Université Joseph Fourier

Les Houches

Session LXXXV

2006

Complex Systems

Lecturers who contributed to this volume

Johannes Berg
Damien Challet
Daniel Fisher
Irene Giardina
Alan Kirman
Satya Majumdar
Rémi Monasson
Andrea Montanari
Mark Newman
Giorgio Parisi
Jim Sethna
Cristina Toninelli
Rudiger Urbanke

ÉCOLE D'ÉTÉ DE PHYSIQUE DES HOUCHES

SESSION LXXXV, 3–28 JULY 2006

ÉCOLE THÉMATIQUE DU CNRS

COMPLEX SYSTEMS

Edited by

Jean-Philippe Bouchaud, Marc Mézard and Jean Dalibard

ELSEVIER

Amsterdam – Boston – Heidelberg – London – New York – Oxford
Paris – San Diego – San Francisco – Singapore – Sydney – Tokyo

Elsevier
Radarweg 29, PO Box 211, 1000 AE Amsterdam, The Netherlands
Linacre House, Jordan Hill, Oxford OX2 8DP, UK

First edition 2007

Library of Congress Cataloging-in-Publication Data
A catalog record for this book is available from the Library of Congress

British Library Cataloguing in Publication Data
A catalogue record for this book is available from the British Library

ISBN: 978-0-444-53006-6
ISSN: 0924-8099

For information on all Elsevier publications
visit our website at books.elsevier.com

Printed and bound in The Netherlands

07 08 09 10 11   10 9 8 7 6 5 4 3 2 1

ÉCOLE DE PHYSIQUE DES HOUCHES

Service inter-universitaire commun
à l'Université Joseph Fourier de Grenoble
et à l'Institut National Polytechnique de Grenoble

Subventionné par le Ministère de l'Éducation Nationale,
de l'Enseignement Supérieur et de la Recherche,
le Centre National de la Recherche Scientifique,
le Commissariat à l'Énergie Atomique

Previous sessions

Publishers:

- Session VIII: Dunod, Wiley, Methuen
- Sessions IX and X: Herman, Wiley
- Session XI: Gordon and Breach, Presses Universitaires
- Sessions XII–XXV: Gordon and Breach
- Sessions XXVI–LXVIII: North Holland
- Session LXIX–LXXVIII: EDP Sciences, Springer
- Session LXXIX–LXXXIV: Elsevier

Organizers

BOUCHAUD Jean-Philippe, SPEC, CEA Saclay, Orme des Merisiers, 91191 Gif sur Yvette, France & Capital Fund Management, 6 Bd Haussmann, 75009 Paris, France

MÉZARD Marc, LPTMS, Bâtiment 100, Université Paris Sud, 91405 Orsay cedex, France

DALIBARD Jean, Laboratoire Kastler Brossel, École normale supérieure, 75231 Paris cedex 05, France

Lecturers

BEN AROUS Gérard, EPFL, CMOS INR 030-Station 14, CH-1015 Lausanne, Switzerland

BERG Johannes, Universität zu Köln, Zülpicher Strasse 77, Institut für Theoretische Physik, 50937 Köln, Germany

CHALLET Damien, Nomura Centre for Quantitative Finance, Mathematical Institute, St Gile's 24-29, Oxford OX1 3LB, UK

FISHER Daniel, Harvard University, Lyman 340, 17 Oxford Street, Cambridge, MA 02138, USA

GIARDINA Irene, Dept. Fisica, Universita La Sapienza, Piazzale A. Moro 2, 00185 Roma, Italia

GRASSBERGER Peter, J. von Neumann Institute for Computing, Forschungzentrum Jülich, 54425 Jülich, Germany

KIRMAN Alan, G.R.E.Q.A.M, 2 rue de la Charité, 13002 Marseille, France

LEIBLER Stanislas, The Rockfeller University, 1230 York Avenue, New York, NY 10021, USA

MAJUMDAR Satya, Bâtiment 100, 15 rue Clémenceau, 91405 Orsay cedex, France

MONASSON Rémi, LPT ENS, 24 rue Lhomond, 75231 Paris cedex 05, France

MONTANARI Andrea, LPT ENS, 24 Rue Lhomond, 75231 Paris cedex 05, France; Electrical Engineering and Statistics Departments, Stanford University, Packard 272, Stanford CA 94305-9510

NEWMAN Mark, Department of Physics, University of Michigan, Randall Laboratory, 450 Church Street, Ann Arbor, MI 48109-1040, USA

PACZUSKI Maya, Department of Physics and Astronomy, SB 605, University of Calgary, University Drive NW, Calgary, Alberta T2N 1N4, Canada

PARISI Giorgio, Dept. Fisica, Universita La Sapienza, Piazzale A. Moro 2, 00185 Roma, Italia

SETHNA Jim, LASSP, Physics Department, Clark Hall, Cornell University, Ithaca, NY 14853-2501, USA

TONINELLI Cristina, LPTMS, Bâtiment 100, 15 rue Clémenceau, 91405 Orsay cedex, France

URBANKE Ruediger, Communication Theory Laboratory, Office inr 116, EPFL-IC- Station 14, 1015 Lausanne, Switzerland

VESPIGNANI Alessandro, Informatics Building, 901 E 10th street, Bloomington, IN 474048-3912, USA

Seminar Speakers

BERTHIER Ludovic, Laboratoire des Colloïdes, Verres et Nanomatériaux, Université Montpellier II, Place E. Bataillon, 34095 Montpellier cedex 5, France

ECKMANN Jean-Pierre, Département de Physique Théorique Université de Genève, 1211 Geneva 4, Switzerland

FRANCESCHELLI Sara, ENS-LSH, 15 Parvis René Descartes, 69007 Lyon, France

MOSSEL Elchanan, Dept. Statistics, 367 Evans Hall, University of California, Berkeley, CA 94720, USA

POTTERS Marc, Capital Fund Management, 6 Bd Haussmann, 75009 Paris, France

Participants

AGLIARI Elena, Dip. Di Fisica, Universita degli Studi di Parma, Viale G.P Usberti 7/a, 43100 Parma, Italy

ALTARELLI Fabrizio, Dip. Di Fisica, Universita la Sapienza, Piazzale A. Moro 2, 00185 Roma, Italy

AMIR Ariel, Weizmann Institute of Science, Herzl Street 76100 Rehovot, Israel

ANDREANOV Alexei, SPhT CEA Saclay, Orme des Merisiers, 91191 Gif sur Yvette, France

BAUKE Heiko, Institut für Theoretische Physik, O. von Guericke Universität, Postfach 4120, 39016 Magdeburg, Germany

BERNHARDSSON Sebastian, Department of Physics, Umea University, 90187 Umea, Sweden

BRITO Carolina, Universidade Federal do Rio Grande do Sul, Ave. Benito Gonçalves 9500, Caixa Postal 15051, 91501-970 Porto Alegre, RS, Brazil

CANDELIER Raphael, SPEC, CEA Saclay, Orme des Merisiers, 91191 Gif sur Yvette, France

CHAUDHURI Pinaki, Laboratoires des Colloïdes Verres et Nanomatériaux, Université Montpellier, 2 Place E. Bataillon, 34095 Montpellier, France

EISLER Zoltan, Budapest University of Technology and Economics, Department Theoretical Physics, Budafoki ut 8, 1111 Budapest, Hungary

ERNEBJERG Morten, Harvard University, Jefferson Physical Laboratory, 17 Oxford Street, Cambridge MA 02138, USA

FRENTZ Zak, Rockfeller University, 1230 York Avenue, New York, NY 10021, USA

GHOSAL Gourab, Department of Physics, University of Michigan, 500E. University Avenue, Ann Arbor, MI 48109, USA

GÖRLICH Andrzej, M. Smoluchowski Instiue of Physics, Jagiellonian University, Ul Remonta 4, 30 059 Krakow, Poland

GUNTHER Stefan, Max Planck Institut for the Physics of Complex Systems, Noethnitzer str. 38, 01187 Dresden Germany

HALLATSCHEK Oskar, Lyman Laboratory of Physics 426, Harvard University, Cambridge MA 02138, USA

KATIFORI Eleni, Physics Department, Jefferson laboratory, Harvard University, 17 Oxford Street, Cambridge MA 02138 USA

KORADA Satish, Babu EPFL-LTHC-IC, Station 14, Lausanne 1015, Switzerland

KUMPULA Jussi, Helsinki University of Technology, Laboratory of Computational Engineering, Tekniikantie 14, PO Box 9203, 02015 TKK, Finland

MANEVA Elitza, IBM Almaden Research Center, 650 Harry Road, San Jose, CA 95120, USA

MELTZER Talya, Street Yehoash 2/8, Rehovot 76448, Israel

MORA Thierry, Université Paris Sud, LPTMS, Bâtiment 100, 15 rue George Clémenceau, 91405 Orsay cedex, France

MEASSON Cyril, LTHC-IC-EPFL, Station 14, 1015 Lausanne, Switzerland

NEHER Richard, A. Sommerfeld Center for Theoretical Physics, LMU Muenchen, Theresienstrasse 37, 80333 Muenchen, Germany

NEUGEBAUER Christoph, Department of Chemistry, Lensfield Road, Cambridge CB2 1EW, UK

OBUSE Hideaki, Condensed Matter Theory laboratory, RIKEN, Hirosawa 2-1 Wako, Saitama 351-0198, Japan

OLLIVIER Julien, Department of Physiology and Centre for Non-Linear Dynamics in Physiology and Medicine, Mc Gill University, 3655 Promenade Sir William Osler, Montreal, Quebec, H3G 1Y6, Canada

OSTOJIC Srdjan, Inst. voor Theoretische Fysica, Universiteit van Amsterdam, Valckenierstraat 65, 1018 XE Amsterdam, The Netherlands

PALEY Christopher, Department of Chemistry, University of Cambridge, Lensfield Road, Cambridge, CB2 1EW, UK

PONSON Laurent, Service de Physique et Chimie des Surfaces et Interfaces, CEA Saclay, 91191 Gif sur Yvette, France

PONSOT Bénédicte, Capital Fund Management, 6 Bd Haussmann, 75009 Paris, France

RAMSTAD Thomas, Department of Physics, Norwegian University of Science and Technology, Hogskoleringen 5 7491 Trondheim, Norway

RAYMOND Jack, Neural Computing Research Group, School of Engineering and Applied Science, Aston University, Aston Triangle, Aston Street, Birmingham, B4 7ET, UK

RENZ Wolfgang, Laborleiter des Labors für Multimediale Systeme, Berliner Tor 7, 20099 Hamburg, Germany

RIVOIRE Olivier, The Rockfeller University, 1230 York Avenue, Box 34, New York, NY 10021, USA

ROUDI Yasser, Gatsby Computational Neuroscience Unit, University College London, Alexandra House, 17 Queen Square, WC1N 3AR, London, UK

ROUQUIER Jean-Baptiste, LIP-ENS, 46 allée d'Italie, 69364 Lyon cedex 07, France

SHREIM Amer, Complex Systems Group, Physics Department, University of Calgary, 2500 University Drive NW, Calgary, Alberta T2N 1N4, Canada

SICILIA Alberto, LPTHE, Tour 24, 5ème étage, 4 place Jussieu, 75252 Paris cedex 05 France

SPORTIELLO Andrea, LPTMS, Université Paris Sud, 15 rue Clémenceau, 91405 Orsay cedex, France

TAMM Mikhail, Physics Dept., Moscow State University, 119992 Vorobyevy gory, Moscow, Russia

TANIMURA Emily, CAMS/EHESS, 54 Boulevard Raspail, 75006 Paris, France

TUMMINELLO Michele, Dip. Di Fisica e Tecnologie Relative, Universita degli Studi di Palermo, Viale delle Scienze, Edificio 18, , 90100 Palermo, Italy

VIVO Pierpaolo, Brunel University, School of Information Systems, Computing and Mathematics, Kingston Lane, Uxbridge-Middlesex UB8 3PH, UK

VICENTE Renato, Escola de Artes Ciencias e Humanidades, Universidade de Sao Paulo, Av. Arlindo Betio 1000, Ermelino Matarazzo, 03828-080 Sao Paulo, Brazil

WALCZAK Aleksandra, Department of Physics, mail code 0354, University of California, 9500 Gilman Drive, La Jolla CA 92093-0354, USA

WINKLER Karen, Ludwig-Maximilians Universitaet Muenchen, Sommerfeld Centre for Theoretical Physics, Theresienstrasse 37, 80333 Muenchen, Germany

WYART Matthieu, DEAS, Harvard University, Room 409, Pierce 29 Oxford Street, Cambridge MA 02138, USA

YAN Koon-Kiu, Department of Physics, Brookhaven National Laboratory, Pennsylvania Street, Upton, NY 11973, USA

ZDEBOROVA Lenka, LPTMS, Bâtiment 100, Université Paris Sud, 15 rue Clémenceau, 91405 Orsay cedex, France

Auditors

BIROLI Guilio, SPhT CEA Saclay, Orme des Merisiers, 91191 Gif sur Yvette, France

BONNEVAY Frédéric, Financial Mathematics, Stanford University, CA 94305, USA

CONTUCCI Pierluigi, Dip. Matematica, Univ. Bologna, Piazza di Porta S. Donato 5, 40126 Bologna, Italy

MOUSTAKAS Aris, Dept. of Physics, Univ. Athens, Panepistimiopolis 15784, Athens, Greece

Preface

Broadly speaking, the study of 'Complex Systems' deals with situations where complex, and often unexpected, collective behaviour emerges from the interaction between many elementary, and relatively simple, constituents. These elementary constituents can be as diverse as spins, atoms, molecules, logical variables, genes, proteins, etc. This is a vast, somewhat ill-defined but rapidly expanding field of research.

In the flurry of activities related to complex systems, we have identified a few lines of research which appeared to us particularly interesting, and at a stage of maturity justifying their presentation during the July 2006 session of the prestigious 'Les Houches' summer school series. Some of the guidelines that we followed in the selection of subjects were: 1) The emergence of well-identified problems on which some clear progress was achieved recently, in particular thanks to the use of Statistical Physics concepts and quantitative methods. 2) The possibility of confronting theoretical ideas to concrete applications on which there exists a corpus of good quality experimental data. 3) The necessary development of multidisciplinary interactions.

While 'Complex Systems Science' often attempts to present itself as a new paradigm, we have rather tried to put forward a more interesting aspect, namely the recent progress brought about by a new, interdisciplinary perspective on problems central to each individual discipline. Two broad themes that looked to us particularly relevant are:

- A) A field of research in the neighborhood of a 'triple point' at which meet the statistical physics of disordered systems, information theory, and some aspects of combinatorial optimization (in particular dealing with the study of typical case complexity, and the development of new types of algorithms for hard computational problems). This field includes, among other topics: error correcting codes, data compression, stochastic optimization algorithms, glassy dynamics, phase transitions in computer science, constraint satisfaction problems and their applications.

- B) The study of collective behaviour in networks of evolving and interacting elements/agents. This theme covers many topics, including population dynamics and the theory of evolution, the study of financial markets and economical organization, and more generally the general mathematical study of networks, the phenomenology and dynamics of social, biological, or computer networks, and game theory.

The first week of the school was dedicated to three long, introductory courses. Gérard Ben Arous lectured on probability theory and stochastic processes. Starting from the universal bulk and tail (extreme) distributions of independent random variables and sums thereof, he gave a precise description of the Poisson-Dirichlet point processes and explained their ubiquitous appearance in physical problems, with examples ranging from random walk with traps to spin glasses; he ended with the mathematical analysis of slow dynamics. Unfortunately, his lecture notes could not be written up for the present volume. The two other lectures introduced topic A above. *Introduction to phase transitions in random optimization problems*, by Rémi Monasson, described the deep connections between statistical physics and optimization. He stressed the occurrence and importance of phase transitions in optimization, which can be either of structural (geometric) nature, or algorithm-dependent. His pedagogical thread was the detailed study of systems of random linear equations between Boolean variables, a topic which is closely related to error-correcting codes as presented by A. Montanari and R. Urbanke (see below). *Mean field theory of Spin Glasses: Statics and Dynamics*, by Giorgio Parisi, proposed an introduction to some of the most sophisticated techniques in the statistical physics of disordered systems, in particular the replica and cavity methods, within the context where they have been first developed, namely the study of spin glasses. These lectures have also dealt with recent issues in the field like the use of stochastic stability and of Ghirlanda-Guerra identities to prove rigorous results.

The second week has been mostly centered on topic B, in particular the study of networks. The long course by Alessandro Vespignani (not reproduced here) on the structure and function of complex networks was an introduction to contemporary research on random networks with particular emphasis on the "small world" structure. Various ensembles of networks, their structure and function, and epidemic propagation problems were investigated. In the same direction, Mark Newman's lectures on *Complex Networks* have shown an interdisciplinary approach to network studies, with applications in social, biological and information sciences. Another long course (not reproduced) was given by Peter Grassberger on numerical methods, with emphasis on modern Monte Carlo methods (Resampling Monte Carlo) and application of information theory concepts to data analysis. Stanislas Leibler gave an insightful introduction to a series of selected problems in biology which are of direct interest to physicists: he covered a broad range of topics ranging from the lac-operon control to bacterial chemotaxis. Three more specialized short courses were given by Maya Paczuski on self-organized criticality, by Irene Giardina on *Metastable States in Glassy Systems*, and by Ludovic Berthier on *The slow dynamics of glassy materials: Insights from computer simulations.*

The third week delved on two main topics developed in long courses. The first one by Andrea Montanari, on *Modern Coding Theory: The Statistical Mechanics Point of View* gave an introduction to all aspects of coding theory, ranging from Shannon's theorems to the most recent Low Density Parity Checks codes, with a special emphasis on the importance of phase transitions (in close connection to the topics presented by Monasson and Parisi) and the results that can be obtained using statistical physics methods. A kind of mirror course was given during the following week by Rudiger Urbanke on *Modern Coding Theory: The Computer Science Point of View*, which also broadened the perspective by presenting several other channel coding problems of great practical importance. They have joined forces to present a unified set of truly interdisciplinary lecture notes in this volume. The long course by Alan Kirman offered a lively introduction to *Economies with Interacting Agents*. His lectures have presented a broad approach to economics, with particular emphasis on the interaction between individual agents, their limited rationality, game theoretic reasoning, and the structure of markets. Three shorter courses of more mathematical nature but great multidisciplinary applications completed the week: Satya Majumdar, in *Random Matrices, The Ulam Problem, Directed Polymer and Growth Models, and Sequence Matching* showed the ubiquitous appearance of the Tracy-Widom distribution in seemingly unrelated problems; Cristina Toninelli, in *Bootstrap and Jamming Percolation* argued about the importance of percolation ideas in describing phase transitions in jamming systems; and finally Jean-Pierre Eckmann offered a glimpse into the mathematical approach to network formation and motif statistics.

Beyond coding and information theory, the fourth week focused on evolution theory, with the seminal lectures by Daniel Fisher on *Evolutionary Dynamics* which outlines a new research program, highlighting – among other things – the importance of finite population sizes and rare events in mutation problems. The lectures by Johannes Berg on *Statistical modelling and analysis of biological networks* reviewed the analysis of biological regulatory networks using information theory concepts. Jim Sethna lectured on *Crackling Noise and Avalanches: Scaling, Critical Phenomena, and the Renormalization Group*, giving a broad perspective on the quantitative analysis of catastrophic collective effects, avalanches, that occur in many physical systems (and beyond). Damien Challet's course on *Minority Games* showed how statistical physics ideas can be applied to game theoretic settings with heterogeneous agents, with applications to economic situations and financial markets. Two shorter presentations by Sara Franceschelli on *Epigenetic landscape and catastrophe theory* and by Marc Potters on his "life in a hedge fund", offered different perspectives (respectively epistemology and practical finance).

The students worked hard within various working groups on topics related to the lectures, and short seminars summarized their results at the end of the session. Some of the results of the group working on '1-in-K SAT' are presented in this volume. The enthusiasm and curiosity of the participants, and the eager attendance of many of the speakers to most lectures, has created a wonderful atmosphere of study and discovery that made this school a very memorable one, that could have easily have extended for a few other weeks (not to mention the extraordinary weather conditions). We thank all the participants for their efforts, in particular the lecturers who have provided us with a written version of their lectures – some of them will most probably soon become classics.

The summer school and the present volume have been made possible by the financial support of the following institutions, whose contribution is gratefully acknowledged:

- La formation permanente du CNRS
- ONCE-CS (programme FET de l'IST de la commission Européenne)
- Commissariat à l'Energie Atomique (DSM & DRECAM)
- Stipco: "Research Training Networks" of the EC, HPRN-CT-2002-00319
- Evergrow: Proactive Initiative "Complex Systems Research" of the EC, FP6, IST Priority
- Dyglagemem: "Research Training Networks" of the EC, HPRN-CT-2002-00319
- Institut des systèmes complexes de l'ENS Lyon
- Capital Fund Management

The staff of the School, especially Brigitte Rousset and Isabelle Lelièvre, have been of great help for the preparation and development of the school, and we would like to thank them warmly on behalf of all students and lecturers.

We also want to thank Raphael Candelier for taking the pictures which illustrate this volume.

J.-P. Bouchaud, M. Mézard and J. Dalibard

CONTENTS

Course 2.   *Modern coding theory: the statistical mechanics and computer science point of view, by Andrea Montanari and Rüdiger Urbanke*   *67*

Contents

Course 1

INTRODUCTION TO PHASE TRANSITIONS IN RANDOM OPTIMIZATION PROBLEMS

R. Monasson

Laboratoire de Physique Théorique de l'ENS
24 rue Lhomond, 75005 Paris, France

J.-P. Bouchaud, M. Mézard and J. Dalibard, eds.
Les Houches, Session LXXXV, 2006
Complex Systems
© *2007 Published by Elsevier B.V.*

Contents

1. Introduction

1.1. Preamble

The connection between the statistical physics of disordered systems and optimization problems in computer science dates back from twenty years at least [43]. After all zero temperature statistical physics is simply the search for the state with minimal energy, while the main problem in combinatorial optimization is to look for the configurations of parameters minimizing some cost function (the length of a tour in the traveling salesman problem (TSP), the number of violated constraints in constrained satisfaction problems, ...) [57]. Yet, despite the beautiful studies of the average properties of the TSP, Graph partitioning, Matching, ..., based on the recently developed mean-field spin glass theory [43], a methodological gap between the fields could not be bridged [30]. In statistical physics statements are usually made on the properties of samples given some quenched disorder distribution such as the typical number of solutions, minimal energy... In optimization, however, one is interested in solving one (or several) particular instances of a problem, and needs efficient ways to do so, that is, requiring a computational effort growing not too quickly with the number of data defining the instance. Knowing precisely the typical properties for a given, academic distribution of instances does not help much to solve practical cases.

At the beginning of the nineties practitioners in artificial intelligence realized that classes of random constrained satisfaction problems used as artificial benchmarks for search algorithms exhibited abrupt changes of behaviour when some control parameter were finely tuned [47]. The most celebrated example was random K-Satisfiability, where one looks for a solution to a set of random logical constraints over a set of Boolean variables. It appeared that, for large sets of variables, there was a critical value of the number of constraints per variable below which there almost surely existed solutions, and above which solutions were absent. An important feature was that search algorithms performances drastically worsened in the vicinity of this critical ratio.

This phenomenon, strongly reminiscent of phase transitions in condensed matter physics, led to a revival of the interface between statistical physics and computer science, which has not vanished yet. The purpose of the present lecture is to introduce the non specialist reader to the concepts and techniques required

5

to understand the literature in the field. For the sake of simplicity the presentation will be limited to one computational problem, namely, linear systems of Boolean equations. A good reason to do so is that this problem concentrates most of the features encountered in other optimization problems, while being technically simpler to study. In addition it is closely related to error-correcting codes in communication theory, see lectures by A. Montanari and R. Urbanke in the present book. Extension to other problems will be mentioned in the conclusions.

The lecture is divided into three parts. Sections 1 and 2 are devoted to the presentation of the model and of elementary concepts related to phase transitions e.g. finite-size scaling, large deviations, critical exponents, symmetry breaking, ... Sections 3 and 4 expose the specific statistical mechanics techniques and concepts developed in disordered systems to deal with highly interacting and random systems, namely the replica and cavity approaches. Finally Section 5 focuses on dynamics and the study of search algorithms.

1.2. Linear systems of Boolean equations

Linear systems of Boolean equations look very much like their well known counterparts for integer-valued variables, except that equalities are defined modulo two. Consider a set of N Boolean variables x_i with indices $i = 1, \ldots, N$. Any variable shall be False (F) or True (T). The sum of two variables, denoted by $+$, corresponds to the logical exclusive OR between these variables defined through,

$$F + T = T + F = T,$$
$$F + F = T + T = F. \tag{1.1}$$

In the following we shall use an alternative representation of the above sum rule. Variables will be equal to 0 or 1, instead of F or T respectively. Then the $+$ operation coincides with the addition between integer numbers modulo two.

The following is a linear equation involving three variables,

$$x_1 + x_2 + x_3 = 1. \tag{1.2}$$

Four among the $2^3 = 8$ assignments of (x_1, x_2, x_3) satisfy the equation: $(1, 0, 0)$, $(0, 1, 0)$, $(0, 0, 1)$ and $(1, 1, 1)$. A Boolean system of equations is a set of Boolean equations that have to be satisfied together. For instance, the following Boolean system involving four variables

$$\begin{cases} x_1 + x_2 + x_3 = 1 \\ x_2 + x_4 = 0 \\ x_1 + x_4 = 1 \end{cases} \tag{1.3}$$

has two solutions: $(x_1, x_2, x_3, x_4) = (1, 0, 0, 0)$ and $(0, 1, 0, 1)$. A system with one or more solutions is called satisfiable. A trivial example of an unsatisfiable Boolean system is

$$\begin{cases} x_1 + x_2 + x_3 = 1 \\ x_1 + x_2 + x_3 = 0 \end{cases}. \tag{1.4}$$

Determining whether a Boolean system admits an assignment of the Boolean variables satisfying all the equations constitutes the XORSAT (exclusive OR Satisfaction) problem. In the following, we shall restrict for some reasons to be clarified in Section 2 to K-XORSAT, a variant of XORSAT where each Boolean equation include K variables precisely.

K-XORSAT belongs to the class P of polynomial problems [57]. Determining whether a system is satisfiable or not can be achieved by the standard Gaussian elimination algorithm in a time (number of elementary operations) bounded from above by some constant times the cube of the number of bits necessary to store the system[1] [57].

If the decision version of K-XORSAT is easy its optimization version is not. Assume you are given a system F, run the Gauss procedure and find that it is not satisfiable. Determining the maximal number $M_S(F)$ of satisfiable equations is a very hard problem. Even approximating this number is very hard. It is known that there is no approximation algorithm (unless P=NP) for XORSAT with ratio $r > \frac{1}{2}$, that is, guaranteed to satisfy at least $r \times M_S(F)$ equations for any F. But $r = \frac{1}{2}$ is achieved, on average, by making a random guess![2]

1.3. Models for random systems

There are many different ways of generating random Boolean systems. Perhaps the simplest one is the following, called *fixed-size ensemble*. To build an equation we pick up uniformly at random K distinct indices among the N ones, say, i_1, i_2 and i_k. Then we consider the equation

$$x_{i_1} + x_{i_2} + \cdots + x_{i_k} = v. \tag{1.5}$$

The second member, v, is obtained by tossing a coin: $v = 0$ or $v = 1$ with equal probabilities (one half) and independently of the indices of the variables in the first member. The process is repeated M times, without correlation between equations to obtain a system with M equations.

---

[1] The storage space is K times the number of equations times the number of bits necessary to label a variable, that is, the logarithm of the number of variables appearing in the system.

[2] Any equation is satisfied by half of the configurations of a variables, so a randomly chosen configuration satisfies on average $\frac{M}{2} \geq \frac{M_S(F)}{2}$ equations.

Another statistical ensemble is the *fixed-probability ensemble*. One scans the set of all $H = 2\binom{N}{K}$ equations one after the other. Each equation is added to the system with probability p, discarded with probability $1 - p$. Then a system with, on average, $p H$ equations (without repetition) is obtained. In practice one chooses $p = \frac{M}{H}$ to have the same (average) number of equations as in the fixed-size ensemble.

The above distributions are not the only possible ones. However they are easy to implement on a computer, are amenable to mathematical studies, and last but not least, lead to a surprisingly rich phenomenology. One of the key quantities which exhibits an interesting behaviour is

$$P_{SAT}(N, \alpha) = \text{Probability that a system of random K-XORSAT with}$$
$$N \text{ variables and } M = \alpha N \text{ equations is satisfiable,}$$

which obviously depends on K and the statistical ensemble. Given N, P_{SAT} is a decreasing function of α. We will see that, in the infinite size limit (and for $K \geq 2$), the decrease is abrupt at some well defined ratio, defining a phase transition between Satisfiable and Unsatisfiable phase [18]. The scope of the lecture is to give some tools to understand this transition and some related phenomena.

2. Basic concepts: overview of static phase transitions in K-XORSAT

In this Section we introduce the basic concepts necessary to the study of random K-XORSAT. It turns out that even the $K = 1$ case, trivial from a computer science point of view (each equation contains a single variable!), can be used as an illustration to important concepts such as scaling and self-averageness. Ideas related to the percolation phase transition and random graphs are illustrated on the $K = 2$ case. Finally the solution space of 3-XORSAT model exemplifies the notion of clusters and glassy states.

2.1. Finite-size scaling (I): scaling function

Figure 1(left) shows the probability P_{SAT} that a randomly extracted 1-XORSAT formula is satisfiable as a function of the ratio α, and for sizes N ranging from 100 to 1000. We see that P_{SAT} is a decreasing function of α and N.

Consider the sub-formula made of the n_i equations with first member equal to x_i. This formula is always satisfiable if $n_i = 0$ or $n_i = 1$. If $n_i \geq 2$ the formula is satisfiable if and only if all second members are equal (to 0, or to 1), an event with probability $(\frac{1}{2})^{n_i - 1}$ decreasing exponentially with the number of equations. Hence we have to consider the following variant of the celebrated

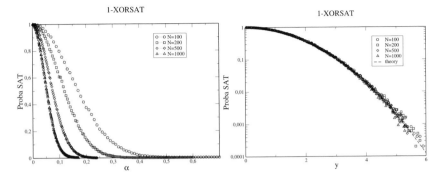

Fig. 1. Left: Probability that a random 1-XORSAT formula is satisfiable as a function of the ratio α of equations per variable, and for various sizes N. Right: same data as in the left panel after the horizontal rescaling $\alpha \to \alpha \times \sqrt{N} = y$; note the use of a log scale for the vertical axis. The dashed line shows the scaling function $\Phi_1(y)$ (2.3).

Birthday problem.[3] Consider a year with a number N of days, how should scale the number M of students in a class to be sure that no two students have the same birthday date?

$$\bar{p} = \prod_{i=0}^{M-1} \left(1 - \frac{i}{N}\right) = \exp\left(-\frac{M(M-1)}{2N} + O(M^3/N^2)\right). \tag{2.1}$$

Hence we expect a cross-over from large to small $\bar{p}$ when M crosses the scaling regime $\sqrt{N}$. Going back to the 1-XORSAT model we expect P_{SAT} to have a non zero limit value when the number of equations and variables are both sent to infinity at a fixed ratio $y = M/\sqrt{N}$. In other words, random 1-XORSAT formulas with N variables, M equations or with, say, $100 \times N$ variables, $10 \times M$ equations should have roughly the same probabilities of being satisfiable. To check this hypothesis we re-plot the data in Figure 1 after multiplication of the abscissa of each point by $\sqrt{N}$ (to keep y fixed instead of α). The outcome is shown in the right panel of Figure 1. Data obtained for various sizes nicely collapse on a single limit curve function of y.

The calculation of this limit function, usually called scaling function, is done hereafter in the fixed-probability 1-XORSAT model where the number of equations is a Poisson variable of mean value $\bar{M} = y\sqrt{N}$. We will discuss the equivalence between the fixed-probability and the fixed-size ensembles later. In the fixed-probability ensemble the numbers n_i of occurrence of each variable x_i are

---

[3]The Birthday problem is a classical elementary probability problem: given a class with M students, what is the probability that at least two of them have the same birthday date? The answer for $M = 25$ is $p \simeq 57\%$, while a much lower value is expected on intuitive grounds when M is much smaller than the number $N = 365$ of days in a year.

independent Poisson variables with average value $\bar{M}/N = y/\sqrt{N}$. Therefore the probability of satisfaction is

$$
\begin{aligned}
P_{SAT}^{p}\left(N, \alpha = \frac{y}{\sqrt{N}}\right) &= \left[e^{-y/\sqrt{N}}\left(1 + \sum_{n \geq 1} \frac{(y/\sqrt{N})^n}{n!}\left(\frac{1}{2}\right)^{n-1}\right)\right]^N \\
&= \left[2e^{-y/(2\sqrt{N})} - e^{-y/\sqrt{N}}\right]^N,
\end{aligned} \tag{2.2}
$$

where the p subscript denotes the use of the fixed-probability ensemble. We obtain the desired scaling function

$$
\Phi_1(y) \equiv \lim_{N \to \infty} \ln P_{SAT}^{p}\left(N, \alpha = \frac{y}{\sqrt{N}}\right) = -\frac{y^2}{4}, \tag{2.3}
$$

in excellent agreement with the rescaled data of Figure 1 (right) [19].

2.2. Self-averageness of energy and entropy

Let us now consider random 1-XORSAT formulas at a finite ratio α, and ask for the distribution of the minimal fraction of unsatisfied equations, hereafter called ground state (GS) energy e_{GS}. For simplicity we work in the fixed-probability ensemble again. The numbers n_i^0, n_i^1 of, respectively, $x_i = 0, x_i = 1$ are independent Poisson variables with mean $\frac{\alpha}{2}$. The minimal number of unsatisfied equations is clearly $\min(n_i^0, n_i^1)$. The GS energy is the sum (divided by M) of N such i.i.d. variables; from the law of large number it almost surely converges towards the average value

$$
e_{GS}(\alpha) = \frac{1}{2}\left(1 - e^{-\alpha}I_0(\alpha) - e^{-\alpha}I_1(\alpha)\right), \tag{2.4}
$$

where I_ℓ denotes the ℓth modified Bessel function. In other words almost all formulas have the same GS energy in the infinite N limit, a property called self-averageness in physics, and concentration in probability.

How many configurations of variables realize have minimal energy? Obviously a variable is free (to take 0 or 1 value) if $n_i^0 = n_i^1$, and is frozen otherwise. Hence the number of GS configurations is $\mathcal{N} = 2^{N_f}$ where N_f is the number of free variables. Call

$$
\rho = \sum_{n \geq 0} e^{-\alpha}\left(\frac{\alpha}{2}\right)^n \frac{1}{(n!)^2} = e^{-\alpha}I_0(\alpha) \tag{2.5}
$$

the probability that a variable is free. Then N_f is a binomial variable with parameter ρ among N; it is sharply concentrated around $\overline{N_f} = \rho N$ with typical fluctuations of the order of $N^{1/2}$. As a consequence, the GS entropy per variable, $s_{GS} = (\log \mathcal{N})/N$, is self-averaging and almost surely equal to its average value $s_{GS} = \rho \log 2$.

Self-averageness is the very useful property. It allows us to study the average value of a random variable, instead of its full distribution. We shall use it in Section 3 and also in the analysis of algorithms of Section 5.5. This property is not restricted to XORSAT but was proved to hold for the GS energy [13] and entropy [52] of other optimization problems.

Not all variables are self-averaging of course. A straightforward example is the number $\mathcal{N}$ of GS configurations itself. Its qth moment reads $\overline{\mathcal{N}^q} = (1 - \rho + \rho 2^q)^N$ where the overbar denotes the average over the formulas. We see that $\overline{\mathcal{N}^q} \gg (\overline{\mathcal{N}})^q$: $\mathcal{N}$ exhibits large fluctuations and is not concentrated around its average. Very rare formulas with atypically large number N_f of free variables contribute more to the qth moment than the vast majority of formulas, and spoil the output. This is the very reason we will need the introduction of the replica approach in Section 3.

2.3. Large deviations for P_{SAT} (I): 1-XORSAT

As we have seen in the previous sections 1-XORSAT formulas with a finite ratio α are unsatisfiable with high probability *i.e.* equal to unity in the infinite N limit. For finite but large N there is a tiny probability that a randomly extracted formula is actually satisfiable. A natural question is to characterize the 'rate' at which P_{SAT} tends to zero as N increases (at fixed α). Answering to such questions is the very scope of large deviation theory (see Appendix A for an elementary introduction). Looking for events with very small probabilities is not only interesting from an academic point of view, but can also be crucial in practical applications. We will see in Section 5.3 that the behaviour of some algorithms is indeed dominated by rare events.

Figure 2 shows minus the logarithm of P_{SAT}, divided by N, as a function of the ratio α and for various sizes N. Once again the data corresponding to different sizes collapse on a single curve, meaning that

$$P_{SAT}(N, \alpha) = e^{-N\omega_1(\alpha) + o(N)}. \tag{2.6}$$

Decay exponent ω_1 is called rate function in probability theory. We can derive its value in the fixed-probability ensemble from (2.2) with $y = \alpha \times \sqrt{N}$, with the immediate result

$$\omega_1^p(\alpha) = \alpha - \ln(2e^{\alpha/2} - 1). \tag{2.7}$$

The agreement with numerics is very good for small ratios, but deteriorates as α increases. The reason is simple. In the fixed-probability ensemble the number M

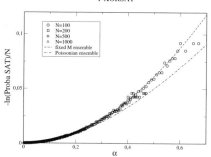

Fig. 2. Same data as Figure1 (left) with: logarithmic scale on the vertical axis, and rescaling by $-1/N$. The scaling functions ω_1 (2.9) and ω_1^p (2.7) for, respectively, the fixed-size and fixed-probability ensembles are shown.

of equations is not fixed but may fluctuate around the average value $\bar{M} = \alpha N$. The ratio $\tilde{\alpha} = M/N$, is with high probability equal to α, but large deviations ($\tilde{\alpha} \neq \alpha$) are possible and described by the rate function,[4]

$$\Omega(\tilde{\alpha}|\alpha) = \tilde{\alpha} - \alpha - \alpha \ln(\alpha/\tilde{\alpha}). \tag{2.8}$$

However the probability that a random 1-XORSAT formula with M equations is satisfiable is also exponentially small in N, with a rate function $\omega_1(\alpha)$ increasing with α. Thus, in the fixed-probability ensemble, a trade-off is found between ratios $\tilde{\alpha}$ close to α (formulas likely to be generated) and close to 0 (formulas likely to be satisfiable). As a result the fixed-probability rate function is

$$\omega_1^p(\alpha) = \min_{\tilde{\alpha}}\big[\omega_1(\tilde{\alpha}) + \Omega(\tilde{\alpha}|\alpha)\big], \tag{2.9}$$

and is smaller than $\omega_1(\alpha)$. It is an easy check that the optimal ratio $\tilde{\alpha}^* = \alpha/(2 - e^{-\alpha/2}) < \alpha$ as expected. Inverting (2.9) we deduce the rate function ω_1 in the fixed-size ensemble, in excellent agreement with numerics (Figure 2). This example underlines that thermodynamically equivalent ensembles have to be considered with care as far as rare events are concerned.

Remark that, when $\alpha \to 0$, $\tilde{\alpha} = \alpha + O(\alpha^2)$, and $\omega_1^p(\alpha) = \omega_1(\alpha) + O(\alpha^3)$. This common value coincides with the scaling function $-\Phi_1(\alpha)$ (2.3). This identity is expected on general basis (Section 2.6) and justifies the agreement between

---

[4] M obeys a Poisson law with parameter $\bar{M}$. Using Stirling formula,

$$e^{-\bar{M}}\frac{\bar{M}^M}{M!} \simeq e^{-\alpha N}(\tilde{\alpha}N)^{\alpha N}\sqrt{2\pi N}\left(\frac{e}{\alpha N}\right)^{\alpha N} = e^{-N\Omega(\tilde{\alpha}|\alpha)+o(N)},$$

where Ω is defined in (2.8).

the fixed-probability scaling function and the numerics based on the fixed-size ensemble (Figure 1, right).

2.4. Percolation in random graphs

Though 1-XORSAT allowed us to understand some general features of random optimization problems it is very limited due to the absence of interactions between variables. A more interesting problem is 2-XORSAT where every equation define a joint constraint on two variables. Formulas of 2-XORSAT can be represented by a graph with N vertices (one for each variable), and αN edges. To each equation of the type $x_i + x_j = e$ corresponds an edge linking vertices i and j, and carrying 0 or 1 label (the value e of the second member). Depending on the input model chosen (Section 1.3) multiple edges are present or not.

As the formula is random so is the graph. Figure 3 shows examples of graphs obtained for various values of α. Notice the qualitative change of structure of graphs when the ratio α varies from low values (graphs are mostly made of small isolated trees) to higher ones (a large part of vertices are now connected together). This change is known as the percolation transition in physics, or the appearance of a giant component in mathematics literature.

Before reviewing some of the aspects of the percolation transition let us mention an important fact on the valency of vertices. As a result of the randomness of the graph generation process, each node share edges with a variable number of neighboring vertices. In the large N limit the degree v of a vertex, *i.e.* the number of its neighbors, is a Poisson variable with mean 2α,

$$\text{Proba}[v] = e^{-2\alpha} \frac{(2\alpha)^v}{v!}. \tag{2.10}$$

For instance the fraction of isolated vertices is $e^{-2\alpha}$. The average degree of a vertex, $c = 2\alpha$, is called connectivity.

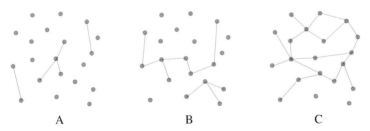

A   B   C

Fig. 3. Examples of random graphs generated at fixed number $M = \alpha N$ of edges (fixed-size model without repetition). All graph include $N = 20$ vertices (grey dots). The average degree of valency, 2α, is equal to 0.5 (**A**), 1 (**B**), and 2 (**C**). The labels of the vertices have been permuted to obtain planar graphs, *i.e.* avoid crossing of edges.

It is natural to decompose the graphs into its connected subgraphs, called components. Erdös and Rényi were able in 1960 to characterize the distribution of sizes of the largest component [12],

• When $c < 1$, the largest component includes $\sim \ln N/(c - 1 - \ln c)$ vertices with high probability. Most components include only a finite number of vertices, and are trees *i.e.* contain no circuit.

• For $c = 1$ the largest component contains $O(N^{2/3})$ vertices.

• When $c > 1$ there is one giant component containing $\sim \gamma(c)N$ vertices; the others components are small *i.e.* look like the components in the $c < 1$ regime. The fraction of vertices in the giant component is the unique positive solution of

$$1 - \gamma = e^{-c\gamma}. \tag{2.11}$$

It is a non analytic function of c, equal to 0 for $c \leq 1$, and positive above, tending to unity when c increases.

The phenomenon taking place at $c = 1$ is an example of (mean-field) percolation transition. We now give a hand-waving derivation of (2.11). Consider a random graph G over N vertices, with connectivity c. Add a new vertex A to the graph to obtain G'. If we want G' to be drawn from the same distribution as G, a number v of edges must be attached to A, where v an integer–valued random number following the Poisson distribution (2.10). After addition of A, some connected components of G will merge in G'. In particular, with some probability p_v, A will be part of the giant component of G'. To estimate p_v, we note that this event will not happen if and only if none of the v neighbors of A in G' belongs to the giant component of G. Thus,

$$1 - p_v = (1 - \gamma)^v, \tag{2.12}$$

where γ is the size (fraction of vertices) of the giant component. Summing both sides of (2.12) over the distribution (2.10) for v, and asserting that the change in size of the giant component between G and G' is $o(1)$ for large N, we obtain (2.11).

The above derivation illustrates an ubiquitous idea in probability and statistical physics, which could be phrased as follows: 'if a system is very large, its statistical properties should be, in some sense, unaffected by a small increase in size'. This idea will be useful, in a more sophisticated context, in Section 4.

2.5. Sat/Unsat transition in 2-XORSAT

Figure 4 shows the probability P_{SAT} that a randomly extracted 2-XORSAT formula is satisfiable as function of α, and for various sizes N. It appears that P_{SAT} drops quickly to zero for large N when α reaches the percolation threshold

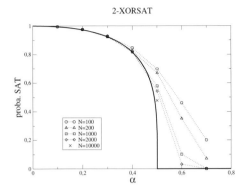

Fig. 4. Probability that a random 2-XORSAT formula is satisfiable as a function of the ratio α of equations per variable, and for various sizes N. The full line is the asymptotic analytical formula (2.18).

$\alpha_c = \frac{1}{2}$. For ratios smaller than α_c the probability of satisfaction is positive, but smaller than unity.

Take $\alpha < \frac{1}{2}$. Then the random graph G associated to a random 2-XORSAT formula is non percolating, and made of many small components. Identical components (differing only by a relabeling of the variables) may appear several times, depending on their topology. For instance consider a connected graph G' made of E edges and V vertices. The average number of times G' appears in G is a function of E and V only,

$$N_{E,V} = \binom{N}{V} \left(\frac{2\alpha}{N} \right)^E \left(1 - \frac{2\alpha}{N} \right)^{\frac{V(V-1)}{2} + V(N-V)} \tag{2.13}$$

since any vertex in G' can establish edges with other vertices in G', but is not allowed to be connected to any of the $N - V$ outside vertices. When N is very large compared to E, V we have

$$N_{E,V} \simeq N^{V-E} \frac{(2\alpha)^E}{V!} e^{-2\alpha V}. \tag{2.14}$$

Three cases should be distinguished, depending on the value of $V - E$:

• $V - E = 1$: this is the largest value compatible with connectedness, and corresponds to the case of trees. From (2.14) every finite tree has of the order of N copies in G.

• $V - E = 0$: this corresponds to trees with one additional edge, that is, to graphs having one cycle (closed loop). The average number of unicyclic graphs is, from (2.14), finite when $N \to \infty$.

- $V - E \leq -1$: the average number of components with more than one cycle vanishes in the large N limit; those graphs are unlikely to be found and can be ignored.[5]

Obviously a 2-XORSAT formula with tree structure is always satisfiable.[6] Hence dangerous sub-formulas, as far as satisfiability is concerned, are associated to unicyclic graphs. A simple thought shows that a unicyclic formula is satisfiable if and only if the number of edges carrying label 1 along the cycle is even. Since the values attached to the edges (second members in the formula) are uncorrelated with the topology of the subgraph (first members) each cycle is satisfiable with probability one half. We end up with the simple formula

$$P_{SAT}(N, \alpha) = \langle 2^{-C(G)} \rangle \tag{2.15}$$

where $C(G)$ denotes the number of cycles in G, and $\langle . \rangle$ the average over G. For a reason which will become clear below let us classify cycles according to their length L. How many cycles of length L can we construct? We have to choose first L vertices among N, and join them one after the other according to some order. As neither the starting vertex nor the direction along the cycle matter, the average number of L-cycles is

$$N_L = \frac{N(N-1)\dots(N-L+1)}{2L} \times \left(\frac{2\alpha}{N}\right)^L \to \Lambda_L = \frac{(2\alpha)^L}{2L}. \tag{2.16}$$

when $N \to \infty$. As the emergence of a cycle between L vertices is a local event (independent of the environment) we expect the number of L-cycles to be Poisson distributed in the large N limit with parameter Λ_L. This statement can actually be proved, and extended to any finite collection of cycles of various lengths [12]: in the infinite size limit, the joint distribution of the numbers of cycles of lengths $1, 2, \dots, L$ is the product of Poisson laws with parameters $\Lambda_1, \Lambda_2, \dots, \Lambda_L$ calculated in (2.16). The probability of satisfaction (2.15) therefore converges to

$$\lim_{N \to \infty} P_{SAT}(N, \alpha) = \prod_{L \geq L_0} \left\{ \sum_{C \geq 0} e^{-\Lambda_L} \frac{(\Lambda_L/2)^C}{C!} \right\} = \prod_{L \geq L_0} e^{-\Lambda_L/2} \tag{2.17}$$

where L_0 is the minimal cycle length. In normal random graphs $L_0 = 3$ since triangles are the shortest cycles. However in our 2-XORSAT model any equation,

---

[5]The probability that such a graph exists is bounded from above by the average number, see Appendix B.

[6]Start from one leaf, assign the attached variable to 0, propagate to the next variable according to the edge value, and so on, up to the completion of the tree.

or more precisely, any first member can appear twice or more, hence $L_0 = 2$. We conclude that [19]

$$\lim_{N \to \infty} P_{SAT}(N, \alpha) = e^{\alpha/2}(1 - 2\alpha)^{\frac{1}{4}} \quad \text{when} \quad \alpha < \alpha_c = \frac{1}{2}. \tag{2.18}$$

The agreement of this result with the large size trend coming out from numerical simulations is visible in Figure 4. As P_{SAT} is a decreasing function of α it remains null for all ratios larger than α_c. The non analyticity of P_{SAT} at α_c locates the Sat/Unsat phase transition of 2-XORSAT.

It is an implicit assumption of statistical physics that asymptotic results of the kind of (2.18), rigorously valid in the $N \to \infty$ limit, should reflect with good accuracy the finite but large N situation. An inspection of Figure 4 shows this is indeed the case. For instance, for ratio $\alpha = .3$, (2.18) cannot be told from the probability of satisfaction measured for formulas with $N = 100$ variables. This statement does not hold for $\alpha = .4$, where the agreement between infinite size theory and numerics sets in when $N = 1000$ at least. It appears that such finite-size effects become bigger and bigger as α gets closer and closer to the Sat/Unsat threshold. This issue, of broad importance in the context of phase transitions and the practical application of asymptotic results, is studied in Section 2.8.

2.6. *Large deviations for P_{SAT} (II): bounds in the Unsat phase of 2-XORSAT*

Consider ratios $\alpha > \alpha_c$. The giant components of the corresponding formulas contain an extensively large number of independent cycles, so we expect from (2.15) that the probability of satisfaction is exponentially small in N, $P_{SAT} = \exp(-N\omega_2(\alpha) + o(N))$. Lower and upper bounds to the rate function ω_2 can be obtained from, respectively, the first and second moment inequalities described in Appendix B. Denoting by $\mathcal{N}$ the number of solutions of a formula, P_{SAT} is the probability that $\mathcal{N} \geq 1$, and is bracketed according to (B.2).

To calculate the first moment of $\mathcal{N}$ remark that an equation is satisfied by one half of the configurations. This result remains true for a restricted set of configurations when we average over the possible choices of (the second member of) the equation. The average number of solutions is thus $2^N/2^M$, from which we get

$$\omega_2(\alpha) \geq (\alpha - 1)\ln 2. \tag{2.19}$$

This lower bound is useless for $\alpha < 1$ since ω_2 is positive by definition. As for the upper bound we need to calculate the second moment $\langle \mathcal{N}^2 \rangle$ of $\mathcal{N}$. As equations are independently drawn

$$\langle \mathcal{N}^2 \rangle = \sum_{X,Y} q(X, Y)^M \tag{2.20}$$

where the sum is carried out over the pairs X, Y of configurations of the N variables, and $q(X, Y)$ is the probability that both X and Y satisfies the same randomly drawn equation. q can be easily expressed in terms of the Hamming distance d between X and Y, defined as the fraction of variables having opposite values in X and Y. The general expression for K-XORSAT is[7]

$$q(d) = \frac{1}{2}\left(1 - (1 - 2d)^K\right) \tag{2.21}$$

and we specialize in this section to $K = 2$. Going back to (2.20) we can sum over Y at fixed X, that is, over the distances d taking multiple values of $\frac{1}{N}$ with the appropriate binomial multiplicity, and then sum over X with the result

$$\langle \mathcal{N}^2 \rangle = 2^N \sum_d \binom{N}{Nd} q(d)^M = \exp\left(N \max_{d \in [0;1]} A(d, \alpha) + o(N)\right) \tag{2.22}$$

in the large N limit, where

$$A(d, \alpha) = (2\alpha - 1)\ln 2 - d \ln d - (1 - d)\ln(1 - d) + \alpha \ln q(d). \tag{2.23}$$

For $\alpha < \frac{1}{2}$ the maximum of A is located in $d^* = \frac{1}{2}$, and equal to $A^* = 0$. When $\alpha > \frac{1}{2}$, A has two global maxima located in $d^*(\alpha) < \frac{1}{2}$ and $1 - d^*(\alpha)$, with equal value $A^*(\alpha) > 0$.

We plot in Figure 5 the lower (2.19) and upper bounds to the rate function,

$$\omega_2(\alpha) \leq 2(1 - \alpha)\ln 2 - \max_{d \in [0;1]} A(d, \alpha) \tag{2.24}$$

from (B.2). At large ratio both bounds asymptotically match, proving that $\omega_2(\alpha) = (\alpha - 1)\ln 2 + O(e^{-2\alpha})$. As the ratio departs from its threshold value by $\epsilon = \alpha - \alpha_c$ the upper bound grows quadratically, $A^*(\alpha_c + \epsilon) \simeq \frac{3}{4}\epsilon^2 + O(\epsilon^3)$. Numerics suggest that the increase of the rate function is slower,

$$\omega_2(\alpha_c + \epsilon) \simeq \Omega \epsilon^3 + O(\epsilon^4), \tag{2.25}$$

for some constant $\Omega \simeq 1$ (Figure 5). We will see in Section 3 that a sophisticated statistical physics technique, called the replica method, actually predict this scaling with $\Omega = \frac{32}{27}$. Actually the rate function can be estimated with the replica approach for any ratio α with the result shown in Figure 5.

---

[7]The equation is satisfied if the number of its variables taking opposite values in Y as in X is even. By definition of d the probability (over its index i) that a variable takes different value in X and Y is d. Hence expression (2.21) for $q(d)$. Beware of the $O(\frac{1}{N})$ corrections to this expression e.g. if variable $x_1 \neq y_1$ (which happens with probability d) then the probability that $x_2 \neq y_2$ is $(dN - 1)/(N - 1) = d + (1 - d)/ - N - 1)$. Those corrections are relevant for the calculation of Gaussian fluctuations around the saddle-point (Appendix C).

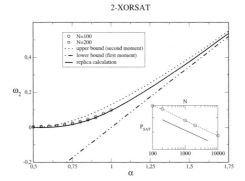

Fig. 5. Rate function $\omega_2(\alpha)$ associated to the probability of satisfaction of 2-XORSAT formulas with ratio α. The dotted line is the upper bound (2.23) and the dot-dashed line the lower bound (2.19). The full line is the output of the replica calculation of Section 3.4, squares and circles represent numerical results for $N = 200, 100$ respectively from 10^6 formulas. Inset: P_{SAT} as a function of the size N at the Sat/Unsat ratio. The slope $-\frac{1}{12}$ (2.35) is shown for comparison.

2.7. Order parameter and symmetry breaking

What is the meaning of the Hamming distance $d^*(\alpha)$ appearing in the calculation of the second moment of the number of solutions? An easy guess would be the average distance between pairs of solutions

$$d_{av}(\alpha) = \lim_{N\to\infty} \left\langle \frac{\sum_{X,Y \text{ solutions of } F} d(X,Y)}{\mathcal{N}(F)^2} \right\rangle_F \tag{2.26}$$

where the average is taken over the satisfiable formulas F with ratio α, and $d(X,Y)$ denotes the (intensive) Hamming distance between two solutions X, Y. However an inspection of the calculation of Section 2.6 shows that

$$d^*(\alpha) = \lim_{N\to\infty} \frac{\langle \sum_{X,Y \text{ solutions of } F} d(X,Y) \rangle_F}{\langle \mathcal{N}(F)^2 \rangle_F} \neq d_{av}(\alpha). \tag{2.27}$$

Actually, though $d^*(\alpha)$ is not the average distance between solutions with the unbiased distribution over formulas, it is the average distance for a biased distribution where each formula is weighted with

$$w(F) = \frac{\mathcal{N}(F)^2}{\sum_{F'} \mathcal{N}(F')^2} \tag{2.28}$$

as can be readily checked upon insertion of $w(F)$ in the numerator of (2.26). We will see in Section (3) how to calculate average properties with the unbiased measure.

Even so definition (2.27) (and (2.26) too) is sloppy. If X is a solution so is $-X$, the configuration where variables values are flipped. Thus the average distance, whatever the weights over formulas, is equal to $\frac{1}{2}$ for any N! The difficulty comes from the ambiguity in how the thermodynamic limit is taken, and is the signature of spontaneous symmetry breaking. In the low temperature phase of the Ising model the magnetization is either $m^* > 0$ or $-m^* < 0$ if an external field h with, respectively, positive or negative vanishing amplitude is added prior to taking the infinite size limit. In the present case what plays the role of the field is a coupling between solutions as is well-known in spin-glass theory [59]. Inserting $\exp[-Nhd(X, Y)]$ in the numerator of (2.27) we obtain, when $N \rightarrow \infty$, d^* if $h \rightarrow 0^+$ and $1 - d^*$ if $h \rightarrow 0^-$. The density μ of probability of distances d between solutions, with the biased measure (2.28), is concentrated below the Sat/Unsat threshold,

$$\mu(d) = \delta\left(d - \frac{1}{2}\right) \quad \text{for} \quad \alpha < \alpha_c, \tag{2.29}$$

and split into two symmetric peaks above the critical ratio,

$$\mu(d) = \frac{1}{2}\delta(d - d^*) + \frac{1}{2}\delta\big(d - (1 - d^*)\big) \quad \text{for} \quad \alpha > \alpha_c. \tag{2.30}$$

The concept of spontaneous symmetry breaking will play a key role in our study of 3-XORSAT (Section 4.3).

2.8. Finite-size scaling (II): critical exponents

Let us summarize what we have found about the probability of satisfying random 2-XORSAT formulas in Section 2.5 and 2.6. Close to the transition we have from (2.18) and (2.25),

$$\ln P_{SAT}(N, \alpha_c + \epsilon) \simeq \begin{cases} \frac{1}{4}\ln(-\epsilon) & \text{when } \epsilon < 0, N \rightarrow \infty \\ -\Omega N\epsilon^3 & \text{when } \epsilon > 0, N \gg 1 \end{cases} . \tag{2.31}$$

The lesson of Section 2.1 is that $\ln P_{SAT}$ may have a non trivial limit when $N \rightarrow \infty$, $\epsilon \rightarrow 0$ provided we keep $y = \epsilon N^\psi$ constant. For 1-XORSAT the exponent ψ was found to be equal to $\frac{1}{2}$, and $\ln P_{SAT}$ to converge to the scaling function $\Phi_1(y)$ (2.3). The situation is similar but slightly more involved for 2-XORSAT. A natural assumption is to look for the existence of a scaling function such that

$$\ln P_{SAT}(N, \epsilon) \simeq N^\rho \Phi_2(\epsilon N^\psi). \tag{2.32}$$

Let us see if (2.32) is compatible with the limiting behaviours (2.31). Fixing $\epsilon < 0$ and sending $N \rightarrow \infty$ we obtain, for $y = \epsilon N^\psi \rightarrow -\infty$, $\frac{1}{4}\ln|y| - \frac{\psi}{4}\ln N$

for the l.h.s, and $N^\rho \times \Phi_2(y)$ for the r.h.s. Hence $\rho = 0$ as in the 1-XORSAT case, but an additive correction is necessary, and we modify scaling Ansatz (2.32) into

$$\ln P_{SAT}(N, \epsilon) \simeq \Phi_2(y = \epsilon N^\psi) - \frac{\psi}{4} \ln N. \tag{2.33}$$

The above equation is now compatible with (2.31) if $\Phi_2(y) \sim \frac{1}{4} \ln |y|$ when $y \to -\infty$. Fixing now $\epsilon > 0$ and sending N to infinity we see that (2.31) is fulfilled if $\Phi_2(y) \sim -\Omega y^3$ when $y \to +\infty$ and

$$\psi = \frac{1}{3}. \tag{2.34}$$

The above value for ψ is expected from the study of random graphs [12] and is related to the size $N^{1-\psi} = N^{\frac{2}{3}}$ of the largest components at the percolation threshold (Section 2.4). ψ is called critical exponent and characterize the width of the critical region of 2-XORSAT. Loosely speaking it means that a formula with N variables and $\frac{N}{2} + \Delta$ equations is 'critical' when $\Delta \sim N^{\frac{2}{3}}$. This information will be useful for the analysis of search algorithms in Section 5.6.

A consequence of (2.33, 2.34) is that, right at the threshold, the probability of satisfaction decays as[8]

$$P_{SAT}(N, \alpha_c) \sim N^{-\frac{1}{12}}. \tag{2.35}$$

This scaling agrees with numerical experiments, though the small value of the decay exponent makes an accurate check delicate (Inset of Figure 5).

2.9. *First and second moments inequalities for the 3-XORSAT threshold*

Figure 6 shows the probability that a random 3-XORSAT formula is satisfiable as a function of α for increasing sizes N. It appears that formulas with ratio $\alpha < \alpha_c \simeq 0.92$ are very likely to be satisfiable in the large N limit, while formulas with ratios beyond this critical value are almost surely unsatisfiable. This behaviour is different from the 2-XORSAT case (Figure 4) in that P_{SAT} seems to tend to unity below threshold.

It is important to realize that, contrary to the 2-XORSAT case, the Sat/Unsat transition is not related to connectivity percolation. Consider indeed a variable, say, x_1. This variable appear, on average, in 3α equations. Each of those equations contain other 2 variables. Hence the 'connectivity' of x_1 is $c = 6\alpha$, which is larger than unity for $\alpha_p = \frac{1}{6}$. In the range $[\alpha_p, \alpha_c]$ the formula is percolating but still satisfiable with high probability. The reason is that cycles do not hinder satisfiability as much as in the 2-XORSAT case.

---

[8]This scaling is correct provided there is no diverging e.g. $O(\ln \ln N)$ corrections to (2.33).

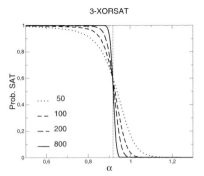

Fig. 6. Probability that a random 3-XORSAT formula is satisfiable as a function of the ratio α of equations per variable, and for various sizes N. The dotted line locates the threshold $\alpha_c \simeq 0.918$.

Use of the first and second moment inequalities (Appendix B) for the number $\mathcal{N}$ of solutions provides us with upper and lower bounds to the Sat/Unsat ratio α_c. The calculation follows the same line as the one of the 2-XORSAT case (Section 2.6). The first moment $\langle \mathcal{N} \rangle = 2^{N(1-\alpha)}$ vanishes for ratios larger than unity, showing that

$$\alpha_c \leq \alpha_1 = 1. \tag{2.36}$$

This upper bound is definitely larger than the true threshold from the numerical findings of Figure 6. We have already encountered this situation in 2-XORSAT: in the $\frac{1}{2} < \alpha < 1$ range formulas are unsatisfiable with probability one (when $N \to \infty$), yet the average number of solutions is exponentially large! The reason is, once more, that the average result is spoiled by rare, satisfiable formulas with many solutions.

As for the second moment expression (2.22, 2.23) still holds with $q(d)$ given by (2.21) with $K = 3$. The absolute maximum of the corresponding function $A(d, \alpha)$ is located in $d^* = \frac{1}{2}$ when $\alpha < \alpha_2 \simeq 0.889$, and $d^* < \frac{1}{2}$ when $\alpha > \alpha_2$. In the latter case $\langle \mathcal{N}^2 \rangle$ is exponentially larger than $\langle \mathcal{N} \rangle^2$, and the second moment inequality (B.2) does not give any information about P_{SAT}. In the former case $\langle \mathcal{N}^2 \rangle$ and $\langle \mathcal{N} \rangle^2$ are equivalent to exponential-in-N order. It is shown in Appendix C that their ratio actually tends to one as $N \to \infty$. We conclude that formulas with ratios of equations per variable less than α_2 are satisfiable with high probability in the infinite size limit, or, equivalently [20],

$$\alpha_c \geq \alpha_2 \simeq 0.889. \tag{2.37}$$

Unfortunately the lower and upper bounds do not match and the precise value of the threshold remains unknown at this stage. We explain in the next section how a simple preprocessing of the formula, before the application of the first and second moment inequalities, can close the gap, and shed light on the structure of the space of solutions.

2.10. Space of solutions and clustering

We start from a simple observation. Assume we have a formula F of 3-XORSAT where a variable, say, x, appears only once, that is, in one equation, say, E : $x + y + z = 0$. Let us call F' the sub-formula obtained from F after removal of equation E. Then the following statement is true: F *is satisfiable if and only if F' is satisfiable.* The proof is obvious: whatever the values of y, z required to satisfy F' equation E can be satisfied by an adequate choice of x, and so can be the whole formula F.

In a random 3-XORSAT formula F with ratio α there are about $N \times 3\alpha\, e^{-3\alpha}$ variables appearing only once in the formula. Removal of those variables (and their equations) produces a shorter formula with $O(N)$ less equations. Furthermore it may happen that variables with multiple occurrences in the original formula have disappeared from the output formula, or appear only once. Hence the procedure can be iterated until no single-occurrence variables are present. We are left with F_2, the largest sub-formula (of the original formula) where every variable appears at least twice.

Many questions can be asked: how many equations are left in F_2? how many variables does it involve? how many solutions does it have? Giving the answers requires a thorough analysis of the removal procedure, with the techniques exposed in Section 5.5 [17, 26, 46]. The outcome depends on the value of the ratio compared to

$$\alpha_d = \min_b -\frac{\log(1-b)}{3b^2} \simeq 0.8184\ldots \tag{2.38}$$

hereafter called clustering threshold. With high probability when $N \to \infty$ F_2 is empty if $\alpha < \alpha_d$, and contains an extensive number of equations, variables when $\alpha > \alpha_d$. In the latter case calculation of the first and second moments of the number of solutions of F_2 shows that this number does not fluctuate around the value $e^{N s_{cluster}(\alpha)+o(N)}$ where

$$s_{cluster}(\alpha) = (b - 3\alpha b^2 + 2\alpha b^3) \ln 2 \tag{2.39}$$

and b is the strictly positive solution of the self-consistent equation

$$1 - b = e^{-3\alpha b^2}. \tag{2.40}$$

Hence F_2 is satisfiable if and only if $\alpha < \alpha_c$ defined through $s_{cluster}(\alpha_c) = 0$, that is,

$$\alpha_c \simeq 0.9179\ldots. \tag{2.41}$$

This value is, by virtue of the equivalence between F and F_2 the Sat/Unsat threshold for 3-XORSAT, in excellent agreement with Figure 6.

How can we reconstruct the solutions of F from the ones of F_2? The procedure is simple. Start from one solution of F_2 (empty string if $\alpha < \alpha_d$). Then introduce back the last equation which was removed since it contained $n \geq 1$ single-occurrence variable. If $n = 1$ we fix the value of this variable in a unique way. If $n = 2$ (respectively $n = 3$) there are 2 (respectively, 4) ways of assigning the reintroduced variables, defining as many solutions from our initial, partial solution. Reintroduction of equations one after the other according to the Last In – First Out order gives us more and more solutions from the initial one, until we get a bunch of solutions of the original formula F. It turns out that the number of solutions created this way is $e^{N s_{in}(\alpha)+o(N)}$ where

$$s_{in}(\alpha) = (1 - \alpha)\ln 2 - s_{cluster}(\alpha). \tag{2.42}$$

The above formula is true for $\alpha > \alpha_d$, and should be intended as $s_{in}(\alpha) = (1 - \alpha)\ln 2$ for $\alpha < \alpha_d$. These two entropies are shown in Figure 7. The total entropy, $s^*(\alpha) = s_{in}(\alpha) + s_{cluster}(\alpha)$, is simply $(1 - \alpha)\ln 2$ for all ratios smaller than the Sat/Unsat threshold. It shows no singularity at the clustering threshold. However a drastic change in the structure of the space of solutions takes place, symbolized in the phase diagram of Figure 8:

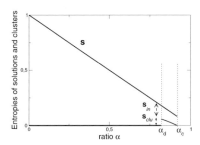

Fig. 7. Entropies (base 2 logarithms divided by size N) of the numbers of solutions and clusters as a function of the ratio α. The entropy of solutions equals $1 - \alpha$ for $\alpha < \alpha_c \simeq 0.918$. For $\alpha < \alpha_d \simeq 0.818$, solutions are uniformly scattered on the N-dimensional hypercube. At α_d the solution space discontinuously breaks into disjoint clusters. The entropies of clusters, $s_{cluster}$, and of solutions in each cluster, s_{in}, are such that $s_{cluster} + s_{in} = s$. At α_c the number of clusters stops being exponentially large ($s_{cluster} = 0$). Above α_c there is almost surely no solution.

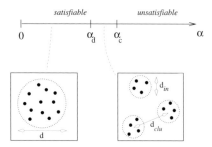

Fig. 8. Phase diagram of 3-XORSAT. A 'geometrical' phase transition takes place in the satisfiable phase at $\alpha_d \simeq 0.818$. At small ratios $\alpha < \alpha_d$ solutions are uniformly scattered on the N-dimensional hypercube, with a typical normalized Hamming distance $d = \frac{1}{2}$. At α_d the solution space discontinuously breaks into disjoint clusters: the Hamming distance $d_{in} \simeq 0.14$ between solutions inside a cluster is much smaller than the typical distance $d_{clu} = \frac{1}{2}$ between two clusters.

• For ratios $\alpha < \alpha_d$ the intensive Hamming distance between two solutions is, with high probability, equal to $d = 1/2$. Solutions thus differ on $N/2 + o(N)$ variables, as if they were statistically unrelated assignments of the N Boolean variables. In addition the space of solutions enjoys some connectedness property. Any two solutions are connected by a path (in the space of solutions) along which successive solutions differ by a bounded number of variables. Loosely speaking one is not forced to cross a big region deprived of solutions when going from one solution to another.

• For ratios $\alpha > \alpha_d$ the space of solutions is not connected any longer. It is made of an exponentially large (in N) number $\mathcal{N}_{clu} = e^{N s_{cluster}}$ of connected components, called clusters, each containing an exponentially large number $\mathcal{N}_{in} = e^{N s_{in}}$ of solutions. Two solutions belonging to different clusters lie apart at a Hamming distance $d_{clu} = 1/2$ while, inside a cluster, the distance is $d_{in} < d_{clu}$. b given by (2.40) is the fraction of variables having the same value in all the solutions of a cluster (defined as the backbone).

We present in Sections 3 and 4 statistical physics tools developed to deal with the scenario of Figure 8.

3. Advanced methods (I): replicas

3.1. From moments to large deviations for the entropy

The analysis of Section 2.6 has shown that the first, and second moments of the number $\mathcal{N}$ of solutions are dominated by rare formulas with a lot of solutions. Let us define the intensive entropy s through $\mathcal{N} = e^{N s}$. As $\mathcal{N}$ is random (at fixed

α, N) so is s. We assume that the distribution of s can be described, in the large size limit, by a rate function $\omega(s)$ (which depends on α). Hence,

$$\langle \mathcal{N}^q \rangle = \int ds \, e^{-N\omega(s)} \times (e^{Ns})^q \sim \exp[N \max_s (qs - \omega(s))] \tag{3.1}$$

using the Laplace method. If we are able to estimate the leading behaviour of the qth moment of the number of solutions when N gets large at fixed α,

$$\langle \mathcal{N}^q \rangle \sim e^{Ng(q)}, \tag{3.2}$$

then ω can be easily calculated by taking the Legendre transform of g. In particular the typical entropy is obtained by $s^* = \frac{dg}{dq}(q \to 0)$. This is the road we will follow below. We will show how $g(q)$ can be calculated when q takes integer values, and then perform an analytic continuation to non integer q. The continuation leads to substantial mathematical difficulties, but is not uncommon in statistical physics e.g. the $q \to 1$ limit of the q-state Potts model to recover percolation, or the $n \to 0$ limit of the $O(n)$ model to describe self-avoiding walks.

To calculate the qth moment we will have to average over the random components of formulas F, that is, the K-uplets of index variables in the first members and the $v = 0, 1$ second members. Consider now homogeneous formulas F_h whose first members are randomly drawn in the same way as for F, but with all second members $v = 0$. The number $\mathcal{N}_h$ of solutions of a homogeneous formula is always larger or equal to one. It is a simple exercise to show that

$$\langle \mathcal{N}^{q+1} \rangle = 2^{N(1-\alpha)} \times \langle (\mathcal{N}_h)^q \rangle, \tag{3.3}$$

valid for any positive integer q.[9] Therefore it is sufficient to calculate the moments of $\mathcal{N}_h = e^{Ng_h(q)}$ since (3.3) gives a simple identity between $g(q + 1)$ and $g_h(q)$. This technical simplification has a deep physical meaning we will comment in Section 4.3.

3.2. Free energy for replicated variables

The qth power of the number of solutions to a homogeneous system reads

$$(\mathcal{N}_h)^q = \left[\sum_X \prod_{\ell=1}^{M} e_\ell(X) \right]^q = \sum_{X^1, X^2, \ldots, X^q} \prod_{\ell=1}^{M} \prod_{a=1}^{q} e_\ell(X^a), \tag{3.4}$$

where $e_\ell(X)$ is 1 if equation ℓ is satisfied by assignment X. The last sum runs over q assignments X^a, with $a = 1, 2, \ldots, q$ of the Boolean variables, called

---

[9] Actually the identity holds for $q = 0$ too, and is known under the name of harmonic mean formula [4].

replicas of the original assignment X. It will turn useful to denote by $\vec{x}_i = (x_i^1, x_i^2, \ldots, x_i^q)$ the q-dimensional vector whose components are the values of variable x_i in the q replicas. To simplify notations we consider the case $K = 3$ only here, but extension to other values of K is straightforward. Averaging over the instance, that is, the triplets of integers labeling the variables involved in each equation ℓ, leads to the following expression for the qth moment,

$$
\langle (\mathcal{N}_h)^q \rangle = \sum_{X^1, X^2, \ldots, X^q} \left\langle \prod_{a=1}^{q} e(X^a) \right\rangle^M
$$

$$
= \sum_{X^1, X^2, \ldots, X^q} \left[\frac{1}{N^3} \sum_{1 \leq i, j, k \leq N} \delta_{\vec{x}_i + \vec{x}_j + \vec{x}_k} + O\left(\frac{1}{N} \right) \right]^M \tag{3.5}
$$

where $\delta_{\vec{x}} = 1$ if the components of $\vec{x}$ are all null mod. 2, and 0 otherwise. We now proceed to some formal manipulations of the above equation (3.5).

First step. Be $\mathcal{X} = \{X^1, X^2, \ldots, X^q\}$ one of the 2^{qN} replica assignment. Focus on variable i, and its attached assignment vector, $\vec{x}_i$. The latter may be any of the 2^q possible vectors e.g. $\vec{x}_i = (1, 0, 1, 0, 0, \ldots, 0)$ if variable x_i is equal to 0 in all but the first and third replicas. The histogram of the assignments vectors given replica assignment $\mathcal{X}$,

$$
\rho(\vec{x}|\mathcal{X}) = \frac{1}{N} \sum_{i=1}^{N} \delta_{\vec{x} - \vec{x}_i}, \tag{3.6}
$$

counts the fraction of assignments vectors $\vec{x}_i$ having value $\vec{x}$ when i scans the whole set of variables from 1 to N. Of course, this histogram is normalized to unity,

$$
\sum_{\vec{x}} \rho(\vec{x}) = 1, \tag{3.7}
$$

where the sum runs over all 2^q assignment vectors. An simple but essential observation is that the r.h.s. of (3.5) may be rewritten in terms of the above histogram,

$$
\frac{1}{N^3} \sum_{1 \leq i, j, k \leq N} \delta_{\vec{x}_i + \vec{x}_j + \vec{x}_k} = \sum_{\vec{x}, \vec{x}'} \rho(\vec{x}) \, \rho(\vec{x}') \, \rho(\vec{x} + \vec{x}'). \tag{3.8}
$$

Keep in mind that ρ in (3.6,3.8) depends on the replica assignment $\mathcal{X}$ under consideration.

Second step. According to (3.8), two replica assignments $\mathcal{X}_1$ and $\mathcal{X}_2$ defining the same histogram ρ will give equal contributions to $\langle(\mathcal{N}_h)^q\rangle$. The sum over replica assignments $\mathcal{X}$ can therefore be replaced over the sum over possible histograms provided the multiplicity $\mathcal{M}$ of the latter is taken properly into account. This multiplicity is also equal to the number of combinations of N elements (the $\vec{x}_i$ vectors) into 2^q sets labeled by $\vec{x}$ and of cardinalities $N\rho(\vec{x})$. We obtain

$$\langle(\mathcal{N}_h)^q\rangle = \sum_{\{\rho\}}^{(norm)} e^{N\mathcal{G}_h(\{\rho\},\alpha)} + o(N), \tag{3.9}$$

where the (*norm*) subscript indicates that the sum runs over histograms ρ normalized according to (3.7), and

$$\mathcal{G}_h(\{\rho\},\alpha) = -\sum_x \rho(x)\ln\rho(x) + \alpha\ln\left[\sum_{\vec{x},\vec{x}'}\rho(\vec{x})\,\rho(\vec{x}')\,\rho(\vec{x}+\vec{x}')\right]. \tag{3.10}$$

In the large N limit, the sum in (3.9) is dominated by the histogram ρ^* maximizing the functional $\mathcal{G}_h$.

Third step. Maximization of function $\mathcal{G}_h$ over normalized histograms can be done within the Lagrange multiplier formalism. The procedure consists in considering the modified function

$$\mathcal{G}_h^{LM}(\{\rho\},\lambda,\alpha) = \mathcal{G}_h(\{\rho\},\alpha) + \lambda\left(1 - \sum_{\vec{x}}\rho(\vec{x})\right), \tag{3.11}$$

and first maximize $\mathcal{G}_h^{LM}$ with respect to histograms ρ without caring about the normalization constraint, and then optimize the result with respect to λ. We follow this procedure with $\mathcal{G}_h$ given by (3.10). Requiring that $\mathcal{G}_h^{LM}$ be maximal provides us with a set of 2^q coupled equations for ρ^*,

$$\ln\rho^*(\vec{x}) + 1 + \lambda - 3\alpha\,\frac{\displaystyle\sum_{\vec{x}'}\rho^*(\vec{x}')\,\rho^*(\vec{x}+\vec{x}')}{\displaystyle\sum_{\vec{x}',\vec{x}''}\rho^*(\vec{x}')\,\rho^*(\vec{x}'')\,\rho^*(\vec{x}'+\vec{x}'')} = 0, \tag{3.12}$$

one for each assignment vector $\vec{x}$. The optimization equation over λ implies that λ in (3.12) is such that ρ^* is normalized. At this point of the above and rather abstract calculation it may help to understand the interpretation of the optimal histogram ρ^*.

3.3. The order parameter

We have already addressed a similar question at the end of the second moment calculation in Section 2.7. The parameter d^* coming out from the calculation was the (weighted) average Hamming distance (2.27) between two solutions of the same random instance. The significance of ρ^* is identical. Consider q' solutions labeled by $a = 1, 2, \ldots, q'$ of the same random and homogeneous instance and a variable, say, x_i. What is the probability, over instances and solutions, that this variable takes, for instance, value 0 in the first and fourth solutions, and 1 in all other solutions? In other words, what is the probability that the assignment vector $\vec{x}_i = (x_i^1, x_i^2, \ldots, x_i^{q'})$ is equal to $\vec{x}' = (0, 1, 1, 0, 1, 1, \ldots, 1)$? The answer is

$$
p(\vec{x}') = \left\langle \frac{1}{(\mathcal{N}_h)^{q'}} \sum_{X^1, X^2, \ldots, X^{q'}} \delta_{\vec{x}_i - \vec{x}} \prod_{l=1}^{M} \prod_{a=1}^{q} e_\ell(X^a) \right\rangle \tag{3.13}
$$

where the dependence on i is wiped out by the average over the instance. The above probability is an interesting quantity; it provides us information about the 'microscopic' nature of solutions. Setting $q' = 1$ gives us the probabilities $p(0)$, $p(1)$ that a variable is false or true respectively, that is, takes the same value as in the null assignment or not. For generic q' we may think of two extreme situations:

• a flat p over assignment vectors, $p(\vec{x}') = 1/2^{q'}$, corresponds to essentially orthogonal solutions;
• on the opposite, a concentrated probability e.g. $p(\vec{x}') = \delta_{\vec{x}'}$ implies that variables are extremely constrained, and that the (almost) unique solution is the null assignment.

The careful reader will have already guessed that our calculation of the qth moment gives access to a weighted counterpart of p. The order parameter

$$
\rho^*(\vec{x}) = \frac{1}{\langle (\mathcal{N}_h)^q \rangle} \sum_{X^1, X^2, \ldots, X^q} \delta_{\vec{x}_i - \vec{x}} \left\langle \prod_{l=1}^{M} \prod_{a=1}^{q} e_\ell(X^a) \right\rangle, \tag{3.14}
$$

is not equal to p even when $q = q'$. However, at the price of mathematical rigor, the exact probability p over vector assignments of integer length q' can be reconstructed from the optimal histogram ρ^* associated to moments of order q when q is real-valued and sent to 0. The underlying idea is the following. Consider (3.14) and an integer $q' < q$. From any assignment vector $\vec{x}$ of length q, we define two assignment vectors $\vec{x}'$, $\vec{x}''$ of respective lengths q', $q - q'$ corresponding to the first q' and the last $q - q'$ components of $\vec{x}$ respectively. Summing (3.14) over the $2^{q-q'}$ assignment vectors $\vec{x}''$ gives,

$$\sum_{\vec{x}''} \rho^*(\vec{x}', \vec{x}'') = \frac{1}{\langle (\mathcal{N}_h)^q \rangle} \sum_{\{X^a\}} \delta_{\vec{x}'_i - \vec{x}'} \left\langle (\mathcal{N}_h)^{q-q'} \prod_{l,a} e_\ell(X^a) \right\rangle. \tag{3.15}$$

As q now appears in the powers of $\mathcal{N}_h$ in the numerator and denominator only, it can be formally send to zero at fixed q', yielding

$$\lim_{q \to 0} \sum_{\vec{x}''} \rho^*(\vec{x}', \vec{x}'') = p(\vec{x}') \tag{3.16}$$

from (3.13). This identity justifies the denomination "order parameter" given to ρ^*.

Having understood the significance of ρ^* helps us to find appropriate solutions to (3.12). Intuitively and from the discussion of the first moment case $q = 1$, p is expected to reflect both the special role of the null assignment (which is a solution to all homogeneous systems) and the ability of other solutions of a random system to be essentially orthogonal to this special assignment. A possible guess is thus

$$p(\vec{x}') = \frac{1 - b}{2^{q'}} + b\delta_{\vec{x}'}, \tag{3.17}$$

where b expresses some degree of 'correlation' of solutions with the null one. Hypothesis (3.17) interpolates between the fully concentrated ($b = 1$) and flat ($b = 0$) probabilities. b measures the fraction of variables (among the N ones) that take the 0 values in all q' solution, and coincides with the notion of backbone introduced in Section 2.10. Hypothesis (3.17) is equivalent, from the connection (3.16) between p and the annealed histogram ρ^* to the following guess for the solution of the maximization condition (3.12),

$$\rho^*(\vec{x}) = \frac{1 - b}{2^q} + b\delta_{\vec{x}}. \tag{3.18}$$

Insertion of Ansatz (3.18) in (3.12) shows that it is indeed a solution provided b is shrewdly chosen as a function of q and α, $b = b^*(q, \alpha)$. Its value can be either found from direct resolution of (3.12), or from insertion of histogram (3.18) in $\mathcal{G}_h$ (3.10) and maximization over b, with the result,

$$g_h(q, \alpha) = \max_{0 \le b \le 1} A_h(b, q, \alpha) \tag{3.19}$$

where

$$A_h(b, q, \alpha) = -\left(1 - \frac{1}{2^q}\right)(1 - b)\ln\left(\frac{1 - b}{2^q}\right)$$
$$-\left(b + \frac{1 - b}{2^q}\right)\ln\left(b + \frac{1 - b}{2^q}\right) + \alpha\ln\left(b^3 + \frac{1 - b^3}{2^q}\right), \tag{3.20}$$

where the maximum is precisely reached in b^*. Notice that, since ρ^* in (3.18) is entirely known from the value of b^*, we shall indifferently call order parameter ρ^*, or b^* itself.

3.4. Results

Numerical investigation of A_h (3.20) shows that: for $\alpha < \alpha_M(q)$ the only local maximum of A_h is located in $b^* = 0$, and $A_h(q, \alpha) = q(1 - \alpha)\ln 2$; when $\alpha_M(q) < \alpha < \alpha^*(q)$, there exists another local maximum in $b > 0$ but the global maximum is still reached in $b^* = 0$; when $\alpha > \alpha^*(q)$, the global maximum is located in $b^* > 0$. This scenario extends to generic q the findings of the second moment calculation carried out in Section 2.6. The α_M and α^* lines divide the q, α plane as shown in Figure 9. Notice that, while the black dots in Figure 9 correspond to integer-valued q, the continuous lines are the output of the implicit analytic continuation to real q done by the replica calculation.

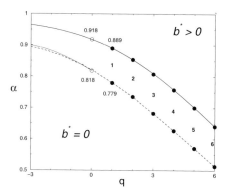

Fig. 9. The q, α plane and the critical lines $\alpha_M(q)$ (dashed), $\alpha^*(q)$ (full tick), and $\alpha_s(q)$ (full thin) appearing in the calculation of the qth moment for homogeneous 3-XORSAT systems. Full dots correspond to integer q values, while continuous curves result from the analytic continuation to real q. The fraction of variables in the backbone, b^*, vanishes below the line $\alpha_M(q)$; the global maximum of A_h in (3.20) is located in $b^* > 0$ for ratios $\alpha > \alpha^*(q)$. Ansatz (3.18) is locally unstable in the hardly visible domain $q < 0$, $\alpha_M(q) < \alpha < \alpha_s(q)$.

Taking the derivative of (3.19) with respect to q and sending $q \to 0$ we obtain the typical entropy of a homogeneous 3-XORSAT system at ratio α,

$$s_h^*(\alpha) = \ln 2 \times \max_{0 \leq b \leq 1} \left[(1 - b)\left(1 - \ln(1 - b)\right) - \alpha(1 - b^3) \right]. \tag{3.21}$$

The optimal value for b coincides with the solution of (2.40). The typical entropy is plotted in Figure 10, and is equal to:
- $(1 - \alpha) \ln 2$ when $\alpha < \alpha_c \simeq 0.918$ (Figure 9); in this range of ratios, homogeneous and full (with random second members) systems have essentially the same properties, with the same cluster Organisation of solutions, and identical entropies of solutions.
- a positive but rapidly decreasing function given by (3.21) when $\alpha > \alpha_c$; above the critical ratio, a full system has no solution any more, while a homogeneous instance still enjoys a positive entropy. The expression for $s_h^*(\alpha)$ coincides with the continuation to $\alpha > \alpha_c$ of the entropy $s_{in}(\alpha)$ (2.42) of solutions in a single cluster for a full system. In other words, a single cluster of solutions, the one with the null solution, survive for ratios $\alpha > \alpha_S$ in homogeneous systems.

Atypical instances can be studied and the large deviation rate function for the entropy can be derived from (3.19) for homogeneous systems, and using equivalence (3.3), for full systems. Minimizing over the entropy we obtain the rate function $\omega_3(\alpha)$ associated to the probability that a random 3-XORSAT system is satisfiable, with the result shown in Figure 10. As expected we find $\omega_3 = 0$ for $\alpha < \alpha_c$ and $\omega_3 > 0$ for $\alpha > \alpha_c$, allowing us to locate the Sat/Unsat threshold.

Notice that the emergence of clustering can be guessed from Figure 9. It coincides with the appearance of a local maximum of A_h (3.20) with a non vanishing

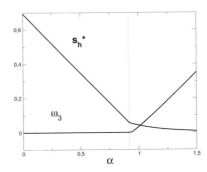

Fig. 10. Rate function ω_3 for the probability of satisfaction of full (bottom curve) and entropy s_h^* of solutions for homogeneous (top curve) 3-XORSAT systems vs. α. The vertical dotted lines indicate the critical Sat/Unsat threshold, $\alpha_c \simeq 0.918$. For $\alpha < \alpha_c$ $\omega_3 = 0$, and $s_h^* = (1 - \alpha) \ln 2$ is the same as for full systems. Above the threshold $\omega^* < 0$. Homogeneous systems are, of course, always satisfiable: the entropy s_h^* is a positive but quickly decreasing function of α.

backbone b. While in the intermediate phase $\alpha_d < \alpha < \alpha_c$, the height of the global maximum equals the total entropy s^*, the height of the local maximum coincides with the entropy of clusters $s_{cluster}$ (2.39).

3.5. Stability of the replica Ansatz

The above results rely on Ansatz (3.18). A necessary criterion for its validity is that ρ^* locates a true local maximum of $\mathcal{G}_h$, and not merely a saddle-point. Hence we have to calculate the Hessian matrix of $\mathcal{G}_h$ in ρ^*, and check that the eigenvalues are all negative [9]. Differentiating (3.10) with respect to $\rho(\vec{x})$ and $\rho(\vec{x}')$ we obtain the Hessian matrix

$$H(\vec{x}, \vec{x}') = -\frac{\delta_{\vec{x}+\vec{x}'}}{\rho^*(\vec{x})} + 6\alpha \frac{\rho^*(\vec{x} + \vec{x}')}{D} - 9\alpha \frac{N(\vec{x})}{D} \frac{N(\vec{x}')}{D}, \qquad (3.22)$$

where $D = \frac{1-b^3}{2^q} + b^3$, $N(\vec{x}) = \frac{1-b^2}{2^q} + b^2 \delta_{\vec{x}}$. We use b instead of b^* to lighten the notations, but it is intended that b is the backbone value which maximizes A_h (3.20) at fixed q, α. To take into account the global constraint over the histogram (3.7) one can express one fraction, say, $\rho(\vec{0})$, as a function of the other fractions $\rho(\vec{x})$, $\vec{x} \neq \vec{0}$. $\mathcal{G}_H$ is now a function of $2^q - 1$ independent variables, with a Hessian matrix $\tilde{H}$ simply related to H,

$$\tilde{H}(\vec{x}, \vec{x}') = H(\vec{x}, \vec{x}') - H(\vec{x}, \vec{0}) - H(\vec{0}, \vec{x}') + H(\vec{0}, \vec{0}). \qquad (3.23)$$

Plugging expression (3.22) into (3.23) we obtain

$$\tilde{H}(\vec{x}, \vec{x}') = \lambda_R \delta_{\vec{x}+\vec{x}'} + \frac{1}{2^q - 1}(\lambda_L - \lambda_R) \quad \text{where}$$

$$\lambda_R = 6\alpha \frac{b}{D} - \frac{2^q}{1-b} \qquad (3.24)$$

$$\lambda_L = 2^q \left(6\alpha \frac{b}{D} - \frac{2^q}{(1-b)(1-b+2^q b)} - 9\alpha(1-2^{-q}) \frac{b^4}{D^2} \right).$$

Diagonalisation of $\tilde{H}$ is immediate, and we find two eigenvalues:
- λ_L (non degenerate). The eigenmode corresponds to a uniform infinitesimal variation of $\rho(\vec{x})$ for all $\vec{x} \neq \vec{0}$, that is, a change of b in (3.18). It is an easy check that

$$\lambda_L = \frac{2^q}{1-2^{-q}} \frac{\partial^2 A_h}{\partial b^2}(b, q, \alpha), \qquad (3.25)$$

where A_h is defined in (3.20). As we have chosen b to maximize A_h this mode, called longitudinal in replica literature [9], is stable.[10]

---

[10]Actually b^* is chosen to *minimize* A_h when $q < 0$, thus λ_L has always the right negative sign.

• λ_R ($2^q - 2$-fold degenerate): the eigenmodes correspond to fluctuations of the order parameter ρ transverse to the replica subspace described by (3.18), and are called replicon in spin-glass theory [9]. Inspection of λ_R as a function of α, q shows that it is always negative when $q > 0$. For $q < 0$ the replicon mode is stable if

$$\alpha > \alpha_s(q) = \frac{1 - b^3 + 2^q b^3}{6b(1 - b)}. \qquad (3.26)$$

which is a function of q only once we have chosen $b = b^*(q, \alpha_s)$.

The unstable region $q < 0, \alpha_M(q) < \alpha < \alpha_s(q)$ is shown in Figure 9 and is hardly visible when $q > -3$. In this region a continuous symmetry breaking is expected [43]. In particular α_s stay below the α^* line for small (in absolute value) and negative q. We conclude that our Ansatz (3.18) defines a maximum of $\mathcal{G}_h$.

Is it the global maximum of $\mathcal{G}_h$? There is no simple way to answer this question. Local stability does not rule out the possibility for a discontinuous transition to another maximum in the replica order parameter space not described by (3.18). A final remark is that a similar calculation can be done for any value of K. The outcome for $K = 2$ is the rate function ω_2 plotted in Figure 5, in good agreement with numerics close to the threshold.

4. Advanced methods (II): cavity

The cavity method, in the context of disordered systems, was historically developed as an alternative to the the replica method [43]. Its application to spin systems on random graphs is extensively explained in [44], and we limit ourselves here to briefly show how it gives back the 3-XORSAT scenario of Section 2.10 [46].

Let us consider a system F involving variables x_i, $i = 1, \ldots, N$. In the following we will indifferently use the variable $x_i = 0, 1$ or its spin representation $S_i = (-1)^{x_i} = \pm 1$ when convenient. Let us define the GS energy $E_i^F(S_i)$ of the system when the ith spin is kept fixed, that is, the minimal number of violated equations in F, taken over the 2^{N-1} configurations. We may always write

$$E_i^F(S_i) = -w_i - h_i S_i, \qquad (4.1)$$

where h_i is called 'field' acting on spin S_i. For a homogeneous system $E_i^F(+1) = 0$, and $E_i^F(-1) = n_i$ for some integer n_i. Hence $h_i = \frac{n_i}{2}$ takes half-integer values.

The above definition can be extended to the case of $\ell > 1$ fixed spins. Let $I \subset \{1, 2, \ldots, N\}$ be a subset of the indices of cardinal $|I| \geq 2$, and S_I denote one of the $2^{|I|}$ configurations of the spins $S_i, i \in I$. The GS energy of F for given S_I can in general be written as

$$E_I^F(S_I) = -w_I - \sum_{i \in I} h_i S_i - \sum_{I' \subset I : |I'| \geq 2} J_{I'} \prod_{i \in I'} S_i \qquad (4.2)$$

where the h_is are the fields and the $J_{I'}$s are effective couplings between subsets of spins.

The basic cavity assumption is that effective couplings are vanishingly small: $J_{I'} = 0$ for every subset I'. This apparently bold hypothesis critically relies on a general property of random graphs (from which our system is built on). Define the distance between two vertices as the minimal number of edges on paths linking these two points. Then vertices in a finite subset are, with high probability when $N \to \infty$, typically at infinite distance from each other.[11] When correlations between variables in GS extinguish with the distance *i.e.* when the correlation length is finite the cavity assumption is correct in the large N limit [43,53]. The assumption will break down when correlations subsist on infinite distance, which happens to be the case in the clustered phase.

4.1. Self-consistent equation for the fields

Under the assumption that couplings between randomly picked up spins are null we are left with the fields only. The goal of this section is to show how to calculate those fields, or more precisely, their probability distribution. The derivation is based on the addition procedure already used in the calculation of the size of the giant component in random graphs (Section 2.4).

Consider a system F over N variables to which we want to add one equation involving one new variable S, and two variables S_1, S_2 appearing in F. The energy function associated to this equation is

$$e(S, S_1, S_2) = \frac{1}{2}(1 - \sigma S S_1 S_2) \qquad (4.3)$$

where $\sigma = +1$, respectively -1, when the second member of the equation is 0, resp. 1. Let us calculate the GS energy of the new system $F' = F +$ added equation when the new variable S is kept fixed,

$$E^{F'}(S) = \min_{S_1, S_2} \left[e(S, S_1, S_2) + E_{1,2}^F(S_1, S_2) \right] = -w - uS. \qquad (4.4)$$

---

[11] An alternative formulation is, for finite size N, that the shortest loops (in extensive number) have lengths of the order of $\log N$ [12].

With the cavity hypothesis the couplings between spins S_1, S_2 is null and the minimization is straightforward. We deduce the following explicit expression for the field acting on S (called bias in the cavity literature [44]),

$$u = \frac{\sigma}{2} \text{sign}(h_1 h_2). \tag{4.5}$$

Suppose we now add $\ell \geq 1$ (and not only one) equations. The above calculation can be easily repeated. The absence of couplings make the total field acting on S a linear combination of the fields coming from each new equation,

$$h = \sum_{j=1}^{\ell} u^j, \tag{4.6}$$

where u^j is calculated from (4.5) and each pair of fields (h_1^j, h_2^j) acting on the spins in the jth equation, $j = 1, \ldots, \ell$.

How many equations should we add for our new system over $N + 1$ variables to have the same statistical features as old one over N variables? First ℓ should be Poisson distributed with parameter 3α. Then, given ℓ, we randomly chose ℓ pairs of variables; for each pair the corresponding bias u can be calculated from (4.5). Assume the output is a set of ℓ independent biases, taking values

$$u = \begin{cases} +\frac{1}{2} & \text{with probability } a_+ \\ 0 & \text{with probability } a_0 \\ -\frac{1}{2} & \text{with probability } a_- \end{cases} \tag{4.7}$$

Obviously $a_+ + a_0 + a_- = 1$. Summing over the equations as in (4.6) we obtain the distribution of the field h acting on the new spin at fixed ℓ,

$$p(h|\ell) = \sum_{\ell_+, \ell_0, \ell_-} \binom{\ell}{\ell_+, \ell_0, \ell_-} a_+^{\ell_+} a_0^{\ell_0} a_-^{\ell_-} \delta_{h - \frac{1}{2}(\ell_+ - \ell_-)}. \tag{4.8}$$

Finally we sum over the Poisson distribution for ℓ to obtain the distribution of fields h,

$$p(h) = e^{-3\alpha(1-a_0)} \sum_{\ell_+, \ell_-} \frac{(3\alpha)^{\ell_+ + \ell_-}}{\ell_+! \ell_-!} a_+^{\ell_+} a_-^{\ell_-} \delta_{h - \frac{1}{2}(\ell_+ - \ell_-)}. \tag{4.9}$$

In turn we calculate the distribution of the biases from the one of the fields through (4.5). The outcome are the values of the probabilities (4.7) in terms of p,

$$a_+ = \sum_{h_1,h_2:h_1h_2>0} p(h_1)p(h_2), \quad a_- = \sum_{h_1,h_2:h_1h_2<0} p(h_1)p(h_2),$$

$$a_0 = \sum_{h_1,h_2:h_1h_2=0} p(h_1)p(h_2) = 2p(0) - p(0)^2. \tag{4.10}$$

The above equations together with (4.9) define three self-consistent conditions for a_0, a_+, a_-. Notice that the free energy can be calculated along the same lines [44].

4.2. Application to homogeneous and full systems

In the case of homogeneous systems ($\sigma = +1$) we expect all the fields to be positive, and look for a solution of (4.10) with $a_- = 0$. Then $p(h)$ (4.9) is a Poisson distribution for the integer-valued variable $2h$, with parameter $3\alpha a_+$. The self-consistent equation (4.10) reads

$$a_0 = 1 - a_+ = 2e^{-3\alpha a_+} - e^{-6\alpha a_+} = 1 - (1 - e^{-3\alpha a_+})^2 \tag{4.11}$$

which coincides with (2.40) with the definition $b = \sqrt{a_+}$. As expected

$$b = \sum_{h\geq\frac{1}{2}} p(h) \tag{4.12}$$

is the fraction of frozen variables (which cannot be flipped from 0 to 1 in GS assignments), in agreement with the notion of backbone of Section 2.10.

The energy is zero at all ratio α by construction. As for the entropy consider adding a new equation to the system F (but with no new variable). With probability $1 - b^3$ at least one of the three variables in the new equation e was not frozen prior to addition, and the number of solutions of the new system $F + e$ is half the one of F. With probability b^3 all three variables are frozen in F (to the zero value) and the number of solutions of $F + e$ is the same as the one of F. Hence the average decrease in entropy is

$$Ns_h^*\left(\alpha + \frac{1}{N}\right) - Ns_h^*(\alpha) \simeq \frac{ds_h^*}{d\alpha} = -(1 - b^3)\ln 2. \tag{4.13}$$

The same differential equation can be obtained by differentiating (3.21). With the limit condition $s_h^*(\alpha \to \infty) = 0$ we obtain back the correct expression for the average entropy of homogeneous systems. The entropy is equal to $(1-\alpha)\ln 2$ at $\alpha = \alpha_c$, and becomes smaller when the ratio decreases. This shows that the solution $b = 0$ must be preferred in this regime to the metastable $b > 0$ solution. We conclude that the cavity assumption leads to sensible results for homogeneous systems at all ratios α.

In full systems the sign σ entering (4.5) takes ± 1 values with equal probabilities. We thus expect $p(h)$ to be an even distribution, and $a_+ = a_- = \frac{1}{2}(1 - a_0)$. Remark that a solution with $a_0 < 1$ cannot exist in the satisfiable phase. It would allow two added equations to impose opposite non zero biases to the new variable *i.e.* to constraint this variable to take opposite values at the same time. Given a_0 we calculate from (4.9) the probability that the field vanishes,

$$p(0) = e^{-3\alpha(1-a_0)} \sum_{\ell=0}^{\infty} \left[\frac{3\alpha}{2}(1 - a_0) \right]^{2\ell} \frac{1}{\ell!^2} \qquad (4.14)$$

and, in turn, derive from (4.10) a self-consistent equation for a_0. Numerical investigations show that $a_0 = 1$ is the unique solution for $\alpha < \alpha_T = 1.167$. When $\alpha > \alpha_T$ there appears another solution with $a_0 < 1$. The clustering and Sat/Unsat transitions are totally absent. This result, incompatible with the exact picture of random 3-XORSAT exposed in Section 2.10, shows that the simple cavity hypothesis does not hold for full systems.

4.3. Spontaneous symmetry breaking between clusters

In the clustered phase variables are known to be strongly correlated and the cavity assumption has to be modified. Actually from what we have done above in the homogeneous case we guess that the independence condition still holds if we can in some way restrict the whole space of solutions to one cluster. To do so we explicitly break the symmetry between clusters as follows [48,59].

Let S_i^*, $i = 1, \ldots, N$ be a reference solution of a full satisfiable system F, and F_h the corresponding homogeneous system. We define the local gauge transform $S_i \rightarrow \hat{S}_i = S_i \times S_i^*$. $\{S\}$ is a solution of F if and only if $\{\hat{S}\}$ is a solution of F_h. As the cavity assumption is correct for the homogeneous system we obtain the distribution of fields $\hat{h}_i \geq 0$ from (4.9). Gauging back to the original spin configuration gives us the fields

$$h_i = S_i^* \times \hat{h}_i. \qquad (4.15)$$

It turns out that the above fields depend only on the cluster to which the reference solution belongs. Indeed for the fraction $1 - b$ of the non frozen spins, $\hat{h}_i = h_i = 0$. For the remaining fraction b of spins in the backbone $\hat{h}_i > \frac{1}{2}$ and S_i^* has a unique value for all solutions in the cluster (Section 2.10). Hence the fields h_i are a function of cluster (c) containing $\{S^*\}$, and will be denoted by $h_i^{(c)}$.

What modification has to be brought to the cavity assumption of Section 4.1 is now clear. Given a subset I of the spins with configuration S_I we define $E_F^{(c)}(S_I)$ as the GS energy over configurations in the cluster (c). Then the cavity assumption is correct (spins in I are uncorrelated) and $E_F^{(c)}$ define the fields $h_i^{(c)}$. How do we perform this restriction in practice? A natural procedure is to break the symmetry between clusters in an explicit manner by adding a small coupling to the reference solution [54, 59]. Remark that symmetry was broken (naturally but explicitly!) in the case of homogeneous systems when we looked for a distribution $p(h)$ with support on positive fields only. It is a remarkable feature of XORSAT (rather unique among disordered systems) that symmetry between disordered clusters can be broken in a constructive and simple way.

The main outcome of the above discussion is that the field attached to variable i is not unique, but depends on the cluster (c). We define the distribution $p_i(h)$ of the fields attached to variable i over the clusters (with uniform weights since all clusters contain the same number of solutions) [50]. The naive cavity assumption corresponds to

$$p_i(h) = \delta_{h-h_i}. \tag{4.16}$$

In presence of many clusters $p_i(h)$ is not highly concentrated. From (4.15) and the fact that $S_i^* = \pm 1$ depending on the cluster from which we pick up the reference solution we find that

$$p_i(h) = \frac{1}{2}\left[\delta_{h-\hat{h}_i} + \delta_{h+\hat{h}_i}\right]. \tag{4.17}$$

As $\hat{h}_i$ is itself randomly distributed we are led to introduce the distribution $\mathcal{P}$ of the field distributions $p_i(h)$. This mathematical object, $\mathcal{P}(p(h))$, is the order parameter of the cavity theory in the clustered phase [44, 50].

4.4. Distribution of field distributions

Let us see how $\mathcal{P}$ can be obtained within the one-more variable approach of Section 4.1. A new equation contains two variables S_i, S_j from F, with fields $h_i^{(c)}$, $h_j^{(c)}$ in each cluster (c). The bias u is a deterministic function of those two fields for each cluster (4.5). We define its distribution over clusters ρ. As u can take three values only and ρ is an even distribution due to the randomness of the second member of the new equation we may write

$$\rho(u) = (1 - \varphi_{ij})\delta_u + \frac{\varphi_{ij}}{2}\left(\delta_{u-\frac{1}{2}} + \delta_{u+\frac{1}{2}}\right). \tag{4.18}$$

The weight φ is a random variable which varies from pair (ij) to pair.

What is the probability distribution $P(\varphi)$ of φ? Either the two variables in the pair belong to the backbone and they are frozen in all clusters; then u will be non zero and $\varphi = 1$. Or one (at least) of the two variables is not frozen and $u = 0$ in all clusters, giving $\varphi = 0$. We may write

$$P(\varphi) = (1 - w)\delta_\varphi + w\delta_{\varphi-1}. \tag{4.19}$$

From the above argument we expect $w = b^2$. Let us derive this result.

Assume we add $\ell \geq 1$ equations to our system. For each one of those equations a bias u^j is drawn randomly according to distribution (4.18). Denote by $m(\leq \ell)$ the number of those equations with parameter $\varphi = 1$; m is binomially distributed with probability w among ℓ. Then $\ell - m$ biases are null, and m biases are not equal to zero. For the formula to remain satisfiable the non-zero biases must be all positive or negative [44], see Section 4.2. Hence the distribution of the field on the new variable is

$$p^m(h) = \frac{1}{2}\left[\delta_{h-\frac{m}{2}} + \delta_{h+\frac{m}{2}}\right], \tag{4.20}$$

in agreement with the expected form (4.17). The upper-script m underlines that field distributions with non zero probability are can be labeled by an integer m; they define a countable set and the distribution $\mathcal{P}$ can be defined as a discrete probability $\mathcal{P}_m$ over the set of positive integers m. The probability $\mathcal{P}_m$ of distribution (4.20) is the convolution of binomial distribution for m at fixed ℓ with the Poisson distribution over ℓ,

$$\mathcal{P}^m = e^{-3\alpha w}\frac{(3\alpha w)^m}{m!}. \tag{4.21}$$

Identities (4.20,4.21) fully determine the distribution of field distributions in term of a single parameter, w.

To close the self-consistency argument consider the two variables in F in, say, the first added equation. Call h, h' their fields, distributed according to $p_m(h)$, $p_{m'}(h')$ for some m, m'. The bias created onto the new variable will be non zero if h and h' may both take non zeros value in some clusters, that is, if m and m' are not equal to zero. This translates into the mathematical identity

$$w = \sum_{m\geq\frac{1}{2},m'\geq\frac{1}{2}} \mathcal{P}_m\mathcal{P}_{m'} = (1 - \mathcal{P}_0)^2 = (1 - e^{-3\alpha w})^2 \tag{4.22}$$

from (4.21). The above equation coincides with (2.40) for $w = b^2$. Notice that w is equal to the probability a_+ that the bias is non zero in the homogeneous case (4.7), in agreement with the discussion of Section 4.3.

It is easy to find back the expressions for the entropies of clusters, $s_{cluster}$, and solutions in a cluster, s_{in}, given in Section 2.10. As for the latter entropy the argument leading to (4.13) can be repeated, with the modification that the second member of the added equation is not necessarily zero but the value it should have for the equation to be satisfied when all three variables are frozen. Hence (4.13) holds with s_h^* replaced with s_{in}. As for the entropy of clusters the same argument again tells us that, on average, half of the clusters will disappear when the three variables are frozen and the second member of the equation is randomly chosen. Therefore

$$\frac{ds_{cluster}}{d\alpha} = -b^3 \ln 2, \tag{4.23}$$

in agreement with equations (2.39,2.40). Summing differential equations (4.23) and (4.13) for $s_{cluster}$ and s_{in} respectively shows that the total entropy of solutions is $(1 - \alpha) \ln 2$ (Section 2.10).

5. Dynamical phase transitions and search algorithms

The deep understanding of the statistical properties of 3-XORSAT makes this problem a valuable benchmark for assessing the performances of various combinatorial search algorithms. At first sight, the idea seems rather odd since 3-XORSAT is a polynomial problem. Interestingly most of the search procedures devised to deal with NP-complete problems e.g. SAT have poor performances *i.e.* take exponentially long average running times on XORSAT above some algorithmic-dependent critical ratio... The purpose of this Section is to present two algorithms exhibiting such a dynamical phase transition, and the techniques required for their analysis.

5.1. Random Walk-SAT (RWSAT): definition, worst-case bound

The first algorithm we consider is the Random Walk-SAT (RWSAT) algorithm introduced by Papadimitriou [58]. RWSAT is based on the observation that a violated equation can be satisfied through negation of one of its variables:

- START FROM A RANDOMLY CHOSEN CONFIGURATION OF THE VARIABLES. CALL ENERGY THE NUMBER E OF UNSATISFIED EQUATIONS.
- WHILE $E \geq 1$;
- PICK UP UNIFORMLY AT RANDOM ONE OF THE E UNSATISFIED EQUATIONS;
- PICK UP UNIFORMLY AT RANDOM ONE OF ITS 3 VARIABLES;
- NEGATE THE VALUE OF THIS VARIABLE, UPDATE E;
- PRINT 'SATISFIABLE', AND HALT.

Notice that, as a result of the negation, some equations that were satisfied may become violated. Therefore the energy is not guaranteed to decrease with the number of steps of the algorithm. RWSAT is able to escape from local minima of the energy landscape, and is *a priori* capable of better performances. On the other hand, RWSAT may run forever... A major question is how long should the algorithm be running before we stop thinking that the studied system has solutions hard to find and get some confidence that there is really no solution.

This question was addressed by Schöning [61], who showed that RWSAT could easily be used as a one-sided randomized algorithm [54].[12] Consider one instance of 3-XORSAT and run RWSAT for $3N$ steps from a randomly chosen configuration of variables. Choose again a random initial configuration and run RWSAT another $3N$ steps, and so on... The probability that no solution has been found after T repetitions of this procedure though the formula is satisfiable is

$$p_{SAT} \leq \exp\left(-T \times \left(\frac{3}{4}\right)^{N+o(N)}\right). \tag{5.1}$$

Hence we obtain a probabilistic proof that the instance is not satisfiable if the algorithm has run unsuccessfully for more than $(\frac{4}{3})^N$ sets of $3N$ steps. It must be clear that this result holds for any instance, no assumption being made on the distribution of formulas. The probability appearing in (5.1) is on the random choices done by RWSAT and the choices of the restart configurations for a fixed formula.

The proof of (5.1) can be sketched as follows. Assume that the formula is satisfiable, and called X^* one of its solutions. Consider now the (extensive) Hamming distance between the solution and the configuration X of variables produced by RWSAT at some instant. After each step only one variable is changed so D changes into $D+1$ (bad move) or $D-1$ (good move). Call x, y, z the variables in the equation which was not satisfied by X. One or three of those variables have opposite values in X^*. In the latter case the flip is always a good move; in the former case the good move happens with probability $\frac{1}{3}$ and a bad move with probability $\frac{2}{3}$. On the overall the probability of a good move is $\frac{1}{3}$ at least.

Think of D as the position of a random walker on the $[0; N]$ segment. Initially the position of the walker is a binomial variable, centered in $\frac{N}{2}$. At each step the walker moves to the left with probability $\frac{1}{3}$, and to the right with probability $\frac{2}{3}$. We look for the probability ρ that the walker is absorbed by the boundary $D = 0$ after S steps. A standard calculation shows that ρ is maximal for $S = 3N$, with the value $\rho \simeq (\frac{3}{4})^N$. After T repetitions the probability of not having been absorbed is $(1-\rho)^T < \exp(-\rho\, T)$, hence (5.1). The proof can be easily extended to K-XORSAT with higher values of K. The number of repetitions necessary to prove unsatisfiability scales as $(\frac{2(K-1)}{K})^N$; it is essentially equal to 2^N for large K, showing that RWSAT does not beat exhaustive search in this limit.

---

[12]Schöning's original work was devoted to the analysis of RWSAT on K-SAT, but his result holds for K-XORSAT too.

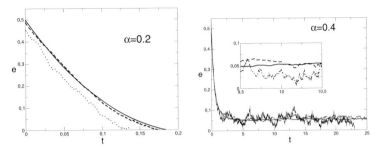

Fig. 11. Fraction e of unsatisfied equations as a function of time t (number of steps divided by N) during the operation of RWSAT on a random 3-XORSAT formula at ratio $\alpha = 0.2$ (left) and $\alpha = 0.4$ (right) with $N = 10^3$ (dotted), 10^4 (dashed), 10^5 (full curve) variables. Note the difference of horizontal scales between the two figures. Inset: blow up of the $t \in [9.5; 10.5]$ region; the amplitude of fluctuations around the plateau decreases with increasing size N.

5.2. *Dynamical transition of RWSAT on random XORSAT instances*

Result (5.1) is true for any instance; what is the typical situation for random systems? Numerical experiments indicate that there is critical value of the ratio of equations per variables, $\alpha_E \simeq 0.33$, hereafter referred to as dynamical threshold, separating two regimes:

• for $\alpha < \alpha_E$, RWSAT generally finds a solution very quickly, namely with a number of flips growing linearly with the number of variables N.[13] Figure 11 shows the plot of the fraction e of unsatisfied clauses as a function of the time (number of steps) T for one randomly drawn system with ratio $\alpha = 0.2$ and $N = 500$ variables. The curve shows a fast decrease from the initial value ($e(T = 0) = \frac{1}{2}$ independently of α for large values of N, but deviations can be found at small sizes, see Figure 11) down to zero on a time scale of the order of N.[14] The resolution time T_{res} depends both on the system of equations under consideration and the choices of the algorithm; its average value scales as

$$\langle T_{res} \rangle = N t_{res} + o(N). \tag{5.2}$$

where t_{res} is an increasing function of α.[15]

• for systems with ratios of equations per variable in the $\alpha_E < \alpha < \alpha_c$ range, the initial relaxation regime taking place on the $O(N)$ time scale does not allow RWSAT to reach a solution (Figure 11B). The fraction e of UNSAT equations

---

[13]A proof of this statement was obtained by [6] for the random SAT model.

[14]This decrease characterizes the overall operation of RWSAT. A precise look at the $e(T)$ curve reveals that the energy e may occasionally increase.

[15]On intuitive grounds, as a step of the algorithm can satisfy $\theta(1)$ equations at a time, we expect the average value of T_{res} to be of the order of the number M of equations at least. Thus t_{res} should grow at least linearly with α. Experiments shows that the growth is in fact more than linear.

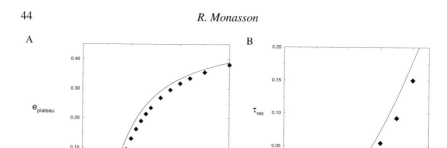

Fig. 12. Fraction $e_{plateau}$ of unsatisfied equations on the plateau (**A**) and logarithm τ_{res} of the average resolution time divided by N (**B**) as a function of the ratio α of equations per variable. Diamonds are the output of numerical experiments, and have been obtained through average of data from simulations over 1,000 systems and runs of RWSAT for various sizes N, and extrapolation to $N \to \infty$ [62]. Full lines are theoretical approximations (5.10),(5.15).

then fluctuates around some plateau value $e_{plateau}$ for a very long time. Fluctuations are smaller and smaller (and the height of the plateau better and better defined) as the size N increases. As a result of fluctuations, the fraction e of unsatisfied equations may temporarily either increase or decrease. When a fluctuation happens to drive RWSAT to $e = 0$, a solution is found and the algorithm stops. The corresponding resolution time, T_{res}, is stochastic; numerical experiments for different sizes N indicate that its expectation value scale as

$$\langle T_{res} \rangle = \exp(N \tau_{res} + o(N)).\tag{5.3}$$

where the coefficient τ_{res} is an increasing function of α. The plateau energy $e_{plateau}$ and the logarithm τ_{res} of the resolution time are shown in Figure 12.

Notice that the dynamical threshold α_E above which the plateau energy is positive is strictly smaller than the critical threshold $\alpha_c \simeq 0.918$, where systems go from satisfiable with high probability to unsatisfiable with high probability. In the intermediate range $\alpha_E < \alpha < \alpha_c$, systems are almost surely satisfiable but RWSAT needs an exponentially large time to prove so. The reason is that RWSAT remains trapped at a high energy level (plateau of Figure 12) for an exponentially large time. The emergence of metastability can be qualitatively studied with simple tools we now expose.

5.3. Approximate theory for the metastable plateau and the escape time

Assume that after T steps of the algorithm the energy (number of unsatisfied equations) is $E_T \geq 1$. Then pick up an unsatisfied equation, say, C, and a variable in C, say, x, and flip it. The energy after the flip is

$$E_{T+1} = E_T - 1 - U + S, \tag{5.4}$$

where S (respectively U) is the number of equations including x which were satisfied (resp. unsatisfied after exclusion of equation C) prior to the flip. S and U are random variables with binomial distributions,

$$\text{Proba}[U] = \binom{E_T - 1}{U} \left(\frac{3}{N}\right)^U \left(1 - \frac{3}{N}\right)^{E_T - 1 - U},$$

$$\text{Proba}[S] = \binom{M - E_T}{S} \left(\frac{3}{N}\right)^S \left(1 - \frac{3}{N}\right)^{M - E_T - S}. \tag{5.5}$$

where the probabilities are intended over the formula content. Taking the average evolution equation (5.4) we obtain

$$\langle E_{T+1} \rangle = \langle E_T \rangle - 1 - \frac{3}{N}(\langle E_T \rangle - 1) + \frac{3}{N}(M - \langle E_T \rangle). \tag{5.6}$$

The above equation is exact. It is now tempting to iterate it with time, from the initial condition $\langle E_{T=0} \rangle = \frac{M}{2}$. This is what we do hereafter but one should realize that this procedure is not correct from a mathematical standpoint. The catch is that one is allowed to average over the formula only once, and certainly not at each time step of the algorithm. Evolution equation (5.6) amounts to redraw randomly the instance at each time step, conditioned to the energy. This approximation nevertheless allows us to write down a simple equation for $\langle E_T \rangle$, which captures much of the true behaviour of RWSAT.

The next step in our analysis is the large size, large time limit. As the energy can typically change by a quantity of the order of unity in one time step we expect the fraction of unsatisfied equations to vary of a time scale of the order of N,

$$\langle E_T \rangle = M e \left(\frac{T}{M} = t\right), \tag{5.7}$$

for some smooth function $e(t)$ of the reduced time t. Finite difference equation (5.6) turns into a differential equation after insertion of (5.7),

$$\frac{de}{dt} = -1 + 3\alpha(1 - 2e), \tag{5.8}$$

with the initial condition $e(0) = \frac{1}{2}$. Clearly (5.8) makes sense as long as $e > 0$; if e vanishes the algorithm stops. Resolution of (5.8) shows the following scenario. If α is smaller than

$$\alpha_E = \frac{1}{3}, \tag{5.9}$$

the fraction e of unsatisfied equations quickly decreases, and vanishes at some time $t_{res}(\alpha)$. This regime corresponds to a successful action of RWSAT in a $O(N)$ number of steps. t_{res} is an increasing function of α which diverges as $\alpha \to \alpha_E$. Above this critical ratio e shows a different behaviour: after a decreasing transient regime e saturates to a positive plateau value

$$e_{plateau}(\alpha) = \frac{1}{2}\left(1 - \frac{\alpha_E}{\alpha}\right). \tag{5.10}$$

The value of the plateau energy is compared to numerics in Figure 12A. The agreement on the location of the dynamical threshold α_E as well as the plateau energy are satisfactory.

The remaining point is to understand how RWSAT finally finds a solution when $\alpha > \alpha_E$. The above theory, based on taking the $N \to \infty$ limit first, washes out the fluctuations of the energy around its metastable value, of crucial importance for resolution [62]. To take into account these fluctuations let us define the probability $Q_{plateau}(E)$ that the energy takes value E in the plateau regime of Figure 11B. A stationary distribution is well defined if we discard the initial transient regime (choose large t) and collect values for E on exponentially large–in–N time scales. The procedure is standard in the study of long-time metastable states.

Within our draw-instance-at-each–step approximation we may write a self-consistent equation for the stationary distribution of energies,

$$Q_{plateau}(E) = \sum_{U,S} \text{Proba}[U]\,\text{Proba}[S]\,Q_{plateau}(E + 1 + U - S) \tag{5.11}$$

where the meaning of U, S was explained right after (5.4). From Section 5.2 we expect fluctuations to decreases sharply with the system size. A reasonable guess for the scaling of the distribution with M is

$$Q_{plateau}(E) = \exp\left[-M\omega\left(\frac{E}{M} = e\right) + o(M)\right] \tag{5.12}$$

where ω is the rate function associated to the fraction of unsatisfied equations. Plugging the above Ansatz into (5.11) and taking the large M limit we find that ω fulfills the following differential equation

$$F\left(\frac{\partial\omega}{\partial e}, e\right) = 0, \tag{5.13}$$

where $F(x, y) = 3\alpha y(e^{-x} - 1) + 3\alpha(1 - y)(e^x - 1) - x$. This equation has to be solved with the condition $\omega(e_{plateau}) = 0$.

An analytical solution can be found for (5.13) when we restrict to the vicinity of the dynamical transition *i.e.* to small values of ω. Expanding F to the second order in its first argument and solving (5.13) we obtain

$$\omega(e) \simeq 2(e - e_{plateau})^2, \tag{5.14}$$

where $e_{plateau}$ is defined in (5.10).

What happens when time increases is now clear. Assume we have run RWSAT up to time $t \sim e^{M\tau}$. Then configurations with energy e such that $\omega(e) < \tau$ have been visited many times and are 'equilibrated' with probability (5.11), (5.14). Configurations with energies outside the band $e_{plateau} \pm \sqrt{\tau/2}$ are not accessible. When the time scales reaches

$$\tau_{res} = \omega(0) \simeq \frac{1}{2}\left(1 - \frac{1}{3\alpha}\right)^2, \tag{5.15}$$

zero energy configurations are encountered, and RWSAT comes to a stop. The agreement between the theoretical estimate (5.15) and the numerical findings (5.3) visible in Figure 12B is acceptable in regard to the crudeness of the approximation done.

5.4. Davis-Putnam-Loveland-Logemann (DPLL) algorithm

The second procedure is the Davis-Putnam-Loveland-Logemann (DPLL) algorithm [22]. Contrary to RWSAT DPLL can provide exact proofs for unsatisfiability. The procedure, widely used in practice, is based on the trial-and-error principle. Variables are assigned according to some heuristic rule (split step), and equations involving those variables simplified. If an equation involving a single variable (unit-equation) appears its variable is chosen accordingly prior to any other heuristic assignment (unit-propagation). If a contradiction is found (two opposite unit-equations) DPLL backtracks to the last heuristically assigned variable, flips it, and resumes the search process. The procedure halts either when all equations have been satisfied (a solution is then found), or when all possible values for the variables have been tried in vane and found to be contradictory (a proof of unsatisfiability is then obtained).

DPLL can be described as a recursive function of the variable assignment A. Given a system S DPLL is first called with the empty assignment $A = \emptyset$:

PROCEDURE DPLL[A]
- LET S_A BE WHAT IS LEFT FROM S GIVEN VARIABLE ASSIGNMENT A;
- IF S_A IS EMPTY, PRINT 'SATISFIABLE'; HALT;
- IF S_A CONTAINS A VIOLATED EQUATION, PRINT 'CONTRADICTION', RETURN; *(backtracking)*

• OTHERWISE, LET U BE THE SET OF UNIT-EQUATIONS IN S_A;
– IF $U \neq \emptyset$, PICK-UP ONE OF THE EQUATIONS IN U, SAY, e, AND CALL DPLL[A∪{e}]; *(unit-propagation)*
– IF $U = \emptyset$, CHOOSE A NOT-YET-ASSIGNED VARIABLE, SAY, x, AND ITS VALUE v ACCORDING TO SOME HEURISTIC RULE, AND CALL DPLL[A∪{$x = v$}], THEN DPLL[A∪{$x = \bar{v}$}]; *(variable splitting)*

Rules for assigning variables in the absence of unit-equations are heuristic in that they aim at doing good assumptions *i.e.* diminishing as much as possible the search process to come from limited information about the current system of equations. Of course, perfect heuristic do exist: trying all possible values for not-yet-assigned variables would ensure that no wrong guess is ever done! But the time required would be exponentially long. In practice, heuristics have to make their decision in polynomial time. Two simple splitting heuristics are:

◊ <u>UC:</u> choose at random and uniformly any unset variable, and assign it to 0 or 1 with equal probabilities ($\frac{1}{2}$).

◊ <u>GUC:</u> choose at random and uniformly any equation with minimal length *i.e.* involving 2 variables if any, or 3 variables otherwise. Pick up at random and uniformly one its variable, and assign it to 0 or 1 with equal probabilities ($\frac{1}{2}$).

UC, which stands for unit-clause [14], amounts to make a random guess and is the simplest possible heuristic. GUC (Generalized UC) is more clever: each time a split is done from an equation with 2 variables, this equation is turned into a unit-equation, and eliminated through unit-propagation. In the following, we call DPLL-UC and DPLL-GUC the variants of DPLL based on the UC and GUC heuristics respectively.

A measure of the computational effort required by DPLL is the number T_{split} of variable splittings. This number varies from system to system (at fixed number N of variables and ratio α), and from run to run of DPLL due to the stochasticity introduced by the heuristic rule. The outcome of numerical experiments for the median number of splits.[16] For a given size N T_{split} shows a maximum located around $\alpha \simeq \alpha_c$. If one fixes α T_{split} is an increasing function of the size N; numerical data support the existence of a dynamical threshold, α_E, separating linear and exponential scalings in N,

$$T_{split} \sim \begin{cases} N t_{split} + o(N) & \text{if } \alpha < \alpha_E \\ \exp(N \tau_{split} + o(N)) & \text{if } \alpha > \alpha_E \end{cases}, \tag{5.16}$$

where t_{split} and τ_{split} are functions of the ratio α. The value of the dynamical threshold can be derived from theoretical calculations shown in Section 5.5 and

---

[16]The median is more representative of the typical value of the number of splits than the expectation value, since the latter may be dominated by huge and unlikely samples, see discussion of Section 2.2.

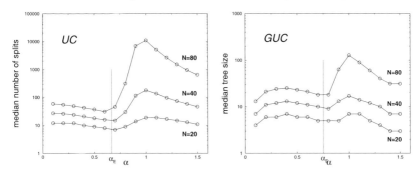

Fig. 13. Median number of splits required by DPLL with the UC (left) and GUC (right) heuristics as a function of the ratio α, and for $N = 20, 40, 60$ variables (from bottom to top). Data have been extracted from the resolution of 10,000 randomly drawn systems; continuous lines are guidelines for the eye. Note the difference of (logarithmic) scale between UC and GUC curves showing that DPLL-GUC is much more efficient than DPLL-UC. The polynomial/exponential transition is located at ratios $\alpha_E = \frac{2}{3}$ and $\alpha_E = 0.7507\ldots$ for UC and GUC respectively.

is equal to $\alpha_E = \frac{2}{3}$ and $\alpha_E \simeq 0.7507\ldots$ for UC and GUC heuristics respectively. Three dynamical regimes are therefore identified [2, 16]:

• *Linear & satisfiable phase* ($\alpha < \alpha_E$): systems with small ratios are solved with essentially no backtracking. A solution is found after $O(N)$ splits.

• *Exponential & satisfiable phase* ($\alpha_E < \alpha < \alpha_c$): systems with ratios slightly below threshold have solutions, but DPLL generally requires an exponential number of splits to find one of them. An explanation for this drastic breakdown of performances will be given in Section 5.5.

• *Exponential & unsatisfiable phase* ($\alpha > \alpha_c$): finally, finding a proof of unsatisfiability typically requires an exponentially large number of splits [15]. Note that, as α gets higher and higher, each variable assignment affects more and more equations (of the order of α), and contradictions are detected earlier and earlier. Rigorous calculations show that $\tau_{split} \sim \frac{1}{\alpha}$ [10], and the computational effort decreases with increasing α (Figure 13). The median number of splits is considerably smaller for DPLL-GUC than for DPLL-UC, a result expected from the advantages of GUC against UC discussed above.

5.5. Linear phase: resolution trajectories in the $2 + p$-XORSAT phase diagram

Action of DPLL on an instance of 3-XORSAT causes changes to the numbers of variables and equations, and thus to the ratio α. Furthermore DPLL turns equations with 3 variables into equations with 2 variables. A mixed $2 + p$-XORSAT distribution, where p is the fraction of 3-equations and α the ratio of the total number of 2- and 3- equations over the number of variables can be used to

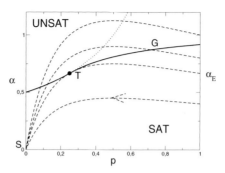

Fig. 14. Phase diagram of $2 + p$-XORSAT and dynamical trajectories of DPLL. The threshold line $\alpha_c(p)$ (bold full line) separates sat from UNSAT phases. Departure points for DPLL trajectories are located on the 3-XORSAT vertical axis with ratios $.4, \frac{2}{3}, .8, 1.$ from bottom to top. The arrow indicates the direction of motion along trajectories parametrized by the fraction t of variables set by DPLL. For small ratios $\alpha < \alpha_E$ ($= \frac{2}{3}$ for the UC heuristic) trajectories remain confined in the sat phase, end in S of coordinates $(0, 0)$, where a solution is found. At α_E the trajectory hits tangentially the threshold line in T of coordinates $(\frac{1}{4}, \frac{2}{3})$. When $\alpha > \alpha_E$ the trajectories intersect the threshold line at some point G (which depends on α), and stops before hitting the $\alpha_D(p)$ dotted line (5.27). After massive backtracking DPLL will find a solution; G corresponds to the highest node in the search tree.

model what remains of the input system.[17] Repeating the calculations of Section 3 for the $2 + p$-XORSAT models we derive the phase diagram of Figure 14. The Sat/Unsat critical line $\alpha_c(p)$ separates the satisfiable from the unsatisfiable phases. For $p \leq p_0 = \frac{1}{4}$ i.e. to the left of point T, the threshold line coincides with the percolation transition as in the 2-XORSAT model, and is given by $\alpha_c(p) = \frac{1}{2(1-p)}$. For $p > p_0$ an intermediate clustered phase is found as in the 3-XORSAT model, and the threshold coincides with the vanishing of the cluster entropy $s_{cluster}$ (Section 2.10).

The phase diagram of 2+p-XORSAT is the natural space in which DPLL dynamic takes place. An input 3-XORSAT instance with ratio α shows up on the right vertical boundary of Figure 14 as a point of coordinates $(p = 1, \alpha)$. Under the action of DPLL the representative point moves aside from the 3-XORSAT axis and follows a trajectory, very much alike real-space renormalization, which depends on the splitting heuristic. Trajectories enjoy two essential features [1]. First the representative point of the system treated by DPLL does not 'leave' the 2+p-XORSAT phase diagram. In other words, the instance is, at any stage of the search process, uniformly distributed from the 2+p-XORSAT distribution con-

---

[17]Equations with a single variable are created too, but are eliminated through unit-propagation. When a heuristic assignment has to be made the system is a mixture of equations with 2 and 3 variables only.

ditioned to its equation per variable ratio α and fraction p of 3-equations. This assumption is not true for all heuristics of split, but holds for UC and GUC [14].[18] Secondly, the trajectory followed by an instance in the course of resolution is a stochastic object, due to the randomness of the instance and of the assignments done by DPLL. In the large size limit ($N \to \infty$) the trajectory becomes self-averaging *i.e.* concentrated around its average locus in the 2+p-XORSAT phase diagram [63]. We will come back below to this concentration phenomenon.

Let α_0 denote the equation per variable ratio of the 3-XORSAT instance to be solved. We call $E_j(T)$ the number of j–equations (including j variables) after T variables have been assigned by the solving procedure. T will be called hereafter 'time', not to be confused with the computational effort. At time $T = 0$ we have $E_3(0) = \alpha_0 N$, $E_2(0) = E_1(0) = 0$. Assume that the variable x assigned at time T is chosen through unit-propagation, that is, independently of the j-equation content. Call $n_j(T)$ the number of occurrences of x in j-equations ($j = 2, 3$). The evolution equations for the populations of 2-,3-equations read

$$E_3(T + 1) = E_3(T) - n_3(T),$$
$$E_2(T + 1) = E_2(T) - n_2(T) + n_3(T). \tag{5.17}$$

Flows n_2, n_3 are of course random variables that depend on the instance under consideration at time T, and on the choice of variable done by DPLL. What are their distributions? At time T there remain $N - T$ untouched variables; x appears in any of the $E_j(T)$ j-equation with probability $p_j = \frac{j}{N-T}$, independently of the other equations. In the large N limit and at fixed fraction of assigned variables, $t = \frac{T}{N}$, the binomial distribution converges to a Poisson law with mean

$$\langle n_j \rangle_T = \frac{j e_j}{1 - t} \quad \text{where} \quad e_j = \frac{E_j(T)}{N} \tag{5.18}$$

is the density of j-equations at time T. The key remark is that, when $N \to \infty$, e_j is a slowly varying and non stochastic quantity and is a function of the fraction $t = \frac{T}{N}$ rather than T itself. Let us iterate (5.17) between times $T_0 = t N$ and $T_0 + \Delta T$ where $1 \ll \Delta T \ll N$ e.g. $\Delta T = O(\sqrt{N})$. Then the change ΔE_3 in the number of 3-equations is (minus) the sum of the stochastic variables $n_j(T)$ for $T = T_0, T_0 + 1, \dots, T_0 + \Delta T$. As these variables are uncorrelated Poisson variables with $O(1)$ mean (5.18) ΔE_3 will be of the order of ΔT, and the change in the density e_3 will be of order of $\Delta T / N \to 0$. Applying central limit theorem $\Delta E_3 / \Delta T$ will be almost surely equal to $-\langle n_3 \rangle_t$ given by (5.18) and with the equation density measured at reduced time t. The argument can

---

[18]Analysis of more sophisticated heuristics e.g. based on the number of occurrences of variables require to handle more complex instance distributions [33].

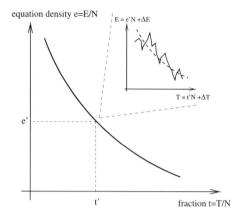

Fig. 15. Deterministic versus stochastic dynamics of the equation population E as a function of the number of steps T of the algorithm. On the slow time scale (fraction $t = T/N$) the density $e = E/N$ of (2- or 3-) equations varies smoothly according to a deterministic law. Blowing up of the dynamics around some point t', e' shows the existence of small and fast fluctuations around this trajectory. Fluctuations are stochastic but their probability distribution depends upon the slow variables t', e' only.

be extended to 2-equations, and we conclude that e_2, e_3 are deterministic (self-averaging) quantities obeying the two coupled differential equations [14]

$$\frac{de_3}{dt}(t) = -\frac{3e_3}{1-t}, \qquad \frac{de_3}{dt}(t) = \frac{3e_3}{1-t} - \frac{2e_2}{1-t}. \qquad (5.19)$$

Those equations, together with the initial condition $e_3(0) = \alpha_0$, $e_2(0) = 0$ can be easily solved,

$$e_3(t) = \alpha_0(1-t)^3, \qquad e_2(t) = 3\alpha_0 t(1-t)^2. \qquad (5.20)$$

To sum up, the dynamical evolution of the equation populations may be seen as a slow and deterministic evolution of the equation densities to which are superimposed fast, small fluctuations. The distribution of the fluctuations adiabatically follows the slow trajectory. This scenario is pictured in Figure 15.

Expressions (5.20) for the equation densities allow us to draw the resolution trajectories corresponding to the action of DPLL on a 3-XORSAT instance. Initially the instance is represented by a point with coordinates ($p = 1, \alpha = \alpha_0$) in Figure 14. As more and more variables are assigned the representative point moves away from the rightmost vertical axis. After a fraction t of variables have been assigned the coordinates of the point are

$$p(t) = \frac{e_3}{e_2 + e_3} = \frac{1-t}{1+2t}, \qquad \alpha(t) = \frac{e_2 + e_3}{1-t} = \alpha_0(1-t)(1+2t). \quad (5.21)$$

Trajectories corresponding to various initial ratios are shown in Figure 14. For small ratios $\alpha_0 < \alpha_E$ trajectories remain confined in the sat phase, end in S of coordinates $(0, 0)$, where a solution is found. At α_E ($= \frac{2}{3}$ for the UC heuristic), the single branch trajectory hits tangentially the threshold line in T of coordinates $(\frac{1}{4}, \frac{2}{3})$. When $\alpha_0 > \alpha_E$ the trajectories enter the Unsat phase, meaning that DPLL has turned a satisfiable instance (if $\alpha_0 < \alpha_c$) into an unsatisfiable one as a result of poor assignments. It is natural to expect that α_E is the highest ratio at which DPLL succeeds in finding a solution without resorting to much backtracking.

5.6. *Dynamics of unit-equations and universality*

The trajectories we have derived in the previous Section are correct provided no contradiction emerges. But contradictions may happen as soon as there are $E_1 = 2$ unit-equations, and are all the more likely than E_1 is large. Actually the set of 1-equations form a 1-XORSAT instance which is unsatisfiable with a finite probability as soon as E_1 is of the order of $\sqrt{N}$ from the results of Section 2.1. Assume now that $E_1(T) \ll N$ after T variables have been assigned, what is the probability ρ_T that no contradiction emerges when the Tth variable is assigned by DPLL? This probability is clearly one when $E_1 = 0$. When $E_1 \geq 1$ we pick up a 1-equation, say, $x_6 = 1$, and wonder whether the opposite 1-equation, $x_6 = 0$, is present among the $(E_1 - 1)$ 1-equations left. As equations are uniformly distributed over the set of $N - T$ untouched variables

$$\rho_T = \left(1 - \frac{1}{2(N - T)}\right)^{\max(E_1(T)-1,0)}. \tag{5.22}$$

The presence of the max in the above equation ensures it remains correct even in the absence of unit-equations ($E_1 = 0$). $E_1(T)$ is a stochastic variable. However from the decoupling between fast and slow time scales sketched in Figure 15 the probability distribution of $E_1(T)$ depends only on the slow time scale t. Let us call $\mu(E_1; t)$ this probability. Multiplying (5.22) over the times $T = 0$ to $T = N-1$ we deduce the probability that DPLL has successfully found a solution without ever backtracking,

$$\rho_{success} = \exp\left(-\int_0^1 \frac{dt}{2(1 - t)} \sum_{E_1 \geq 1} \mu(E_1; t)(E_1 - 1)\right) \tag{5.23}$$

in the large N limit.

We are left with the calculation of μ [29]. Figure 16 sketches the stochastic evolution of the number E_1 during one step. The number of 1-equations produced

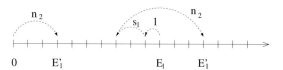

Fig. 16. Evolution of the number E_1 of 1-equations as one more variable is assigned. n_2 denotes the number of 2-equations reduced to 1-equations, s_1 the number of 1-equations satisfied. If $E_1 \geq 1$ a variable is fixed through unit-propagation: E_1 decreases by one plus s_1, and increases by n_2. In the absence of unit-equation ($E_1 = 0$) the number of 1-equations after the assignment is simply $E'_1 = n_2$.

from 2-equations, n_2, is a Poisson variable with average value, from (5.20),

$$d(t) = \frac{2e_2(t)}{1-t} = 6\alpha_0 t (1-t) \tag{5.24}$$

when $N \to \infty$. The number of satisfied 1-equations, s_1, is negligible as long as E_1 remains bounded. The probability that the number of 1-equations goes from E_1 to E'_1 when $T \to T+1$ defines the entry of the transition matrix

$$M(E'_1, E_1; t) = \sum_{n_2 \geq 0} e^{-d(t)} \frac{d(t)^{n_2}}{n_2!} \delta_{E'_1 - (E_1 + n_2 - \delta_{E_1})}. \tag{5.25}$$

from which a master equation for the probability of E_1 at time T may be written. On time scales $1 \ll \Delta T \ll N$ this master equation converges to the equilibrium distribution μ [16,29], conveniently expressed in terms of the generating function

$$G(x; t) = \sum_{E_1 \geq 0} \mu(E_1; t) x^{E_1} = \frac{(1 - d(t))(x - 1)}{x \, e^{d(t)(1-x)} - 1}. \tag{5.26}$$

The above is a sensible result for $d(t) \leq 1$ but does not make sense when $d(t) > 1$ since a probability cannot be negative! The reason is that we have derived (5.26) under the implicit condition that no contradiction was encountered. This assumption cannot hold when the average rate of 1-equation production, $d(t)$, is larger that one, the rate at which 1-equations are satisfied by unit-propagation. From (5.24) we see, when $\alpha > \alpha_E = \frac{2}{3}$, the trajectory would cross the

$$\alpha_D(p) = \frac{1}{2(1-p)} \tag{5.27}$$

on which $d = 1$ for some time $t_D < 1$. A contradiction is very likely to emerge before the crossing.

When $\alpha < \alpha_E$ d remains smaller than unity at any time. In this regime the probability of success reads, using (5.23) and (5.26),

$$\rho_{success} = \exp\left(\frac{3\alpha}{4} - \frac{1}{2}\sqrt{\frac{3\alpha}{2-3\alpha}} \, \tanh^{-1}\left[\sqrt{\frac{3\alpha}{2-3\alpha}}\right]\right). \tag{5.28}$$

$\rho_{success}$ is a decreasing function of the ratio α, down from unity for $\alpha = 0$ to zero for $\alpha = \alpha_E$. The present analysis of the UC heuristic can be easily transposed to the GUC heuristic. Details are not given here but can be found in [2, 16]. The result is an expression for $\rho_{success}$ larger than its UC counterpart (5.28), and vanishing in $\alpha_E \simeq 0.7507$. Interestingly the way $\rho_{success}$ vanishes when α reaches α_E,

$$-\ln \rho_{success}(\alpha_E - \epsilon) \sim \epsilon^{-\frac{1}{2}} \qquad (\epsilon \to 0^+) \tag{5.29}$$

is the same for both heuristics. This similarity extends to a whole class of heuristics which can be described by the flow of equation densities only and based on unit-propagation [25]. The probability that DPLL finds a solution without backtracking to a 3-XORSAT instance of size N satisfies finite-size scaling at the dynamical critical point,

$$-\ln \rho_{success}(\alpha_E - \epsilon, N) \sim N^{\frac{1}{6}} \, \Phi\left(\epsilon N^{\frac{1}{3}}\right), \tag{5.30}$$

where the scaling function Φ is independent of the heuristics and can be calculated exactly [25]. The exponent characterizing the width of the critical region is the one associated to percolation in random graphs (2.34). A consequence of (5.30) is that, right at α_E, $\rho_{success} \sim \exp(-Cst \times N^{\frac{1}{6}})$ decreases as a stretched exponential of the size. The value of the exponent, and its robustness against the splitting heuristics can be understood from the following argument [25].

Let us represent 1- and 2- equations by a graph G over the set of $N-T$ vertices (one for each variable x_i) with E_1 marked vertices (one for each unit-equation $x_i = 0, 1$), and E_2 signed edges ($x_i + x_j = 0, 1$), see Section 2.4. d is simply the average degree of vertices in G. Unit-propagation corresponds to removing a marked vertex (and its attached edges), after having marked its neighbours; the process is iterated until the connected component is entirely removed (no vertex is marked). Meanwhile, new edges have been created from the reduction of 3-equations into 2-equations. Then a vertex is picked up according to the heuristic and marked, and unit-propagation resumes. The success/failure transition coincides with the percolation transition on G: $d = 1$ as expected. From random graph theory [12] the percolation critical window is of width $|d - 1| \sim N^{-1/3}$. As d is proportional to the ratio α_0 (5.24) we find back $\psi = \frac{1}{3}$. The time spent by resolution trajectories in the critical window is $\Delta t \sim \sqrt{|d - 1|} \sim N^{-1/6}$, corresponding to $\Delta T = N\Delta t \sim N^{5/6}$ eliminated variables. As the largest components have size $S \sim N^{2/3}$ the number of such components eliminated is $C = \Delta T/S \sim N^{1/6}$. What is the probability q that a large component is removed without encountering a contradiction? During the removal of the component the number of marked vertices 'freely' diffuses, and reaches $E_1 \sim \sqrt{S} \sim N^{1/3}$. The probability that no contradiction occurs is, from (5.22), $q \sim (1 - \frac{Cst}{N})^{E_1 \times S}$, a finite quantity. Thus $\rho_{success} \sim q^C \sim \exp(-N^{1/6})$. The presence of numerous, smaller components does not affect this scaling.

5.7. Exponential phase: massive backtracking

For ratios $\alpha_0 > \alpha_E$ DPLL is very likely to find a contradiction. Backtracking enters into play, and is responsible for the drastic slowing down of the algorithm (Figure 13).

The history of the search process can be represented by a search tree, where the nodes represent the variables assigned, and the descending edges their values (Figure 17). The leaves of the tree correspond to solutions (S), or to contradictions (C). The analysis of the $\alpha < \alpha_E$ regime leads us to the conclusion that search trees look like Figure 17A at small ratios.[19] Consider now the case of unsatisfiable formulas ($\alpha_0 > \alpha_c$) where all leaves carry contradictions after DPLL halts (Figure 17C). DPLL builds the tree in a sequential manner, adding nodes and edges one after the other, and completing branches through backtracking steps. We can think of the same search tree built in a parallel way [16]. At time (depth T) our tree is composed of $L(T) \leq 2^T$ branches, each carrying a partial assignment over T variables. Step T consists in assigning one more variable to each branch, according to DPLL rules, that is, through unit-propagation or split. Possible consequences are: emergence of a contradiction and end of the branch, simplification of the attached formulas and the branch keeps growing.

The number of branches $L(T)$ is a stochastic variable. Its average value can be calculated as follows [51]. Let us define the average number $L(\vec{E}; T)$ of branches with equation populations $\vec{E} = (E_1, E_2, E_3)$ at depth T. Initially $L(\vec{E}; 0) = 1$ for $\vec{E} = (0, 0, \alpha_0 N)$, 0 otherwise. Call $M(\vec{E}', \vec{E}; T)$ the average number of branches with population $\vec{E}'$ generated from a branch with population $\vec{E}$ once the Tth variable is assigned. Transition matrix M is an extension of (5.25) to the whole population vector $\vec{E}$ and not only E_1. We have $0 \leq M \leq 2$, the extreme values corresponding to a contradiction and to a split respectively. We claim that

$$L(\vec{E}'; T+1) = \sum_{\vec{E}} M(\vec{E}', \vec{E}; T) \, L(\vec{E}; T). \tag{5.31}$$

---

[19] A small amount of backtracking may be necessary to find the solution since $\rho_{success} < 1$ [29], but the overall picture of a single branch is not qualitatively affected.

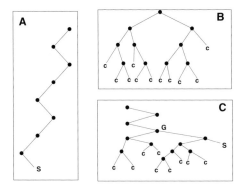

Fig. 17. Search trees in three regimes of Section 5.4: **A.** linear, satisfiable ($\alpha < \alpha_E$); **B.** exponential, satisfiable ($\alpha_E < \alpha < \alpha_c$); **C.** exponential, unsatisfiable ($\alpha > \alpha_c$). Leaves are marked with S (solutions) or C (contradictions). G is the highest node to which DPLL backtracks, see Figure 14.

Evolution equation (5.31) is somewhat suspicious since it looks like the approximation (5.11) we have done in the analysis of RWSAT. Yet a major difference exists which makes (5.31) exact [51]. Drawing randomly many times the same instance, as we are doing, is in principle forbidden but not along *one* branch for the very reason the analysis of Section 5.5 was correct. Actually what we have done in Section 5.5 is to draw randomly at time T the equations containing the Tth variable. But this is correct since those equations are immediately simplified into shorter equations and their remaining content remains unknown [33]. The situation seems more complicated in the case of the whole tree since the same equation can appear at different depth along distinct branches. Indeed the number of branches produced from two distinct branches after assignment of one variable are correlated variables. But thanks to the linearity of expectation those correlations do not matter and (5.31) is correct.

Transition matrix M can be explicitly written down. It is more convenient to write (5.31) for the generating function of the number of branches, $B(\vec{x}; T) = \sum_{\vec{E}} L(\vec{E}; T) x_1^{E_1} x_2^{E_2} x_3^{E_3}$, with the result

$$B(\vec{x}; T+1) = \frac{1}{f_1} B(\vec{f}; T) + \left(2 - \frac{1}{f_1}\right) B(0, f_2, f_3; T) - 2B(\vec{0}; T), \quad (5.32)$$

where $\vec{f}$ is the vector with components

$$f_1 = x_1 + \frac{\frac{1}{2} - x_1}{N - T}, f_2 = x_2 + \frac{2(x_2 - x_1)}{N - T}, f_3 = x_3 + \frac{3(x_2 - x_3)}{N - T}. \quad (5.33)$$

The three terms on the r.h.s. of (5.32) correspond, from left to right: unit-propagation (the branch keeps growing), variable splitting (2 branches are created from the previous one), branches carrying empty instances (satisfied instance). Equation (5.32) together with the initial condition $B(\vec{x}; 0) = x_3^{\alpha_0 N}$ completely defines the average dynamics of the search tree. We sketch the main steps of its resolution below [16]:

1. To count the number of branches irrespectively of the number of unit-equations we should consider the value $x_1 = 1$. However, as long as branches grow the number E_1 of unit-equations cannot be large, and remains bounded. We can therefore choose $x_1 = \frac{1}{2}$ which simplifies (5.32) without affecting the large size scaling of L and B. This technical trick is reminiscent of Knuth's kernel method [34].

2. For large N it is reasonable to expect that the number of branches grows exponentially with the depth, or, equivalently,

$$\sum_{E_1} L(E_1, E_2, E_3; T) \sim e^{N\lambda(e_2, e_2; t) + o(N)} \tag{5.34}$$

where e_2, e_3 are the densities of equations as usual. From point 1 the Legendre transform of λ

$$\gamma(x_2, x_3; t) = \max_{e_2, e_3}\left[\lambda(e_2, e_2; t) + e_2 \ln x_2 + e_3 \ln x_3\right] \tag{5.35}$$

fulfills the partial differential equation (PDE)

$$\frac{\partial \gamma}{\partial t} = \ln 2 + \frac{1 - 2x_2}{1 - t}\frac{\partial \gamma}{\partial x_2} + \frac{3(x_2 - x_3)}{1 - t}\frac{\partial \gamma}{\partial x_3}. \tag{5.36}$$

with the initial condition $\gamma(x_2, x_3; t) = \alpha_0 \ln x_3$.

3. The first order PDE can be solved exactly with the characteristic method. The output, after Legendre inversion through (5.35), is the entropy $\lambda(e_2, e_3; t)$ of branches at reduced depth t. Let us call $\lambda^*(t)$ the maximum value of λ over the equation densities for a fixed fraction t of assigned variables.

4. $\lambda^*(t)$ is a function growing from $\lambda^* = 0$ at $t = 0$, reaching a maximum value λ_M^* in t_M, and decreasing for larger times $t \leq 1$. t_M is the depth in the tree of Figure 17C where most contradictions are found; the number of C leaves is, to exponential order, $e^{N\lambda_M^*}$. We conclude that the size of the tree we were looking for is

$$\tau_{split} = \lambda_M^*, \tag{5.37}$$

compare with (5.16). For large $\alpha \gg \alpha_c$ one finds $\tau_{split} \sim \ln 2/(6\alpha)$ in agreement with [10]. The calculation can be extended to higher values of K.

The above calculation holds for the unsatisfiable, exponential phase. How can we understand the satisfiable but exponential regime $\alpha_E < \alpha_0 < \alpha_c$? The resolution trajectory crosses the Sat/Unsat critical line at some point G shown in Figure 14. Immediately after G the instance left by DPLL is unsatisfiable. A subtree with all its leaves carrying contradictions will develop below G (Figure 17B). The size τ_{split}^G of this subtree can be easily calculated from the above theory. The only change is the initial condition over γ: $\gamma(x_2, x_3; 0) = \alpha_G(p_G \ln x_3 + (1 - p_G) \ln x_2)$ where (p_G, α_G) are the coordinates of G which can be calculated from α_0 and the knowledge of the critical Sat/Unsat line. Once this subtree has been built DPLL backtracks to G, flips the attached variable and will finally end up with a solution. Hence the (log of the) number of splits necessary will be typically equal to $\tau_{split} = (1 - t_G)\tau_{split}^G$ [16].

6. Conclusions

Previous Sections have allowed us to illustrate rather general techniques and ideas to deal with random systems. It does not come as a surprise that other problems than XORSAT e.g. the satisfaction of Boolean constraints, graph coloring, the covering of vertices, ... have been successfully studied with these tools. Many of those problems, when given an input distribution based on random graphs, actually share a lot of common features with XORSAT. The reader is referred to [3, 11, 36, 40, 45, 49] (satisfiability), [36, 55] (coloring), [31, 64] (vertex cover), ... for entry points to the literature. Let us also mention that many other interesting optimization problems, not directly related to random graphs, have been studied with the techniques of Sections 4 and 5, and the results sometimes rigorously proved e.g. matching [5, 41, 56], traveling salesman [42], number partitioning [38, 39], graph partitioning [30], ... Finally, from a historical point of view, one should not forget that statistical mechanics tools have found numerous and beautiful applications in the study of the learning and storage properties of neural networks [8, 27], all the more so the random satisfiability problem can be recast as an Ising perceptron problem [35].

The study of random optimization problems is obviously interesting from a probabilistic point of view. As far as computer science is concerned they can be seen as useful benchmarks for testing and improving resolution procedures. A successful example is the translation of the cavity equations of Section 5 into an algorithm for solving given instances of the satisfiability problem [45]. This algorithm, called Survey Propagation, extends to the clustered phase the Belief Propagation procedure of wide-spread use in statistical inference, and is a very efficient procedure to find solutions to 3-Satisfiability slightly below threshold. Another application of statistical physics ideas is the conception of new heuristics for DPLL capable of proving the unsatisfiability of formulas with 700 hundreds variables at threshold [24].

Despite those successes important question remain open. First is there a relationship between clustering and hardness of resolution? This question is reminiscent of a very general issue in statistical physics, namely the relationship between dynamical and static properties of disordered or glassy systems [21]. The onset of clustering, or more precisely of strong correlations between variables over the space of solutions drastically worsens the performances of sampling algorithms e.g. Monte Carlo procedures [36, 53]. However, in practical applications, one looks for a solution rather than for the sampling of the solution space... From this point of view knowing whether solutions are clustered or not does not seem to be of crucial relevance. Actually a local and polynomial search strategy capable of finding solutions well above the clustering threshold has been explicitly found for various optimizations problems [37].

Another open question is what happens at large K, that is, when constraints involve more and more variables. The performances of all known algorithms, be they local search procedures or DPLL solvers, seem to deteriorate. Worst-case bound indicate that the large K case is very difficult [32]. From statistical mechanics point of view problems look like more and more the random energy model [43] as K increases, but can we beat the worst-case bounds on average? Finally let us mention a recent work by Feige [28] which, for the first time, showed that the complexity of solving random SAT (or XORSAT) model had a fundamental interest in worst-case approximation theory. Consider 3-SAT instances with ratio $\alpha \gg \alpha_c$. Most of them have GS energy close to $\alpha N/2$, but a very tiny fraction of those instances have energy smaller than, say, ϵN where $\epsilon \ll \alpha$ is fixed. Is there a polynomial algorithm capable of recognizing all such atypical formulas from the vast majority of typical instances? Insights from statistical physics suggest that, the answer is positive for SAT (if we want most satisfiable instances to be detected and not all of them) while XORSAT seems to be much harder [7]! Actually, to the knowledge of the author, no local search algorithm (based on random walk, variable assignment, Monte Carlo, message-passing, cooling procedure, ...) is efficient for solving XORSAT. This makes the study of this problem even more valuable from a computer science point of view.

Appendix A. A primer on large deviations

Large deviation theory is the field of probability which deals with very unlikely events [23]. You are given a fair (unbiased) coin and toss it N times. The number

H of head draws has probability

$$p_N(H) = \frac{1}{2^N}\binom{N}{H}.$$ (A.1)

When N gets large H is highly concentrated around $H^* = N/2$ with small relative fluctuations of the order of $O(\sqrt{N})$. Yet we can ask for the probability of observing a fraction $h = H/N$ equal to say, 25%, of heads, far away from the likely value $h^* = 50\%$. To calculate this probability we use Stirling's asymptotic expression for the binomial coefficient in (A.1) to obtain

$$p_N(H = hN) = e^{-N\omega(h)+o(N)},$$ (A.2)

where

$$\omega(h) = \ln 2 + h \ln h + (1-h)\ln(1-h)$$ (A.3)

is called rate function. The meaning of (A.2) is that events with value of $h \neq h^*$ are exponentially rare in N, and $\omega(h)$ give the decay (rate) exponent. The answer to our question is $e^{-N\omega(.25)} \sim e^{-0.13N}$ when N is large. Some comments are:
• $\omega(h)$ is strictly positive, except in $h = h^* = \frac{1}{2}$ where it vanishes. This is the only value for the fraction of head draws with non exponentially small–in–N probability.
• Let $h = h^* + \delta h$ where δh is small. Using $\omega(h^*) = \omega'(h^*) = 0$ we have

$$P_N\big(H = (h^* + \delta h)N\big) = \exp\left[-N\frac{1}{2}\omega''(h^*)(\delta h)^2 + \ldots\right],$$ (A.4)

that is, δh is Gaussianly distributed with zero mean and variance $(N\omega''(h^*))^{-1} = (4N)^{-1}$. Hence central limit theorem is found back from the parabolic behaviour of the rate function around its minimum.[20]
• ω is here a convex function of its argument. This property is true rate functions describing independent events. Indeed, suppose we have H positive (according to some criterion e.g. being a head for a coin) events among a set of N events, then another set of N' events among which H' are positive. If the two sets are uncorrelated

$$p_{N+N'}(H + H') \geq p_N(H) \times p_{N'}(H')$$ (A.5)

since the same total number $H + H'$ of positive events could be observed in another combination of $N + N'$ events. Taking the logarithm and defining $h =$

---

[20]Non standard behaviour e.g. fluctuations of the order of N^ν with $\nu \neq \frac{1}{2}$ as found in Levy flights correspond to non-analyticities of ω in h^* or the vanishing of the second derivative.

$H/N, h' = H'/N, u = N/(N + N')$ we obtain

$$\omega(uh + (1 - u)h') \leq u\omega(h) + (1 - u)\omega(h'),\qquad\qquad\text{(A.6)}$$

for any $u \in [0; 1]$. Hence the representative curve of ω lies below the chord joining any two points on this curve, and ω is convex. Non-convex rate functions are found in presence of strong correlations.[21]

Appendix B. Inequalities of first and second moments

Let $\mathcal{N}$ be a random variable taking values on the positive integers, and call $p_{\mathcal{N}}$ its probability. We denote by $\langle \mathcal{N} \rangle$ and $\langle \mathcal{N}^2 \rangle$ the first and second moments of $\mathcal{N}$ (assumed to be finite), and write

$$p(\mathcal{N} \geq 1) = \sum_{\mathcal{N}=1,2,3,\dots} p_{\mathcal{N}} = 1 - p_0\qquad\qquad\text{(B.1)}$$

the probability that $\mathcal{N}$ is not equal to zero. Our aim is to show the inequalities

$$\frac{\langle \mathcal{N} \rangle^2}{\langle \mathcal{N}^2 \rangle} \leq p(\mathcal{N} \geq 1) \leq \langle \mathcal{N} \rangle.\qquad\qquad\text{(B.2)}$$

The right inequality, call 'first moment inequality', is straightforward:

$$\langle \mathcal{N} \rangle = \sum_{\mathcal{N}} \mathcal{N} p_{\mathcal{N}} = \sum_{\mathcal{N} \geq 1} \mathcal{N} p_{\mathcal{N}} \geq \sum_{\mathcal{N} \geq 1} p_{\mathcal{N}} = p(\mathcal{N} \geq 1).\qquad\qquad\text{(B.3)}$$

Consider now the linear space made of vectors $\mathbf{v} = (v_0, v_1, v_2, \dots\}$ whose components are labeled by positive integers, with the scalar product

$$\mathbf{v} \cdot \mathbf{v}' = \sum_{\mathcal{N}} p_{\mathcal{N}} v_{\mathcal{N}} v'_{\mathcal{N}}.\qquad\qquad\text{(B.4)}$$

Choose now $v_{\mathcal{N}} = \mathcal{N}$, and $v'_0 = 0, v'_{\mathcal{N}} = 1$ for $\mathcal{N} \geq 1$. Then

$$\mathbf{v} \cdot \mathbf{v} = \langle \mathcal{N}^2 \rangle, \quad \mathbf{v} \cdot \mathbf{v}' = \langle \mathcal{N} \rangle, \quad \mathbf{v}' \cdot \mathbf{v}' = p(\mathcal{N} \geq 1).\qquad\qquad\text{(B.5)}$$

The left inequality in (B.2) is simply the Cauchy-Schwarz inequality for $\mathbf{v}, \mathbf{v}'$:
$(\mathbf{v} \cdot \mathbf{v}')^2 \leq (\mathbf{v} \cdot \mathbf{v}) \times (\mathbf{v}' \cdot \mathbf{v}')$.

---

[21] Consider the following experiment. You are given three coins: the first one is fair (coin A), the second and third coins, respectively denoted by B and C, are biased and give head with probabilities, respectively, $\frac{1}{4}$ and $\frac{3}{4}$. First draw coin A once. If the outcome is head pick up coin B, otherwise pick up coin C. Then draw your coin N times. What is the rate function associated to the fraction h of heads?

Appendix C. Corrections to the saddle-point calculation of $\langle \mathcal{N}^2 \rangle$

In this Appendix we show that $\langle \mathcal{N}^2 \rangle$ is asymptotically equivalent to $\langle \mathcal{N} \rangle^2$, where $\mathcal{N}$ is the number of solutions of a 3-XORSAT formula with ratio $\alpha < \alpha_2 \simeq 0.889$. This requires to take care of the finite-size corrections around the saddle-point calculations of Section 2.6. Let $Z = (z_1, z_2, \ldots, z_N)$ denotes a configuration of variables at distance d from the zero configuration *i.e.* dN variables z_i are equal to 1, the other $(1 - d)N$ variables are null. Let $q(d, N)$ be the probability that Z satisfies the equation $z_i + z_j + z_k = 0$ where (i, j, k) is a random triplet of distinct integers (unbiased distribution):

$$q(d, N) = \frac{1}{\binom{N}{3}} \left[\binom{(1 - d)N}{3} + (1 - d)N \binom{dN}{2} \right] \tag{C.1}$$

$$= q(d) \left(1 + \frac{h(d)}{N} \right) + \ldots \quad \text{where} \quad h(d) = \frac{6d(2d - 1)}{3d^2 + (1 - d)^2},$$

and $q(d)$ is defined in (2.21) with $K = 3$. Terms of the order of N^{-2} have been discarded.

Using formula (2.22) with $q(d)$ substituted with $q(d, N)$ and the Stirling formula for the asymptotic behaviour of combinatorial coefficients we have

$$\langle \mathcal{N}^2 \rangle \sim \sum_{d=0, \frac{1}{N}, \frac{2}{N}, \ldots} \frac{\sqrt{2\pi N}}{\sqrt{2\pi Nd} \sqrt{2\pi N(1 - d)}} e^{NA(d, \alpha) + \alpha h(d)} \tag{C.2}$$

where $A(d, \alpha)$ is defined in (2.23), and $\sim$ indicates a true asymptotic equivalence (no multiplicative factor omitted). The r.h.s. of (C.2) is the Riemann sum associated to the integral

$$\langle \mathcal{N}^2 \rangle \sim \int_0^1 \frac{N \, dd}{\sqrt{2\pi Nd(1 - d)}} e^{NA(d, \alpha) + \alpha h(d)}. \tag{C.3}$$

We now estimate the integral through the saddle-point method. For $\alpha < \alpha_2 \simeq 0.889$ the dominant contribution to the integral comes from the vicinity of $d^* = \frac{1}{2}$. There are quadratic fluctuations around this saddle-point, with a variance equal to N times the inverse of (the modulus of) the second derivative A_{dd} of A with respect to d. Carrying out the Gaussian integral over those fluctuations we obtain

$$\langle \mathcal{N}^2 \rangle \sim \frac{N \, e^{NA(d^*, \alpha) + \alpha h(d^*)}}{\sqrt{2\pi Nd^*(1 - d^*)}} \sqrt{\frac{2\pi}{N |A_{dd}(d^*, \alpha)|}} \sim \langle \mathcal{N} \rangle^2 \tag{C.4}$$

since $h(d^*) = 0$, $A_{dd}(d^*, \alpha) = -4$. Therefore, from the second moment inequality, $P_{SAT} \to 1$ when $N \to \infty$ at ratios smaller than α_2.

References

[1] D. Achlioptas, Theor. Comp. Sci. **265** (2001) 159.

[2] D. Achlioptas, P. Beame, and M. Molloy, Journal of Computer and System Sciences **68** (2004) 238.

[3] D. Achlioptas, A. Naor and Y. Perez, Nature **435** (2005) 759.

[4] D.J. Aldous, Discrete Math. **76** (1989) 167.

[5] D.J. Aldous, Rand. Struct. Algo. **48** (2001) 381.

[6] M. Alekhnovich and E. Ben-Sasson, *Analysis of the Random Walk Algorithm on Random 3-CNFs*, preprint (2002).

[7] F. Altarelli, R. Monasson and F. Zamponi, J. Phys. A **40** (2007) 867.

[8] D.J. Amit, *Modeling Brain Function*, (Cambridge University Press, Cambridge, 1989).

[9] J.R.L. de Almeida and D.J. Thouless, J. Phys. A **11** (1978) 983.

[10] P. Beame, R. Karp, T. Pitassi, and M. Saks, (Proceedings of the ACM Symp. on Theory of Computing, 1998, pp. 561).

[11] G. Biroli, R. Monasson and M. Weigt, Eur. Phys. J. B **14** (2000) 551.

[12] B. Bollobas, *Random Graphs* (Cambridge University Press, Cambridge, 2001).

[13] A.Z. Broder, A.M. Frieze and E. Upfal, (Proceedings of Symposium of Discrete Algorithms (SODA), Austin, 1993).

[14] M.T. Chao and J. Franco, Information Science **51** (1990) 289; SIAM Journal on Computing **15** (1986) 1106.

[15] V. Chvàtal and E. Szmeredi, Journal of the ACM **35** (1988) 759.

[16] S. Cocco and R. Monasson, Phys. Rev. Lett. **86** (2001) 1658; Eur. Phys. J. B **22** (2002) 505.

[17] S. Cocco, O. Dubois, J. Mandler and R. Monasson, Phys. Rev. Lett. **90** (2003) 047205.

[18] N. Creignou and H. Daudé, Discrete Applied Mathematics **96-97** (1999) 41.

[19] N. Creignou and H. Daudé, RAIRO: Theoretical Informatics and Applications **37** (2003) 127.

[20] N. Creignou, H. Daudé and O. Dubois, Combinatorics, Probability and Computing **12** (2003) 113.

[21] L. Cugliandolo and J. Kurchan, Phys. Rev. Lett. **71** (1993) 173.

[22] M. Davis and H. Putnam, J. Assoc. Comput. Mach. **7** (1960) 201; M. Davis, G. Logemann and D. Loveland, Communications of the ACM **5** (1962) 394.

[23] A. Dembo and O. Zeitouni, *Large deviations techniques and applications* (Springer-Verlag, New York, 1993).

[24] G. Dequen and O. Dubois, (Proceedings of Theory and Applications of Satisfiability Testing, 6th International Conference, SAT 2003. Santa Margherita Ligure, 2003, pp. 486).

[25] C. Deroulers and R. Monasson, Eur. Phys. J. B **49** (2006) 339.

[26] O. Dubois and J. Mandler, (Proc. of the 43rd annual IEEE symposium on Foundations of Computer Science, Vancouver, 2002).

[27] A. Engel and C. Van den Broeck, *Statistical Mechanics of Learning*, (Cambridge University Press, Cambridge, 2001).

[28] U. Feige, (Proceedings of 34th STOC conference, 2002, pp. 534).

[29] A. Frieze and S. Suen, Journal of Algorithms **20** (1996) 312.

[30] Y. Fu and P.W. Anderson, J. Phys. A **19** (1986) 1605.

[31] A.K. Hartmann and M. Weigt, Theor. Comp. Sci. **265** (2001) 199.

[32] R. Impagliazzo and R. Paturi, (Proceedings of the IEEE Conference on Computational Complexity, 1999, pp. 237).

[33] A.C. Kaporis, L.M. Kirousis, and Y.C. Stamatiou, *How to prove conditional randomness using the principle of deferred decisions*, technical report, Computer technology Institute, Patras (2002).

[34] D. Knuth, *The Art of Computer Programming; vol 1: fundamental algorithms, section 2.2.1*, (Addison-Wesley, Ney York, 1968).

[35] W. Krauth and M. Mézard, J. Phys. (France) **50** (1989) 3057.

[36] F. Krzakala *et al. Gibbs States and the Set of Solutions of Random Constraint Satisfaction Problems* (preprint, 2006).

[37] F. Krzakala and J. Kurchan, *A landscape analysis of constraint satisfaction problems* (preprint, 2007).

[38] S. Mertens, Phys. Rev. Lett. **81** (1998) 4281; Phys. Rev. Lett. **84** (2000) 1347.

[39] S. Mertens, Theor. Comp. Sci. **265** (2001) 79.

[40] M. Mertens, M. Mézard and R Zecchina, Rand. Struct. Algo. **28** (2006) 340.

[41] M. Mézard and G. Parisi, J. Phys. (Paris) **48** (1987) 1451.

[42] M. Mézard and G. Parisi, J. Phys. (Paris) **47** (1986) 1285.

[43] M. Mézard, G. Parisi and M. Virasoro, *Spin glasses and beyond* (World Scientific, Singapore, 1987).

[44] M. Mézard and G. Parisi, Eur. Phys. J. B **20** (2001) 217; J. Stat. Phys **111** (2003) 111.

[45] M. Mézard and R Zecchina, Phys. Rev. E **56** (2002) 066126.

[46] M. Mézard, F. Ricci-Tersenghi, and R. Zecchina, J. Stat. Phys. **111** (2003) 505.

[47] D. Mitchell, B. Selmann and H. Levesque, Proc. of the Tenth Natl. Conf. on Artificial Intelligence (AAAI-92), (1992) 440.

[48] R. Monasson, Phys. Rev. Lett. **75** (1995) 2847.

[49] R. Monasson and R. Zecchina, Phys. Rev. E **56** (1997) 1357.

[50] R. Monasson, J. Phys. A **31** (1998) 513.

[51] R. Monasson, Lecture Notes in Computer Science **3624** (2005) 402.

[52] A. Montanari and D. Shah, (Proceedings of Symposium of Discrete Algorithms (SODA), New Orleans, 2007).

[53] A. Montanari and G. Semerjian, Phys. Rev. Lett. **94** (2005) 247201.

[54] R. Motwani and P. Raghavan, *Randomized algorithms* (Cambridge University Press, Cambridge, 1995).

[55] R. Mulet, A. Pagnani, M. Weigt and R Zecchina, Phys. Rev. Lett. **89** (2002) 268701.

[56] H. Orland, J. Phys. (Paris) Lett. **46** (1985) L763; M. Mézard and G. Parisi, J. Phys. (Paris) Lett. **46** (1985) L771.

[57] C. Papadimitriou and K. Steiglitz, *Combinatorial Optimization: Algorithms and Complexity* (Dover, 1998).

[58] C.H. Papadimitriou, (Proceedings of the 32nd Annual IEEE Symposium on Foundations of Computer Science, 1991, pp. 163).

[59] G. Parisi and M. Virasoro, J. Phys. (Paris) **50** (1986) 3317.

[60] F. Ricci-Tersenghi, M. Weigt and R. Zecchina, Phys. Rev. E **63** (1999) 026702.

[61] U. Schöning, Algorithmica **32** (2002) 615.

[62] G. Semerjian and R. Monasson, Phys. Rev. E **67** (2003) 066103; W. Barthel, A. Hartmann and M. Weigt, Phys. Rev. E **67** (2003) 066104.

[63] N. Wormald, The Annals of Applied Probability **5** (1995) 1217.

[64] M. Weigt and A.K. Hartmann Phys. Rev. Lett. **84** (2000) 6118; Phys. Rev. Lett. **86** (2001) 1658.

Course 2

MODERN CODING THEORY: THE STATISTICAL
MECHANICS AND COMPUTER SCIENCE POINT
OF VIEW

Andrea Montanari[1] and Rüdiger Urbanke[2]

[1]Electrical Engineering and Statistics Departments, Stanford University, Packard 272, Stanford CA
94305-9510
[2] Communication Theory Laboratory, Office inr 116, EPFL-IC-Station 14, 1015 Lausanne,
Switzerland

J.-P. Bouchaud, M. Mézard and J. Dalibard, eds.
Les Houches, Session LXXXV, 2006
Complex Systems

Contents

1. Introduction and outline

The last few years have witnessed an impressive convergence of interests between disciplines which are *a priori* well separated: coding and information theory, statistical inference, statistical mechanics (in particular, mean field disordered systems), as well as theoretical computer science. The underlying reason for this convergence is the importance of probabilistic models and/or probabilistic techniques in each of these domains. This has long been obvious in information theory [53], statistical mechanics [10], and statistical inference [45]. In the last few years it has also become apparent in coding theory and theoretical computer science. In the first case, the invention of Turbo codes [7] and the re-invention of Low-Density Parity-Check (LDPC) codes [30, 28] has motivated the use of random constructions for coding information in robust/compact ways [50]. In the second case (theoretical computer science) the relevance of randomized algorithms has steadily increased (see for instance [41]), thus motivating deep theoretical developments. A particularly important example is provided by the Monte Carlo Markov Chain method for counting and sampling random structures.

Given this common probabilistic background, some analogies between these disciplines is not very surprising nor is it particularly interesting. The key new ingredient which lifts the connections beyond some superficial commonalities is that one can name specific problems, questions, and results which lie at the intersection of these fields while being of central interest for each of them. The set of problems and techniques thus defined can be somewhat loosely named "theory of large graphical models." The typical setting is the following: a large set of random variables taking values in a finite (typically quite small) alphabet with a "local" dependency structure; this local dependency structure is conveniently described by an appropriate graph.

In this lecture we shall use "modern" coding theory as an entry point to the domain. There are several motivations for this: (*i*) theoretical work on this topic is strongly motivated by concrete and well-defined practical applications; (*ii*) the probabilistic approach mentioned above has been quite successful and has substantially changed the field (whence the reference to *modern* coding theory); (*iii*) a sufficiently detailed picture exists illustrating the interplay among different view points.

We start in Section 2 with a brief outline of the (channel coding) problem. This allows us to introduce the standard definitions and terminology used in this field. In Section 3 we introduce ensembles of codes defined by sparse random graphs and discuss their most basic property – the weight distribution. In Section 4 we phrase the decoding problem as an inference problem on a graph and consider the performance of the efficient (albeit in general suboptimal) message-passing decoder. We show how the performance of such a combination (sparse graph code and message-passing decoding) can be analyzed and we discuss the relationship of the performance under message-passing decoding to the performance of the optimal decoder. In Section 5 we briefly touch on some problems beyond coding, in order to show as similar concept emerge there. In particular, we discuss how message passing techniques can be successfully used in some families of counting/inference problems. In Section 6 we show that several of the simplifying assumptions (binary case, symmetry of channel, memoryless channels) are convenient in that they allow for a simple theory but are not really necessary. In particular, we discuss a simple channel with memory and we see how to proceed in the asymmetric case. Finally, we conclude in Section 7 with a few fundamental open problems.

To readers who would like to find current contributions on this topic we recommend the *IEEE Transactions on Information Theory*. A considerably more in-depth discussion can be found in the two upcoming books *Information, Physics and Computation* [36] and *Modern Coding Theory* [50]. Standard references on coding theory are [6, 9, 26] and very readable introductions to information theory can be found in [12, 20]. Other useful reference sources are the book by Nishimori [44] as well as the book by MacKay [29].

2. Background: the channel coding problem

The central problem of communications is how to transmit information reliably through a noisy (and thus unreliable) communication channel. Coding theory aims at accomplishing this task by adding a properly designed redundancy to the transmitted message. This redundancy is then used at the receiver to reconstruct the original message despite the noise introduced by the channel.

2.1. The problem

In order to model the situation described above we shall assume that the noise is random with some known distribution.[1] To keep things simple we shall assume

---

[1] It is worth mentioning that an alternative approach would be to consider the noise as 'adversarial' (or worst case) under some constraint on its intensity.

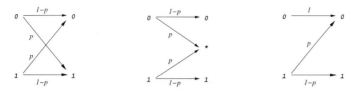

Fig. 1. Schematic description of three simple binary memoryless channels. From left to right: binary symmetric channel BSC(p), binary erasure channel BEC(ϵ), and Z channel ZC(p).

that the communication channel admits as input binary symbols $x \in \{0, 1\}$, while the output belongs to some finite alphabet $\mathcal{A}$. We denote the probability of observing the output $y \in \mathcal{A}$ given that the input was $x \in \{0, 1\}$ by $Q(y|x)$. The channel model is defined by the transition probability matrix

$$Q = \big\{ Q(y|x) : x \in \{0, 1\}, \ y \in \mathcal{A} \big\}. \tag{2.1}$$

Of course, the entries of this matrix must be non-negative and normalized in such a way that $\sum_y Q(y|x) = 1$. It is convenient to have a few simple examples in mind. We refer to Fig. 1 for an illustration of the channel models which we introduce in the following three examples.

Example 1: The *binary symmetric channel* BSC(p) is defined by letting $\mathcal{A} = \{0, 1\}$ and $Q(0|0) = Q(1|1) = 1 - p$; the normalization then enforces $Q(1|0) = Q(0|1) = p$. In words, the channel "flips" the input bit with probability $p \in [0, 1]$. Since flips are introduced for each bit independently we say that the channel is *memoryless*. Except for an example in Section 6.2 all channels which we consider are memoryless.

Example 2: The *binary erasure channel* BEC(ϵ) is defined by $\mathcal{A} = \{0, 1, *\}$ and $Q(0|0) = Q(1|1) = 1 - \epsilon$ while $Q(*|0) = Q(*|1) = \epsilon$. In words, the channel input is erased with probability ϵ and it is transmitted correctly otherwise.

Example 3: The *Z-channel* ZC(p) has an output alphabet $\mathcal{A} = \{0, 1\}$ but acts differently on input 0 (that is transmitted correctly) and 1 (that is flipped with probability p). We invite the reader to write the transition probability matrix.

Since in each case the input is binary we speak of a *binary-input* channel. Since further in all models each input symbol is distorted independently from all other ones we say that the channels are *memoryless*. It is convenient to further restrict our attention to *symmetric* channels: this means that there is an involution on $\mathcal{A}$ (i.e. a mapping $\iota : \mathcal{A} \to \mathcal{A}$ such that $\iota \circ \iota = 1$) so that $Q(y|0) = Q(\iota(y)|1)$. (E.g., if $\mathcal{A} = \mathbb{R}$ then we could require that $Q(y|0) = Q(-y|1)$.) This condition is satisfied by the first two examples above but not by the third one. To summarize these three properties one refers to such models as BMS channels.

In order to complete the problem description we need to formalize the information which is to be transmitted. We shall model this probabilistically as well and assume that the transmitter has an information source that provides an infinite stream of i.i.d. fair coins: $\{z_i; i = 0, 1, 2, \dots\}$, with $z_i \in \{0, 1\}$ uniformly at random. The goal is to reproduce this stream faithfully after communicating it over the noisy channel.

Let us stress that, despite its simplification, the present setting contains most of the crucial and challenges of the channel coding problem. Some of the many generalizations are described in Section 6.

2.2. Block coding

The (general) coding strategy we shall consider here is *block coding*. It works as follows:

• The source stream $\{z_i\}$ is chopped into blocks of length L. Denote one such block by $\underline{z}$, $\underline{z} = (z_1, \dots, z_L) \in \{0, 1\}^L$.

• Each block is fed into an *encoder*. This is a map $\mathsf{F} : \{0, 1\}^L \to \{0, 1\}^N$, for some fixed $N > L$ (the *blocklength*). In words, the encoder introduces redundancy in the source message. Without loss of generality we can assume F to be injective. It this was not the case, even in the absence of noise, we could not uniquely recover the transmitted information from the observed codeword.

• The image of $\{0, 1\}^L$ under the map F is called the *codebook*, or sometimes the *code*, and it will be denoted by $\mathfrak{C}$. The code contains $|\mathfrak{C}| = 2^L$ strings of length N called *codewords*. These are the possible channel inputs. The codeword $\underline{x} = \mathsf{F}(\underline{z})$ is sent through the channel, bit by bit.

• Let $\underline{y} = (y_1, \dots, y_N) \in \mathcal{A}^N$ be the channel output. Conditioned on $\underline{x}$ the y_i, $i = 1, \dots, L$, are independent random variables with distribution $y_i \overset{\mathrm{d}}{=} Q(\cdot | x_i)$ (here and below $\overset{\mathrm{d}}{=}$ denotes identity in distribution and $x \overset{\mathrm{d}}{=} P(\cdot)$ means that x is a random variable with distribution $P(\cdot)$).

• The channel output is fed into a *decoder*, which is a map $\widehat{\mathsf{F}} : \mathcal{A}^N \to \{0, 1\}^L$. It is the objective of the decoder to reconstruct the source $\underline{z}$ from the noisy channel output $\underline{y}$.

The flow chart describing this coding scheme is shown in Fig. 2. It is convenient to slightly modify the above scheme. Notice that, under the hypothesis that the encoder is injective, the codebook is in one-to-one correspondence with the source sequences. Since these are equiprobable, the transmitted codewords are equiprobable as well. We can therefore equivalently assume that the trans-

Fig. 2. Flow chart of a block coding scheme.

mitter picks a codeword uniformly at random and transmits it. Every reference to the source stream can be eliminated if we redefine the decoder to be a map $\widehat{F} : A^N \rightarrow \{0, 1\}^N$, i.e., the decoder aims to reconstruct the transmitted codeword. If $\widehat{F}(y) \notin \mathcal{C}$ we declare an error.[2] In the following we shall also use the notation $\widehat{F}(\underline{y}) = \underline{\widehat{x}}(\underline{y}) = (\widehat{x}_1(\underline{y}), \ldots, \widehat{x}_N(\underline{y}))$.

One crucial parameter of a code is its *rate*: it quantifies how many bits of information are transmitted per channel use,

$$R \equiv \frac{L}{N} = \frac{1}{N} \log_2 |\mathcal{C}|. \tag{2.2}$$

Two fundamental performance parameters are the *bit* (or 'symbol') and *block* (or 'word') *error rates*. The block error rate is the probability that the input codeword is not recovered correctly at the end of the process,

$$P_B \equiv \mathbb{P}\{\underline{\widehat{x}}(\underline{y}) \neq \underline{x}\}. \tag{2.3}$$

The bit error rate is the expected fraction of bits that are not recovered correctly,

$$P_b \equiv \frac{1}{N} \sum_{i=1}^{N} \mathbb{P}\{\widehat{x}_i(\underline{y}) \neq x_i\}. \tag{2.4}$$

It should not be too surprising that one can trade-off rate and error probability. We want to achieve a high rate and achieve a low probability of error. However, increasing the rate decreases the redundancy built into the codeword, thus inducing a higher error probability. The aim of coding theory is to choose the code $\mathcal{C}$ and the decoding function $\underline{\widehat{x}}(\cdot)$ in a way to optimize this trade-off.

2.3. Decoding

Given the code there is a simple (although in general not computationally efficient) prescription for the decoder. If we want to minimize the block error rate, we must chose the most likely codeword,

$$\underline{\widehat{x}}^B(\underline{y}) \equiv \arg\max_{\underline{x}} \mathbb{P}\{\underline{X} = \underline{x} | \underline{Y} = \underline{y}\}. \tag{2.5}$$

---

[2]More precisely, if we are interested only in the block probability of error, i.e., the frequency at which the whole block of data is decoded correctly, then indeed any one-to-one mapping between information word and codeword performs identical. If, on the other hand, we are interested in the fraction of *bits* that we decode correctly then the exact mapping from information word to codeword does come into play. We shall ignore this somewhat subtle point in the sequel.

To minimize the bit error rate we must instead return the sequence of most likely bits,

$$\widehat{x}_i^b(\underline{y}) \equiv \arg\max_{x_i} \mathbb{P}\{X_i = x_i | \underline{Y} = \underline{y}\}. \tag{2.6}$$

The reason of these prescriptions is the object of the next exercise.

Exercise 1: Let (U, V) be a pair of discrete random variables. Think of U as a 'hidden' variable and imagine you observe $V = v$. We want to understand what is the optimal estimate for U given $V = v$. Show that the function $v \mapsto \widehat{u}(v)$ that minimizes the error probability $\mathsf{P}(\widehat{u}) \equiv \mathbb{P}\{U \neq \widehat{u}(V)\}$ is given by

$$\widehat{u}(v) = \arg\max_{u} \mathbb{P}\{U = u | V = v\}. \tag{2.7}$$

It is instructive to explicitly write down the conditional distribution of the channel input given the output. We shall denote it as $\mu_{\mathfrak{C},y}(\underline{x}) = \mathbb{P}\{\underline{X} = \underline{x} | \underline{Y} = \underline{y}\}$ (and sometimes drop the subscripts $\mathfrak{C}$ and y if they are clear from the context). Using Bayes rule we get

$$\mu_{\mathfrak{C},y}(\underline{x}) = \frac{1}{Z(\mathfrak{C}, y)} \prod_{i=1}^{N} Q(y_i | x_i)\, \mathbb{I}_{\mathfrak{C}}(\underline{x}), \tag{2.8}$$

where $\mathbb{I}_{\mathfrak{C}}(\underline{x})$ denotes the code membership function ($\mathbb{I}_{\mathfrak{C}}(\underline{x}) = 1$ if $\underline{x} \in \mathfrak{C}$ and $= 0$ otherwise).

According to the above discussion, decoding amounts to computing the marginals (for symbol MAP) or the mode[3] (for word MAP) of $\mu(\cdot)$. More generally, we would like to understand the properties of $\mu(\cdot)$: is it concentrated on a single codeword or spread over many of them? In the latter case, are these close to each other or very different? And what is their relationship with the transmitted codeword?

The connection to statistical mechanics emerges in the study of the decoding problem [56, 51]. To make it completely transparent we rewrite the distribution $\mu(\cdot)$ in Boltzmann form

$$\mu_{\mathfrak{C},y}(\underline{x}) = \frac{1}{Z(\mathfrak{C}, y)} e^{-E_{\mathfrak{C},y}(\underline{x})}, \tag{2.9}$$

$$E_{\mathfrak{C},y}(\underline{x}) = \begin{cases} -\sum_{i=1}^{N} \log Q(y_i | x_i), & \text{if } \underline{x} \in \mathfrak{C}, \\ +\infty, & \text{otherwise.} \end{cases} \tag{2.10}$$

---

[3] We recall that the mode of a distribution with density $\mu(\cdot)$ is the value of x that maximizes $\mu(x)$.

The word MAP and bit MAP rule can then be written as

$$\widehat{\underline{x}}^{\text{B}}(\underline{y}) = \arg\min_{\underline{x}} E_{\mathcal{C},y}(\underline{x}), \tag{2.11}$$

$$\widehat{x}_i^{\text{b}}(\underline{y}) = \arg\max_{x_i} \sum_{x_j : j \neq i} \mu_{\mathcal{C},y}(\underline{x}). \tag{2.12}$$

In words, word MAP amounts to computing the ground state of a certain energy function, and bit MAP corresponds to computing the expectation with respect to the Boltzmann distribution. Notice furthermore that $\mu(\cdot)$ is itself random because of the randomness in y (and we shall introduce further randomness in the choice of the code). This is analogous to what happens in statistical physics of disordered systems, with $\underline{y}$ playing the role of quenched random variables.

2.4. Conditional entropy and free energy

As mentioned above, we are interested in understanding the properties of the (random) distribution $\mu_{\mathcal{C},y}(\cdot)$. One possible way of formalizing this idea is to consider the entropy of this distribution.

Let us recall that the (Shannon) entropy of a discrete random variable X (or, equivalently, of its distribution) quantifies, in a very precise sense, the 'uncertainty' associated with X.[4] It is given by

$$H(X) = -\sum_{x} \mathbb{P}(x) \log \mathbb{P}(x). \tag{2.13}$$

For two random variables X and Y one defines the conditional entropy of X given Y as

$$H(X|Y) = -\sum_{x,y} \mathbb{P}(x, y) \log \mathbb{P}(x|y)$$

$$= \mathbb{E}_y \left\{ -\sum_{x} \mathbb{P}(x|Y) \log \mathbb{P}(x|Y) \right\}. \tag{2.14}$$

This quantifies the remaining uncertainty about X when Y is observed.

Considering now the coding problem. Denote by $\underline{X}$ the (uniformly random) transmitted codeword and by $\underline{Y}$ the channel output. The right-most expression in Eq. (2.14) states that $H(\underline{X}|\underline{Y})$ is the expectation of the entropy of the conditional distribution $\mu_{\mathcal{C},y}(\cdot)$ with respect to $\underline{y}$.

---

[4]For a very readable account of information theory we recommend [12].

Let us denote by $\nu_{\mathcal{C}}(x)$ the probability that a uniformly random codeword in $\mathcal{C}$ takes the value x at the i-th position, averaged over i. Then a straightforward calculation yields

$$H(\underline{X}|\underline{Y}) = -\frac{1}{|\mathcal{C}|} \sum_{\underline{x},\underline{y}} \prod_{i=1}^{N} Q(y_i|x_i) \log \left\{ \frac{1}{Z(\mathcal{C}, \underline{y})} \prod_{i=1}^{N} Q(y_i|x_i) \right\}, \qquad (2.15)$$

$$= -N \sum_{x} \nu_{\mathcal{C}}(x) Q(y|x) \log Q(y|x) + \mathbb{E}_{\underline{y}} \log Z(\mathcal{C}, \underline{y}). \qquad (2.16)$$

The 'type' $\nu_{\mathcal{C}}(x)$ is usually a fairly straightforward characteristic of the code. For most of the examples considered below we can take $\nu_{\mathcal{C}}(0) = \nu_{\mathcal{C}}(1) = 1/2$. As a consequence the first of the terms above is trivial to compute (it requires summing over $2|\mathcal{A}|$ terms).

On the other hand the second term is highly non-trivial. The reader will recognize the expectation of a free energy, with $\underline{y}$ playing the role of a quenched random variable.

The conditional entropy $H(\underline{X}|\underline{Y})$ provides an answer to the question: how many codewords is $\mu_{\mathcal{C},\underline{y}}(\cdot)$ spread over? It turns out that about $e^{H(\underline{X}|\underline{Y})}$ of them carry most of the weight.

2.5. Shannon Theorem and random coding

As mentioned above, there exists an obvious trade-off between high rate and low error probability. In his celebrated 1948 paper [53], Shannon derived the optimal error probability-vs-rate curve in the limit of large blocklengths. In particular, he proved that if the rate is larger than a particular threshold, then the error probability can be made arbitrarily small. The threshold depends on the channel and it is called the channel *capacity*. The capacity of a BMS channel (measured in bits per channel use) is given by the following elementary expression,

$$\mathrm{C}(Q) = H(X) - H(X|Y)$$
$$= 1 + \sum_{y} Q(y|0) \log_2 \left\{ \frac{Q(y|0)}{Q(y|0) + Q(y|1)} \right\}.$$

For instance, the capacity of a BSC(p) is $\mathrm{C}(p) = 1 - \mathfrak{h}_2(p)$, (where $\mathfrak{h}_2(p) = -p \log_2 p - (1-p) \log_2(1-p)$ is the entropy of a Bernoulli random variable of parameter p) while the capacity of a BEC(ϵ) is $\mathrm{C}(\epsilon) = 1 - \epsilon$. As an illustration, the capacity of a BSC(p) with flip probability $p \approx 0.110028$ is $\mathrm{C}(p) = 1/2$: such a channel can be used to transmit reliably $1/2$ bit of information per channel use.

Theorem 2.1 (Channel Coding Theorem). *For any BMS channel with transition probability Q and $R < C(Q)$ there exists a sequence of codes $\mathfrak{C}_N$ of increasing blocklength N and rate $R_N \to R$ whose block error probability $P_B^{(N)} \to 0$ as $N \to \infty$.*

Vice versa, for any $R > C(Q)$ the block error probability of a code with rate at least R is bounded away from 0.

The prove of the first part ('achievability') is one of the first examples of the so-called 'probabilistic method'. In order to prove that there exists an object with a certain property (a code with small error probability), one constructs a probability distribution over all potential candidates (all codes of a certain blocklength and rate) and shows that a random element has the desired property with non-vanishing probability. The power of this approach is in the (meta-mathematical) observation that random constructions are often much easier to produce than explicit, deterministic ones.

The distribution over codes proposed by Shannon is usually referred to as the *random code* (or, *Shannon*) *ensemble*, and is particularly simple. One picks a code $\mathfrak{C}$ uniformly at random among all codes of blocklength N and rate R. More explicitly, one picks 2^{NR} codewords as uniformly random points in the hypercube $\{0, 1\}^N$. This means that each codeword is a string of N fair coins $\underline{x}^{(\alpha)} = (x_1^{(\alpha)}, \ldots, x_N^{(\alpha)})$ for[5] $\alpha = 1, \ldots, 2^{NR}$.

Once the ensemble is defined, one can estimate its average block error probability and show that it vanishes in the blocklength for $R < C(Q)$. Here we will limit ourselves to providing some basic 'geometric' intuition of why a random code from the Shannon ensemble performs well with high probability.[6]

Let us consider a particular codeword, say $\underline{x}^{(0)}$, and try to estimate the distance (from $\underline{x}^{(0)}$) at which other codewords in $\mathfrak{C}$ can be found. This information is conveyed by the *distance enumerator*

$$\mathcal{N}_{\underline{x}^{(0)}}(d) \equiv \#\{\underline{x} \in \mathfrak{C}\backslash\underline{x}^{(0)} \text{ such that } d(\underline{x}, \underline{x}^{(0)}) = d\}, \tag{2.17}$$

where $d(\underline{x}, \underline{x}')$ is the Hamming distance between $\underline{x}$ and $\underline{x}'$ (i.e., the number of positions in which $\underline{x}$ and $\underline{x}'$ differ). The expectation of this quantity is the number of codewords different from $\underline{x}^{(0)}$ (that is $(2^{NR} - 1)$) times the probability that any

---

[5]The reader might notice two imprecisions with this definition. First, 2^{NR} is not necessarily an integer: one should rather use $\lceil 2^{NR} \rceil$ codewords, but the difference is obviously negligible. Second, in contradiction with our definition, two codewords may coincide if they are independent. Again, only an exponentially small fraction of codewords will coincide and they can be neglected for all practical purposes.

[6]Here and in the rest of the lectures, the expression *with high probability* means 'with probability approaching one as $N \to \infty$'

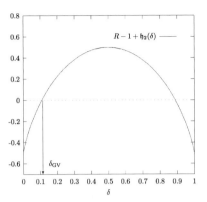

Fig. 3. Exponential growth rate for the expected distance enumerator $\mathbb{E}\,\mathcal{N}_{\underline{x}^{(0)}}(n\delta)$ within the random code ensemble.

given codeword has distance d from $\underline{x}^{(0)}$. Since each entry is independent and different with probability $1/2$, we get

$$\mathbb{E}\,\mathcal{N}_{\underline{x}^{(0)}}(d) = (2^{NR} - 1)\,\frac{1}{2^N}\binom{N}{d} \doteq 2^{N[R-1+\mathfrak{h}_2(\delta)]}, \qquad (2.18)$$

where $\delta = d/N$ and $\doteq$ denotes equality to the leading exponential order.[7]

The exponent $R - 1 + \mathfrak{h}_2(\delta)$ is plotted in Fig. 3. For δ sufficiently small (and $R < 1$) this exponent is negative. Its first zero, to be denoted as $\delta_{GV}(R)$, is called the Gilbert-Varshamov distance. For any $\delta < \delta_{GV}(R)$ the expected number of codewords of distance at most $N\delta$ from $\underline{x}^{(0)}$ is exponentially small in N. It follows that the probability to find *any codeword* at distance smaller than $N\delta$ is exponentially small in N.

Vice-versa, for $d = N\delta$, with $\delta > \delta_{GV}(R)$, $\mathbb{E}\,\mathcal{N}_{\underline{x}^{(0)}}(d)$ is exponentially large in N. Indeed, $\mathcal{N}_{\underline{x}^{(0)}}(d)$ is a binomial random variable, because each of the $2^{NR} - 1$ codewords is at distance d independently and with the same probability. As a consequence, $\mathcal{N}_{\underline{x}^{(0)}}(d)$ is exponentially large as well with high probability.

The bottom line of this discussion is that, for any given codeword $\underline{x}^{(0)}$ in $\mathfrak{C}$, the closest other codeword is, with high probability, at distance $N(\delta_{GV}(R) \pm \varepsilon)$. A sketch of this situation is provided in Fig. 4.

Let us assume that the codeword $\underline{x}^{(0)}$ is transmitted through a BSC(p). Denote by $\underline{y} \in \{0, 1\}^N$ the channel output. By the law of large numbers $d(\underline{x}, \underline{y}) \approx Np$ with high probability. The receiver tries to reconstruct the transmitted codeword

---

[7]Explicitly, we write $f_N \doteq g_N$ if $\frac{1}{N}\log f_N/g_N \to 0$.

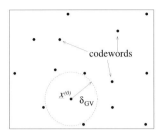

Fig. 4. Pictorial description of a typical code from the random code ensemble.

from $\underline{y}$ using word MAP decoding. Using Eq. (2.10), we see that the 'energy' of a codeword $\underline{x}^{(\alpha)}$ (or, in more conventional terms, its log-likelihood) is given by

$$E(\underline{x}^{(\alpha)}) = -\sum_{i=1}^{N} \log Q(y_i | x_i)$$

$$= -\sum_{i=1}^{N} \{ \mathbb{I}(y_i = x_i^{(\alpha)}) \log(1-p) + \mathbb{I}(y_i \neq x_i^{(\alpha)}) \log p \} \qquad (2.19)$$

$$= N A(p) + 2 B(p) d(\underline{x}^{(\alpha)}, \underline{y}), \qquad (2.20)$$

where $A(p) \equiv -\log p$ and $B(p) \equiv \frac{1}{2} \log(1-p)/p$. For $p < 1/2$, $B(p) > 0$ and word MAP decoding amounts to finding the codeword $\underline{x}^{(\alpha)}$ which is closest in Hamming distance to the channel output $\underline{y}$. By the triangle inequality, the distance between $\underline{y}$ and any of the 'incorrect' codewords is $\gtrsim N(\delta_{GV}(R) - p)$. For $p < \delta_{GV}(R)/2$ this is with high probability larger than the distance from $\underline{x}^{(0)}$.

The above argument implies that, for $p < \delta_{GV}(R)/2$, the expected block error rate of a random code from Shannon's ensemble vanishes as $N \to \infty$. Notice that the channel coding theorem promises instead vanishing error probability whenever $R < 1 - \mathfrak{h}_2(p)$, that is (for $p < 1/2$) $p < \delta_{GV}(R)$. The factor 2 of discrepancy can be recovered through a more careful argument.

Without entering into details, it is interesting to understand the basic reason for the discrepancy between the Shannon Theorem and the above argument. This is related to the geometry of high dimensional spaces. Let us assume for simplicity that the minimum distance between *any two* codewords in $\mathfrak{C}$ is at least $N(\delta_{GV}(R) - \varepsilon)$. In a given random code, this is the case for most codeword pairs. We can then eliminate the pairs that do not satisfy this constraint, thus modifying the code rate in a negligible way (this procedure is called *expurgation*). The resulting code will have *minimum distance* (the minimum distance among any two codewords in $\mathfrak{C}$) $d(\mathfrak{C}) \approx N \delta_{GV}(R)$.

Imagine that we use such a code to communicate through a BSC and that exactly n bits are flipped. By the triangular inequality, as long as $n < d(\mathfrak{C})/2$, the word MAP decoder will recover the transmitted message for *all* error patterns. If on the other hand $n > d(\mathfrak{C})/2$, there are error patterns involving n bits such that the word-MAP decoder does not return the transmitted codeword. If for instance there exists a single codeword $\underline{x}^{(1)}$ at distance $d(\mathfrak{C}) = 2n - 1$ from $\underline{x}^{(0)}$, any pattern involving n out of the $2n - 1$ such that $x_i^{(0)} \neq x_i^{(1)}$, will induce a decoding error. However, it might well be that *most* error patterns with the same number of errors can be corrected.

Shannon's Theorem points out that this is indeed the case until the number of bits flipped by the channel is roughly equal to the minimum distance $d(\mathfrak{C})$.

3. Sparse graph codes

Shannon's Theorem provides a randomized construction to find a code with 'essentially optimal' rate vs error probability trade-off. In practice, however, one cannot use random codes for communications. Just storing the code $\mathfrak{C}$ requires a memory which grows exponentially in the blocklength. In the same vein the optimal decoding procedure requires an exponentially increasing effort. On the other hand, we can not use very short codes since their performance is not very good. To see this assume that we transmit over the BSC with parameter p. If the blocklength is N then the standard deviation of the number of errors contained in a block is $\sqrt{Np(1 - p)}$. Unless this quantity is very small compared to Np we have to either over-provision the error correcting capability of the code so as to deal with the occasionally large number of errors, waisting transmission rate most of the time, or we dimension the code for the typical case, but then we will not be able to decode when the number of errors is larger than the average. This means that short codes are either inefficient or unreliable (or both).

The general strategy for tackling this problem is to introduce more structure in the code definition, and to hope that such structure can be exploited for the encoding and the decoding. In the next section we shall describe a way of introducing structure that, while preserving Shannon's idea of random codes, opens the way to efficient encoding/decoding.

There are two main ingredients that make *modern coding* work and the two are tightly connected. The first important ingredient is to use codes which can be described by *local* constraints only. The second ingredient is to use a local algorithm instead of an high complexity global one (namely symbol MAP or word MAP decoding). In this section we describe the first component.

3.1. Linear codes

One of the simplest forms of structure consists in requiring $\mathfrak{C}$ to be a linear subspace of $\{0, 1\}^N$. One speaks then of a *linear code*. For specifying such a code it is not necessary to list all the codewords. In fact, any linear space can be seen as the kernel of a matrix:

$$\mathfrak{C} = \left\{ \underline{x} \in \{0, 1\}^N : \mathbb{H}\underline{x} = \underline{0} \right\}, \tag{3.1}$$

where the matrix vector multiplication is assumed to be performed modulo 2. The matrix $\mathbb{H}$ is called the *parity-check matrix*. It has N columns and we let $M < N$ denote its number of rows. Without loss of generality we can assume $\mathbb{H}$ to have maximum rank M. As a consequence, $\mathfrak{C}$ is a linear space of dimension $N - M$. The rate of $\mathfrak{C}$ is

$$R = 1 - \frac{M}{N}. \tag{3.2}$$

The a-th line in $\mathbb{H}\underline{x} = \underline{0}$ has the form (here and below $\oplus$ denotes modulo 2 addition)

$$x_{i_1(a)} \oplus \cdots \oplus x_{i_k(a)} = 0. \tag{3.3}$$

It is called a *parity check*.

The parity-check matrix is conveniently represented through a *factor graph* (also called *Tanner graph*). This is a bipartite graph including two types of nodes: M function nodes (corresponding to the rows of $\mathbb{H}$, or the parity-check equations) and N variable nodes (for the columns of $\mathbb{H}$, or the variables). Edges are drawn whenever the corresponding entry in $\mathbb{H}$ is non-vanishing.

Example 4: In Fig. 5 we draw the factor graph corresponding to the parity-check matrix (here $N = 7$, $M = 3$)

$$\mathbb{H} = \begin{bmatrix} 1 & 0 & 1 & 0 & 1 & 0 & 1 \\ 0 & 1 & 1 & 0 & 0 & 1 & 1 \\ 0 & 0 & 0 & 1 & 1 & 1 & 1 \end{bmatrix}. \tag{3.4}$$

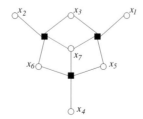

Fig. 5. Factor graph for the parity-check matrix (3.4).

In the following we shall use indices $i, j, \ldots$ for variable nodes and $a, b, \ldots$ for check nodes. We shall further denote by ∂i (respectively, ∂a) the set of nodes that are adjacent to variable node i (to factor node a).

Remarkably, introducing the linear space structure does not deteriorate the performances of the resulting code. Let us introduce Shannon's *parity-check ensemble*: it is defined by letting the parity-check matrix $\mathbb{H}$ be a uniformly random matrix with the prescribed dimensions. Explicitly, each of the NM entries H_{ai} is an independent Bernoulli random variable of mean $1/2$. Probabilistic arguments similar to the ones for the random code ensemble can be developed for the random parity-check ensemble. The conclusion is that random codes from this ensemble allow to communicate with arbitrarily small block error probability at any rate $R < \mathsf{C}(Q)$, where $\mathsf{C}(Q)$ is the capacity of the given BMS channel.

Unfortunately, linearity is not sufficient to guarantee that a code admits a low-complexity decoding algorithm. In particular, the algorithm which we discuss in the sequel works well only for codes that can be represented by a sparse parity-check matrix $\mathbb{H}$ (i.e. a parity check matrix with $O(N)$ non-vanishing entries). Notice that a given code $\mathfrak{C}$ has more than one representation of the form (3.1). A priori one could hope that, given a uniformly random matrix $\mathbb{H}$, a new matrix $\mathbb{H}'$ could be built such that $\mathbb{H}'$ is sparse and that its null space coincides with the one of $\mathbb{H}$. This would provide a sparse representation of $\mathfrak{C}$. Unfortunately, this is the case only for a vanishing fraction of matrices $\mathbb{H}$, as shown by the exercise below.

Exercise 2: Consider a linear code $\mathfrak{C}$, with blocklength N, and dimension $N - M$ (as a linear space). Prove the following sequence of arguments.

(i) The total number of binary $N \times M$ parity-check matrices is 2^{NM}.

(ii) Each code $\mathfrak{C}$ has $2^{\binom{M}{2}} \prod_{i=1}^{M} \left(2^i - 1\right)$ distinct $N \times M$ parity-check matrices $\mathbb{H}$.

(iii) The number of such matrices with at most aN non-zero entries is $\sum_{i=0}^{aN} \binom{NM}{i} \leq 2^{NM\,\mathfrak{h}_2(a/(N-M))}$.

(iv) Conclude from the above that, for any given a, the fraction of parity-check matrices $\mathbb{H}$ that admit a sparse representation in terms of a matrix $\mathbb{H}'$ with at most aN ones, is of order $e^{-N\gamma}$ for some $\gamma > 0$.

With an abuse of language in the following we shall sometimes use the term 'code' to denote a pair code/parity-check matrix.

3.2. Low-density parity-check codes

Further structure can be introduced by restricting the ensemble of parity-check matrices. Low-density parity-check (LDPC) codes are codes that have at least one sparse parity-check matrix.

Rather than considering the most general case let us limit ourselves to a particularly simple family of LDPC ensembles, originally introduced by Robert Gallager [19]. We call them 'regular' ensembles. An element in this family is characterized by the blocklength N and two integer numbers k and l, with $k > l$. We shall therefore refer to it as the (k, l) regular ensemble). In order to construct a random Tanner graph from this ensemble, one proceeds as follows:

1. Draw N variable nodes, each attached to l half-edges and $M = Nl/k$ (we neglect here the possibility of Nl/k not being an integer) check nodes, each with k half edges.

2. Use an arbitrary convention to label the half edges form 1 to Nl, both on the variable node side as well as the check node side (note that this requires that $Mk = Nl$).

3. Choose a permutation π uniformly at random among all permutations over Nl objects, and connect half edges accordingly.

Notice that the above procedure may give rise to multiple edges. Typically there will be $O(1)$ multiple edges in a graph constructed as described. These can be eliminated easily without effecting the performance substantially. From the analytical point of view, a simple choice consists in eliminating all the edges (i, a) if (i, a) occurs an even number of times, and replacing them by a single occurrence (i, a) if it occurs an odd number of times.

Neglecting multiple occurrences (and the way to resolve them), the parity-check matrix corresponding to the graph constructed in this way does include k ones per row and l ones per column. In the sequel we will keep l and k fixed and consider the behavior of the ensemble as $N \to \infty$. This implies that the matrix has only $O(N)$ non-vanishing entries. The matrix is *sparse*.

For practical purposes it is important to maximize the rate at which such codes enable one to communicate with vanishing error probability. To achieve this goal, several more complex ensembles have been introduced. As an example, one simple idea is to consider a generic row/column weight distribution (the weight being the number of non-zero elements), cf. Fig. 6 for an illustration. Such ensembles are usually referred to as 'irregular', and were introduced in [27].

3.3. Weight enumerator

As we saw in Section 2.5, the reason of the good performance of Shannon ensemble (having vanishing block error probability at rates arbitrarily close to the capacity), can be traced back to its minimum distance properties. This is indeed only a partial explanation (as we saw errors could be corrected well beyond half its minimum distance). It is nevertheless instructive and useful to understand the geometrical structure (and in particular the minimum distance properties) of typical codes from the LDPC ensembles defined above.

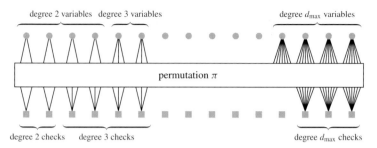

degree 2 variables degree 3 variables   degree d_{max} variables

permutation π

degree 2 checks   degree 3 checks   degree d_{max} checks

Fig. 6. Factor graph of an irregular LDPC code. Variable nodes and function nodes can have any degree between 2 and d_{max}. Half edges on the two sides are joined through a uniformly random permutation.

Let us start by noticing that, for linear codes, the distance enumerator does not depend upon the reference codeword. This is a straightforward consequence of the observation that, for any $\underline{x}^{(0)} \in \mathfrak{C}$ the set $\underline{x}^{(0)} \oplus \mathfrak{C} \equiv \{\underline{x}^{(0)} \oplus \underline{x} : \underline{x} \in \mathfrak{C}\}$ coincides with $\mathfrak{C}$. We are therefore led to consider the distance enumerator with respect to the all-zero codeword $\underline{0}$. This is also referred to as the *weight enumerator*,

$$\mathcal{N}(w) = \#\left\{\underline{x} \in \mathfrak{C} : w(\underline{x}) = w\right\}, \tag{3.5}$$

where $w(\underline{x}) = d(\underline{x}, \underline{0})$ is the number of non-zero entries in $\underline{x}$.

Let us compute the expected weight enumerator $\overline{\mathcal{N}}(w) \equiv \mathbb{E}\mathcal{N}(w)$. The final result is

$$\overline{\mathcal{N}}(w) = \frac{(lw)!(F - lw)!}{F!}\binom{N}{w}\,\mathsf{coeff}[q_k(z)^M, z^{lw}]. \tag{3.6}$$

Here, $F = Nl = Mk$ denotes the number of edges in the Tanner graph, $q_k(z) \equiv \frac{1}{2}[(1 + z)^k + (1 - z)^k]$, and, given a polynomial $p(z)$ and an integer n, $\mathsf{coeff}[p(z), z^n]$ denotes the coefficient of z^n in the polynomial $p(z)$.

Let us proved Eq. (3.6). Let $\underline{x} \in \{0, 1\}^N$ be a binary word of length N and weight w. Notice that $\mathbb{H}\underline{x} = 0$ if and only if the corresponding factor graph has the following property. Consider all variable nodes i such that $x_i = 1$, and color in red all edges incident on these nodes. Color in blue all the other edges. Then all the check nodes must have an even number of incident red edges. A little thought shows that $\overline{\mathcal{N}}(w)$ is the number of 'colored' factor graphs having this property, divided by the total number of factor graphs in the ensemble.

A valid colored graph must have wl red edges. It can be constructed as follows. First choose w variable nodes. This can be done in $\binom{N}{w}$ ways. Assign to each node in this set l red sockets, and to each node outside the set l blue sockets. Then, for each of the M function nodes, color in red an even subset of its sockets in such a way that the total number of red sockets is $E = wl$. The number of ways of doing this is[8] $\text{coeff}[q_k(z)^M, z^{lw}]$. Finally we join the variable node and check node sockets in such a way that colors are matched. There are $(lw)!(F - lw)!$ such matchings out of the total number of $F!$ corresponding to different elements in the ensemble.

Let us compute the exponential growth rate $\phi(\omega)$ of $\overline{\mathcal{N}}(w)$. This is defined by

$$\overline{\mathcal{N}}(w = N\omega) \doteq e^{N\phi(\omega)}. \tag{3.7}$$

In order to estimate the leading exponential behavior of Eq. (3.6), we set $w = N\omega$ and estimate the $\text{coeff}[\dots, \dots]$ term using the Cauchy Theorem,

$$\begin{aligned}
\text{coeff}\left[q_k(z)^M, z^{wl}\right] &= \oint \frac{q_k(z)^M}{z^{lw+1}} \frac{dz}{2\pi i} \\
&= \oint \exp\left\{N\left[\frac{l}{k}\log q_k(z) - l\omega \log z\right]\right\} \frac{dz}{2\pi i}.
\end{aligned} \tag{3.8}$$

Here the integral runs over any path encircling the origin in the complex z plane. Evaluating the integral using the saddle point method we finally get $\overline{\mathcal{N}}(w) \doteq e^{N\phi}$, where

$$\phi(\omega) \equiv (1 - l)\mathfrak{h}(\omega) + \frac{l}{k}\log q_k(z) - \omega l \log z, \tag{3.9}$$

and z is a solution of the saddle point equation

$$\omega = \frac{z}{k} \frac{q_k'(z)}{q_k(z)}. \tag{3.10}$$

The typical result of such a computation is shown in Fig. 7. As can be seen, there exists $\omega_* > 0$ such that $\phi(\omega) < 0$ for $\omega \in (0, \omega_*)$. This implies that a typical code from this ensemble will not have any codeword of weight between 0 and $N(\omega_* - \varepsilon)$. By linearity the minimum distance of the code is at least $\approx N\omega_*$. This implies in particular that such codes can correct any error pattern over the binary symmetric channel of weight $\omega_*/2$ or less.

Notice that $\phi(\omega)$ is an 'annealed average', in the terminology of disordered systems. As such, it can be dominated by rare instances in the ensemble. On the other hand, since $\log \mathcal{N}_N(N\omega) = \Theta(N)$ is an 'extensive' quantity, we expect it to be *self averaging* in the language of statistical physics. In mathematics terms one says that it should *concentrate in probability*. Formally, this means that there

---

[8]This is a standard generating function calculation, and is explained in Appendix A.

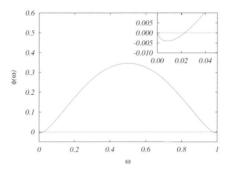

Fig. 7. Logarithm of the expected weight enumerator for the $(3, 6)$ ensemble in the large blocklength limit. Inset: small weight region. Notice that $\phi(\omega) < 0$ for $\omega < \omega_* \approx 0.02$: besides the 'all-zero' word there is no codeword of weight smaller than $N\omega_*$ in the code with high probability.

exists a function $\Phi_N(\omega)$ that is non-random (i.e., does not depend upon the code) and such that

$$\lim_{N \to \infty} \mathbb{P}\{|\log \mathcal{N}_N(N\omega) - \Phi_N(\omega)| \geq N\delta\} = 0. \tag{3.11}$$

Further we expect that $\Phi_N(\omega) = N\phi_{\mathrm{q}}(\omega) + o(N)$ as $N \to \infty$. Despite being rather fundamental, both these statements are open conjectures.

The coefficient $\phi_{\mathrm{q}}(\omega)$ is the growth rate of the weight enumerator for typical codes in the ensembles. In statistical mechanics terms, it is a 'quenched' free energy (or rather, entropy). By Jensen inequality $\phi_{\mathrm{q}}(\omega) \leq \phi(\omega)$. A statistical mechanics calculation reveals that the inequality is strict for general (irregular) ensembles. On the other hand, for regular ensembles as the ones considered here, $\phi_{\mathrm{q}}(\omega) = \phi(\omega)$: the annealed calculation yields the correct exponential rate. This claim has been supported rigorously by the results of [47, 3, 31].

Let us finally comment on the relation between distance enumerator and the Franz-Parisi potential [18], introduced in the study of glassy systems. In this context the potential is used to probe the structure of the Boltzmann measure. One considers a system with energy function $E(x)$, a reference configuration x_0 and some notion of distance between configurations $d(x, x')$. The constrained partition function is then defined as

$$Z(x_0, w) = \int e^{-E(x)} \delta(d(x_0, x) - w) \, dx. \tag{3.12}$$

One then defines the potential $\Phi_N(\omega)$ as the typical value of $\log Z(x_0, w)$ when x_0 is a random configuration with the same Boltzmann distribution and $w = N\omega$.

Self averaging is expected to hold here too:

$$\lim_{N \to \infty} \mathbb{P}_{x_0}\left\{\left|\log Z(x_0, N\omega) - \Phi_N(\omega)\right| \geq N\delta\right\} = 0. \tag{3.13}$$

Here N may denote the number of particles or the volume of the system and $\mathbb{P}_{x_0}\{\cdots\}$ indicates probability with respect to x_0 distributed with the Boltzmann measure for the energy function $E(x_0)$.

It is clear that the two ideas are strictly related and can be generalized to any joint distribution of N variables $(x_1, \ldots, x_N)$. In both cases the structure of such a distribution is probed by picking a reference configuration and restricting the measure to its neighborhood.

To be more specific, the weight enumerator can be seen as a special case of the Franz-Parisi potential. It is sufficient to take as Boltzmann distribution the uniform measure over codewords of a linear code $\mathfrak{C}$. In other words, let the configurations be binary strings of length N, and set $E(\underline{x}) = 0$ if $\underline{x} \in \mathfrak{C}$, and $= \infty$ otherwise. Then the restricted partition function is just the distance enumerator with respect to the reference codeword, which indeed does not depend on it.

4. The decoding problem for sparse graph codes

As we have already seen, MAP decoding requires computing either marginals or the mode of the conditional distribution of $\underline{x}$ being the channel input given output $\underline{y}$. In the case of LDPC codes the posterior probability distribution factorizes according to underlying factor graph G:

$$\mu_{\mathfrak{C}, \underline{y}}(\underline{x}) = \frac{1}{Z(\mathfrak{C}, \underline{y})} \prod_{i=1}^{N} Q(y_i | x_i) \prod_{a=1}^{M} \mathbb{I}(x_{i_1(a)} \oplus \cdots \oplus x_{i_k(a)} = 0). \tag{4.1}$$

Here $(i_1(a), \ldots, i_k(a))$ denotes the set of variable indices involved in the a-th parity check (i.e., the non-zero entries in the a-th row of the parity-check matrix $\mathbb{H}$). In the language of spin models, the terms $Q(y_i | x_i)$ correspond to an external random field. The factors $\mathbb{I}(x_{i_1(a)} \oplus \cdots \oplus x_{i_k(a)} = 0)$ can instead be regarded as hard core k-spins interactions. Under the mapping $\sigma_i = (-1)^{x_i}$, such interactions depend on the spins through the product $\sigma_{i_1(a)} \cdots \sigma_{i_k(a)}$. The model (4.1) maps therefore onto a k-spin model with random field.

For MAP decoding, minimum distance properties of the code play a crucial role in determining the performances. We investigated such properties in the previous section. Unfortunately, there is no known way of implementing MAP decoding efficiently. In this section we discuss two decoding algorithms that exploit the sparseness of the factor graph to achieve efficient decoding. Although such strategies are sub-optimal with respect to word (o symbol) MAP decoding, the graphical structure can be optimized itself, leading to state-of-the-art performances.

After briefly discussing bit-flipping decoding, most of this section will be devoted to message passing that is the approach most used in practice. Remarkably, both bit flipping as well as message passing are closely related to statistical mechanics.

4.1. Bit flipping

For the sake of simplicity, let us assume that communication takes place over a binary symmetric channel. We receive the message $\underline{y} \in \{0, 1\}^N$ and try to find the transmitted codeword $\underline{x}$ as follows:

```
Bit-flipping decoder
```

0. Set $\underline{x}(0) = \underline{y}$.

1. Find a bit belonging to more unsatisfied than satisfied parity checks.

2. If such a bit exists, flip it: $x_i(t+1) = x_i(t) \oplus 1$. Keep the other bits: $x_j(t+1) = x_j(t)$ for all $j \neq i$. If there is no such bit, return $\underline{x}(t)$ and halt.

3. Repeat steps 2 and 3.

The bit to be flipped is usually chosen uniformly at random among the ones satisfying the condition at step 1. However this is irrelevant for the analysis below.

In order to monitor the bit-flipping algorithm, it is useful to introduce the function:

$$U(t) \equiv \#\{\text{parity-check equations not satisfied by } \underline{x}(t)\}. \tag{4.2}$$

This is a non-negative integer, and if $U(t) = 0$ the algorithm is halted and it outputs $\underline{x}(t)$. Furthermore, $U(t)$ cannot be larger than the number of parity checks M and decreases (by at least one) at each cycle. Therefore, the algorithm complexity is $O(N)$ (this is a commonly regarded as the ultimate goal for many communication problems).

It remains to be seen if the output of the bit-flipping algorithm is related to the transmitted codeword. In Fig. 8 we present the results of a numerical experiment. We considered the (5, 10) regular ensemble and generated about 1000 random code and channel realizations for each value of the noise level p in some mesh. Then, we applied the above algorithm and traced the fraction of successfully decoded blocks, as well as the residual energy $U_* = U(t_*)$, where t_* is the

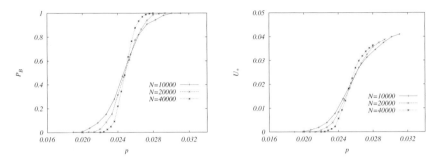

Fig. 8. Numerical simulations of bit-flipping decoding of random codes from the (5, 10) regular ensemble. On the left: block error rate achieved by this scheme. On the right: fraction of unsatisfied parity checks in the word found by the algorithm.

total number of iterations of the algorithm. The data suggests that bit-flipping is able to overcome a finite noise level: it recovers the original message with high probability when less than about 2.5% of the bits are corrupted by the channel. Furthermore, the curves for the block error probability P_B^{bf} under bit-flipping decoding become steeper and steeper as the system size is increased. It is natural to conjecture that asymptotically, a phase transition takes place at a well defined noise level p_{bf}: $P_B^{bf} \to 0$ for $p < p_{bf}$ and $P_B^{bf} \to 1$ for $p > p_{bf}$. Numerically $p_{bf} = 0.025 \pm 0.005$.

This threshold can be compared with the one for word MAP decoding, that we will call p_c: The bounds in [60] state that $0.108188 \le p_c \le 0.109161$ for the (5, 10) ensemble, while a statistical mechanics calculation yields $p_c \approx 0.1091$. Bit-flipping is significantly sub-optimal, but it is still surprisingly good, given the extreme simplicity of the algorithm.

These numerical findings can be confirmed rigorously [55].

Theorem 4.1. *Consider a regular (l, k) LDPC ensemble and let $\mathfrak{C}$ be chosen uniformly at random from the ensemble. If $l \ge 5$ then there exists $\varepsilon > 0$ such that, with high probability,* Bit-flipping *is able to correct any pattern of at most $N\varepsilon$ errors produced by a binary symmetric channel.*

Given a generic word $\underline{x}$ (i.e., a length N binary string that is not necessarily a codeword), let us denote, with a slight abuse of notation, by $U(\underline{x})$ the number of parity-check equations that are not satisfied by $\underline{x}$. The above result, together with the weight enumerator calculation in the previous section, suggests the following picture of the function $U(\underline{x})$. If $\underline{x}^{(0)} \in \mathfrak{C}$, than $U(\underline{x}^{(0)}) = 0$. Moving away from $\underline{x}^{(0)}$, $U(\underline{x})$ will become strictly positive. However as long as $d(\underline{x}, \underline{y})$ is small enough, $U(\underline{x})$ does not have any local minimum distinct from $\underline{x}^{(0)}$. A greedy procedure with a starting point within such a Hamming radius is able to recon-

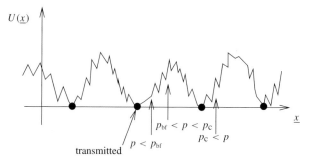

Fig. 9. Sketch of the cost function $U(\underline{x})$ (number of unsatisfied parity checks) for a typical random LDPC code. Filled circles correspond to codewords, and arrows to received messages in various possible regimes.

struct $\underline{x}^{(0)}$. As we move further away, $U(\underline{x})$ stays positive (no other codewords are encountered) but local minima start to appear. Bit flipping gets trapped in such minima. Finally, for $d(\underline{x}^{(0)}, \underline{x}) \geq N\omega_*$ new codewords, i.e., minima with $U(\underline{x}) = 0$, are encountered.

4.2. Message passing

Message-passing algorithms are iterative and have low complexity. Unlike the bit-flipping procedure in the previous section, the basic variables are now associated to directed edges in the factor graph. More precisely, for each edge (i, a) (corresponding to a non-zero entry in the parity-check matrix at row a and column i), we introduce two messages $v_{i \to a}$ and $\widehat{v}_{a \to i}$. Messages are elements of some set (the message alphabet) that we shall denote by M. Depending on the specific algorithm, M can have finite cardinality, or be infinite, for instance $\mathsf{M} = \mathbb{R}$. Since the algorithm is iterative, it is convenient to introduce a time index $t = 0, 1, 2, \dots$ and label the messages with the time at which they are updated: $v_{i \to a}^{(t)}$ and $\widehat{v}_{a \to i}^{(t)}$ (but we will sometimes drop the label below).

The defining property of message-passing algorithms is that the message flowing from node u to v at a given time is a function of messages entering u from nodes w distinct from v at the previous time step. Formally, the algorithm is defined in terms of two sets of functions $\Phi_{i \to a}(\cdot)$, $\Psi_{a \to i}(\cdot)$, that define the update operations at variable and function nodes as follows

$$v_{i \to a}^{(t+1)} = \Phi_{i \to a}\big(\{\widehat{v}_{b \to i}^{(t)}; b \in \partial i \backslash a\}; y_i\big),$$
$$\widehat{v}_{a \to i}^{(t)} = \Psi_{a \to i}\big(\{v_{j \to a}^{(t)}; j \in \partial a \backslash i\}\big). \tag{4.3}$$

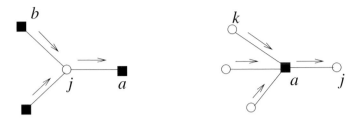

Fig. 10. Graphical representation of message passing updates.

Notice that messages are updated in parallel and that the time counter is incremented only at variable nodes. Alternative scheduling schemes can be considered but we will stick to this for the sake of simplicity. After a pre-established number of iterations, the transmitted bits are estimated using *all* the messages incoming at the corresponding nodes. More precisely, the estimate at function i is defined through a new function

$$\widehat{x}_i^{(t)}(\underline{y}) = \Phi_i(\{\widehat{v}_{b \to i}^{(t)}; b \in \partial i\}; y_i). \tag{4.4}$$

A graphical representation of message passing updates is provided in Fig. 10.

A specific message-passing algorithm requires the following features to be specified:

1. The message alphabet M.

2. The initialization $\{v_{i \to a}^{(0)}\}, \{\widehat{v}_{i \to a}^{(0)}\}$.

3. The update functions $\{\Phi_{i \to a}(\cdot)\}, \{\Psi_{a \to i}(\cdot)\}$.

4. The final estimate functions $\{\Phi_i(\cdot)\}$.

The most prominent instance of a message-passing algorithm is the Belief Propagation (BP) algorithm. In this case the messages $v_{i \to a}^{(t)}(x_i)$ and $\widehat{v}_{a \to i}^{(t)}(x_i)$ are distributions over the bit variables $x_i \in \{0, 1\}$. The message $\widehat{v}_{a \to i}^{(t)}(x_i)$ is usually interpreted as the *a posteriori* distributions of the bit x_i given the information coming from edge $a \to i$. Analogously, $v_{i \to a}(x_i)$ is interpreted as the *a posteriori* distribution of x_i, given all the information collected through edges distinct from (a, i). Since the messages normalization (explicitly $v_{i \to a}(0) + v_{i \to a}(1) = 1$) can be enforced at any time, we shall neglect overall factors in writing down the relation between to messages (and correspondingly, we shall use the symbol $\propto$).

BP messages are updated according to the following rule, whose justification we will discuss in the next section

$$v_{i \to a}^{(t+1)}(x_i) \propto Q(y_i|x_i) \prod_{b \in \partial i \setminus a} \widehat{v}_{b \to i}^{(t)}(x_i), \tag{4.5}$$

$$\widehat{v}_{a \to i}^{(t)}(x_i) \propto \sum_{\{x_j\}} \mathbb{I}(x_i \oplus x_{j_1} \oplus \cdots \oplus x_{j_{k-1}} = 0) \prod_{j \in \partial a \setminus i} v_{j \to a}^{(t)}(x_j), \tag{4.6}$$

where we used $(i, j_1, \ldots, j_{k-1})$ to denote the neighborhood ∂a of factor node a. After any number of iterations the single bit marginals can be estimated as follows

$$v_i^{(t+1)}(x_i) \propto Q(y_i|x_i) \prod_{b \in \partial i} \widehat{v}_{b \to i}^{(t)}(x_i). \tag{4.7}$$

The corresponding MAP decision for bit i (sometimes called 'hard decision', while $v_i(x_i)$ is the 'soft decision') is

$$\widehat{x}_i^{(t)} = \arg \max_{x_i} v_i^{(t)}(x_i). \tag{4.8}$$

Notice that the above prescription is ill-defined when $v_i(0) = v_i(1)$. It turns out that it is not really important which rule to use in this case. To preserve the $0 - 1$ symmetry, we shall assume that the decoder returns $\widehat{x}_i^{(t)} = 0$ or $= 1$ with equal probability.

Finally, as initial condition one usually takes $\widehat{v}_{a \to i}^{(-1)}(\cdot)$ to be the uniform distribution over $\{0, 1\}$ (explicitly $\widehat{v}_{a \to i}^{(-1)}(0) = \widehat{v}_{a \to i}^{(-1)}(1) = 1/2$).

Since for BP the messages are distributions over binary valued variables, they can be described by a single real number, that is often chosen to be the bit log-likelihood:[9]

$$h_{i \to a} = \frac{1}{2} \log \frac{v_{i \to a}(0)}{v_{i \to a}(1)}, \qquad u_{a \to i} = \frac{1}{2} \log \frac{\widehat{v}_{a \to i}(0)}{\widehat{v}_{a \to i}(1)}. \tag{4.9}$$

We refer to Fig. 11 for a pictorial representation of these notations. We further introduce the channel log-likelihoods

$$B_i = \frac{1}{2} \log \frac{Q(y_i|0)}{Q(y_i|1)}. \tag{4.10}$$

---

[9]The conventional definition of log-likelihoods does not include the factor $1/2$. We introduce this factor here for uniformity with the statistical mechanics convention (the h's and u's being analogous to effective magnetic fields).

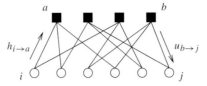

Fig. 11. Factor graph of a regular LDPC code, and notation for the belief propagation messages.

The BP update equations (4.5), (4.6) read in this notation

$$h_{i \to a}^{(t+1)} = B_i + \sum_{b \in \partial i \backslash a} u_{b \to i}^{(t)}, \qquad u_{a \to i}^{(t)} = \operatorname{atanh} \left\{ \prod_{j \in \partial a \backslash i} \tanh h_{j \to a}^{(t)} \right\}. \qquad (4.11)$$

In this language the standard message initialization would be $u_{a \to i}^{(-1)} = 0$. Finally, the overall log-likelihood at bit i is obtained by combining *all* the incoming messages in agreement with Eq. (4.7). One thus gets the decision rule

$$\widehat{x}_i^{(t)} = \begin{cases} 0 & \text{if } B_i + \sum_{b \in \partial i} u_{b \to i}^{(t)} > 0, \\ 1 & \text{if } B_i + \sum_{b \in \partial i} u_{b \to i}^{(t)} < 0. \end{cases} \qquad (4.12)$$

Notice that we did not commit to any special decision if $B_i + \sum_{b \in \partial i} u_{b \to i}^{(t)} = 0$. To keep complete symmetry we'll establish that the decoder returns 0 or 1 with equal probability in this case.

4.3. Correctness of Belief Propagation on trees

The justification for the BP update equations (4.5), (4.6) lies in the observation that, whenever the underlying factor graph is a tree, the estimated marginal $v_i^{(t)}(x_i)$ converges after a finite number of iterations to the correct one $\mu_i(x_i)$. In particular, under the tree assumption, and for any t sufficiently large, $\widehat{x}_i^{(t)}(\underline{y})$ coincides with the symbol MAP decision.

In order to prove this statement, consider a tree factor graph G. Given a couple of adjacent nodes u, v, denote by $G(u \to v)$ the subtree rooted at the directed edge $u \to v$ (this contains all that can be reached from v through a non-reversing path whose first step is $v \to u$). If i is a variable index and a a parity-check index, let $\mu_{i \to a}(\cdot)$ be the measure over $\underline{x} = \{x_j : j \in G(i \to a)\}$, that is obtained by retaining in Eq. (4.1) only those terms that are related to nodes in $G(i \to a)$:

$$\mu_{i \to a}(\underline{x}) = \frac{1}{Z(i \to a)} \prod_{j \in G(i \to a)} Q(y_i | x_i)$$
$$\times \prod_{b \in G(i \to a)} \mathbb{I}(x_{i_1(b)} \oplus \cdots \oplus x_{i_k(b)} = 0). \tag{4.13}$$

The measure $\widehat{\mu}_{a \to i}(\cdot)$ is defined analogously for the subtree $G(a \to i)$. The marginals $\mu_{i \to a}(x_i)$ (respectively $\widehat{\mu}_{a \to i}(x_i)$) are easily seen to satisfy the recursions

$$\mu_{i \to a}(x_i) \propto Q(y_i | x_i) \prod_{b \in \partial i \setminus a} \widehat{\mu}_{b \to i}(x_i), \tag{4.14}$$

$$\widehat{\mu}_{a \to i}(x_i) \propto \sum_{\{x_j\}} \mathbb{I}(x_i \oplus x_{j_1} \oplus \cdots \oplus x_{j_{k-1}} = 0) \prod_{j \in \partial a \setminus i} \mu_{j \to a}(x_j), \tag{4.15}$$

which coincide, apart from the time index, with the BP recursion (4.5), (4.6). That such recursions converges to $\{\mu_{i \to a}(x_i), \widehat{\mu}_{a \to i}(x_i)\}$ follows by induction over the tree depth.

In statistical mechanics equations similar to (4.14), (4.15) are often written as recursions on the constrained partition function. They allow to solve exactly models on trees. However they have been often applied as mean-field approximation to statistical models on non-tree graphs. This is often referred to as the *Bethe-Peierls approximation* [8].

The Bethe approximation presents several advantages with respect to 'naive-mean field' [61] (that amounts to writing 'self-consistency' equations for expectations over single degrees of freedom). It retains correlations among degrees of freedom that interact directly, and is exact on some non-empty graph (trees). It is often asymptotically (in the large size limit) exact on locally tree-like graphs. Finally, it is quantitatively more accurate for non-tree like graphs and offers a much richer modeling palette.

Within the theory of disordered systems (especially, glass models on sparse random graphs), Eqs. (4.14) and (4.15) are also referred to as the *cavity equations*. With respect to Bethe-Peierls, the cavity approach includes a hierarchy of ('replica symmetry breaking') refinements of such equations that aim at capturing long range correlations [37]. This will briefly described in Section 5.

We should finally mention that several improvements over Bethe approximation have been developed within statistical physics. Among them, Kikuchi's cluster variational method [24] is worth mentioning since it motivated the development of a 'generalized belief propagation' algorithm, which spurred a lot of interest within the artificial intelligence community [61].

4.4. Density evolution

Although BP converges to the exact marginals on tree graphs, this says little about its performances on practical codes such as the LDPC ensembles introduced in Section 3. Fortunately, a rather precise picture on the performance of LDPC ensembles can be derived in the large blocklength limit $N \to \infty$. The basic reason for this is that the corresponding random factor graph is locally tree-like with high probability if we consider large blocklengths.

Before elaborating on this point, notice that the performance under BP decoding (e.g., the bit error rate) is independent on the transmitted codeword. For the sake of analysis, we shall hereafter assume that the all-zero codeword $\underline{0}$ has been transmitted.

Consider a factor graph G and let (i, a) be one of its edges. Consider the message $v_{i \to a}^{(t)}$ sent by the BP decoder in iteration t along edge (i, a). A considerable amount of information is contained in the distribution of $v_{i \to a}^{(t)}$ with respect to the channel realization, as well as in the analogous distribution for $\widehat{v}_{a \to i}^{(t)}$. To see this, note that under the all-zero codeword assumption, the bit error rate after t iterations is given by

$$P_{\mathrm{b}}^{(t)} = \frac{1}{n} \sum_{i=1}^{n} \mathbb{P}\left\{ \Phi_i(\{\widehat{v}_{b \to i}^{(t)}; b \in \partial i\}; y_i) \neq 0 \right\}. \tag{4.16}$$

Therefore, if the messages $\widehat{v}_{b \to i}^{(t)}$ are independent, then the bit error probability is determined by the distribution of $\widehat{v}_{a \to i}^{(t)}$.

Rather than considering one particular graph (code) and a specific edge, it is much simpler to take the average over all edges and all graph realizations. We thus consider the distribution $\mathsf{a}_t^{(N)}(\cdot)$ of $v_{i \to a}^{(t)}$ with respect to the channel, the edges, *and* the graph realization. While this is still a quite difficult object to study rigorously, it is on the other hand possible to characterize its large blocklength limit $\mathsf{a}_t(\cdot) = \lim_N \mathsf{a}_t^{(N)}(\cdot)$. This distribution satisfies a simple recursion.

It is convenient to introduce the *directed neighborhood* of radius r of the directed edge $i \to a$ in G, call it $\mathsf{B}_{i \to a}(r; G)$. This is defined as the subgraph of F that includes all the variable nodes that can be reached from i through a non-reversing path of length at most r, whose first step *is not* the edge (i, a). It includes as well all the function nodes connected *only* to the above specified variable nodes. In Fig. 13 we reproduce an example of a directed neighborhood of radius $r = 3$ (for illustrative purposes we also include the edge (i, a)) in a $(2, 3)$ regular code.

If F is the factor graph of a random code from the (k, l) LDPC ensemble, then $\mathsf{B}_{i \to a}(r; F)$ is with high probability a depth-r regular tree with degree l at

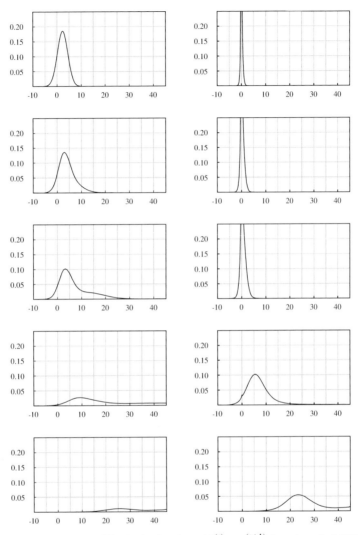

Fig. 12. Evolution of the probability density functions of $h^{(t)}$ an $u^{(t+1)}$ for an irregular LDPC code used over a gaussian channel. From top to bottom $t = 0, 5, 10, 50,$ and 140.

variable nodes and degree k at check nodes (as in Fig. 13 where $l = 2$ and $k = 3$). The basic reason for this phenomenon is rather straightforward. Imagine to explore the neighborhood progressively, moving away from the root, in a breadth first fashion. At any finite radius r, about c^r/N vertices have been visited (here

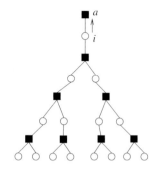

Fig. 13. A radius 3 directed neighborhood $B_{i \to a}(3; G)$.

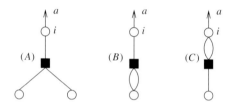

Fig. 14. The three possible radius–1 directed neighborhoods in a random factor graph from the regular (2, 3) graph ensemble.

$c = (k - 1)(l - 1)$). The vertices encountered at the next layer will be 'more or less' uniformly random among all the ones not visited so far. As a consequence they will be distinct with high probability, and $B_{i \to a}(r + 1; G)$ will be a tree as well. This argument breaks down when the probability that two of the $\Theta(c^r)$ new vertices coincide, that is for $c^{2r} = \Theta(N)$.[10] This is equivalent to $r \simeq \frac{1}{2} \log_c N$.

The skeptical reader is invited to solve the following exercise.

Exercise 3: In order to illustrate the above statement, consider the example of a random code from the regular (2, 3) ensemble (each variable has degree 2 and each check has degree 3). The three possible radius-1 neighborhoods appearing in the associated factor graph are depicted in Fig. 14.

(a) Show that the probability that a given edge (i, a) has neighborhoods as in (B) or (C) is $O(1/N)$.

(b) What changes for a generic radius r?

---

[10]This is the famous birthday problem. The probability that two out of a party of n peoples were born on the same day of the year, scales like n^2/N for $n^2 \ll N$ (N is here the number of days in a year).

For illustrative reasons, we shall occasionally add a 'root edge' to $\mathsf{B}_{i \to a}(r; G)$, as for $i \to a$ in Fig. 13.

Now consider the message $v_{i \to a}^{(t)}$. This is a function of the factor graph G and of the received message $\underline{y}$. However, a moment's thought shows that it will depend on G only through its directed neighborhood $\mathsf{B}_{i \to a}(t + 1; G)$, and only on the received symbols y_j, $j \in \mathsf{B}_{i \to a}(t; G)$.

In view of the above discussion, let us consider the case in which $\mathsf{B}_{i \to a}(t + 1; G)$ is a (k, l)-regular tree. We further assume that the received symbols y_j are i.i.d. with distribution $Q(y|0)$, and that the update rules (4.3) do not depend on the edge we are considering (i.e., $\Phi_{i \to a}(\cdot) = \Phi(\cdot)$ and $\Psi_{i \to a}(\cdot) = \Psi(\cdot)$ independent of i, a).

Let $v^{(t)}$ be the message passed through the root edge of such a tree after t BP iterations. Since the actual neighborhood $\mathsf{B}_{i \to a}(t + 1; G)$ is with high probability a tree, $v_{i \to a}^{(t)} \overset{d}{\to} v^{(t)}$ as $N \to \infty$. The symbol $\overset{d}{\to}$ denotes convergence in distribution. In other words, for large blocklengths, the message distribution after t iterations is asymptotically the same that we would have obtained if the graph were a tree.

Consider now a (k, l)-regular tree, and let $j \to b$ an edge directed towards the root, at distance d from it. It is not hard to realize that the message passed through it after $r - d - 1$ (or more) iterations is distributed as $v^{(r-d-1)}$. Furthermore, if $j_1 \to b_1$ and $j_2 \to b_2$ are both directed upwards and none belongs to the subtree rooted at the other one, then the corresponding messages are independent. Together with Eq. (4.3), these observation imply that

$$v^{(t+1)} \overset{d}{=} \Phi\big(\widehat{v}_1^{(t)}, \ldots, \widehat{v}_{l-1}^{(t)}; y\big), \qquad \widehat{v}^{(t)} \overset{d}{=} \Psi\big(v_1^{(t)}, \ldots, v_{k-1}^{(t)}\big). \qquad (4.17)$$

Here $\widehat{v}_1^{(t)}, \ldots, \widehat{v}_{l-1}^{(t)}$ are i.i.d. copies of $\widehat{v}^{(t)}$, and $v_1^{(t)}, \ldots, v_{k-1}^{(t)}$ i.i.d. copies of $v^{(t)}$. Finally, y is a received symbol independent from the previous variables and distributed according to $Q(y|0)$.

Equations (4.17), or the sequence of distributions that they define, are usually referred to as *density evolution*. The name is motivated by the identification of the random variables with their densities (even if these do not necessarily exist). They should be parsed as follows (we are refer here to the first equation in (4.17); an analogous phrasing holds for the second): pick $l - 1$ i.i.d. copies $\widehat{v}^{(t)}$ and y with distribution $Q(y|0)$, compute $\Phi(\widehat{v}_1^{(t)}, \ldots, \widehat{v}_{l-1}^{(t)}; y)$. The resulting quantity will have distribution $v^{(t+1)}$. Because of this description, they are also called 'recursive distributional equations'.

Until this point we considered a generic message passing procedure. If we specialize to BP decoding, we can use the parametrization of messages in terms of log-likelihood ratios, cf. Eq. (4.9), and use the above arguments to characterize the limit random variables $h^{(t)}$ and $u^{(t)}$. The update rules (4.11) then imply

$$h^{(t+1)} \stackrel{\mathrm{d}}{=} B + u_1^{(t)} + \cdots + u_{l-1}^{(t)}, \qquad u^{(t)} \stackrel{\mathrm{d}}{=} \operatorname{atanh}\{\tanh h_1^{(t)} \cdots \tanh h_{k-1}^{(t)}\}.$$

$$(4.18)$$

Here $u_1^{(t)}, \ldots, u_{l-1}^{(t)}$ are i.i.d. copies of $u^{(t)}$, $h_1^{(t)}, \ldots, h_{k-1}^{(t)}$ are i.i.d. copies of $h^{(t)}$, and $B = \frac{1}{2} \log \frac{Q(y|0)}{Q(y|1)}$, where y is independently distributed according to $Q(y|0)$. It is understood that the recursion is initiated with $u^{(-1)} = 0$.

Physicists often write distributional recursions explicitly in terms of densities. For instance, the first of the equations above reads

$$\mathsf{a}_{t+1}(h) = \int \prod_{b=1}^{l-1} \mathrm{d}\widehat{\mathsf{a}}_t(u_b) \, \mathrm{d}\mathsf{p}(B) \; \delta\left(h - B - \sum_{b=1}^{l-1} u_b\right), \qquad (4.19)$$

where $\widehat{\mathsf{a}}_t(\cdot)$ denotes the density of $u^{(t)}$, and $\mathsf{p}(\cdot)$ the density of B. We refer to Fig. 12 for an illustration of how the densities $\mathsf{a}_t(\cdot)$, $\widehat{\mathsf{a}}_t(\cdot)$ evolve during the decoding process.

In order to stress the importance of density evolution notice that, for any continuous function $f(x)$,

$$\lim_{N\to\infty} \mathbb{E}\left\{\frac{1}{N} \sum_{i=1}^{N} f\left(h_{i\to a}^{(t)}\right)\right\} = \mathbb{E}\{f(h^{(t)})\}, \qquad (4.20)$$

where the expectation is taken with respect to the code ensemble. Similar expressions can be obtained for functions of several messages (and are particularly simple when such message are asymptotically independent). In particular,[11] if we let $\mathsf{P}_{\mathsf{b}}^{(N,t)}$ be the expected (over an LDPC ensemble) bit error rate for the decoding rule (4.12), and let $\mathsf{P}_{\mathsf{b}}^{(t)} = \lim_{N\to\infty} \mathsf{P}_{\mathsf{b}}^{(N,t)}$ be its large blocklength limit. Then

$$\mathsf{P}_{\mathsf{b}}^{(t)} = \mathbb{P}\{B + h_1^{(t)} + \cdots + h_l^{(t)} < 0\} + \frac{1}{2}\,\mathbb{P}\{B + h_1^{(t)} + \cdots + h_l^{(t)} = 0\},$$

$$(4.21)$$

where $h_1^{(t)}, \ldots, h_l^{(t)}$ are i.i.d. copies of $h^{(t)}$.

---

[11] The suspicious reader will notice that this is not exactly a particular case of the previous statement, because $f(x) = \mathbb{I}(x < 0) + \frac{1}{2}\mathbb{I}(x = 0)$ is not a continuous function.

4.5. The Belief Propagation threshold

Density evolution would not be such an useful tool if it could not be simulated efficiently. The idea is to estimate numerically the distributions of the density evolution variables $\{h^{(t)}, u^{(t)}\}$. As already discussed this gives access to a number of statistics on BP decoding, such as the bit error rate $P_b^{(t)}$ after t iterations in the large blocklength limit.

A possible approach consists in representing the distributions by samples of some fixed size. Within statistical physics this is sometimes called the *population dynamics algorithm* (and made its first appearance in the study of the localization transition on Cayley trees [46]). Although there exist more efficient alternatives in the coding context (mainly based on Fourier transform, see [49, 48]), we shall describe population dynamics because it is easily programmed.

Let us describe the algorithm within the setting of a general message passing decoder, cf. Eq. (4.17). Given an integer $\mathcal{N} \gg 1$, one represent the messages distributions with two samples of size $\mathcal{N}$: $\mathfrak{P}^{(t)} = \{v_1^{(t)}, \ldots, v_{\mathcal{N}}^{(t)}\}$, and $\widehat{\mathfrak{P}}^{(t)} = \{\widehat{v}_1^{(t)}, \ldots, \widehat{v}_{\mathcal{N}}^{(t)}\}$. Such samples are used as proxy for the corresponding distributions. For instance, one would approximate an expectation as

$$\mathbb{E}f(v^{(t)}) \approx \frac{1}{\mathcal{N}} \sum_{i=1}^{\mathcal{N}} f(v_i^{(t)}). \tag{4.22}$$

The populations are updated iteratively. For instance $\mathfrak{P}^{(t+1)}$ is obtained from $\widehat{\mathfrak{P}}^{(t)}$ by generating $v_1^{(t+1)}, \ldots, v_{\mathcal{N}}^{(t+1)}$ independently as follows. For each $i \in [\mathcal{N}]$, draw indices $b_1(i), \ldots, b_l(i)$ independently and uniformly at random from $[\mathcal{N}]$, and generate y_i with distribution $Q(y|0)$. Then compute $v_i^{(t+1)} = \Phi(\{\widehat{v}_{b_n(i)}^{(t)}\}; y_i)$ and store it in $\mathfrak{P}^{(t+1)}$.

An equivalent description consists in saying that we proceed as if $\widehat{\mathfrak{P}}^{(t)}$ exactly represents the distribution of $u^{(t)}$ (which in this case would be discrete). If this was the case, the distribution of $h^{(t+1)}$ would be composed of $|\mathcal{A}| \cdot \mathcal{N}^{l-1}$ Dirac deltas. In order not to overflow memory, the algorithm samples $\mathcal{N}$ values from such a distribution. Empirically, estimates of the form (4.22) obtained through population dynamics have systematic errors of order $\mathcal{N}^{-1}$ and statistical errors of order $\mathcal{N}^{-1/2}$ with respect to the exact value.

In Fig. 15 we report the results of population dynamics simulations for two different LDPC ensembles, with respect to the BSC. We consider two performance measures: the bit error rate $P_b^{(t)}$ and the bit conditional entropy $H^{(t)}$. The latter is defined as

$$H^{(t)} = \lim_{N \to \infty} \frac{1}{N} \sum_{i=1}^{N} \mathbb{E} \, H\big(X_i | \overline{v}_i^{(t)}\big), \tag{4.23}$$

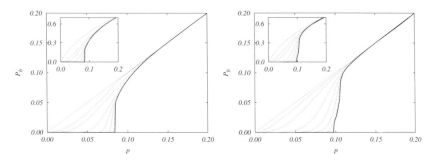

Fig. 15. The performances of two LDPC ensembles as predicted by a numerical implementation of density evolution. On the left, the $(3,6)$ regular ensemble. On the right, an optimized irregular ensemble. Dotted curves refer (from top to bottom) to $t = 0, 1, 2, 5, 10, 20, 50$ iterations, and bold continuous lines to the limit $t \to \infty$. In the inset we plot the expected conditional entropy $\mathbb{E}\,H(X_i|\overline{v}_i^{(t)})$.

and encodes the uncertainty about bit x_i after t BP iterations. It is intuitively clear that, as the algorithm progresses, the bit estimates improve and therefore $P_b^{(t)}$ and $H^{(t)}$ should be monotonically decreasing functions of the number of iterations. Further, they are expected to be monotonically increasing functions of the crossover probability p. Both statement can be easily checked on the above plots, and can be proved rigorously as well.

Since $P_b^{(t)}$ is non-negative and decreasing in t, it has a finite limit

$$P_b^{BP} \equiv \lim_{t \to \infty} P_b^{(t)}, \tag{4.24}$$

which is itself non-decreasing in p. The limit curve P_b^{BP} is estimated in Fig. 15 by choosing t large enough so that $P_b^{(t)}$ is independent of t within the numerical accuracy.

Since $P_b^{BP} = P_b^{BP}(p)$ is a non-decreasing function of p, one can define the *BP threshold*

$$p_d \equiv \sup\left\{ p \in [0, 1/2] : P_b^{BP}(p) = 0 \right\}. \tag{4.25}$$

Analogous definitions can be provided for other channel families such as the BEC(ϵ). In general, the definition (4.25) can be extended to any family of BMS channels BMS(p) indexed by a real parameter $p \in I$, $I \subseteq \mathbb{R}$ being an interval (obviously the sup will be then taken over $p \in I$). The only condition is that the family is 'ordered by physical degradation'. We shall not describe this concept formally, but limit ourselves to say that that p should be an 'honest' noise parameter, in the sense that the channel worsen as p increases.

Table 1
Belief propagation thresholds for a few regular LDPC ensembles.

l	k	R	p_d	Shannon limit
3	4	1/4	0.1669(2)	0.2145018
3	5	2/5	0.1138(2)	0.1461024
3	6	1/2	0.0840(2)	0.1100279
4	6	1/3	0.1169(2)	0.1739524

Analytical upper and lower bounds can be derived for p_d. In particular it can be shown that it is strictly larger than 0 (and smaller than $1/2$) for all LDPC ensembles with minimum variable degree at least 2. Numerical simulation of density evolution allows to determine it numerically with good accuracy. In Table 1 we report the results of a few such results.

Let us stress that the threshold p_d has an important practical meaning. For any $p < p_d$ one can achieve arbitrarily small bit error rate with high probability by just picking one random code from the ensemble LDPC and using BP decoding and running it for a large enough (but independent of the blocklength) number of iterations. For $p > p_d$ the bit error rate is asymptotically lower bounded by $P_b^{BP}(p) > 0$ for any fixed number of iterations. In principle it could be that after, let's say n^a, $a > 0$ iterations a lower bit error rate is achieved. However simulations show quite convincingly that this is not the case.

In physics terms the algorithm undergoes a phase transition at p_d. At first sight, such a phase transition may look entirely dependent on the algorithm definition and not 'universal' in any sense. As we will discuss in the next section, this is not the case. The phase transition at p_d is somehow intrinsic to the underlying measure $\mu(\underline{x})$, and has a well studied counterpart in the theory of mean field disordered spin models.

Apart from the particular channel family, the BP threshold depends on the particular code ensemble, i.e. (for the case considered here) on the code ensemble. It constitutes therefore a primary measure of the 'goodness' of such a pair. Given a certain design rate R, one would like to make p_d as large as possible. This has motivated the introduction of code ensembles that generalize the regular ones studied here (starting from 'irregular' ones). Optimized ensembles have been shown to allow for exceptionally good performances. In the case of the erasure channel, they allowed to saturate Shannon's fundamental limit [27]. This is an important approach to the design of LDPC ensembles.

Let us finally mention that the BP threshold was defined in Eq. (4.25) in terms of the bit error rate. One may wonder whether a different performance parameter may yield a different threshold. As long as such parameter can be written in the form $\frac{1}{N} \sum_i f(h_i^{(t)})$ this is not the case. More precisely

$$p_{\mathrm{d}} = \sup\{p \in I : h^{(t)} \overset{\mathrm{d}}{\to} +\infty\}, \tag{4.26}$$

where, for the sake of generality we assumed the noise parameter to belong to an interval $I \subseteq \mathbb{R}$. In other words, for any $p < p_{\mathrm{d}}$ the distribution of BP messages becomes a delta at plus infinity.

4.6. Belief Propagation versus MAP decoding

So far we have seen that detailed predictions can be obtained for the performance of LDPC ensembles under message passing decoding (at least in the large blocklength limit). In particular the threshold noise for reliable communication is determined in terms of a distributional recursion (density evolution). This recursion can in turn be efficiently approximated numerically, leading to accurate predictions for the threshold.

It would be interesting to compare such predictions with the performances under optimal decoding strategies. Throughout this section we shall focus on symbol MAP decoding, which minimizes the bit error rate, and consider a generic channel family $\{\mathrm{BMS}(p)\}$ ordered[12] by the noise parameter p.

Given an LDPC ensemble, let $\mathrm{P}_{\mathrm{b}}^{(N)}$ be the expected bit error rate when the blocklength is N. The *MAP threshold* p_{c} for such an ensemble can be defined as the largest (or, more precisely, the supremum) value of p such that $\lim_{N\to\infty} \mathrm{P}_{\mathrm{b}}^{(N)} = 0$. In other words, for any $p < p_{\mathrm{c}}$ one can communicate with an arbitrarily small error probability, by using a random code from the ensemble, provided N is large enough.

By the optimality of MAP decoding, $p_{\mathrm{d}} \leq p_{\mathrm{c}}$. In coding theory some techniques have been developed to prove upper and lower bounds on p_{c} [19, 52]. In particular it is easy to find ensembles for which there exist a gap between the two thresholds (namely $p_{\mathrm{d}} < p_{\mathrm{c}}$ strictly). Consider for instance (k, l) regular ensembles with a fixed ratio $l/k = 1 - R$. It is then possible to show that, as $k, l \to \infty$, the BP threshold goes to 0 while the MAP threshold approaches the Shannon limit.

This situation is somewhat unsatisfactory. The techniques used to estimate p_{d} and p_{c} are completely different. This is puzzling since the two thresholds can be extremely close and even coincide for some ensembles. Furthermore, we know that $p_{\mathrm{d}} \leq p_{\mathrm{c}}$ by a general argument (optimality of MAP decoding), but this inequality is not 'built in' the corresponding derivations. Finally, it would be interesting to have a sharp estimate for p_{c}.

---

[12]Such that the channel worsen as p increases. Examples are the binary symmetric or binary erasure channels.

It turns out that a sharp characterization of p_c can be obtained through statistical mechanics techniques [42, 58, 38]. The statistical mechanics result has been proved to be a rigorous upper bound for general code ensembles, and it is conjectured to be tight [39, 35].

The starting point is to consider the conditional entropy of the channel input $\underline{x}$ given the output $\underline{y}$, $H_N(\underline{X}|\underline{Y})$. As shown in Eq. (2.16) this is given by the expectation of the log partition function appearing in Eq. (4.1) (apart from a trivial additive factor).

Let $f_N = \mathbb{E}H_N(\underline{X}|\underline{Y})/N$ denote the entropy density averaged over the code ensemble. Intuitively speaking, this quantity allows to estimate the typical number of inputs with non-negligible probability for a given channel output. If f_N is bounded away from 0 as $N \to \infty$, the typical channel output corresponds to an exponential number of (approximately) equally likely inputs. If on the other hand $f_N \to 0$, the correct input has to be searched among a sub-exponential number of candidates. This leads us to identify[13] the MAP threshold as the largest noise level such that $f_N \to 0$ as $N \to \infty$.

The Bethe free energy provides a natural way to approximate log-partition functions on sparse graphs. It is known to be exact if the underlying graph is a tree and its stationary points are in correspondence with the fixed points of BP. In statistical physics terms, it is the correct variational formulation for the Bethe Peierls approximation. In random systems which are locally tree like, it is normally thought to provide the correct $N \to \infty$ limit unless long range correlations set in. These are in turn described through 'replica symmetry breaking' (see below).

As many mean field approximations, the Bethe approximation can be thought of as a way of writing the free energy as a function of a few correlation functions. More specifically, one considers the single-variable marginals $\{b_i(x_i) : i \in \{1, \ldots, N\}\}$, and the joint distributions of variables involved in a common check node $\{b_a(\underline{x}_a) : a \in \{1, \ldots, M\}\}$. In the present case the Bethe free energy reads

$$F_B(\underline{b}) = -\sum_{i=1}^{N}\sum_{x_i} b_i(x_i) \log Q(y_i|x_i) \tag{4.27}$$

$$+ \sum_{a=1}^{M}\sum_{\underline{x}_a} b_a(\underline{x}_a) \log b_a(\underline{x}_a) - \sum_{i=1}^{N}(|\partial i| - 1)\sum_{x_i} b_i(x_i) \log_2 b_i(x_i).$$

The marginals $\{b_i(\cdot)\}$, $\{b_a(\cdot)\}$ are regarded as variables. They are constrained to be probability distributions (hence non-negative) and to satisfy the marginal-

---

[13] A rigorous justification of this identification can be obtained using Fano's inequality.

ization conditions

$$\sum_{x_j,\ j\in\partial a\backslash i} b_a(\underline{x}_a) = b_i(x_i) \quad \forall i \in \partial a, \qquad \sum_{x_i} b_i(x_i) = 1 \quad \forall i. \tag{4.28}$$

Further, in order to fulfill the parity-check constraints $b_a(\underline{x}_a)$ must be forced to vanish unless $x_{i_a(1)}\oplus\cdots\oplus x_{i_a(k)} = 0$ (as usual we use the convention $0\log 0 = 0$). Since they do not necessarily coincide with the actual marginals of $\mu(\cdot)$, the $\{b_a\}$, $\{b_i\}$ are sometimes called *beliefs*.

Approximating the log-partition function $-\log Z(\underline{y})$ requires minimizing the Bethe free energy $F_{\mathrm{B}}(\underline{b})$. The constraints can be resolved by introducing Lagrange multipliers, that are in turn expressed in terms of two families of real valued messages $\underline{u} \equiv \{u_{a\to i}\}$, $\underline{h} = \{h_{i\to a}\}$. If we denote by $P_u(x)$ the distribution of a bit x whose log likelihood ratio is u (in other words $P_u(0) = 1/(1+e^{-2u})$, $P_u(1) = e^{-2u}/(1+e^{-2u})$), the resulting beliefs read

$$b_a(\underline{x}_a) = \frac{1}{z_a}\mathbb{I}_a(\underline{x}) \prod_{j\in\partial a} P_{h_{j\to a}}(x_j),$$

$$b_i(x_i) = \frac{1}{z_i} Q(y_i|x_i) \prod_{a\in\partial i} P_{u_{a\to i}}(x_i), \tag{4.29}$$

where we introduced the shorthand $\mathbb{I}_a(\underline{x})$ to denote the indicator function for the a-th parity check being satisfied. Using the marginalization conditions (4.28) as well as the stationarity of the Bethe free energy with respect to variations in the beliefs, one obtains the fixed point BP equations

$$h_{i\to a} = B_i + \sum_{b\in\partial i\backslash a} u_{b\to i}, \qquad u_{a\to i} = \operatorname{atanh}\left\{\prod_{j\in\partial a\backslash i} \tanh h_{j\to a}\right\}. \tag{4.30}$$

These in turn coincide for with the fixed point conditions for belief propagation, cf. Eqs. (4.11).

The Bethe free energy can be written as a function of the messages by plugging the expressions (4.29) into Eq. (4.27). Using the fixed point equations, we get

$$\begin{aligned}
F_{\mathrm{B}}(\underline{u},\underline{h}) =\ & \sum_{(ia)\in E} \log\left[\sum_{x_i} P_{u_{a\to i}}(x_i) P_{h_{i\to a}}(x_i)\right] \\
& - \sum_{i=1}^{N} \log\left[\sum_{x_i} Q(y_i|x_i)\prod_{a\in\partial i} P_{u_{a\to i}}(x_i)\right] \\
& - \sum_{a=1}^{M} \log\left[\sum_{\underline{x}_a} \mathbb{I}_a(\underline{x})\prod_{i\in\partial a} P_{h_{i\to a}}(x_i)\right]. \tag{4.31}
\end{aligned}$$

We are interested in the expectation of this quantity with respect to the code and channel realization, in the $N \to \infty$ limit. We assume that messages are asymptotically identically distributed $u_{a \to i} \overset{\mathrm{d}}{=} u$, $h_{i \to a} \overset{\mathrm{d}}{=} h$, and that messages incoming in the same node along distinct edges are asymptotically independent. Under these hypotheses we get the limit

$$\lim_{N \to \infty} \frac{1}{N} \mathbb{E}\, F_{\mathrm{B}}(\underline{u}, \widehat{\underline{u}}) = -\phi_{u,h} + \sum_y Q(y|0) \log_2 Q(y|0), \tag{4.32}$$

where

$$
\begin{aligned}
\phi_{u,h} \equiv\ & -l\, \mathbb{E}_{u,h} \log_2 \left[\sum_x P_u(x) P_h(x) \right] \\
& + \mathbb{E}_y \mathbb{E}_{\{u_i\}} \log_2 \left[\sum_x \frac{Q(y|x)}{Q(y,0)} \prod_{i=1}^{l} P_{u_i}(x) \right] \\
& + \frac{l}{k} \mathbb{E}_{\{h_i\}} \log_2 \left[\sum_{x_1 \ldots x_k} \mathbb{I}_a(x) \prod_{i=1}^{k} P_{h_i}(x_i) \right].
\end{aligned}
\tag{4.33}
$$

Notice that the random variables u, h are constrained by Eq. (4.30), which must be fulfilled in distributional sense. In other words u, h must form a fixed point of the density evolution recursion (4.19). Given this proviso, if the above assumptions are correct and the Bethe free energy is a good approximation for the log partition function one expects the conditional entropy per bit to be $\lim_{N \to \infty} f_N = \phi_{u,h}$. This guess is supported by the following rigorous result.

Theorem 4.2. *If u, h are symmetric random variables satisfying the distributional identity $u \overset{\mathrm{d}}{=} \mathrm{atanh}\{\prod_{i=1}^{k-1} \tanh h_i\}$, then*

$$\lim_{N \to \infty} f_N \geq \phi_{u,h}. \tag{4.34}$$

It is natural to conjecture that the correct limit is obtained by optimizing the above lower bound, i.e.

$$\lim_{N \to \infty} f_N = \sup_{u,h} \phi_{u,h}, \tag{4.35}$$

where, once again the sup is taken over the couples of symmetric random variables satisfying $u \overset{\mathrm{d}}{=} \mathrm{atanh}\{\prod_{i=1}^{k-1} \tanh h_i\}$. In fact it is easy to show that, on the fixed point, the distributional equation $h \overset{\mathrm{d}}{=} B + \sum_{a=1}^{l-1} u_a$ must be satisfied as well. In other words the couple u, h must be a density evolution fixed point.

This conjecture has indeed been proved in the case of communication over the binary erasure channel for a large class of **LDPC** ensembles (including, for instance, regular ones).

The expression (4.35) is interesting because it bridges the analysis of BP and MAP decoding. For instance, it is immediate to show that it implies $p_d \leq p_c$.

Exercise 4: This exercise aims at proving the last statement.

(*a*) Recall that $u, h = +\infty$ constitute a density evolution fixed point for any noise level. Show that $\phi_{h,u} = 0$ on such a fixed point.

(*b*) Assume that, if any other fixed point exists, then density evolution converges to it (this can indeed be proved in great generality).

(*c*) Deduce that $p_d \leq p_c$.

Evaluating the expression (4.35) implies an a priori infinite dimensional optimization problem. In practice good approximations can be obtained through the following procedure:

1. Initialize h, u to a couple of symmetric random variables $h^{(0)}, u^{(0)}$.

2. Implement numerically the density evolution recursion (4.19) and iterate it until an approximate fixed point is attained.

3. Evaluate the functional $\phi_{u,h}$ on such a fixed point, after enforcing $u \overset{d}{=}$ atanh$\left\{ \prod_{i=1}^{k-1} \tanh h_i \right\}$ exactly.

The above procedure can be repeated for several different initializations $u^{(0)}$, $h^{(0)}$. The largest of the corresponding values of $\phi_{u,h}$ is then picked as an estimate for $\lim_{N \to \infty} f_N$.

While his procedure is not guaranteed to exhaust all the possible density evolution fixed points, it allows to compute a sequence of lower bounds to the conditional entropy density. Further, one expects a small finite number of density evolution fixed points. In particular, for regular ensembles and $p > p_d$, a unique (stable) fixed point is expected to exist apart from the no-error one $u, h = +\infty$. In Table 2 we present the corresponding MAP thresholds for a few regular ensembles.

Table 2

MAP thresholds for a few regular LDPC ensembles and communication over the BSC(p).

l	k	R	p_c	Shannon limit
3	4	1/4	0.2101(1)	0.2145018
3	5	2/5	0.1384(1)	0.1461024
3	6	1/2	0.1010(2)	0.1100279
4	6	1/3	0.1726(1)	0.1739524

For further details on these results, and complete proofs, we refer to [39]. Here we limit ourselves to a brief discussion why the conjecture (4.35) is expected to hold from a statistical physics point of view.

The expression (4.35) corresponds to the 'replica symmetric ansatz' from the present problem. This usually breaks down if some form of long-range correlation ('replica symmetry breaking') arises in the measure $\mu(\cdot)$. This phenomenon is however not expected to happen in the case at hand. The technical reason is that the so-called Nishimori condition holds for $\mu(\cdot)$ [38]. This condition generally holds for a large family of problems arising in communications and statistical inference. While Nishimori condition does not provide an easy proof of the conjecture (4.35), it implies a series of structural properties of $\mu(\cdot)$ that are commonly regarded as incompatible with replica symmetry breaking.

Replica symmetry breaking is instead necessary to describe the structure of 'metastable states' [17]. This can be loosely described as very deep local minima in the energy landscape introduced in Section 4.1. Here 'very deep' means that $\Theta(N)$ bit flips are necessary to lower the energy (number of unsatisfied parity checks) when starting from such minima. As the noise level increases, such local minima become relevant at the so called 'dynamic phase transition'. It turns out that the critical noise for this phase transition coincides with the BP threshold p_d. In other words the double phase transition at p_d and p_c is completely analogous to what happens in the mean field theory of structural glasses (see for instance Parisi's lectures at this School). Furthermore, this indicates that p_d has a 'structural' rather than purely algorithmic meaning.

5. Belief Propagation beyond coding theory

The success of belief propagation as an iterative decoding procedure has spurred a lot of interest in its application to other statistical inference tasks.

A simple formalization for this family of problems is provided by factor graphs. One is given a factor graph $G = (V, F, E)$ with variable nodes V, function nodes F, and edges E and considers probability distributions that factorize accordingly

$$\mu(\underline{x}) = \frac{1}{Z} \prod_{a \in F} \psi_a(\underline{x}_{\partial a}). \tag{5.1}$$

Here the variables x_i take values in a generic finite alphabet $\mathcal{X}$, and the *compatibility functions* $\psi_a : \mathcal{X}^{\partial a} \to \mathbb{R}_+$ encode dependencies among them. The prototypical problem consists in computing marginals of the distribution $\mu(\cdot)$, e.g.,

$$\mu_i(x_i) \equiv \sum_{\underline{x}_{\sim i}} \mu(\underline{x}). \tag{5.2}$$

Belief propagation can be used to accomplish this task in a fast and distributed (but not necessarily accurate) fashion. The general update rules read

$$v_{i \to a}^{(t+1)}(x_i) \propto \prod_{b \in \partial i \setminus a} \widehat{v}_{b \to i}^{(t)}(x_i), \qquad \widehat{v}_{a \to i}^{(t)}(x_i) \propto \sum_{\{x_j\}} \psi_a(\underline{x}_{\partial a}) \prod_{j \in \partial a \setminus i} v_{j \to a}^{(t)}(x_j).$$

$$(5.3)$$

Messages are then used to estimate local marginals as follows

$$\bar{v}_i^{(t+1)}(x_i) \propto \prod_{b \in \partial i} \widehat{v}_{b \to i}^{(t)}(x_i).$$

$$(5.4)$$

The basic theoretical question is of course to establish a relation, if any between $\mu_i(\cdot)$ and $\bar{v}_i(\cdot)$.

As an example, we shall consider *satisfiability* [23]. Given N Boolean variables x_i, $i \in \{1, \ldots, N\}$, $x_i \in \{\text{True, False}\}$, a formula is the logical expression obtained by taking the AND of M *clauses*. Each clause is the logical OR of a subset of the variables or their negations. As an example, consider the formula (here $\bar{x}_i$ denotes the negation of x_i)

$$(\bar{x}_1 \vee \bar{x}_2 \vee \bar{x}_4) \wedge (x_1 \vee \bar{x}_2) \wedge (x_2 \vee x_4 \vee x_5) \wedge (x_1 \vee x_2 \vee \bar{x}_5) \wedge (x_1 \vee \bar{x}_3 \vee x_5).$$

$$(5.5)$$

An assignment of the N variables satisfies the formula if, for each of the clause, at least one of the involved *literals* (i.e. either the variables or their negations) evaluates to True.

A satisfiability formula admits a natural factor graph representation where each clause is associated to a factor node and each variable to a variable node. An example is shown in Fig. 16. Admitting that the formula has at least one satisfying assignment, it is natural to associate to a $\mathcal{F}$ the uniform measure over such assignments $\mu_{\mathcal{F}}(\underline{x})$. It is easy to realize that such a distribution takes the

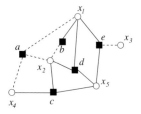

Fig. 16. Factor graph representation of the satisfiability formula (5.5). Circle correspond to variables and squares to clauses. Edges are represented as dashed line if the variable is negated in the corresponding clause.

form (5.1) where the compatibility function $\psi_a(\underline{x}_{\partial a})$ takes value 1 if the assignment $\underline{x}$ satisfies clause a and 0 otherwise.

Satisfiability, i.e., the problem of finding a solution to a satisfiability formula or proving that it is unsatisfiable, is one of the prototypical NP-complete problems. Given a satisfiability formula, computing marginals with respect to the associated distribution $\mu_{\mathcal{F}}(\cdot)$ is relevant for tasks such as counting the number of solutions of $\mathcal{F}$ or sampling them uniformly. These are well known #-P complete problems.[14]

The currently best algorithms for solving random instances for the K-SAT problem are based on a variant of BP, which is called *survey propagation* [11, 33, 11].

5.1. Proving correctness through correlation decay

A simple trick to bound the error incurred by BP consists in using the correlation decay properties [1, 57] of the measure $\mu(\cdot)$. Let $i \in \{1, \dots, N\}$ be a variable index and denote by $\underline{x}_t$ the set of variables sitting on nodes at distance t from i. Further, denote by $\underline{x}_{>t}$ the set of variables whose distance from i is *at least t*. Then the local structure of the probability distribution (5.1)

$$\mu_i(x_i) = \sum_{\underline{x}_{\geq t}} \mu(x_i|\underline{x}_{\geq t})\mu(\underline{x}_{\geq t}) = \sum_{\underline{x}_t} \mu(x_i|\underline{x}_t)\mu(\underline{x}_t). \tag{5.6}$$

Let $\mathsf{B}_i(t)$ denote the subgraph induced by nodes whose distance from i is *at most* t, and $\partial \mathsf{B}_i(t)$ its boundary (nodes whose distance from i is exactly t). Further, for any $j \in \partial \mathsf{B}_i(t)$ let $a(j)$ be the unique function nodes inside $\mathsf{B}_i(t)$ that is adjacent to j. It is intuitively clear that belief propagation computes the marginal at i *as if* the graph did not extend beyond $\mathsf{B}_i(t)$. More precisely, if the initial condition $v_{i\to a}^{(0)}(x_i)$ is properly normalized, then we have the exact expression

$$\overline{v}_i^{(t)}(x_i) = \sum_{\underline{x}_t} \mu(x_i|\underline{x}_t) \prod_{j \in \partial \mathsf{B}_i(t)} v_{j\to a(j)}^{(0)}(x_j). \tag{5.7}$$

As a consequence of Eq. (5.6) and (5.7) we have

$$\left|\mu_i(x_i) - \overline{v}_i^{(t)}(x_i)\right| \leq \sup_{\underline{x}_t, \underline{x}_t'} \left|\mu(x_i|\underline{x}_t) - \mu(x_i|\underline{x}_t')\right|. \tag{5.8}$$

This provides an upper bound on the error incurred by BP when computing the marginal of x_i base on the local structure of the underlying graph in terms of the influence of far away variables. To make things fully explicit, assume that the graph has *girth*[15] g and that $\sup_{\underline{x}_t, \underline{x}_t'} |\mu(x_i|\underline{x}_t) - \mu(x_i|\underline{x}_t')| \leq \exp(-\kappa t)$ for some

---

[14] The notation #-P refers to the hardness classification for counting problems.

[15] Recall that the girth of a graph is the length of its shortest cycle.

positive κ. This implies

$$\left| \mu_i(x_i) - \overline{v}_i^{(t)}(x_i) \right| \leq e^{-\kappa g/2}. \tag{5.9}$$

As an example of such error estimates, we shall consider *random k-satisfiability* [16]. This is a standard model to generate 'synthetic' satisfiability formulae. It amounts to picking a formula uniformly at random among all the ones including N variables and $M = N\alpha$ k-clauses (a k-clause is a clause that involve *exactly* k distinct variables). We shall of course limit to $k \geq 2$, the case $k = 1$ being trivial.

Consider a uniformly random variable node in the factor graph associated to a random formula, and its depth-t neighborhood $\mathsf{B}_i(t)$. Proceeding as in the previous section it is not hard to show that, for any fixed t, $\mathsf{B}_i(t)$ is with high probability (as $N \to \infty$) a tree. An appropriate model the distribution of such a tree, is given by the tree ensemble $\mathsf{T}_*(t)$ described as follows. For $t = 0$, it is the graph containing a unique variable node. For any $t \geq 1$, start by a single variable node (the root) and add $l \stackrel{\mathrm{d}}{=} \mathsf{Poisson}(k\alpha)$ clauses, each one including the root, and $k - 1$ new variables (first generation variables). For each one of the l clauses, the corresponding literals are non-negated or negated independently with equal probability. If $t \geq 2$, generate an independent copy of $\mathsf{T}_*(t - 1)$ for each variable node in the first generation and attach it to them.

Assume that, for a typical random tree formula $\mathsf{T}(t)$, the marginal distribution of the variable at the root is weakly dependent on the values assigned at the boundary. Following the above lines, one can use this fact to prove that BP computes good approximations for the marginals in a random k-SAT formula. In fact it turns out that an estimate of the form[16]

$$\mathbb{E}_{\mathsf{T}(t)} \sup_{\underline{x}_t, \underline{x}_t'} \left| \mu(x_i | \underline{x}_t) - \mu(x_i | \underline{x}_t') \right| \leq e^{-\kappa t} \tag{5.10}$$

can be proved if the clause density α stays below a threshold $\alpha_{\mathrm{u}}(k)$ that is estimated to behave as $\alpha_{\mathrm{u}}(k) = \frac{2 \log k}{k}[1 + o_k(1)]$.

While we refer to the original paper [40] for the details of the proof we limit ourselves to noticing that the left hand side of Eq. (5.10) can be estimated efficiently using a density evolution procedure. This allows to estimate the threshold $\alpha_{\mathrm{u}}(k)$ numerically. Consider in fact the log-likelihood (here we are identifying {True, False} with {+1, −1})

$$h^{(t)}(\underline{x}_t) = \frac{1}{2} \log \frac{\mu(+1 | \underline{x}_t)}{\mu(-1 | \underline{x}_t)}. \tag{5.11}$$

---

[16]Here the sup is taken over assignments $\underline{x}_t, \underline{x}_t'$ that can be extended to solutions of $\mathsf{T}(t)$.

This quantity depends on the assignment of the variables on the boundary, $\underline{x}_t$. Since we are interested on a *uniform* bound over the boundary, let us consider the extreme cases

$$\overline{h}^{(t)} = \max_{\underline{x}_t} h^{(t)}(\underline{x}_t), \qquad \underline{h}^{(t)} = \min_{\underline{x}_t} h^{(t)}(\underline{x}_t). \tag{5.12}$$

It is then possible to show that the couple $(\overline{h}^{(t)}, \underline{h}^{(t)})$ obeys a recursive distributional equation that, as mentioned, can be efficiently implemented numerically.

6. Belief Propagation beyond the binary symmetric channel

So far we have considered mainly the case of transmission over BMS channels, our reference example being the BSC. There are many other channel models that are important and are encountered in practical situations. Fortunately, it is relatively straightforward to extend the previous techniques and statements to a much larger class, and we review a few such instances in this section.

6.1. Binary memoryless symmetric channels

In order to keep the notation simple, we assumed channel output to belong to a finite alphabet $\mathcal{A}$. In our main example, the BSC, we had $\mathcal{A} = \{0, 1\}$. But in fact all results are valid for a wider class of *binary memoryless symmetric (BMS) channels*. One can prove that there is no loss of generality in assuming the output alphabet to be the real line $\mathbb{R}$ (eventually completed with $\overline{\mathbb{R}} = \mathbb{R} \cup \{\pm\infty\}$).

Let $\underline{y} = (y_1, \dots, y_N)$ be the vector of channel outputs on input $\underline{x} = (x_1, \dots, x_N)$. For a BMS the input is binary, i.e. $\underline{x} \in \{0, 1\}^N$. Further the channel is *memoryless*, i.e. the probability density of getting $\underline{y} \in \overline{\mathbb{R}}^N$ at output when the input is $\underline{x}$, is

$$Q(\underline{y}|\underline{x}) = \prod_{t=1}^{N} Q(y_t|x_t).$$

Finally, the *symmetry* property can be written without loss of generality, as $Q(y_t|x_t = 1) = Q(-y_t|x_t = 0)$.

One of the most important elements in this class is the *additive white Gaussian noise* (AWGN) channel, defined by

$$y_t = x_t + z_t, \quad t \in \{1, \dots, N\},$$

where the sequence $\{z_t\}$ is i.i.d. consisting of Gaussian random variables with mean zero and variance σ^2. It is common in this setting to let x_i take values in

$\{+1, -1\}$ instead of $\{0, 1\}$ as we have assumed so far. The AWGNC transition probability density function is therefore

$$Q(y_t|x_t) = \frac{1}{\sqrt{2\pi\sigma^2}} \, e^{-\frac{(y-x)^2}{2\sigma^2}}.$$

The AWGNC is the basic model of transmission of an electrical signal over a cable (here the noise is due to thermal noise in the receiver) and it is also a good model of a wireless channel in free space (e.g., transmission from a satellite).

Although the class of BMS channels is already fairly large, it is important in practice to go beyond it. The extension to the non-binary case is quite straightforward and so we will not discuss it in detail. The extension to channels with memory or the asymmetric case are more interesting and so we present them in the subsequent two sections.

6.2. Channels with memory

Loosely speaking, in a memoryless channel the channel acts on each transmitted bit independently. In a channel with memory, on the other hand, the channel acts generally on the whole block of input bits together. An important special case of a channel with memory is if the channel can be modeled as a Markov chain, taking on a sequence of "channel states." Many physical channels posses this property and under this condition the message-passing approach can still be applied. For channels with memory there are two problems. First, we need to determine the capacity of the channel. Second, we need to devise efficient coding schemes that achieve rates close to this capacity. It turns out that both problems can be addressed in a fairly similar framework. Rather than discussing the general case we will look at a simple but typical example.

Let us start by computing the information rate/capacity of channels with memory, assuming that the channel has a Markov structure. As we discussed in Section 2.5 in the setting of BMS channels, the channel capacity can be expressed as the difference of two entropies, namely as $H(X) - H(X|Y)$. Here, X denotes the binary input and Y denotes the observation at the output of the channel whose input is X. Given two random variables X, Y, this entropy difference is called the *mutual information* and is typically denoted by $I(X; Y) = H(X) - H(X; Y)$.

A general channel, is defined by a channel transition probability $Q(y_1^N|x_1^N)$ (here and below x_1^N denotes the vector $(x_1, \ldots, x_N)$). In order to define a joint distribution of the input and output vectors, we have to prescribe a distribution on the channel input, call it $p(x_1^N)$. The channel capacity is obtained by maximizing the mutual information over all possible distributions of the input, and eventually taking the $N \to \infty$ limit. In formulae

$$C(Q) = \lim_{N \to \infty} \sup_{p(\cdot)} I\left(X_1^N; Y_1^N\right)/N.$$

For BMS channels it is possible to show that the maximum occurs for the uniform prior: $p(x_1^N) = 1/2^N$. Under this distribution, $I(X_1^N; Y_1^N)/N$ is easily seen not to depend on N and we recover the expression in Sec. 2.5.

For channels with memory we have to maximize the mutual information over all possible distributions over $\{0, 1\}^N$ (a space whose dimension is exponential in N), and take the limit $N \to \infty$. An easier task is to choose a convenient input distribution $p(\cdot)$ and then compute the corresponding mutual information in the $N \to \infty$ limit:

$$I = \lim_{N \to \infty} I\left(X_1^N; Y_1^N\right)/N. \tag{6.1}$$

Remarkably, this quantity has an important operational meaning. It is the largest rate at which we can transmit reliably across the channel using a coding scheme such that the resulting input distribution matches $p(\cdot)$.

To be definite, assume that the channel is defined by a state sequence $\{\sigma_t\}_{t \geq 0}$, taking values in a finite alphabet, such that the joint probability distribution factors in the form

$$p\left(x_1^n, y_1^n, \sigma_0^n\right) = p(\sigma_0) \prod_{i=1}^n p(x_i, y_i, \sigma_i \mid \sigma_{i-1}). \tag{6.2}$$

We will further assume that the transmitted bits $(x_1, \cdot, x_N)$ are iid uniform in $\{0, 1\}$. The factor graph corresponding to (6.2) is shown in Fig. 17. It is drawn in a somewhat different way compared to the factor graphs we have seen so far. Note that in the standard factor graph corresponding to this factorization all variable nodes have degree two. In such a case it is convenient not to draw the factor graph as a bipartite graph but as a standard graph in which the nodes correspond to the factor nodes and the edges correspond to the variable nodes (which have degree two and therefore connect exactly two factors). Such a graphical representation is also known as *normal* graph or as Forney-style factor graph (FSFG), in honor of Dave Forney who introduced them [15]. Let us now look at a concrete example.

Example 5:[Gilbert-Elliott Channel] The Gilbert-Elliot channel is a model for a *fading* channel, i.e., a channel where the quality of the channel is varying over time. In this model we assume that the channel quality is evolving according to a Markov chain. In the simplest case there are exactly two states, and this is the original Gilbert-Elliott channel (GEC) model. More precisely, consider the two-state Markov chain depicted in Fig. 18. Assume that $\{X_t\}_{t \geq 1}$ is i.i.d., taking values in $\{\pm 1\}$ with uniform probability.

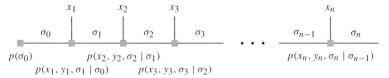

Fig. 17. The FSFG corresponding to (6.2).

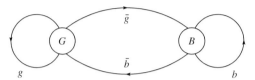

Fig. 18. The Gilbert-Elliott channel with two states.

The channel is either in a *good* state, denote it by G, or in a *bad* state, call it B. In either state the channel is a BSC. Let the crossover probability in the good state be ϵ_G and in the bad state be ϵ_B, with $0 \leq \epsilon_G < \epsilon_B \leq 1/2$. Let P be the 2×2 matrix

$$P = \begin{pmatrix} g & \bar{b} \\ \bar{g} & b \end{pmatrix}$$

which encodes the transition probabilities between the states (the columns indicate the present state and the rows the next state). Define the *steady state* probability vector $p = (p_G, p_B)$, i.e., the vector which fulfills $Pp^T = p^T$. This means that in steady state the system spends a fraction p_G of the time in state G and a fraction p_B of the time in state B. If we consider e.g. the state G then the detailed balance condition reads $p_G \bar{g} = p_B \bar{b}$. From this we get $p = (\bar{b}/(\bar{g} + \bar{b}), \bar{g}/(\bar{g} + \bar{b}))$. More generally, let us assume that we have s states, $s \in \mathbb{N}$, and that the channel in state i, $i \in [s]$, is the $BSC(\epsilon_i)$. Let P be the $s \times s$ matrix encoding the transition probabilities between these states. Let p denote the steady-state probability distribution vector. If $(I - P^T + E)$ is invertible then a direct check shows that $p = e(I - P^T + E)^{-1}$, where e is the all-one vector of length s, I is the $s \times s$ identity matrix and E is the $s \times s$ all-one matrix.

Note that the state sequence is ergodic as long as the Markov chain is irreducible (i.e. there is a path of strictly positive probability from any state to any other state) and aperiodic (i.e. there there exists such a path for any number of steps large enough). In the original Gilbert-Elliot model this is true as long as $0 < g, b < 1$.

Consider the computation of the maximal rate at which we can transmit reliably. We have

$$I(X_1^N; Y_1^N) = H(Y_1^N) - H(Y_1^N|X_1^N).$$

Let us see how we can compute $\lim_{N \to \infty} H(Y_1^N)/N$. Because of the ergodicity assumption on the state sequence, $-\frac{1}{N} \log p(y_1^N)$ converges with probability one to $\lim_{N\to\infty} H(Y_1^N)/N$. It follows that if we can compute $-\frac{1}{N}\log p(y_1^N)$ for a very large sequence, then with high probability the value will be close to the desired entropy rate. Instead of computing $p(y_1^N)$, let us compute $p(\sigma_N, y_1^N)$. From this we trivially get our desired quantity by summing,

$$p(y_1^N) = \sum_{\sigma_N} p(\sigma_N, y_1^N).$$

Note that

$$p(\sigma_N, y_1^N) = \sum_{x_N, \sigma_{N-1}} p(x_N, \sigma_{N-1}, \sigma_N, y_1^N)$$

$$= \sum_{x_N, \sigma_{N-1}} \underbrace{p(x_N, \sigma_N, y_N \mid \sigma_{N-1})}_{\text{kernel}} \underbrace{p(\sigma_{N-1}, y_1^{N-1})}_{\text{message}}. \tag{6.3}$$

From this we see that $p(\sigma_N, y_1^N)$ can be computed recursively. In fact this recursion corresponds to running the BP message-passing rules on the factor graph depicted in Fig. 17 (which is a tree): denote the message which is passed along the edge labeled by σ_N by $\nu_N(\sigma_N)$. Then according to the BP message-passing rules we have

$$\nu_N(\sigma_N) = \sum_{x_N, \sigma_{N-1}} p(x_N, \sigma_N, y_N \mid \sigma_{N-1}) \nu_{N-1}(\sigma_{N-1}).$$

If we compare this to the recursion stated in (6.3) we see that these two recursions are identical. In other words, $\nu_N(\sigma_N) = p(\sigma_N, y_1^N)$, so that

$$\lim_{N\to\infty} H(Y_1^N)/N = -\lim_{N\to\infty} \log\left(\sum_{\sigma_N} \nu_N(\sigma_N)\right)/N. \tag{6.4}$$

From a practical perspective it is typically more convenient to pass *normalized* messages $\tilde{\nu}_N(\sigma_N)$ so that $\sum_\sigma \tilde{\nu}_N(\sigma_N) = 1$. The first message $\nu_0(\sigma_0) = p(\sigma_0)$ is already a probability distribution and, hence, normalized, $\tilde{\nu}_0(\sigma_0) = \nu_0(\sigma_0)$. Compute $\nu_1(\sigma_1)$ and let $\lambda_1 = \sum_{\sigma_1} \nu_1(\sigma_1)$. Define $\tilde{\nu}_1(\sigma_1) = \nu_1(\sigma_1)/\lambda_1$. Now note that by definition of the message-passing rules all subsequent messages in the case of rescaling differ from the messages which are sent in the unscaled case only by this scale factor. Therefore, if λ_i denotes the normalization constant by which we have to divide at step i so as to normalize the message then $\tilde{\nu}_N(\sigma_N) = $

$v_N(\sigma_N)/(\prod_{i=1}^{N} \lambda_i)$. It follows that

$$\lim_{N\to\infty} H(Y_1^N)/N = -\lim_{N\to\infty} \log\left(\sum_{\sigma_N} \alpha_N(\sigma_N)\right)/N$$

$$= -\lim_{N\to\infty} \log\left(\left(\prod_{i=1}^{N} \lambda_i\right) \sum_{\sigma_N} \tilde{\alpha}_N(\sigma_N)\right)/N$$

$$= \lim_{N\to\infty} \left(\sum_{i=1}^{N} \log(\lambda_i)\right)/N.$$

It remains to compute $H(Y_1^N \mid X_1^N)$. We write $H(Y_1^N \mid X_1^N)/N = H(Y_1^N, X_1^N)/N - H(X_1^N)/N$. The second part is trivial since the inputs are i.i.d. by assumption so that $H(X_1^N)/N = 1$. For the term $H(Y_1^N, X_1^N)/N$ we use the same technique as for the computation of $H(Y_1^N)/N$. Because of the ergodicity assumption on the state sequence, $-\frac{1}{N}\log p(y_1^N, x_1^N)$ converges with probability one to $\lim_{N\to\infty} H(Y_1^N, X_1^N)/N$. We write $p(y_1^N, x_1^N) = \sum_{\sigma_N} p(\sigma_N, y_1^N, x_1^N)$ and use the factorization

$$p(\sigma_N, y_1^N, x_1^N) = \sum_{\sigma_{N-1}} p(\sigma_{N-1}, \sigma_N, y_1^N, x_1^N)$$

$$= \sum_{\sigma_{N-1}} \underbrace{p(x_N, \sigma_N, y_N \mid \sigma_{N-1})}_{\text{kernel}} \cdot \underbrace{p(\sigma_{N-1}, y_1^{N-1}, x_1^{N-1})}_{\text{message}}.$$

In words, we generate a random instance X_1^N and Y_1^N and run the BP algorithm on the FSFG shown in Fig. 17 assuming that *both* Y_1^N and X_1^N are 'quenched.' Taking the logarithm, multiplying by minus one and normalizing by $1/N$ gives us an estimate of the desired entropy.

Now that we can compute the maximal rate at which we can transmit reliably, let us consider coding. The symbol MAP decoder is

$$\widehat{x}_i(\underline{y}) = \text{argmax}_{x_i}\, p(x_i \mid y_1^N)$$

$$= \text{argmax}_{x_i} \sum_{\{x_j, j\neq i\}} p(x_1^N, y_1^N, \sigma_0^N)$$

$$= \text{argmax}_{x_i} \sum_{\{x_j, j\neq i\}} p(\sigma_0) \prod_{j=1}^{N} p(x_j, y_j, \sigma_j \mid \sigma_{j-1}) \mathbb{I}_{\mathfrak{C}}(x_1^N).$$

In words, the FSFG in Fig. 17 describes also the factorization for the message-passing decoder if we add to it the factor nodes describing the definition of the code. As always, this factor graph together with the initial messages stemming from the channel completely specify the message-passing rules, except for the message-passing schedule. Let us agree that we alternate one round of decoding with one round of channel estimation. No claim as to the optimality of this scheduling rule is made.

Notice that the correlations induced by the Markovian structure of the channel are in general short ranged in time. This is analogous to what happens with a one-dimensional spin model, whose correlation length is always finite (at nonzero temperature). A good approximation to the above message passing schedule is therefore obtained by a 'windowed' decodes. This means that the state at time t is estimated only of the basis of observations between time $t - R$ and $t + R$, for some finite R.

Assuming windowed decoding for channel estimation, it is not hard to show that after a fixed number of iterations, the decoding neighborhood is again asymptotically tree-like. In the case of the GEC the channel symmetry can be used to reduce to the all-zero codeword. Therefore, we can employ the technique of density evolution to determine thresholds and to optimize the ensembles.

Example 6:[GEC: State Estimation]   For the case of transmission over the GEC the iterative decoder implicitly also estimates the state of the channel. Let us demonstrate this by means of the following example. We pick a GEC with three states. Let

$$
P = \begin{pmatrix} 0.99 & 0.005 & 0.02 \\ 0.005 & 0.99 & 0.02 \\ 0.005 & 0.005 & 0.96 \end{pmatrix},
$$

which has a steady state probability vector of $p \approx (0.4444, 0.4444, 0.1112)$. Finally, let the channel parameters of the BSC in these three states be $(\epsilon_1, \epsilon_2, \epsilon_3) \approx (0.01, 0.11, 0.5)$. This corresponds to an *average* error probability of $\epsilon_{\text{avg}} = \sum_{i=1}^{3} p_i \epsilon_i \approx 0.108889$. Using the methods described above, the capacity of this channel (assuming uniform inputs) can be computed to be equal to $C \approx 0.583$ bits per channel use. This is markedly higher than $1 - h(\epsilon_{\text{avg}}) \approx 0.503444$, which is the capacity of the BSC(ϵ_{avg}). The last channel is the channel which we experience if we ignore the Markov structure.

Fig. 19 shows the evolution of the densities for an optimized ensemble of rate $r \approx 0.5498$. The pictures on the right correspond to the messages which are passed from the part of the factor graph which estimates the state towards the part of the factor graph which describes the code. These messages therefore can be interpreted as the current estimate of the state the channel is in at a given point in time. Note that after 10 iterations 5 clear peaks emerge. These peaks are at $\pm \log(0.99/0.01) \approx \pm 4.595$, $\pm \log(0.9/0.1) \approx \pm 2.197$, $\pm \log(0.5/0.5) = 0$. They correspond to the received likelihoods in the three possible channel states. In other words, the emergence of the peaks shows that at this stage the system has identified the channel states with high reliability. This is quite pleasing. Although the channel state is not known to the receiver and can not be observed directly, in the region where the iterative decoder works reliably it also automatically estimates the channel state with high confidence.

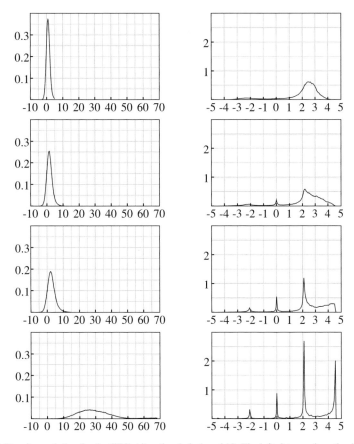

Fig. 19. Density evolution for the GEC at iteration 1, 2, 4, and 10. The left pictures show the densities of the messages which are passed from the code towards the part of the FSFG which estimates the channel state. The right hand side shows the density of the messages which are the estimates of the channel state and which are passed to the FSFG corresponding to the code.

Although we only looked a very particular example it is quite typical of the general situation: as long as the channel memory can be described by a Markov chain the factor graph approach applies and we can use message-passing schemes to construct efficient coding schemes [14, 4, 21, 22, 25, 43].

6.3. *Asymmetric channels – the Z channel*

Let us now consider the second generalization, namely the case of *non-symmetric* channels.

Consider the channel depicted on the right of Fig. 1. For obvious reasons it is called the Z channel (ZC). This channel has binary input and it is memoryless but it is *not* symmetric. Nevertheless, essentially the same type of analysis which we performed in Section 4 can be applied to this case as well. Symmetry is therefore a *nice* property to have but it is *not essential*.

Consider the capacity of this channel. Since the channel is not symmetric the capacity is not necessarily given by the mutual information between channel input and channel output for a uniform input distribution of the input. We must instead maximize the mutual information over the input distribution. Since the channel is memoryless, it can be assumed that the input is given by a sequence of i.i.d. Bernoulli variables. Assuming that $p(x_i = 0) = \alpha$, the output distribution is

$$\big(p(y_i = 0), p(y_i = 1)\big) = (\alpha\bar{p}, 1 - \alpha\bar{p}),$$

so that the mutual information $I_\alpha(X; Y)$ for a fixed α is equal to

$$I_\alpha(X; Y) = H(Y) - H(Y \mid X) = h(\alpha\bar{p}) - \alpha h(p). \qquad (6.5)$$

Some calculus reveals that the optimal choice of α is

$$\alpha(p) = \frac{p^{p/\bar{p}}}{1 + \bar{p}p^{p/\bar{p}}}, \qquad (6.6)$$

so that

$$C_{ZC(p)} = h(\alpha(p)\bar{p}) - \alpha(p)h(p).$$

Fig. 20 compares $C_{ZC(p)}$ with $I_{\alpha=\frac{1}{2}}(X; Y)$, i.e., it compares the capacity with the transmission rate which is achievable with *uniform* input distribution. This is important and surprising – only little is lost by insisting on an uniform input

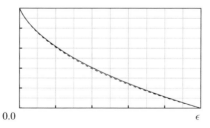

0.0 ϵ

Fig. 20. Comparison of $C_{ZC(p)}$ (solid curve) with $I_{\alpha=\frac{1}{2}}(X; Y)$ (dashed curve), both measured in bits.

distribution: the rate which is achievable by using a uniform input distribution is at least a fraction $\frac{1}{2} e \ln(2) \approx 0.924$ of capacity over the entire range of p (with equality when p approaches one). Even more fortunate, from this perspective the Z channel is the extremal case [32, 54]: the information rate of any binary-input memoryless channel when the input distribution is the uniform one is at least a fraction $\frac{1}{2} e \ln(2)$ of its capacity. From the above discussion we conclude that, when dealing with asymmetric channels, not much is lost if we use a binary linear coding scheme (inducing a uniform input distribution).

Consider the density evolution analysis. Because of the lack of symmetry we can no longer make the all-one codeword assumption. Therefore, it seems at first that we have to analyze the behavior of the decoder with respect to each codeword. Fortunately this is not necessary. First note that, since we consider an ensemble average, only the *type* of the codeword matters. More precisely, let us say that a codeword has type τ if the fraction of zeros is τ. For $\underline{x} \in \mathfrak{C}$, let $\tau(\underline{x})$ be its type. Let us assume that we use an LDPC ensemble whose dominant type is one-half. This means that "most" codewords contain roughly as many zeros as one. Although it is possible to construct degree-distributions which violate this constraint, "most" degree distributions do fulfill it. Under this assumption there exists some strictly positive constant γ such that

$$\mathbb{P}\left\{ \tau(\underline{x}) \notin [1/2 - \delta/\sqrt{n}, 1/2 + \delta/\sqrt{n}] \right\} \leq e^{-\delta^2 \gamma}, \tag{6.7}$$

where the probability is with respect to a uniformly random codeword $\underline{x}$. We can therefore analyze the performance of such a system in the following way: determine the error probability assuming that the type of the transmitted codeword is "close" to the typical one. Since sub-linear changes in the type do not figure in the density analysis, this task can be accomplished by a straightforward density evolution analysis. Now add to this the probability that the type of a random codeword deviates significantly from the typical one. The second term can be made arbitrarily small (see right hand side of (6.7)) by choosing δ sufficiently large.

We summarize: if we encounter a non-symmetric channel and we are willing to sacrifice a small fraction of capacity then we can still use standard LDPC ensembles (which impose a uniform input distribution) to transmit at low complexity. If it is crucial that we approach capacity even closer, a more sophisticated approach is required. We can combine LDPC ensembles with non-linear mappers which map the uniform input distribution imposed by linear codes into a non-uniform input distribution at the channel input in order to bring the mutual information closer to capacity. For a detailed discussion on coding for the Z-channel we refer the reader to [34, 59, 5].

7. Open problems

Let us close by reviewing some of the most important open challenges in the channel coding problem.

7.1. Order of limits

Density evolution computes the limit

$$\lim_{t \to \infty} \lim_{N \to \infty} \mathbb{E}\big[P_b^{(N,t)}\big].$$

In words we determined the limiting performance of an ensemble under a *fixed* number of iterations as the blocklength tends to infinity and then let the number of iterations tend to infinity. As we have seen, this limit is relatively easy to compute. What happens if the order of limits is exchanged, i.e., how does the limit

$$\lim_{N \to \infty} \lim_{t \to \infty} \mathbb{E}\big[P_b^{(N,t)}\big]$$

behave? This limit is closer in spirit to the typical operation in practice: for each fixed length the BP decoder continues until no further progress is achieved. We are interested in the limiting performance as the blocklength tends to infinity.

For the BEC it is known that the two limits coincide. If we combine this with the fact that for the BEC the performance is a monotone function in the number of iterations (any further iteration can only make the result better) then we get the important observation that regardless of how we take the limit (jointly or sequentially), as long as both the blocklength as well as the number of iterations tend to infinity we get the same result. From a practical perspective this is comforting to know: it shows that we can expect a certain robustness of the performance with respect to the details of the operation.

It is conjectured that the same statement holds for general BMS channels. Unfortunately, no proof is known.

7.2. Finite-length performance

The threshold gives an indication of the asymptotic performance: for channel parameters which are better than the threshold sufficiently long codes allow transmission at an arbitrarily low probability of bit error. If, on the other hand, we transmit over a channel which has a parameter that is worse than the threshold then we can not hope to achieve a low probability of error. This is an important insight but from a practical perspective we would like to know how fast the finite-length performance approaches this asymptotic limit. There can be many

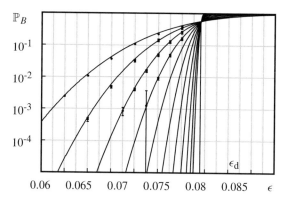

Fig. 21. Performance of the BP decoder for the $(3, 6)$-regular ensemble when transmission takes place over the BSC. The block lengths are $n = 2^i$, $i = 10, \ldots, 20$. The dots correspond to simulations. For most simulation points the 95% confidence intervals are smaller than the dot size. The lines correspond to the analytic approximation based on scaling laws.

different ensembles that all have the same asymptotic performance but that might have a substantially different finite-length behavior. Can we predict which one we should choose a priori without having to resort to simulations? The typical convergence of the performance to the asymptotic limit is shown in Fig. 21. The points correspond to simulation results whereas the solid curves correspond to a general scaling conjecture [2]. Roughly speaking, this scaling conjecture states that around the threshold the error probability behaves as follows: Let the channel be parameterized by ϵ with increasing ϵ indicating a worsening of the channel. Let ϵ_d be the BP threshold, and define $z = \sqrt{n}(\epsilon - \epsilon_d)$. Then for z fixed and n tending to infinity we have

$$P_B(n, \epsilon) = \Phi(z/\alpha)\big(1 + o(1)\big),$$

where $\Phi(\,\cdot\,)$ is the error function (i.e. $\Phi(x)$ is the probability that a standard normal random variable is smaller than x), and α is a constant which depends on the channel as well as on the channel.

For the BEC this scaling law has been shown to be correct [2]. In fact, even a refined version is known [13]: define $z = \sqrt{n}(\epsilon - \epsilon_d + \beta n^{-\frac{2}{3}})$ where β is a constant depending on the ensemble. Then for z fixed and n tending to infinity we have

$$P_B(n, \epsilon) = Q(z/\alpha)\big(1 + O(n^{-1/3})\big).$$

For general channels on the other hand the problem is largely open. If proved to be correct, finite length scaling laws could be used as a tool for an efficient finite-length optimization.

7.3. Capacity-achieving codes

For the most part we have taken the point of view that we are given an ensemble of codes and a family of channels and we would like to determine the performance of this combination. For instance, the most fundamental question is to determine the threshold noise for such a code/decoder combination. Hopefully this threshold is close to the best possible as determined by Shannon's capacity formula.

But we can take a more active point of view. Given a family of channels how should we choose the ensemble in order for the threshold noise to be as high as possible. In other words, can we approach the capacity of the channel?

For the BSC this question has been answered by Luby, Mitzenmacher, Shokrollahi, Spielman, and Steman in [27]. These authors showed that by a suitable choice of the degree distribution one can approach the capacity arbitrarily closely. More precisely, in order to approach capacity up to a fraction δ the average degree has to grow like $\log(1/\delta)$ and this is the best possible. For general channels it is not known whether capacity can be achieved. Although the resolution of this problem will most likely only have a small practical implication it is without doubt the most important open theoretical problem in this area.

Appendix A. A generating function calculation

We want to compute the number of ways of picking n distinct objects from M groups each containing k objects, in such a way that the number of selected elements in each group is even. Let us denote this number by C_n. The result of this computation is used in Section 3.3, where we claimed the result to be $C_n = \text{coeff}[q_k(z)^M, z^n]$ (in that case $n = lw$).

First notice that given $m_1, \ldots, m_M$ all even, the number of ways of choosing m_1 objects from the first group, m_2 from the second, etc is

$$
\binom{k}{m_1}\binom{k}{m_2}\cdots\binom{k}{m_M}. \tag{A.1}
$$

The desired number C_n is the sum of this quantity over the even numbers m_1, $\ldots, m_n$ such that $m_1, \ldots, m_M \in \{0, \ldots k\}$ and $m_1 + \cdots + m_M = n$. The corresponding generating function is

$$C(z) \equiv \sum_{n=0}^{kM} C_n z^n = \sum_{m_1 \ldots m_n}^{\text{even}} \binom{k}{m_1}\binom{k}{m_2}\cdots\binom{k}{m_M} z^{m_1}\cdots z^{m_N}, \tag{A.2}$$

the sum being restricted over even integers $m_1, \ldots, m_M \in \{0, \ldots k\}$. We now notice that the sum factorizes yielding

$$C(z) = \left\{ \sum_{m_{\text{even}}} \binom{k}{m} z^m \right\}^M = q_k(z)^M, \tag{A.3}$$

which proves the claim.

Acknowledgments

We would like to thank Jean-Philippe Bouchaud and Marc Mézard for organizing the 2006 Summer School on Complex Systems in Les Houches. We would also like to thank Cyril Méasson and Christine Neuberg with their help in creating Fig. 19.

The work of A. Montanari was partially supported by the European Union under the project EVERGROW. The work of R. Urbanke was partially supported by the NCCR-MICS, a center supported by the Swiss National Science Foundation under grant number 5005-67322.

References

[1] D. ALDOUS AND J. M. STEELE, *The Objective Method: Probabilistic Combinatorial Optimization and Local Weak Convergence*, in Probability on discrete structures, H. Kesten, ed., Springer, New York, 2004.

[2] A. AMRAOUI, A. MONTANARI, T. RICHARDSON, AND R. URBANKE, *Finite-length scaling for iteratively decoded LDPC ensembles*, in Proc. 41th Annual Allerton Conference on Communication, Control and Computing, Monticello, IL, 2003.

[3] O. BARAK AND D. BURSHTEIN, *Lower bounds on the spectrum and error rate of LDPC code ensembles*, in Proc. of the IEEE Int. Symposium on Inform. Theory, Adelaide, Australia, September 4–9 2005, pp. 42–46.

[4] G. BAUCH, H. KHORRAM, AND J. HAGENAUER, *Iterative equalization and decoding in mobile communications systems*, in Proceedings of GLOBECOM, 1997.

[5] A. BENNATAN AND D. BURSHTEIN, *Iterative decoding of LDPC codes over arbitrary discrete-memoryless channels*, in Proceedings of the 41-st Allerton Conference on Communication, Control, and Computing, Monticello, IL, Oct. 2003, pp. 1416–1425.

[6] E. R. BERLEKAMP, *Algebraic Coding Theory*, Aegean Park Press, 1984.

[7] C. BERROU, A. GLAVIEUX, AND P. THITIMAJSHIMA, *Near Shannon limit error-correcting coding and decoding*, in Proceedings of ICC'93, Geneve, Switzerland, May 1993, pp. 1064–1070.

[8] H. A. BETHE, *Statistical theory of superlattices*, Proc. Roy. Soc. London A, 150 (1935), pp. 552–558.

[9] R. E. BLAHUT, *Theory and Practice of Error Control Codes*, Addison-Wesley, 1983.

[10] L. E. BOLTZMANN, *Vorlesungen über Gastheorie*, J.A. Barth, Leipzig, 1896.

[11] A. BRAUNSTEIN, M. MÉZARD, AND R. ZECCHINA, *Survey propagation: algorithm for satisfiability*. arXiv:cond-math/cond-mat/0212002.

[12] T. M. COVER AND J. A. THOMAS, *Elements of Information Theory*, Wiley, New York, 1991.

[13] A. DEMBO AND A. MONTANARI, *Finite size scaling for the core of large random hypergraphs*. Xarch:math.PR/0702007, 2007.

[14] C. DOUILLARD, A. PICART, M. JÉZÉQUEL, P. DIDIER, C. BERROU, AND A. GLAVIEUX, *Iterative correction of intersymbol interference: Turbo-equalization*, European Trans. on Commun., 6 (1995), pp. 507–511.

[15] G. D. FORNEY, JR., *Codes on graphs: Normal realizations*, IEEE Trans. Inform. Theory, 47 (2001), pp. 520–548.

[16] J. FRANCO AND M. PAULL, *Probabilistic analysis of the Davis-Putnam procedure for solving satisfiability*, Discrete Appl. Math, 5 (1983), pp. 77–87.

[17] S. FRANZ, M. LEONE, A. MONTANARI, AND F. RICCI-TERSENGHI, *Dynamic phase transition for decoding algorithms*, Phys. Rev. E, 22 (2002), p. 046120.

[18] S. FRANZ AND G. PARISI, *Recipes for metastable states in spin glasses*, J. Physique I, 5 (1995), p. 1401.

[19] R. G. GALLAGER, *Low-density parity-check codes*, IRE Transactions on Information Theory, 8 (1962), pp. 21–28.

[20] ———, *Information theory and reliable communication*, Wiley, 1968.

[21] J. GARCIA-FRIAS AND J. D. VILLASENOR, *Combining hidden Markov source models and parallel concatenated codes*, IEEE Communications Letters, 1 (1997), pp. 111–113.

[22] ———, *Turbo decoders for Markov channels*, IEEE Commun. Lett., 2 (1998), pp. 257–259.

[23] M. R. GAREY AND D. S. JOHNSON, *Computers and Intractability: A Guide to the Theory of NP-Completeness*, W. H. Freeman & Co., New York, 1979.

[24] R. KIKUCHI, *A theory of cooperative phenomena*, Phys. Rev., 81 (1951), pp. 988–1003.

[25] F. R. KSCHISCHANG AND A. W. ECKFORD, *Low-density parity-check codes for the gilbert-elliot channel*, in Proc. 41th Annual Allerton Conference on Communication, Control and Computing, Monticello, IL, 2003.

[26] S. LIN AND D. J. COSTELLO, JR., *Error Control Coding*, Prentice-Hall, 2nd ed., 2004.

[27] M. LUBY, M. MITZENMACHER, A. SHOKROLLAHI, D. A. SPIELMAN, AND V. STEMANN, *Practical loss-resilient codes*, in Proceedings of the 29th annual ACM Symposium on Theory of Computing, 1997, pp. 150–159.

[28] D. J. C. MACKAY, *Good error correcting codes based on very sparse matrices*, IEEE Trans. Inform. Theory, 45 (1999), pp. 399–431.

[29] D. J. C. MACKAY, *Information Theory, Inference & Learning Algorithms*, Cambridge University Press, Cambridge, 2002.

[30] D. J. C. MACKAY AND R. M. NEAL, *Good codes based on very sparse matrices*, in Cryptography and Coding. 5th IMA Conference, C. Boyd, ed., no. 1025 in Lecture Notes in Computer Science, Springer, Berlin, 1995, pp. 100–111.

[31] N. MACRIS, *Correlation inequalities: a useful tool in the theory of LDPC codes*, in Proc. of the IEEE Int. Symposium on Inform. Theory, Adelaide, Australia, Sept. 2005, pp. 2369–2373.

[32] E. E. MAJANI AND H. RUMSEY, JR., *Two results on binary-input discrete memoryless channels*, in Proc. of the IEEE Int. Symposium on Inform. Theory, June 1991, p. 104.

[33] E. MANEVA, E. MOSSEL, AND M. J. WAINWRIGHT, *A new look at survey propagation and its generalizations*, in SODA, Vancouver, Canada, 2005.

[34] R. J. MCELIECE, *Are turbo-like codes effective on nonstandard channels?*, IEEE inform. Theory Soc. Newslett., 51 (2001), pp. 1–8.

[35] C. MÉASSON, A. MONTANARI, T. RICHARDSON, AND R. URBANKE, *The Generalized Area Theorem and Some of its Consequences*. submitted to IEEE IT, 2005.

[36] M. MEZARD AND A. MONTANARI, *Information, Physics and Computation*, Clarendon Press - Oxford, 2007. To be published.

[37] M. MÉZARD AND G. PARISI, *The bethe lattice spin glass revisited*, Eur. Phys. J. B, 20 (2001), p. 217.

[38] A. MONTANARI, *The glassy phase of Gallager codes*, Eur. Phys. J. B, 23 (2001), pp. 121–136.

[39] ———, *Tight bounds for LDPC and LDGM codes under MAP decoding*, IEEE Trans. Inform. Theory, 51 (2005), pp. 3221–3246.

[40] A. MONTANARI AND D. SHAH, *Counting good truth assignments of random k-sat formulae*, in SODA, New Orleans, USA, Jan. 2007, pp. 1255–1264.

[41] R. MOTWANI AND P. RAGHAVAN, *Randomized Algorithms*, Cambridge University Press, Cambridge, 1995.

[42] T. MURAYAMA, Y. KABASHIMA, D. SAAD, AND R. VICENTE, *Statistical physics of regular low-density parity-check error-correcting codes*, Phys. Rev. E, 62 (2000), p. 1577.

[43] C. NEUBERG, *Gilbert-Elliott channel and iterative decoding*. EPFL, Semester Project (Supervisor: Cyril Méason), 2004.

[44] H. NISHIMORI, *Statistical Physics of Spin Glasses and Information Processing*, Oxford University Press, Oxford, 2001.

[45] J. PEARL, *Fusion, propagation, and structuring in belief networks*, Artificial Intelligence, 29 (1998), pp. 241–288.

[46] P. W. A. R. ABOU-CHACRA AND D. J. THOULESS, *A selfconsistent theory of localization*, J. Phys C, 6 (1973), p. 1734.

[47] V. RATHI, *On the asymptotic weight and stopping set distribution of regular LDPC ensembles*, IEEE Trans. Inform. Theory, 52 (2006), pp. 4212–4218.

[48] T. RICHARDSON AND H. JIN, *A new fast density evolution*, in Proc. of the IEEE Inform. Theory Workshop, Monte Video, Uruguay, Feb. 2006. pp. 183–187.

[49] T. RICHARDSON AND R. URBANKE, *The capacity of low-density parity check codes under message-passing decoding*, IEEE Trans. Inform. Theory, 47 (2001), pp. 599–618.

[50] ———, *Modern Coding Theory*, Cambridge University Press, 2007. To be published.

[51] P. RUJAN, *Finite temperature error-correcting codes*, Phys. Rev. Lett., 70 (1993), pp. 2968–2971.

[52] I. SASON AND S. SHAMAI, *Performance Analysis of Linear Codes under Maximum-Likelihood Decoding: A Tutorial*, vol. 3 of Foundations and Trends in Communications and Information Theory, NOW, Delft, the Netherlands, July 2006.

[53] C. E. SHANNON, *A mathematical theory of communication*, Bell System Tech. J., 27 (1948), pp. 379–423, 623–656.

[54] N. SHULMAN AND M. FEDER, *The uniform distribution as a universal prior*, IEEE Trans. Inform. Theory, 50 (2004), pp. 1356–1362.

[55] M. Sipser and D. A. Spielman, *Expander codes*, IEEE Trans. Inform. Theory, 42 (1996), pp. 1710–1722.

[56] N. Sourlas, *Spin-glass models as error-correcting codes*, Nature, 339 (1989), pp. 693–695.

[57] S. Tatikonda and M. Jordan, *Loopy belief propagation and Gibbs measures*, in Proc. Uncertainty in Artificial Intell., Alberta, Canada, Aug. 2002, pp. 493–500.

[58] J. van Mourik, D. Saad, and Y. Kabashima, *Critical noise levels for LDPC decoding*, Physical Review E, 66 (2002).

[59] C.-C. Wang, S. R. Kulkarni, and H. V. Poor, *Density evolution for asymmetric memoryless channels*, IEEE Trans. Inform. Theory, 51 (2005), pp. 4216–4236.

[60] G. Wiechman and I. Sason, *On the parity-check density and achievable rates of LDPC codes for memoryless binary-input output-symmetric channels*, in Proc. of the Allerton Conf. on Commun., Control and Computing, Monticello, IL, USA, September 28–30 2005, pp. 1747–1758.

[61] J. S. Yedidia, W. T. Freeman, and Y. Weiss, *Constructing free energy approximations and generalized belief propagation algorithms*, IEEE Trans. Info.Theory, 51 (2005), pp. 2282–2313.

Course 3

MEAN FIELD THEORY OF SPIN GLASSES: STATICS AND DYNAMICS

Giorgio Parisi

Dipartimento di Fisica, Sezione INFN, SMC and UdRm1 of INFM,
Università di Roma "La Sapienza",
Piazzale Aldo Moro 2, I-00185 Rome (Italy)
giorgio.parisi@roma1.infn.It

J.-P. Bouchaud, M. Mézard and J. Dalibard, eds.
Les Houches, Session LXXXV, 2006
Complex Systems
© *2007 Published by Elsevier B.V.*

Contents

Preamble

In these lecture I will review some theoretical results that have been obtained for spin glasses. I will concentrate my attention on the presentation of the mean field approach and on its numerical and experimental verifications. I will present in a modern the various hypothesis at the basis of the theory and I will discuss their mathematical and physical status.

1. Introduction

Spin glasses have been intensively studied in the last thirty years. They are very interesting for many reasons:
- Spin glasses are the simplest example of glassy systems. There is a highly non-trivial mean field approximation where one can study phenomena that have been discovered for the first time in this context, e.g. the existence of many equilibrium states and the ultrametricity relation among these states. In this framework it is possible to derive some of the main properties of a generic glassy systems, e.g. history dependent response [1–3]; this property, in the context of mean field approximation, is related to the existence of many *equilibrium states*.[1]
- The study of spin glasses opens a very important window for studying off-equilibrium behavior. Aging [9] and the related violations of the equilibrium fluctuation dissipation relations emerge in a natural way and they can be studied in this simple setting [10–13]. Many of the ideas developed in this context can be used in other physical fields like fragile glasses, colloids, granular materials, combinatorial optimization problems and for other complex systems [14].
- The theoretical concepts and the tools developed in the study of spin glasses are based on two logically equivalent, but very different, methods: the algebraic broken replica symmetry method and the probabilistic cavity approach [2, 3]. They have a wide domain of applications. Some of the properties that appear in the mean field approximation, like ultrametricity, are unexpected and counterintuitive.

---

[1] This sentence is too vague: one should discuss the its precise mathematical meaning; although we will present in these lecture a physically reasonable definition, for a careful discussion see ref. [4–8].

• Spin glasses also provide a testing ground for a more mathematical inclined probabilistic approach: the rigorous proof of the correctness of the solution of the mean field model came out after twenty years of efforts where new ideas (e.g. stochastic stability [15–17]) and new variational principles [18, 19] were at the basis of a recent rigorous proof [20] of the correctness of the mean field approximation for in the case of the infinite range Sherrington-Kirkpatrick model [21].

In these lectures I will present a short review of some of the results that have been obtained using the probabilistic cavity approach. I will mostly discuss the mean field approximation for the infinite range Sherrington-Kirkpatrick. I will only mention *en passant* how to extend these results to finite connectivity model and to finite dimensional systems. The very interesting applications of these techniques to other problems coming from physics (e.g. glasses [22]) and other disciplines [14] will not be discussed.

2. General considerations

The simplest spin glass Hamiltonian is of the form:

$$H = \sum_{i,k=1,N} J_{i,k} \sigma_i \sigma_k, \tag{2.1}$$

where the J's are *quenched* (i.e. time independent) random variables located on the links connecting two points of the lattice and the σ's are Ising variables (i.e. they are equal to ± 1). The total number of points is denoted with N and it goes to infinity in the thermodynamic limit.

We can consider four models of increasing complexity:

• The SK model [21]: All J's are random and different from zero, with a Gaussian or a bimodal distribution with variance $N^{-1/2}$. The coordination number (z) is $N - 1$ and it goes to infinity with N. In this case a *simple* mean field theory is valid in the infinite N limit [2, 3].

• The Bethe lattice model [23–25]: The spins live on a random lattice and only $Nz/2$ J's are different from zero: they have variance $z^{-1/2}$. The average coordination number is finite (i.e. z). In this case a modified mean field theory is valid.

• The large range Edwards Anderson model [26]: The spins belong to a finite dimensional lattice of dimension D. Only nearest spins at a distance less than R interact and the variance of the J's is proportional to $1/R^{D/2}$. If R is large, the corrections to mean field theory are small for thermodynamic quantities, although they may change the large distance behavior of the correlations functions and the nature of the phase transition, which may disappear.

• The Edwards Anderson model [27]: The spins belong to a finite dimensional lattice of dimensions D: Only nearest neighbor interactions are different from zero and their variance is $D^{-1/2}$. In this case finite corrections to mean field theory are present, that are certainly very large in one or two dimensions, where no transition is expected. The Edwards Anderson model corresponds to the limit $R = 1$ of the the large range Edwards Anderson model; both models are expected to belong to the same universality class. The large range model Edwards Anderson provides a systematic way to interpolate between the mean field results and the short range model.[2]

As far as the free energy is concerned, one can prove the following rigorous results:

$$\lim_{z \to \infty} \text{Bethe}(z) = \text{SK},$$

$$\lim_{D \to \infty} \text{Edwards Anderson}(D) = \text{SK},$$

$$\lim_{R \to \infty} \text{finite range EA}(R) = \text{SK}. \tag{2.2}$$

The Sherrington Kirkpatrick model is thus a good starting point for studying also the finite dimensional case with short range interaction, that is the most realistic and difficult case. This starting point becomes worse and worse when the dimension decreases, e.g. it is not any more useful in the limiting case where $D = 1$.

3. Mean field theory

3.1. General considerations

Let us start again with the Hamiltonian in eq. (2.1) and let us proceed in the most naive way. Further refinement will be added at the end.

We consider the local magnetizations $m_i = \langle \sigma_i \rangle$ and we write the simplest possible form of the free energy as function of the magnetization. Using the standard arguments that can be found in the books [5], we get:

$$F[m] = \sum_{i,k=1,N} J_{i,k} m_i m_k - T \sum_i S(m_i), \tag{3.1}$$

where $S(m)$ is the usual binary entropy:

$$-S(m) = \left(\frac{1+m}{2} \log\left(\frac{1+m}{2} \right) + \frac{1-m}{2} \log\left(\frac{1-m}{2} \right) \right). \tag{3.2}$$

---

[2]In alternative approach one introduces ν spins per site that are coupled to all the spins in the nearest points of the lattice. It is possible formally to construct an expansion in the parameter $g \equiv 1/\nu$, the so called loop expansion.

If we look to the minimum of the free energy we get the following equations (that are valid at any stationary point of the free energy):

$$m_i = \tanh(h_i^{eff}), \qquad h_i^{eff} = \sum_k J_{i,k}\sigma_k. \tag{3.3}$$

This is well known and fifty years ago we could have stopped here. However now we understand that mean field approximation involves uncontrolled approximations and therefore we need to work in a more controlled framework.

In a more modern approach one consider a model and one tries to write down an expression for the free energy that is exact for that particular model. In this way one is sure that the range of validity of the formulae one is writing is not void (and what is more important, if no technical mistakes have been done, there could be no contradictions). As far as the exact model is often obtained as limit of a more realistic model when some parameter goes to infinity, it is possible to estimate the corrections to these asymptotic results.

A very interesting case, where usually mean field exact formula are valid, happens when the coordination number (i.e. the number of spins that are connected to a given spin) goes to infinity. Let us consider the following construction. We assume that for a given i, $J_{i,k}$ is different from zero only for z different values of k (for simplicity let us also assume that the J's take only the values ± 1 with equal probability).

We are interested to study the limit where z goes to infinity. One can immediately see that in the random case a finite result is obtained (at least in the high temperature phase) only if J goes to zero as $z^{-1/2}$ (a similar result can be obtained also in the low temperature phase). This result is in variance with the ferromagnetic case where the J's are all positive and they should go zero as z, in order to have a finite result in the low temperature phase. Indeed a good thermodynamic scaling is present when h^{eff} is of O(1). In the ferromagnetic case the z terms in the expression for h^{eff} add coherently and therefore each of them must be of order z^{-1}; on the contrary in the spin glass case, if they are uncorrelated (this is true in a first approximation), h^{eff} is the sum of z random terms and the result is of O(1) only if each term is of order $z^{-1/2}$.

If one makes a more careful computation and we look to the corrections to the mean field expression that survive in this limit, one obtains:

$$F_{TAP}[m] = \sum_{i,k=1,N} \left(\frac{1}{2} J_{i,k} m_i m_k - \frac{1}{4}\beta J_{i,k}^2 (1 - m_i^2)(1 - m_k^2) \right)$$
$$- T \sum_i S(m_i), \tag{3.4}$$

where $\beta = 1/T$ (we put the Boltzmann-Drude constant α to the value 3/2).

This free energy has an extra term with respect to the previous free energy eq. (3.3) and is called the TAP free energy [28], because it firstly appeared in a preprint signed by Thouless, Anderson, Lieb and Palmer.[3] The extra term can be omitted in the ferromagnetic case, because it gives a vanishing contribution.

We must add a word of caution: the TAP free energy has not been derived from a variational equation: the magnetization should satisfy the the stationary equation:

$$\frac{\partial F}{\partial m_i} = 0, \tag{3.5}$$

but there is no warranty that the correct value of the magnetization is the absolute minimum of the free energy. Indeed in many cases the paramagnetic solution (i.e. $m_i = 0$) is the absolute minimum at all temperatures, giving a value of the free energy density that would be equal to $-\frac{1}{4}\beta$. The corresponding internal energy would be equal to $-\frac{1}{2}\beta$, that would give an result divergent at $T = 0$, that in most of the cases would be unphysical.

Let us be more specific (although these consideration are quite generic) and let us consider the case of the Sherrington Kirkpatrick model, where all spins are connect and $z = N - 1$.

Here one could be more precise and compute the corrections to the TAP free energy: an explicit computation shows that these corrections are not defined in some regions of the magnetization space [1, 29]. When the corrections are not defined, the TAP free energy does not make sense. In this way (after a detailed computation) one arrives the conclusion that one must stay in the region where the following condition is satisfied:

$$\beta^2 \mathrm{Av}\big((1 - m_i^2)^2\big) \leq 1, \tag{3.6}$$

where Av stands for the average over all the points i of the sample. When the previous relation is not valid, one finds that the correlations function, that were assumed to be negligible, are divergent [2] and the whole computation fails. The correct recipe is to find the absolute minimum of the TAP free energy only in this region [1]. Of course the paramagnetic solution is excluded as soon as $T < 1$.

Let us as look to the precise expression of the TAP stationarity equations: they are

$$m_i = \tanh(\beta h_i^{eff}), \qquad h_i^{eff} = \sum_k \left(J_{i,k} m_k - \beta m_i J_{i,k}^2 (1 - m_k^2) \right). \tag{3.7}$$

---

[3]There are other possible terms, for example $\sum_{i,k,l} J_{i,k}^2 J_{k,l}^2 J_{l,i}^2$, but they do vanish in this limit.

For large N these equations can be simplified to

$$h_i = \sum_k \left(J_{i,k}m_k - \beta m_i(1 - q_{EA})\right), \tag{3.8}$$

where

$$q_{EA} = \mathrm{Av}\left(m_i^2\right). \tag{3.9}$$

Apparently the TAP equations are more complex that the naive ones (3.3); in reality their analysis is simpler. Indeed, using perturbation expansion with respect to the term $J_{i,k}$ that are of order $O(N^{1/2})$, we can rewrite the effective field as

$$h_i^{eff} = \sum_k J_{i,k}m_k^c, \tag{3.10}$$

where the cavity magnetizations m_k^c are the values of the magnetizations at the site k in a system with $N - 1$ spins, where the spin at i has been removed [2] (the correct notation would be $m_{k;i}^c$ in order to recall the dependence of the cavity magnetization on the spin that has removed, but we suppress the second label i when no ambiguities are present). The previous equations are also called Bethe's equations because they were the starting point of the approximation Bethe in his study of the two dimensional Ising model.

The crucial step is based on the following relation:

$$m_k = m_k^c + \beta m_i J_{i,k}\left(1 - (m_{k;i}^c)^2\right). \tag{3.11}$$

This relation can be proved by using the fact that the local susceptibility $\partial m_i / \partial h_i$ is given by $\beta(1 - m_i^2)$ in the mean field approximation. Although the difference between $m(k)$ and m_k^c is small, i.e. $O(N^{-1/2})$, one obtains that the final effect on h_{eff} is of $O(1)$.

The validity of eq. (3.10) is rather fortunate, because in this formula $J_{i,k}$ and m_k^c are uncorrelated. Therefore the central limit theorem implies that, when one changes the $J_{i,k}$, the quantity h_i^{eff} has a Gaussian distribution with variance q_{EA}. Therefore the probability distribution of h^{eff} is given by

$$P(h)dh = (2\pi)^{-1/2}\exp(-q^2/2)dh \equiv d\mu_q(h) \tag{3.12}$$

However this result is valid only under the hypothesis that there is a one to one correspondence of the solutions of the TAP equation at N and at $N - 1$, a situation that would be extremely natural if the number of solutions of the TAP equations would be a fixed number (e..g 3, as happens in the ferromagnetic case at low temperature). As we shall see, this may be not the case and this difference brings in all the difficulties and the interesting features of the models.

3.2. The cavity method: the naive approach

Given the fact that all the points are equivalent, it is natural to impose the condition that the statistical properties of the spin at i, when the $J_{i,k}$ change, are i independent. For a large system this statistical properties coincide with those obtained by looking to the properties of the other $N - 1$ spins of the system. This condition leads to the cavity method, where one compare the average properties of the spins in similar systems with slightly different size.

The cavity method is a direct approach that in principle can be used to derive in an probabilistic way all the results that have been obtained with the replica method.[4] The replica approach is sometimes more compact and powerful, but it is less easy to justify, because the working hypothesis that are usually done cannot be always translated in a transparent way in physical terms.[5]

The idea at the basis of the cavity method is simple: we consider a system with N spins ($i = 1, N$) and we construct a new system with $N + 1$ spins by adding an extra spin (σ_0). We impose the consistency condition that the average properties of the new spin (the average being done with respect to $J_{0,i}$) are the same of that of the old spins [2, 24].

In this way, if we assume that there is only one non-trivial solution to the TAP equations,[6] we get that for that particular solution

$$\overline{m_0^2} = \text{Av}(m_i^2), \tag{3.13}$$

where the bar denotes the average over all the $J_{0,i}$.

The Hamiltonian of the new spin is:

$$\sigma_0 \sum_{i=1,N} J_{0,i}\sigma_i. \tag{3.14}$$

If we suppose that the spins σ_i have vanishing correlations and we exploit the fact that each individual term of the Hamiltonian is small, we find that

$$m_0 \equiv \langle \sigma_0 \rangle = \tanh(\beta h^{\textit{eff}}).$$
$$h^{\textit{eff}} = \sum_{i=1,N} J_{0,i}m_i^c, \tag{3.15}$$

---

[4]The replica method is based on a saddle point analysis of some n dimensional integral in the limit where n goes to zero.

[5]Not all the results of the replica approach have been actually derived using the cavity approach.

[6]We neglect the fact the if m_i is a particular solution of the TAP equations, also $-m_i$ is a solution of the TAP equations. This degeneracy is removed if we add an infinitesimal magnetic field or we restrict ourselves only the half space of configurations with positive magnetization.

where m_i^c (for $i \neq 0$) denotes the magnetization of the spin σ_i *before* we add the spin 0. In this way we have re-derived the TAP equations for the magnetization in a direct way, under the crucial hypothesis of absence of long range correlation among the spins (a result that should be valid if we consider the expectation values taken inside one equilibrium state). It should be clear why the whose approach does not make sense when the condition eq. (3.6) is not satisfied.

We have already seen that when the variables $J_{0,i}$ are random, the central limit theorem implies that h is a Gaussian random variable with variance

$$\overline{h^2} = q_{EA} \equiv \frac{\sum_{i=1,N}(m_i^c)^2}{N}, \tag{3.16}$$

and the distribution probability of h is given by eq. (3.12). The same result would be true also when the variables m_i^c are random.

If we impose the condition that the average magnetization squared of the new point is equal to average over the old points, we arrive to the consistency equation:

$$q_{EA} = \overline{m_0^2} = \int d\mu_{q_{EA}}(h) \tanh^2(\beta h), \tag{3.17}$$

where $d\mu_{q_{EA}}(h)$ denotes a normalized Gaussian distribution with zero average and variance q_{EA}. It is easy to check [2] that the h-dependent increase in the total free energy of the system (when we go from N to $N+1$) is

$$\beta \Delta F(h) \equiv -\ln(\cosh(\beta h)). \tag{3.18}$$

It is appropriate to add a comment. The computation we have presented relates the magnetization a spin of the systems with $N+1$ spins to the magnetizations of the system with N spins: they are not a closed set of equations for a given system. However we can also write the expression of magnetization at the new site as function of the magnetizations of the system with $N+1$ spins, by computing the variation of the magnetizations in a perturbative way. Let us denote by m the magnetization of the old system (N spins) and by m' the magnetization of the new system ($N+1$ spins). Perturbation theory tells us that

$$m_i' \approx m_i + J_{0,i}m_0'\frac{\partial m_i}{\partial h_i} = m_i + J_{0,i}m_0'\beta\left(1 - (m_i')^2\right). \tag{3.19}$$

It is not surprising that using the previous formula we get again the TAP equations [2, 28]:

$$m'_0 = \tanh(\beta h)$$

$$h = \sum_{i=1,N} J_{0,i} m_i \approx \sum_{i=1,N} J_{0,i} m'_i - m'_0 \sum_i J^2_{0,i} \beta (1 - m'_i)$$

$$\approx \sum_{i=1,N} J_{0,i} m'_i - m'_0 \beta (1 - q_{EA}) \tag{3.20}$$

where $(N+1)q_{EA} = \sum_{i=0,N} m'_i$ (we have used the relation $\overline{J^2_{0,i}} = N^{-1}$).

A detailed computation show that the free energy corresponding to a solution of the TAP equations is given by the TAP free energy. The computation of the total free energy is slightly tricky. Indeed we must take care than if we just add one spin the coordination number z change and we have to renormalize the J's. This gives an extra term that should be taken into account.

The explicit computation of the free energy could be avoided by computing the internal energy density and verifying a *a posteriori* the guessed form of the free energy is correctly related to the internal energy. Here we can use a general argument: when the J's are Gaussian distributed the internal energy density e is given

$$e = \frac{1}{2} \overline{\sum_i J_{0,i} \langle \sigma_0 \sigma_i \rangle} = \frac{1}{2N} \overline{\sum_i \frac{\partial}{\partial J_{0,i}} \langle \sigma_0 \sigma_i \rangle}$$

$$= \beta \frac{1}{2N} \overline{\sum_i \left(\langle (\sigma_0 \sigma_i)^2 \rangle - \langle \sigma_0 \sigma_i \rangle^2 \right)} = \frac{1}{2} \beta (1 - q^2). \tag{3.21}$$

In the first step we have integrated by part with respect to the Gaussian distribution of the J, in the last step we have used the assumption that connect correlation are small and therefore

$$\langle \sigma_0 \sigma_i \rangle^2 \approx (m_0 m_i)^2. \tag{3.22}$$

In both ways one obtain the expression for the TAP free energy, where the only parameter free is $q \equiv q_{EA}$; we can call this quantity $f(q)$; its explicit expression is

$$\beta f(q) = -\frac{1}{4} \beta^2 (1 - q^2) + \int d\mu_q(h) \ln(\cosh(\beta h)) \tag{3.23}$$

The value of q can be computed in two mathematical equivalent ways:
- We use the equation (3.17) to find a non-linear equation for q.
- We find the stationary point of the free energy respect to q by solving the equation $\partial f / \partial q = 0$.

An integration by part is the needed step to prove the equivalence of the two approaches.

An explicit computation shows that the non-trivial stationary point (i.e $q \neq 0$), that should be relevant at $T < 1$, is no not a minimum of the free energy, as it should be a maximum for reasons that would be clear later. In this way one finds the form of the function $q(T)$. which turns out to be rather smooth:

- $q(T)$ is zero for $T > 1$.
- $q(T)$ it is proportional to $1 - T$ for T slightly below 1.
- $q(T)$ behave as $1 - AT$, with positive A when T goes to zero.

When this solution was written down [21], it was clear that something was wrong. The total entropy was negative at low temperature. This was particular weird, because in usual mean field models the entropy is just the average over the sites of the local entropy that is naturally positive, The *villain* is the extra term that we have added to the naive free energy in order to get the TAP free energy: indeed one finds that the entropy density is given by

$$S = \text{Av}\big(S(m_i)\big) - \frac{(1-q)^2}{4T^2}. \tag{3.24}$$

The average of the local term goes like T at low temperatures, while the extra term goes to $-A^2/4$, i.e. -0.17. It is important to note that the extra term is needed to obtain that for $T > 1$, $e = -\beta/2$, a results that can be easily checked by constructing the high temperature expansion.

On the contrary numerical simulations [21] showed a non-negative entropy (the opposite would be an extraordinary and unexpected event) very similar to the analytic computation for $T > 0.4$ and going to zero at T^2 (the analytic result was changing sign around $T = 0.3$).

Also the value of the internal energy was wrong: the analytic computation gave $e = -0.798$ while the numerical computations gave $e = -0.765 \pm 0.01$ (now numerical simulations give $e = -0.7633 \pm 0.0001$).

It is clear that we are stacked, the only way out is to leave out the hypothesis of the existence of only one solution of the TAP equations (or equivalently of only one equilibrium state) and to introduce this new ingredient in the approach. This will be done in the next section.

4. Many equilibrium states

4.1. The definition of equilibrium states

The origin of the problem firstly became clear when De Almeida and Thouless noticed that in the whole region $T < 1$, the inequality (3.6) was violated:

$$\beta^2 \text{Av}\big((1 - m_i^2)^2\big) = 1 + B_0(1 - T)^2, \tag{4.1}$$

with B_0 positive.

The problem became sharper when it was rigorously proved that the *wrong* naive result was a necessary consequence of the existence of a only one equilibrium state, i.e. of the assumption that connected correlations functions are zero (in an infinitesimal magnetic field) and therefore

$$\langle \sigma_i \sigma_k \rangle \approx \langle \sigma_i \rangle \langle \sigma_k \rangle. \tag{4.2}$$

In order to develop a new formalism that overcomes the pitfalls of the naive replica approach, it is necessary to discuss the physical meaning of these results and to consider the situation where many states are present. The first step consists in introducing the concept of *pure states in a finite volume* [2,6]. This concept is crystal clear from a physical point of view, however it is difficult to state it in a rigorous way. We need to work at finite, but large volumes, because the infinite volume limit for local observables is somewhat tricky and we have a chaotic dependence on the size [4,6].

At this end we consider a system with a total of N spins. We partition the equilibrium configuration space in regions, labeled by α, and each region must carry a finite weight. We define averages restricted to these regions [6–8]: these regions will correspond to our finite volume pure states or phases. It is clear that in order to produce something useful we have to impose sensible constraints on the form of these partitions. We require that the restricted averages on these regions of phase space are such that connected correlation functions are small when the points are different in an infinite range model. This condition is equivalent to the statement that the fluctuation of intensive quantities[7] vanishes in the infinite volume limit inside a given phase.

The phase decomposition depends on the temperature. In a ferromagnet at low temperature two regions are defined by considering the sign of the total magnetization. One region includes configurations with a positive total magnetization, the second selects negative total magnetization. There are ambiguities for those configurations that have exactly zero total magnetization, but the probability that these configurations are present at equilibrium is exponentially small at low temperature.

In order to present an interpretation of the results we assume that such decomposition exists also each instance of our problem. Therefore the *finite* volume Boltzmann-Gibbs measure can be decomposed in a sum of such finite volume pure states and inside each state the connect correlations are small. The states of the system are labeled by α: we can write

$$\langle \cdot \rangle = \sum_\alpha w_\alpha \langle \cdot \rangle_\alpha, \tag{4.3}$$

---

[7]Intensive quantities are defined in general as $b = \frac{1}{N} \sum_{i=1}^{N} B_i$, where the functions B_i depend only on the value of σ_i or from the value of the nearby spins.

with the normalization condition

$$\sum_{\alpha} w_{\alpha} = 1 - \epsilon, \tag{4.4}$$

where $\epsilon \to 0$ when $N \to \infty$.

The function $P_J(q)$ for a particular sample is given by

$$P_J(q) = \sum_{\alpha,\beta} w_{\alpha} w_{\beta} \delta(q_{\alpha,\beta} - q), \tag{4.5}$$

where $q_{\alpha,\gamma}$ is the overlap among two generic configurations in the states α and γ:

$$q_{\alpha,\gamma} = \frac{\sum_{l=1,N} m_{\alpha}(i) m_{\gamma}(i)}{N}, \tag{4.6}$$

and $m_{\alpha}(i)$ and $m_{\gamma}(i)$ are respectively the magnetizations within the state α and γ.

For example, if we consider two copies (i.e. σ and τ) of the same system, we have;

$$q[\sigma, \tau] = \frac{\sum_{l=1,N} \sigma(i) \tau(i)}{N} \tag{4.7}$$

and

$$\langle g(q[\sigma, \tau]) \rangle_J = \int g(q) P_J(q) dq, \tag{4.8}$$

where $g(q)$ is a generic function of q.

Given two spin configurations (σ and τ) we can introduce the natural concept of distance by

$$d^2(\sigma, \tau) \equiv \frac{1}{2N} \sum_{i=1}^{N} (\sigma_i - \tau_i)^2, \tag{4.9}$$

that belongs to the interval [0–2], and is zero only if the two configurations are equal (it is one if they are uncorrelated). In the thermodynamical limit, i.e. for $N \to \infty$, the distance of two configurations is zero if the number of different spins remains finite. The percentage of different σ's, not the absolute number, is relevant in this definition of the distance. It is also important to notice that at a given temperature β^{-1}, when N goes to infinity, the number of configurations inside a state is extremely large: it is proportional to $\exp(NS(\beta))$, where $S(\beta)$ is the entropy density of the system).

In a similar way we can introduce the distance between two states defined as the distance among two generic configurations of the states:

$$d^2_{\alpha,\gamma} = 1 - q_{\alpha,\gamma} \qquad (4.10)$$

In the case where the self-overlap

$$q_{\alpha,\alpha} = \frac{\sum_{1=1,N} m_\alpha(i)^2}{N}, \qquad (4.11)$$

does not depend on α and it is denoted by q_{EA}, we have that the distance of a state with itself is not zero:

$$d^2_{\alpha,\alpha} = 1 - q_{EA} \qquad (4.12)$$

We expect that finite volume pure states will enjoy the following properties that hopefully characterizes the finite volume pure states [6,7]:
• When N is large each state includes an exponentially large number of configurations.[8]
• The distance of two different generic configurations C_α and C_γ (the first belonging to state α and the second to state γ) does not depend on the C_α and C_γ, but only on α and β. The distance $d_{\alpha,\gamma}$ among the states α and γ, is the distance among two generic configurations in these two states. The reader has already noticed that with this definition the distance of a state with itself is not zero. If we want we can define an alternative distance:

$$D_{\alpha,\gamma} \equiv d_{\alpha,\gamma} - \frac{1}{2}\left(d_{\alpha,\alpha} + d_{\gamma,\gamma}\right), \qquad (4.13)$$

in such a way that the distance of a state with itself is zero ($D_{\alpha,\alpha} = 0$).
• The distance between two configurations belonging to the same state α is strictly smaller than the distance between one configuration belonging to state α and a second configuration belonging to a different state γ. This last property can be written as

$$d_{\alpha,\alpha} < d_{\alpha,\gamma} \text{ if } \alpha \neq \gamma. \qquad (4.14)$$

This property forbids to have different states such that $D_{\alpha,\gamma} = 0$, and it is crucial in avoiding the possibility of doing a too fine classification.[9]

---

[8] We warn the reader that in the case of a glassy system it may be not possible to consider $N \to \infty$ limit of a given finite volume pure state: there could be no one to one correspondence among the states at N and those at $2N$ due to the chaotic dependence of the statistical expectation values with the size of the system [4].

[9] For example if in a ferromagnet (in witless mood) we would classify the configurations into two states that we denote by e and o, depending if the total number of positive spins is even or odd, we would have that $d_{e,e} = d_{e,o} = d_{o,o}$.

• The classification into states is the finest one that satisfies the three former properties.

The first three conditions forbid a too fine classification, while the last condition forbids a too coarse classification.

For a given class of systems the classification into states depends on the temperature of the system. In some case it can be rigorously proved that the classification into states is possible and unique [30, 31] (in these cases all the procedures we will discuss lead to the same result). In usual situations in Statistical Mechanics the classification in phases is not very rich. For usual materials, in the generic case, there is only one state. In slightly more interesting cases there may be two states. For example, if we consider the configurations of a large number of water molecules at zero degrees, we can classify them as water or ice: here there are two states. In slightly more complex cases, if we tune carefully a few external parameters like the pressure or the magnetic field, we may have coexistence of three or more phases (a tricritical or multicritical point).

In all these cases the classification is simple and the number of states is small. On the contrary in the mean field approach to glassy systems the number of states is very large (it goes to infinity with N), and a very interesting nested classification of states is possible. We note "en passant" that this behavior implies that the Gibbs rule[10] is not valid for spin glasses.

We have already seen the only way out from the inconsistent result of the naive approach is to assume the existence of many equilibrium states. A more precise analysis will show that we need not only a few equilibrium states but an infinite number and which are the properties of these states.

4.2. The description of a many states system

The next step consists in describing a system with many equilibrium state and to find out which is the most appropriate way to code the relevant information,

Let us assume that a system has many states. In equation

$$\langle \cdot \rangle = \sum_\alpha w_\alpha \langle \cdot \rangle_\alpha, \tag{4.15}$$

there is a large number of w's that are non-zero.

How are we are supposed to describe such a system? First of all we should give the list of the w_α. In practice it is more convenient to introduce the free energy of a state F_α defined by

$$w_\alpha \propto \exp(-\beta F_\alpha), \tag{4.16}$$

---

[10]The Gibbs rule states that in order to have coexistence of K phases (K-critical point), we must tune K parameters. In spin glasses no parameters are tuned and the number of coexisting phases is infinite!

where the proportionality constant is fixed by the condition of normalization of the w's, i.e. eq. (4.4). Of course the free energies are fixed modulo an overall addictive constant,

The second problem it to describe the states themselves. In this context we are particularly interested in describing the relations among the different states. At this end we can introduce different kinds of overlaps:

- The spin overlap, which we have already seen, i.e.

$$q_{\alpha,\gamma} = N^{-1} \sum_i \langle \sigma_i \rangle_\alpha \langle \sigma_i \rangle_\gamma. \tag{4.17}$$

- The link (or energy) overlap, which is defined as:

$$q_{\alpha,\gamma}^L = N^{-1} \sum_{i,k} J_{i,k} \langle \sigma_i \sigma_k \rangle_\alpha J_{i,k} \langle \sigma_i \sigma_k \rangle_\gamma. \tag{4.18}$$

- We could introduce also more fancy overlaps like

$$q_{\alpha,\gamma}^{(l)} = N^{-1} \sum_{i,k} J_{i,k}^{(l)} \langle \sigma_i \sigma_k \rangle_\alpha J_{i,k}^{(l)} \langle \sigma_i \sigma_k \rangle_\gamma, \tag{4.19}$$

where $J^{(l)}$ denotes the l-th power of the matrix J.
- More generally, in the same way as the distances we can introduce overlaps that depend on the local expectation value of an operator $O(i)$

$$q_{\alpha,\gamma}^{(O)} = N^{-1} \sum_i \langle O(i) \rangle_\alpha \langle O(i) \rangle_\gamma, \tag{4.20}$$

In principle the description of the system should be given by the list of all F_α and by the list of all possible overlaps a between all the pairs α, γ. However it may happens, as it can be argued to happen in the SK model, that all overlap are function of the same overlap q. For example in the SK model we have

$$q^L = q^2. \tag{4.21}$$

More complex formulae holds for the other overlaps.

In the case in short range model we may expect that

$$q^L = l(q) \tag{4.22}$$

where $l(q)$ is a model (and temperature) dependent function [32,33].

If this *reductio ad unum* happens, we say that we have *overlap equivalence* and this is a property that has far reaching consequences. The simplest way to state this properties is the following. We consider a system composed by two replicas (σ and τ) and let us denote by $\langle \cdot \rangle_q$ the usual expectation Boltzmann-Gibbs expectation value with the constraint that

$$q[\sigma, \tau] = q. \tag{4.23}$$

Overlap equivalence states that in this restricted ensemble of the connected correlation functions of the overlap (and of the generalized overlaps) are negligible in the infinite range case (they would decay with the distance in a short range model).

There is no *a priori* compulsory need for assuming overlap equivalence; however it is by far the simplest situation and before considering other more complex cases it is better to explore this simple scenario, that it is likely to be valid at least in the SK model.

If overlap equivalence holds, in order to have a macroscopic description of the system, we must know only the list of all F_α and the list of al $q_{\alpha,\gamma}$. As we have already mentioned, we can also assume that for large N $q_{\alpha,\alpha}$ does not depend on α and it is equal to q_{EA} (this property is a consequence that the states cannot be distinguished one form the other using intrinsic properties). At the end the high level description of the system is given by [7, 19]:

$$\mathcal{D} = \left\{ q_{EA}, \{F_\alpha\}, \{q_{\alpha,\gamma}\} \right\}, \tag{4.24}$$

where the indices α and γ goes from 1 to $\Omega(N)$, where $\Omega(N)$ is a function that goes to infinity with N. The precise way in which $\Omega(N)$ increases with N is irrelevant because all high free energy states give a very small contribution to statistical sums.

In a random system the descriptor $\mathcal{D}$ is likely to depend on the system,[11] therefore we cannot hope to compute it analytically. The only thing that we can do is to compute the probability distribution of the descriptor $\mathcal{P}(D)$. Some times one refer to this task as high level statistical mechanics [7] because one has to compute the probability distribution of the states of the system an not of the configurations, as it is done in the old *bona fide* statistical mechanics (as it was done in the low level statistical mechanics).

4.3. A variational principle

Low level and high level statistical mechanics are clearly intertwined, as it can be see from the following considerations, where one arrives to computing the free energy density of the SK model using a variational principle.

---

[11] In a glassy non-random system it is quite possible that there is a strong dependence of $\mathcal{D}$ on the size N of the system.

Here we sketch only the main argument (the reader should read the original papers [18, 19]). Here for simplicity we neglect the terms coming for the necessity of renormalize the J when we change N; these terms can be added at the price of making the formalism a little heavier.

Let us consider an SK system with N particles with a descriptor $\mathcal{D}_N$. If we add a a new site we obtain a descriptor $\mathcal{D}_{N+1}$ where the overlaps do not change and the new free energies are given by

$$F_\alpha(N+1) = F_\alpha(N) + \Delta F_\alpha, \qquad \beta \Delta F_\alpha = -\ln\big(\cosh(\beta h_\alpha)\big) \tag{4.25}$$

where

$$h_\alpha = \sum_i J_{0,i} m_\alpha(i). \tag{4.26}$$

Also in the case where the h_α are Gaussian random uncorrelated, the fact that the new free energy depends on h_α implies that at fixed value of the free energy $F_\alpha(N+1)$ the corresponding value of the old free energy $F_\alpha(N)$ is correlated with h_α. It is a problem of conditioned probability. If we look at the distribution of h_α, conditioned at the value of the old free energy $F_\alpha(N)$, we find a Gaussian distribution, but if we look to the same quantity, conditioned at the value of $F_\alpha(N+1)$, the distribution is no more Gaussian. Therefore if we extract a state with a probability proportional to $w_\alpha \propto \exp(-\beta F_\alpha(N+1))$ the distribution of h_α will be not Gaussian.

Now it is clear that the h_α (conditioned to the old values of the free energy) are random Gaussian correlated variables because

$$\overline{h_\alpha h_\gamma} = q_{\alpha,\gamma}. \tag{4.27}$$

This correlation induces correlations among the ΔF_α.

Let us suppose for simplicity that we have succeeded in finding a choice of $\mathcal{P}(\mathcal{D})$ that is self-reproducing, i.e. a probability distribution such that

$$\mathcal{P}_{N+1}(\mathcal{D}) = \mathcal{P}_N(\mathcal{D}). \tag{4.28}$$

We have neglect an overall shift $\Delta F(\mathcal{P})$ of the free energies, that obviously depend on $\mathcal{P}$. The self-reproducing property is non-trivial as we shall see later. Notice than in going from N to $N+1$ the q's do not change, so the crucial point are the correlations among the q's and the F's that may change going form N to $N+1$.

A very interesting variational principle [19] tell us that the true free energy density f is given by

$$f = \max_{\mathcal{P}} \Delta F(\mathcal{P}), \tag{4.29}$$

which automatically implies that

$$f \geq \Delta F(\mathcal{P}) \quad \forall \mathcal{P}. \tag{4.30}$$

The reader should notice that there are no typos in the previous formulae. The free energy is the maximum, and not the minimum, with respect to all possible probabilities of the descriptor. It is difficult to explain why these inequalities are reversed with respect to the usual ones.

We could mumble that in random system we have to do the average over the logarithm of the free energy and the logarithm has the opposite convexity of the exponential. This change of sign is very natural also using the replica approach, but it is not trivial to give a compulsory physical explanation. The only suggestion we can give is *follow the details of the proof.*

The mathematical proof is not complex, there is a simplest form of this inequality that has been obtained by an interpolating procedure by Guerra [18] that is really beautiful and quite simple (it involves only integration by part and the notion that a square in non-negative).

We must add a remark: although the theorem tells us that the correct value of the free energy is obtained for a precise choice of the probability $\mathcal{P}^*(\mathcal{D})$ that maximize eq. (4.29), there is no warranty that the actual form of the probability of the descriptors is given by $\mathcal{P}^*(\mathcal{D})$. Indeed the theorem itself does not imply that the concept of descriptor is mathematical sound; all the heuristic interpretations of the descriptor and of the meaning of the theorem is done at risk of the physical reader.

While the lower bound is easy to use, for particular cases it coincides with Guerra's one [18], it is particularly difficult to find in this formalism that a $\mathcal{P}(\mathcal{D})$ such that it is the actual maximum. The only thing that we are able to do is to prove by other means [20] which is the value of f and to find out a $\mathcal{P}^*(\mathcal{D})$ that it saturate the bounds [2]. It uniqueness (if restricted to the class of functions that satisfy eq. (4.28)), is far from being established. It is clear that we miss some crucial ingredient in order to arrive to a fully satisfactory situation.

5. The explicit solution of the Sherrington Kirkpatrick model

5.1. Stochastic stability

The property of stochastic stability was introduced in a different context [15–17], however it is interesting to discuss them in this perspective. Stochastic stability stability is crucial because it strongly constrains the space of all possible descriptors and it is the hidden responsible of most of the unexpected cancellations or simplifications that are present in the computation of the explicit solution of the Sherrington Kirkpatrick model.

Let us sketch the definition of stochastic stability in this framework. Let us consider a distribution $\mathcal{P}(\mathcal{D})$ and let us consider a new distribution $\mathcal{P}'(\mathcal{D})$ where

$$F'_\alpha = F_\alpha + r_\alpha. \tag{5.1}$$

Here the r's are random variables such that the correlation among r_α and r_γ is a function of only $q_{\alpha,\gamma}$. We say that the distribution is stochastically stable iff the new distribution is equal to the previous one, after averaging over the r's, apart from an overall shift or the F's, i.e iff

$$\mathcal{P}(\mathcal{D}) = \mathcal{P}'(\mathcal{D}). \tag{5.2}$$

It is evident that, if a distribution is stochastically stable, it is automatically self-reproducing because the the shifts in free energy can be considered random correlated variables. This strongly suggest to consider only stochastic stable $cP(\mathcal{D})$. Moreover general arguments, that where at the origin of the first version of stochastic stability [15–17] imply that $\mathcal{P}(\mathcal{D})$ should be stochastically stable.

More precisely the original version of stochastic stability implies the existence of a large set of sum rules, which can be derived in this context if we assume that the equation (5.2) is valid. These sum rules give relations among the joint average of the overlap of more *real replicas* (i.e. a finite number of copies of the system in the same realization of the quenched disorder). One can show [34] that under very general assumptions of continuity that also valid in finite dimensional models for the appropriate quantities.

These relations connects the expectation values of products of various functions $P_J(q)$ to the expectation value of the function $P_J(q)$ itself.

An example of these relations is the following

$$P(q) = \overline{P_J(q)},$$

$$P(q_{1,2}, q_{3,4}) \equiv \overline{P_J(q_{1,2})P_J(q_{3,4})}$$
$$= \frac{2}{3}P(q_{1,2})P(q_{3,4}) + \frac{1}{3}P(q_{1,2})\delta(q_{1,2} - q_{3,4}). \tag{5.3}$$

where we have denoted by an bar the average over the J's.

This relation is difficult to test numerically, as far as in finite volume systems we do not have exact delta functions, therefore very often one test simplified versions of these equations. For example let us consider four replicas of the system (all with the same Hamiltonian) and we denote by

$$E(\ldots) \equiv \overline{\langle\ldots\rangle} \tag{5.4}$$

the global average, taken both over the thermal noise and over the quenched disorder. The previous equation implies that

$$E\left(q_{1,2}^2 q_{3,4}^2\right) = \frac{2}{3}E\left(q_{1,2}^2\right)^2 + \frac{1}{3}E\left(q_{1,2}^4\right),\tag{5.5}$$

This relation is very interesting: it can be rewritten (after some algebra), in a slightly different notation as

$$\overline{\left(\langle A\rangle_J - \overline{\langle A\rangle_J}\right)^2} = \frac{1}{3}\overline{\langle\left(A - \langle A\rangle_J\right)^2\rangle_J},\tag{5.6}$$

with $A = q^2$. This relation implies that the size of sample to sample fluctuations of an observable (l.h.s.) is quantitatively related to the average of the fluctuations of the same observable inside a given sample (r.h.s.). This relation is very interesting because it is not satisfied in most of the non-stochastically stable models and it is very well satisfied in three dimensional models [6].

An other relation of the same kind is:

$$E\left(q_{1,2}^2 q_{2,3}^2\right) = \frac{1}{2}E\left(q_{1,2}^2\right)^2 + \frac{1}{2}E\left(q_{1,2}^4\right).\tag{5.7}$$

One could write an infinite set of similar, but more complex sum rules.

Stochastic stability is an highly non-trivial requirement (notice that the union of two stochastically stable systems is not stochastically stable).

Therefore it is interesting to present here the simplest non trivial case of stochastically stable system, where all the overlaps are equal

$$q_{\alpha,\gamma} = q_0 < q_{EA}, \quad \forall\,\alpha,\gamma\;.\tag{5.8}$$

In this case (which is usually called one-step replica-symmetry breaking) we have only to specify the the weight of each state that (as usually) is given by

$$w_\alpha \propto \exp(-\beta F_\alpha).\tag{5.9}$$

Let us consider the case where the joint probability of the F's is such that the F' are random independent variable: the probability of finding an F in the interval $[F; F + dF]$ is given by

$$\rho(F)dF.\tag{5.10}$$

The total number of states is infinite and therefore the function $\rho(F)$ has a divergent integral. In this case it is easy to prove that stochastic stability implies that the function ρ must be of the form

$$\rho(F) \propto \exp(\beta m F).\tag{5.11}$$

At this end let us consider the effect of a perturbation of strength ϵ on the free energy of a state, say α. The unperturbed value of the free energy is denoted by F_α. The new value of the free energy G_α is given by $G_\alpha = F_\alpha + \epsilon r_\alpha$ where r_α are identically distributed uncorrelated random numbers. Stochastic stability implies that the distribution $\rho(G)$ is the same as $\rho(F)$. Expanding to second order in ϵ we see that this condition leads to

$$d\rho/dF \propto d^2\rho/dF^2. \tag{5.12}$$

The only physical solution is given by eq. (5.11) with an appropriate value of m. The parameter m must satisfy the condition $m < 1$ if the sum

$$\sum_\alpha \exp(-\beta F_\alpha) \tag{5.13}$$

is convergent.

We have seen that stochastic stability fixes the form of the function ρ and therefore connects in an inextricable way the low and the high free energy part of the function ρ.

In this case an explicit computation show that the function $P(q)$ is given by

$$P(q) = m\delta(q - q_0) + (1 - m)\delta(q - q_{EA}). \tag{5.14}$$

It is interesting that the same parameter m enters both in the form of the function $\rho(F)$ at large values of F and in the form of the function $P(q)$ which is dominated by the lowest values of F, i.e. those producing the largest w's. This result is deeply related to the existence of only one family of stochastic stable systems with uncorrelated F's.

This relation is interesting from the physical point of view, because one could argue that the off-equilibrium dynamics depends on the behavior of the function $\rho(\Delta F)$ at very large argument, and in principle it could be not related to the static property that depend on the function ρ for small values of the argument. However stochastic stability forces the function $\rho(\Delta F)$ to be of the form (5.11), in all the range of ΔF. Consequently the high F and low F behavior are physically entwined [13].

Stochastic stable distribution have remarkable properties, an example of it can be found in the appendix I, taken from [24].

Stochastic stability in finite dimensional system has many interesting applications. For example it is possible, by analyzing the properties of a single large system, to reconstruct the properties of the whole $\mathcal{P}(\mathcal{D})$ averaged over the ensemble of systems to which the individual system naturally belong [38,39]: in a similar way the fluctuation dissipation relations tells us that we can reconstruct the $P(q)$ function from the analysis of the response and the correlations during aging [13].

5.2. The first non trivial approximation

We start by presenting here the first non trivial approximation, i.e. the so called one step replica symmetry breaking. Let us assume that the probability distribution of the descriptor is given by Eqs. (5.8) and (5.11). We need to compute both the shift in free energy and joint probability distribution of the effective field and of the free energy.

The computation goes as follows. We suppose that in the system with N spins we have a population of states whose total free energies F_α are distributed (when F_α is not far from a given reference value F^* that for lighten the notation we take equal to zero) as

$$\mathcal{N}_N(F_N) \propto \exp(\beta m F_N). \tag{5.15}$$

When we add the new spin, we will find a value of the field h that depends on the state α. We can now consider the joint probability distribution of the new free energy and of the magnetic field. We obtain

$$\mathcal{N}_{N+1}(F, h) = \int dh\, P_{q_{EA}}(h) \int dF_N \mathcal{N}_N(F_N) \delta\big(F - F_N - \Delta F(h)\big), \tag{5.16}$$

where $P_{q_{EA}}(h)$ is the probability distribution of the effective magnetic field produced on the spin at 0, for a generic state and it is still given by $d\mu_{q_{EA}}(h)$. It is crucial to take into account that the new free energy will differs from the old free energy by an energy shift that is h dependent. If we integrate over F_N and we use the explicit exponential form for $\mathcal{N}_N(F_N)$ we find that

$$\mathcal{N}_{N+1}(F, h) \propto \exp(\beta m F) \int d\mu_{q_{EA}}(h) \exp\big(\beta m \Delta F(h)\big)$$
$$\propto \exp(\beta m F) P_{N+1}(h). \tag{5.17}$$

The probability distribution of the field at *fixed* value of the new free energy is given by

$$P_{N+1}(h) \propto \mu_{q_{EA}}(h) \exp\big(\beta m \Delta F(h)\big) = \mu_{q_{EA}}(h) \cosh^m(\beta h). \tag{5.18}$$

It is clear that $P_{N+1}(h)$ different from $\mu_{q_{EA}}(h)$ as soon as $m \neq 0$. In this way we find the consistency equation of the replica approach for q_{EA}:

$$q_{EA} = \frac{\int dh\, \mu_{q_{EA}}(h) \cosh^m(\beta h) \tanh(\beta h)^2}{\int dh\, \mu_{q_{EA}}(h) \cosh^m(\beta h)}. \tag{5.19}$$

If we put everything together taking care of also the other terms coming for the renormalization of the $j's$ we find

$$\beta f(q_1, m) = -\beta^2(1 - (1 - m)q_1^2) + m^{-1} \ln\left(\int d\mu_{q_1}(h) \cosh^m(\beta h)\right), \quad (5.20)$$

where we have used the notation q_1 at the place of of q_{EA}.

This result is obtained in the case where $q_0 = 0$. Otherwise we have

$$\beta f(q_0, q_1, m) = \frac{1}{4}\beta^2(1 - (1 - m)q_1^2 - mq_0^2)$$
$$+ m^{-1}\int d\mu_{q_0}h_0 \ln\left(\int d\mu_{q_1-q_0}(h_1) \cosh^m\big(\beta(h_0 + h_1)\big)\right). \quad (5.21)$$

The value of the parameter and of the free energy can be found by solving the equations

$$\frac{\partial f}{\partial q_0} = \frac{\partial f}{\partial q_1} = \frac{\partial f}{\partial m}. \quad (5.22)$$

A few comments are in order:
• The probability distribution of h at fixed value of the free energy of the N spins system ($P_s(h)$) is *not* the probability distribution of h at fixed value of the free energy of the $N + 1$ spins system $P_{N+1}(h)$: the two free energies differs by an h dependent addictive factor and they do not have a flat distribution (as soon as $m \neq 0$). The probability distribution of h at a fixed value of the free energy of the N spins system is Gaussian, but the probability distribution of h at fixed value of the free energy of the $N + 1$ spins system is not a Gaussian.
• Only in the case were $\mathcal{N}_N(F_N)$ is an exponential distribution, $\mathcal{N}_{N+1}(F, h)$ factorizes into the product of an F and an h dependent factor and the $\mathcal{N}_{N+1}(F)$ has the same form of $\mathcal{N}_N(F)$. Self-consistency can be reached only in the case of an exponential distribution for $\mathcal{N}_N(F_N)$. This fact is a crucial consequence of stochastic stability.
• The self consistency equations for the q do not fix the value of m. This is natural because m is an independent variational parameter and we must maximize over q_0, q_1 and m. If one looks to the distribution of the metastable of having a free energy density higher that the ground state (problem that we cannot discuss here for reason of space [35,36]) one finds that the distribution of the free energies inside the metastable states is characterized by a different value of m than the ground state. In this different context the equation $\partial f/\partial m = 0$ determines which is the free energy of the metastable states that are true equilibrium states.

If one analyze the solution of these equation one finds that the entropy is still negative at low temperature, but only by of a tiny amount and that the value of B_0 in eq. (4.1) is smaller of a factor 9 with respect to the previous computation. We are on the write track, but we have not reached our goal.

5.3. Ultrametricity

Let us consider two generic spin configuration at equilibrium and let us use the label c to denote it and the label α to denote the state to which they belong.

Their overlap satisfies the following properties (usually called one step replica symmetry breaking):
- If $c = c'$ the overlap is equal to 1,
- If $c \neq c'$, $\alpha = \alpha'$ the overlap is equal to q_1,
- If $c \neq c'$, $\alpha \neq \alpha'$ the overlap is equal to q_0.

We see that we have the beginning of an hierarchical scheme. It is clear that the simplest generalization will assume that states are grouped into clusters, that are labeled by γ and the overlaps are given by
- If $c = c'$ the overlap is equal to 1,
- If $c \neq c'$, $\alpha = \alpha'$ the overlap is equal to q_2,
- If $c \neq c'$, $\alpha \neq \alpha'$ and $\gamma = \gamma'$ the overlap is equal to q_1.
- If $c \neq c'$, $\alpha \neq \alpha'$ and $\gamma \neq \gamma'$ the overlap is equal to q_0.

This case is called two step replica symmetry breaking.

In order to specify the descriptor of the system we have to define the probability distribution of the weight w's or equivalently of the free energy F's.

The following ansatz is the only one compatible with stochastic stability [37]

$$F_{\alpha,\gamma} = F_\alpha + F_\gamma,$$
$$\rho(F_\alpha) = \exp(\beta m_2 F_\alpha) \quad \rho(F_\gamma) = \exp(\beta m_1 F_\gamma), \tag{5.23}$$

where we have

$$m_1 < m_2. \tag{5.24}$$

This condition implies that the number of states increases faster that number of clusters.[12]

If we compute the function $P(q)$ it turns out to be given

$$m_1 \delta(q - q_0) + (m_2 - m_1)\delta(q - q_1) + (1 - m_1)\delta(q - q_2) \tag{5.25}$$

---

[12]Sometimes one may consider the case where this condition is violated: in this case one speaks of an inverted tree. The physical meaning of this inverted tree is clarified in [41]; an inverted tree for metastable states is discussed in [40], however the physical implications are less clear. It seems to me that there are some hidden properties that we do not fully understand.

The free energy as a more complex expression, that can be written as a

$$\beta f(q_0, q_1, q_2, m_1, m_2)$$
$$= \frac{1}{4}\beta^2\left(1 - (1 - m_2)q_2^2 + (m_2 - m_1)q_1^2 + m_1 q_0^2\right)$$
$$+ m_1^{-1}\int d\mu_{q_0}(h_0)\ln\left(\int d\mu_{q_1-q_0}(h_1)I(h_0, h_1)^{m_2/m_1}\right),$$
$$I(h_0, h_1) = \int d\mu_{q_2-q_1}(h_2)\cosh^{m_2}\left(\beta(h_0 + h_1 + h_2)\right). \tag{5.26}$$

At the end progress has been done, the value of A_0 in equation (4.1) is much smaller and the zero temperature entropy is nearly equal to zero (e.g. $S(0) = -.003$). The results are nearly consistent but not completely consistent. Moreover the zero temperature energy seems to be convergent: it was $-.798$ in the naive approach, $-.765$ in the one step approach and $-.7636$ in the two step approach (the correct value is $-.7633$).

It is clear that we can go on introducing more steps of replica symmetry breaking: at the third steps we will to assume that clusters are grouped into families, that are labeled by δ and the overlaps are given by

- If $c = c'$ the overlap is equal to 1,
- If $c \neq c'$, $\alpha = \alpha'$ the overlap is equal to q_3,
- If $c \neq c'$, $\alpha \neq \alpha'$ and $\gamma = \gamma'$ the overlap is equal to q_2.
- If $c \neq c'$, $\alpha \neq \alpha'$, $\gamma \neq \gamma'$ and $\delta = \delta'$ the overlap is equal to q_1.
- If $c \neq c'$, $\alpha \neq \alpha'$, $\gamma =\neq \gamma'$ and $\delta \neq \delta'$ the overlap is equal to q_0.

In the same way as before stochastic stability implies that

$$F_{\alpha,\gamma,\delta} = F_\alpha + F_\gamma + F_\delta,$$
$$\rho(F_\alpha) = \exp(\beta m_3 F_\alpha),$$
$$\rho(F_\gamma) = \exp(\beta m_2 F_\gamma),$$
$$\rho(F_\delta) = \exp(\beta m_1 F_\delta), \tag{5.27}$$

where we have

$$m_1 < m_2 < m_3. \tag{5.28}$$

In this case the function $P(q)$ is given by

$$m_1\delta(q - q_0) + (m_2 - m_1)\delta(q - q_1)$$
$$+ (m_3 - m_2)\delta(q - q_2) + (1 - m_3)\delta(q - q_3) \tag{5.29}$$

G. Parisi

This hierarchical construction of the probability distribution of the descriptor is called ultrametric, because if we use the distance defined in eq. (4.10) the following inequality holds [2]:

$$d_{\alpha,\beta} < \max(d_{\alpha,\gamma}, d_{\gamma,\beta}) \quad \forall \gamma \tag{5.30}$$

If the the ultrametricity is satisfied, we can associate to the set of states a tree, formed by a root, by the nodes of the the tree and by the leaves are the states. States branching from the same node belongs to the same cluster. In the two step breaking all the nodes are the clusters. In the three step breaking the nodes are both the clusters and the families. In order from above we have the root, the families, the clusters and the states. In order to fix the probability distribution of the descriptor, we have to give the self distance of the states, the distance of the states belonging to the same clusters, the distance of the states belonging to different clusters, but to the same family, and the distance among the states belonging to different families [2, 42, 43]. As we have already remarked the probability distribution of the leaves is fixed by stochastic stability,

At the present moment it is not rigorously know which of the following two alternatives are correct:

• Ultrametricity is a necessary condition, i.e. a consequence of overlap equivalence and of stochastic stability,

• Ultrametricity is an independent hypothesis.

However there are strong arguments that suggest that the first statement is true [44] (although a rigorous proof is missing) and personally I am convinced that this is the case.

The question if overlap equivalence (and consequently ultrametricity) is a consequence of stochastic stability it is a more difficult question. Opposite arguments can be presented:

• The fact that there is no known example of a stochastically stable system where we can define two non-equivalent overlaps, can be considered a strong suggestion of the non-existence of such systems.

• The construction of non-trivial stochastically stable system is rather difficult; in some sense it was done by chance using the algebraic replica method. It is likely that the construction of stochastically stable systems with two non-equivalent overlaps will be rather complex (if it is non-void) and it may take a very strong effort to exhibit an explicit case.

It is clear that further research in this direction is needed: it would be very interesting to discover a stochastically stable systems with two non-equivalent overlaps.

Generally speaking stochastic stability may have unexpected consequences: for example for a stochastically stable system (in the original sense [15–17]), without doing any assumption on the decomposition into states, it is possible to prove [45] that the support of the function $P(q)$ is discrete, i.e., if N is large, we can find n very small intervals such that the integral of the function $P(q)$ outside these intervals goes to zero with n, at least as n^{-2}.

5.4. *Continuous Symmetry Breaking*

From the discussion of the previous section is completely evident that we can go on and introduce more and more levels of symmetry breaking. The problem is how to control the limit where the number of levels is going to infinity.

At this end we can introduce the function $q(x)$ on the interval $[0 : 1]$ that is given by

$$q(x) = q_l \quad \text{if} \quad m_l < x < m_{l+1}, \tag{5.31}$$

where the index l runs from zero to K and we have used the convention that $m_0 = 0$ and $m_{K+1} = 1$. This function is interesting for many reasons.

- In the case of K-step symmetry breaking the function $q(x)$ is piecewise constant and takes $K + 1$ values.
- The inverse function $x(q)$ is discontinuous (we assume that the q''s are increasing and we use the convention $x(0) = 0$). We have the very simple relation

$$P(q) = \frac{dx}{dq}. \tag{5.32}$$

- In the limit $K \to \infty$ the function $q(x)$ may become a continuous function and in this case we can speak of continuous replica symmetry breaking.
- The limit $K \to \infty$ is rather smooth for any reasonable observable.

From one side it is possible to construct in this way an infinite tree with branches at all the levels by a painful explicit construction. A priori it is not obvious that this construction has a well defined limit, however it is possible to check the all the properties of the tree can be computed and explicit formulae can be obtained [2, 42, 43, 46] for all the possible quantities. It is possible to define this tree in a rigorous directly in the continuous limit [47] by stating the properties of resulting measure.

It crucial that although branching may happens at any level, we can reduce ourself to consider trees with only a finite number of branches. Indeed the number of equilibrium states is diverging when the volume goes to infinity, but most of the states carry a very small weight (i.e. they are high free energy states). An explicit computation [43, 47] shows that the total weight of the states with weight less than ϵ goes to zero when ϵ goes to zero, therefore removing an infinite number of states, that carry negligible weight, we remain with a finite number of them.

G. Parisi

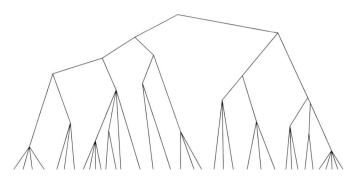

Fig. 1. An example of a taxonomic tree of states in the case of an hierarchical breaking of the replica symmetry.

In this construction one finds the the only free quantity is the function $P(q)$ itself; its form determines analytically the full $\mathcal{P}(\mathcal{D})$ [37].

The inequality (5.30) implies that the most relevant states form an ultrametric tree, i.e. they can be put on a tree in such a way that the states are on the branches of the tree and the distance among the states is the maximum level one has to cross for going from one states to an other. In other words the states of a system can be classified in a taxonomic way. The most studies possibilities are the following:

• All different states are at given distance d one from the other and the taxonomy is rather trivial (*one step replica symmetry breaking*).

• The states have a continuous distribution and branching points exist at any level (*continuous replica symmetry breaking*). A suggestive drawing of how the tree may look like, if we draw only the finite number of branches that have weight larger than ϵ, is shown in fig. (1). A careful analysis of the properties of the tree can be found in [31,43].

Other distributions are possible, but they are less common or at least less studied.

5.5. *The final result*

In order to compute the free energy is useful to write down a formula for its value that is valid also in the case where there is a finite number of steps. After some thinking, we can check that the following expression (5.33) is a compact way to write the previous formulae for the free energy in terms of the function $x(q)$ (or $q(x)$). The expression for the free energy is a functional (a functional of the function $q[x]$) and it is given by:

$$\beta f[q(x)] = \frac{\beta^2}{4}\left(1 - \int_0^1 q^2(x)\,dx\right) + f(0,0) \tag{5.33}$$

where the function $f(q, h)$ is defined in the strip $0 \le q \le q_{EA} \equiv q_{K+1}$ and it obeys the non linear equation:

$$\dot{f} = -\left(f'' + x(q)(f')^2\right), \tag{5.34}$$

where dots and primes mean respectively derivatives with respect to q and h. The initial condition is

$$f(q, y) = \cosh(\beta m_K h). \tag{5.35}$$

Now the variational principle tell us that the true free energy is greater than the maximum with respect to all function $q(x)$ of the previous functional. It is clear that a continuous function may be obtained as limit of a stepwise continuous function and this shows how to obtain the result for a continuous tree in a smooth way.

Talagrand [20] is using on a cleaver the generalization of the bounds by Guerra [18] to systems composed by a two identical copies: he finally arrives to write also an upper bound for the free energy and to prove that they coincide. In this way after 24 years the following formula for the free energy of the SK model was proved to be true:

$$f = \max_{q(x)} f\big[q(x)\big]. \tag{5.36}$$

We have already remarked that the mathematical proof does not imply that the probability distribution of the descriptors is the one predicted by the physical theory. The main ingredient missing is the proof that the ultrametricity properties is satisfied. This is the crucial point, because ultrametricity and stochastic stability (which is known to be valid) completely fix the distribution. It is likely that the needed step would be to transform the estimates of [48] into rigorous bounds, however this is a rather difficult task. Maybe we have to wait other 24 years for arriving to the rigorous conclusion that the heuristic approach gave the correct results, although I believe that there are no serious doubts on its full correctness.

6. Bethe lattices

6.1. An intermezzo on random graphs

It is convenient to recall here the main properties of random graphs [49].

There are many variants of random graphs: fixed local coordination number, Poisson distributed local coordination number, bipartite graphs... They have the same main topological structures in the limit where the number (N) of nodes goes to infinity.

We start by defining the random Poisson graph in the following way: given N nodes we consider the ensemble of *all* possible graphs with $M = \alpha N$ edges (or links). A random Poisson graph is a generic element of this ensemble.

The first quantity we can consider for a given graph is the local coordination number z_i, i.e. the number of nodes that are connected to the node i. The average coordination number z is the average over the graph of the z_i:

$$z = \frac{\sum_{i=1,N} z_i}{N}. \tag{6.1}$$

In this case it is evident that

$$z = 2\alpha. \tag{6.2}$$

It takes a little more work to show that in the thermodynamic limit ($N \to \infty$), the probability distribution of the local coordination number is a Poisson distribution with average z.

In a similar construction two random points i and k are connected with a probability that is equal to $z/(N-1)$. Here it is trivial to show that the probability distribution of the z_i is Poisson, with average z. The total number of links is just $zN/2$, apart from corrections proportional to $\sqrt{N}$. The two Poisson ensembles, i.e. fixed total number of links and fluctuating total number of links, cannot be distinguished locally for large N and most of the properties are the same.

Random lattices with fixed coordination number z can be easily defined; the ensemble is just given by all the graphs with $z_i = z$ and a random graph is just a generic element of this ensemble.

One of the most important facts about these graphs is that they are locally a tree, i.e. they are locally cycle-less. In other words, if we take a generic point i and we consider the subgraph composed by those points that are at a distance less than d on the graph,[13] this subgraph is a tree with probability one when N goes to infinity at fixed d. For finite N this probability is very near to 1 as soon as

$$\ln(N) > A(z)\, d, \tag{6.3}$$

$A(z)$ being an appropriate function. For large N this probability is given by $1 - O(1/N)$.

If $z > 1$ the nodes percolate and a finite fraction of the graph belongs to a single giant connected component. Cycles (or loops) do exist on this graph, but they have typically a length proportional to $\ln(N)$. Also the diameter of the graph, i.e. the maximum distance between two points of the same connected component

---

[13] The distance between two nodes i and k is the minimum number of links that we have to traverse in going from i to k.

is proportional to $\ln(N)$. The absence of small loops is crucial because we can study the problem locally on a tree and we have eventually to take care of the large loops (that cannot be seen locally) in a self-consistent way. i.e. as a boundary conditions at infinity.

Before applying the problem will be studied explicitly in the next section for the ferromagnetic Ising model.

6.2. The Bethe Approximation in $D = 2$

Random graphs are sometimes called Bethe lattices, because a spin model on such a graph can be solved exactly using the Bethe approximation. Let us recall the Bethe approximation for the two dimensional Ising model.

In the standard mean field approximation, one writes a variational principle assuming the all the spins are not correlated [5]; at the end of the computational one finds that the magnetization satisfies the well known equation

$$m = \tanh(\beta J z m), \tag{6.4}$$

where $z = 4$ on a square lattice ($z = 2d$ in d dimensions) and J is the spin coupling ($J > 0$ for a ferromagnetic model). This well studied equation predicts that the critical point (i.e. the point where the magnetization vanishes) is $\beta_c = 1/z$. This result is not very exiting in two dimensions (where $\beta_c \approx .44$) and it is very bad in one dimensions (where $\beta_c = \infty$). On the other end it becomes more and more correct when $d \to \infty$.

A better approximation can be obtained if we look to the system locally and we compute the magnetization of a given spin (σ) as function of the magnetization of the nearby spins (τ_i, $i = 1, 4$). If we assume that the spins τ are uncorrelated, but have magnetization m, we obtain that the magnetization of the spin σ (let us call it m_0) is given by:

$$m_0 = \sum_\tau P_m[\tau]\tanh\left(\beta J \sum_{i=1,4} \tau_1\right), \tag{6.5}$$

where

$$P_m[\tau] = \prod_{i=1,4} P_m(\tau_i), \qquad P_m(\tau) = \frac{1+m}{2}\delta_{\tau,1} + \frac{1-m}{2}\delta_{\tau,-1}. \tag{6.6}$$

The sum over all the 2^4 possible values of the τ can be easily done.

If we impose the self-consistent condition

$$m_0(m) = m, \tag{6.7}$$

we find an equation that enables us to compute the value of the magnetization m.

This approximation remains unnamed (as far as I know) because with a little more work we can get the better and simpler Bethe approximation. The drawback of the previous approximation is that the spins τ cannot be uncorrelated because they interact with the same spin σ: the effect of this correlation can be taken into account ant this leads to the Bethe approximation.

Let us consider the system where the spin σ has been removed. There is a cavity in the system and the spins τ are on the border of this cavity. We assume that in this situation these spins are uncorrelated and they have a magnetization m_C. When we add the spin σ, we find that the probability distribution of this spin is proportional to

$$\sum_\tau P_{m_C}[\tau] \exp\left(\beta J \sigma \sum_{i=1,4} \tau_i\right). \tag{6.8}$$

The magnetization of the spin σ can be computed and after some simple algebra we get

$$m = \tanh\{z \, \mathrm{atanh}[\tanh(\beta J)m_C]\}, \tag{6.9}$$

with $z = 4$.

This seems to be a minor progress because we do not know m_C. However we are very near the final result. We can remove one of the spin τ_i and form a larger cavity (two spins removed). If in the same vein we assume that the spins on the border of the cavity are uncorrelated and they have the same magnetization m_C, we obtain

$$m_C = \tanh\{(z-1)\mathrm{atanh}[\tanh(\beta J)m_C]\}. \tag{6.10}$$

Solving this last equation we can find the value of m_C and using the previous equation we can find the value of m.

It is rather satisfactory that in 1 dimensions ($z = 2$) the cavity equations become

$$m_C = \tanh(\beta J)m_C. \tag{6.11}$$

This equation for finite β has no non-zero solutions, as it should be.

The internal energy can be computed in a similar way: the energy density per link is given by

$$E_{link} = \frac{\tanh(\beta J) + m_C^2}{1 + \tanh(\beta J)m_C^2}. \tag{6.12}$$

We can obtain the free energy by integrating the internal energy as function of β.

In a more sophisticated treatment we write the free energy as function of m_C:

$$\frac{\beta F(m_C)}{N} = F_{site}(m_C) - \frac{z}{2} F_{link}(m_C), \qquad (6.13)$$

where $F_{link}(m_C)$ and $F_{site}(m_C)$ are appropriate functions [50]. This free energy is variational, in other words the equation

$$\frac{\partial F}{\partial m_C} = 0 \qquad (6.14)$$

coincides with the cavity equation (6.10).

6.3. Bethe lattices and replica symmetry breaking

It should be now clear why the Bethe approximation is correct for random lattices. If we remove a node of a random lattice, the nearby nodes (that were at distance 2 before) are now at a very large distance, i.e. $O(\ln(N))$. In this case we can write

$$\langle \tau_{i_1} \tau_{i_2} \rangle \approx m_{i_1} m_{i_2} \qquad (6.15)$$

and everything seems easy.

This is actually easy in the ferromagnetic case where in absence of magnetic field at low temperature the magnetization may take only two values ($\pm m$). In more complex cases, (e.g. antiferromagnets) there are many different possible values of the magnetization because there are many equilibrium states and everything become complex (as it should) because the cavity equations become equations for the probability distribution of the magnetizations [2].

I will derive the TAP cavity equations on a random Bethe lattice [24,26,28,51] where the average number of neighbors is equal to z and each point has a Poisson distribution of neighbors. In the limit where z goes to infinity we recover the SK model.

Let us consider a node i; we denote by ∂i the set of nodes that are connected to the point i. With this notation the Hamiltonian can be written

$$H = \frac{1}{2} \sum_i \sum_{k \in \partial i} J_{i,k} \sigma(i) \sigma(k). \qquad (6.16)$$

We suppose that for a given value of N the system is in a state labeled by α and we suppose such a state exists also for the system of $N - 1$ spins when the spin i is removed. Let us call $m(i)_\alpha$ the magnetization of the spin k and $m(k; i)_\alpha$ the magnetization of the spin k when the site i is removed. Two generic spins are, with probability one, far on the lattice: they are at an average distance of order $\ln(N)$; it is reasonable to assume that in a given state the correlations of two generic spins are small (a crucial characteristic of a state is the absence of infinite-range correlations). Therefore the probability distribution of two generic spins is factorized (apart from corrections vanishing in probability when N goes to infinity) and it can be written in terms of their magnetizations.

We have already seen that the usual strategy is to write the equation directly for the cavity magnetizations. We obtain (for $l \in \partial i$):

$$m(i; l) = \tanh\left(\sum_{k \in \partial i; k \neq l} \text{atanh}(\tanh(\beta J_{i,k}) m(k; i)) \right). \tag{6.17}$$

Following this strategy we remain with Nz equations for the Nz cavity magnetizations $m(i; k)$; the true magnetizations ($m(i)$) can be computed at the end using equation (6.17).

In the SK limit (i.e. $z \approx N$), it is convenient to take advantage of the fact that the J's are proportional to $N^{-1/2}$ and therefore the previous formulae may be partially linearized; using the the law of large numbers and the central limit theorem one recovers the previous result for the SK model.

It is the cavity method, where one connects the magnetizations for a system of N spins with the magnetizations for a system with $N - 1$ spins. In absence of the spin i, the spins $\sigma(k)$ ($k \in \partial i$) are independent from the couplings $J_{i,k}$.

We can now proceed exactly as in SK model. We face a new difficulty already at the replica symmetric level: the effective field

$$h_{\text{eff}} = \sum_{k \in \partial i; k \neq l} \text{atanh}\left(\tanh(\beta J_{i,k}) m(k; i) \right) \tag{6.18}$$

is the sum of a finite number of terms and it no more Gaussian also for Gaussian J's. This implies that the distribution probability of h_{eff} depends on the full probability distribution of the magnetizations and not only on its second moment. If one writes down the equations, one finds a functional equation for the probability distribution of the magnetizations ($P(m)$).

When replica symmetry is broken, also at one step level, in each site we have a probability distribution of the magnetization over the different states $Q(m)$ and the description of the probability distribution of the magnetization is now given by the probability distribution $P[Q(m)]$, that is a functional of the probabilities $Q(m)$ [24, 25, 51–53].

The computation becomes much more involved, but at least in the case of one step replica breaking (an sometimes at the two steps level) they can be carried up to the end. The one step approximation is quite interesting in this context, because there are optimization models in which it gives the exact results [14, 52].

A complete discussion of the stability of the one step replica approach can be found in [54].

7. Finite dimensions

7.1. General considerations

It is well known that mean field theory is correct only in the infinite dimensional limit. Usually in high dimensions it gives the correct results. Below the upper critical dimensions (that is 6 in the case of spin glasses) the critical exponents at the transition point change and they can be computed near to the upper critical dimensions using the renormalization group.

Sometimes the renormalization group is needed to study the behavior in the low temperature phase (e.g. in the $O(n)$ spin models), if long range correlations are present. A careful and complete analysis of spin glasses in the low temperature phase is still missing due to the extreme complexity of the corrections to the mean field approximation. Many very interesting results have been obtained [55], but the situation is still not clear. May be one needs a new approach, e.g starting to compute the corrections to mean field theory at zero temperature.

There are two different problems that we would like to understand much better:

• The value of the lower critical dimension.

• If the predictions of the mean field theory at least approximately are satisfied in three dimensions

These two problems will be addressed in the next two sections

7.2. The lower critical dimension

The calculation of free energy increase due to an interfaces is a well known method to compute the lower critical dimension in the case of spontaneous symmetry breaking. I will recall the basic points and I will try to apply it to spin glasses.

Generally speaking in the simplest case we can consider a system with two possible coexisting phases (A and B), with different values of the order parameter,

For standard ferromagnets we may have:

• A: spins up.

• B: spins down.

We will study what happens in a finite system in dimensions D of size $M^d L$ with $d = D - 1$.

We put the system in phase A at $z = 0$ and in phase B at $z = L$. The free energy of the interface is the increase in free energy due to this choice of boundary conditions with respect to choosing the same phase at $z = 0$ and $z = L$. In many cases we have that the free energy increase $\delta F(M, L)$ behaves for large M and L as:

$$\delta F(L, M) = M^d / L^{\omega}, \tag{7.1}$$

where ω is independent from the dimension.

There is a lower critical dimension where the free energy of the interface is finite:

$$D_c = \omega + 1, \tag{7.2}$$

and

$$\delta F(L, L) = L^{(D - D_c)}. \tag{7.3}$$

Heuristic arguments, which sometimes can be made rigorous, tell us that when $D = \omega + 1$, (the lowest critical dimension) the two phases mix in such a way that symmetry is restored.

In most cases the value of ω from mean field theory is the exact one and therefore we can calculate in this way the value of the lower critical dimension. The simplest examples are the ferromagnetic Ising model $\omega = 0$ and the ferromagnetic Heisenberg model $\omega = 1$.

For spin glasses the order parameter is the overlap q and all values of q in the interval $[-q_{EA}, q_{EA}]$ are allowed. We consider two replicas of the same system described by a Hamiltonian:

$$H = H[\sigma] + H[\tau], \tag{7.4}$$

where H is the Hamiltonian of a a single spin glass. We want to compute the free energy increase corresponding to imposing an expectation value of q equal to q_1 at $z = 0$ and q_2 at $z = L$.

A complex computation gives [66] (for small $|q_1 - q_2|$)

$$\delta F \propto ML(|q_1 - q_2|/L)^{5/2} \propto L^{(D-5/2)}. \tag{7.5}$$

As a consequence, the *naive* prediction of mean field theory for the lower critical dimension for spontaneous replica symmetry breaking is $D_c = 2.5$. I stress that these predictions are *naive*; corrections to the mean field theory are neglected. In known cases these kind of computations give the correct result,

A direct check of this prediction has never been done, however it quite remarkable that there are numerical results that strongly suggest that the lower critical dimension is really 2.5. Indeed a first method based on finding the dimension where the transition temperature is zero [56, 57], gives $D_c = 2.491$. A more accurate method based on interpolating the value of a critical exponent [57] gives

$$D_c = 2.4986, \tag{7.6}$$

with an error of order 10^{-3} (my estimate).

It seem that this prediction for the lower critical dimension is well satisfied. It success implies that tree dimensional systems should be describe in the low temperature phase by the mean field approximation, although the correction could slowly vanish with the system size and the critical exponents could be quite different from the naive one, due to the fact three dimensions are only 0.5 dimensions above the lower critical dimensions. What happens in three dimension will be discussed in the next section.

7.3. The three dimensional case

Let us describe some of the equilibrium properties that has be computed or measured in the there dimensional case. We have to make a very drastic selection of the very vast literature, for a review see [6, 58].

All the simulations done with systems up 24^3 [59] find that at low temperature at equilibrium a large spin glass system remains for an exponentially large time in a small region of phase space, but it may jump occasionally in a relatively short time to an other region of phase space (it is like the theory of *punctuated equilibria*: long periods of *stasis*, punctuated by fast changes). Example of the function $P_J(q)$, i.e. the probability distribution of the overlap among two equilibrium configurations, is shown in fig. (2) for two three dimensional samples.

We have defined $P(q) \equiv \overline{P_J(q)}$, where the average is done over the different choices of the couplings J, see fig. (3). We have seen that average is needed

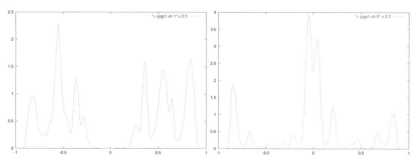

Fig. 2. The function $P_J(q)$ for two samples (i.e two choices of J) for $D = 3$, $L = 16$ (16^3 spins) from [58].

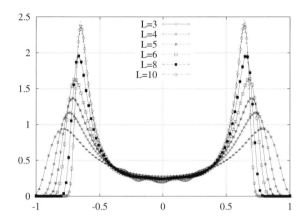

Fig. 3. The function $P(q) = \overline{P_J(q)}$ after average over many samples (L=3...10) from [60].

because the theory predicts (and numerical simulations also in three dimensions do confirm) that the function $P_J(q)$ changes dramatically from system to system. It is clear that the function remains non-trivial in the infinite volume limit.

In the mean field approximation the function $P(q)$ (and its fluctuations from system to system) can be computed analytically together with the free energy: at zero magnetic field $P(q)$ has two delta functions at $\pm q_{EA}$, with a flat part in between. The shape of the three dimensional function $P(q)$ is not very different from the one of the mean fields models.

The validity of the sum rules derived from stochastic stability (Eqs. (5.5, 5.7)) has been rather carefully verified [6]. This is an important test of the theory because they are derived in the infinite volume limit and there is no reason whatsoever that they should be valid in a finite system, if the system would be not sufficient large to mimic the behavior of the infinite volume system. There are also indication that both overlap equivalence [32, 33] and ultrametricity [67, 68] are satisfied, with corrections that goes to zero slowly by increasing the system size.

Very interesting phenomena happen when we add a very small magnetic field. They are very important because the magnetic properties of spin glass can be very well studied in real experiment.

From the theoretical point of view we expect that order of the states in free energy is scrambled when we change the magnetic field [2]: their free energies differ of a factor $O(1)$ and the perturbation is of order N. Different results should be obtained if we use different experimental protocols:

• If we add the field at low temperature, the system remains for a very large time in the same state, only asymptotically it jumps to one of the lower equilibrium states of the new Hamiltonian.

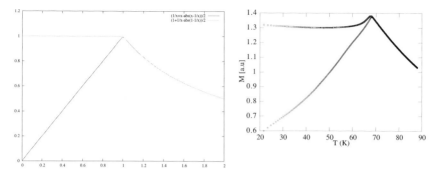

Fig. 4. The two susceptibilities; upper panel, the analytic results in the mean field approximation [2]; lower panel, the experimental results for a metallic spin glass [61].

• If we cool the system from high temperature in a field, we likely go directly to one of the good lowest free energy states.

Correspondingly there are two susceptibilities that can be measured also experimentally:

• The so called linear response susceptibility χ_{LR}, i.e. the response within a state, that is observable when we change the magnetic field at fixed temperature and we do not wait too much. This susceptibility is related to the fluctuations of the magnetization inside a given state.

• The true equilibrium susceptibility, χ_{eq}, that is related to the fluctuation of the magnetization when we consider also the contributions that arise from the fact that the total magnetization is slightly different (of a quantity proportional to $\sqrt{N}$) in different states. This susceptibility is very near to χ_{FC}, the field cooled susceptibility, where one cools the system in presence of a field.

The difference of the two susceptibilities is the hallmark of replica symmetry breaking. In fig. (4) we have both the analytic results for the SK model [2] and the experimental data on metallic spin glasses [61]. The similarities between the two panels are striking.

Which are the differences of this phenomenon with a well known effect; i.e. hysteresis?

• Hysteresis is due to defects that are localized in space and produce a finite barrier in free energy. The mean life of the metastable states is finite and it is roughly $\exp(\beta \Delta F)$ where ΔF is a number of order 1 in natural units.

• In the mean field theory of spin glasses the system must cross barriers that correspond to rearrangements of arbitrary large regions of the system. The largest value of the barriers diverge in the thermodynamic limit.

• In hysteresis, if we wait enough time the two susceptibilities coincide, while they remain always different in this new framework if the applied magnetic field is small enough (non linear susceptibilities are divergent).

The difference between hysteresis and the this new picture (replica symmetry breaking) becoming clearer as we consider fluctuation dissipation relations during aging [10–13, 69, 70], but this would be take to far from our study of equilibrium properties.

8. Some other applications

Some of the physical ideas that have been developed for spin glasses have also been developed independently by people who were working in the study of structural glasses. However in the fields of structural glasses there were no soluble models that displayed interesting behavior so most of the analytic tools and of the corresponding physical insight were firstly developed for spin glasses (for a review see for example [22]).

The two fields remained quite separate one from the other also because there was a widespread believe that the presence of quenched disorder (absent in structural glasses) was a crucial ingredient of the theory developed for spin glasses. Only in the middle of the 90's it became clear that this was a misconception [71, 72] and it was realized that the spin glass theory may be applied also to systems where non-intrinsic disorder is present (e.g. hard spheres).

This became manifest after the discovery of models where there is a transition from a high temperature *liquid* phase to a low temperature glassy phase (some of these models at low temperature have two phases, a disorder phase and an ordered, *crystal* phase). Spin glass theory was able to predict the correct behavior both at high and at low temperature [71–73].

There are many other problems where these ideas have been used. In the past there has been a great interest in applying them to problems with a biological flavor, like the computation of the optimal performance of associative memories based on neural networks [74]. Also learning abilities of simpler objects like the perceptrons can be computed.

In recent years they have been widely used in the field of combinatorial optimization problems and constraint satisfaction, but this will be discussed elsewhere in this volume [14].

9. Conclusions

From the retrospective point of view it may seem extraordinary the amount of new ideas that has been necessary in order to understand in depth the basic properties of spin glasses and other glassy system. A strong effort should be done in order to have better quantitative predictions and a more precise comparison between theory and experiments.

I am convinced that there are still many very interesting surprises that will come from a careful study of spin glasses (and related problem) and it is quite likely that the most exciting result will come from an area I am not able to indicate. I am very curious to see the future of the field.

Appendix I: A useful little theorem

In this appendix we want to prove a useful little theorem [24], which may be useful in many contests.

Theorem: consider a set of $M (\gg 1)$ iid random free energies f_α, $(\alpha \in \{1 \ldots M\})$ distributed with the exponential density eq. (5.11), and a set of M positive numbers a_α. Then, neglecting terms which go to zero when M goes to infinity, the following relation holds:

$$\left\langle \ln \left(\frac{\sum_\alpha a_\alpha \exp(-\beta f_\alpha)}{\sum_\alpha \exp(-\beta f_\alpha)} \right) \right\rangle_f \equiv \left\langle \ln \left(\sum_\alpha w_\alpha a_\alpha \right) \right\rangle_f = \frac{1}{x} \ln \left(\frac{1}{M} \sum_\alpha a_\alpha^x \right),$$
(9.1)

where $\langle \cdot \rangle_f$ denotes an average over the distribution of f.

Corollary 1: In the same conditions as the theorem, for any set of M real numbers b_α, one has:

$$\left\langle \frac{\sum_\alpha a_\alpha b_\alpha \exp(-\beta f_\alpha)}{\sum_\alpha a_\alpha \exp(-\beta f_\alpha)} \right\rangle_f = \frac{\sum_\alpha a_\alpha^x b_\alpha}{\sum_\alpha a_\alpha^x}.$$
(9.2)

Corollary 2: If the numbers a_α are M iid positive random variables, such that the average of a^x exists, which are uncorrelated with the f_α, then one has:

$$\left\langle \ln \left(\frac{\sum_\alpha a_\alpha \exp(-\beta f_\alpha)}{\sum_\alpha \exp(-\beta f_\alpha)} \right) \right\rangle_f \equiv \left\langle \ln \left(\sum_\alpha w_\alpha a_\alpha \right) \right\rangle_f = \frac{1}{x} \ln \left(\langle a_\alpha^x \rangle_a \right),$$
(9.3)

where $\langle \cdot \rangle_a$ denotes an average over the distribution of a.

Remark: We notice that in the two limits $x \to 0$ and $x \to 1$ the equations can be simply understood:

- In the limit $x = 0$, in a typical realization of the random free energies, only one weight w is equal to one and all the others are zero. Averaging over the realizations of free energies amounts to spanning uniformly the set of indices of this special non zero weight.
- In the limit $x = 1$ the number of relevant w goes to infinity and each individual contribution goes to zero. An infinite number of term is present in the l.h.s. of eq. (9.1) and the r.h.s. of the eq. (9.1) becomes $\ln[(1/M)\sum_\alpha a_\alpha]$, as it should.

References

[1] P.W. Anderson (1983), in *Ill condensed matter*, eds. R. Balian, R. Maynard and G. Toulouse (North-Holland, Berlin), pp. 159–259.

[2] M. Mézard, G. Parisi and M.A. Virasoro (1987), *Spin Glass Theory and Beyond* (World Scientific, Singapore).

[3] G. Parisi (1992), *Field Theory, Disorder and Simulations* (World Scientific, Singapore).

[4] C.M. Newman and D.L. Stein (1996), *J. Stat. Phys.* **82**, 1113.

[5] See for example: G. Parisi (1987), *Statistical Field Theory* (Academic Press, New York).

[6] E. Marinari, G. Parisi, F. Ricci-Tersenghi, J. Ruiz-Lorenzo and F. Zuliani (2000), *J. Stat. Phys.* **98**, 973.

[7] G. Parisi (1987), *Physica Scripta* **35**, 123.

[8] G. Parisi (1995), *Phil. Mag. B* **71**, 471.

[9] J.-P. Bouchaud (1992), *J. Phys. France* **2** 1705.

[10] L.F. Cugliandolo and J. Kurchan (1993), *Phys. Rev. Lett.* **71**, 173.

[11] L.F. Cugliandolo and J. Kurchan (1994), *J. Phys. A: Math.* **27**, 5749.

[12] S. Franz and M. Mézard (1994), *Europhys. Lett.* **26**, 209.

[13] S. Franz, M. Mézard, G. Parisi and L. Peliti (1999), *J. Stat. Phys.* **97**, 459.

[14] See for example the lectures by R. Monasson and A. Montanari at this school.

[15] F. Guerra (1997), *Int. J. Phys. B* **10**, 1675.

[16] M. Aizenman and P. Contucci (1998), *J. Stat. Phys.* **92**, 765.

[17] G. Parisi (2004), *Int. Jou. Mod. Phys. B* **18**, 733–744.

[18] F. Guerra (2002), *Comm. Math. Phys.* **233** 1.

[19] M. Aizenman, R. Sims and S.L. Starr (2003), *Phys. Rev. B* **68**, 214403.

[20] M. Talagrand (2006), *Ann. of Math.* **163**, 221.

[21] D. Sherrington and S. Kirkpatrick (1975), *Phys. Rev. Lett.* **35**, 1792.

[22] G. Parisi (2003), in *Les Houches Summer School – Session LXXVII: Slow relaxation and non equilibrium dynamics in condensed matter*, ed. by J.-L. Barrat, M.V. Feigelman, J. Kurchan, and J. Dalibard (Elsevier).

[23] L. Viana and A.J. Bray (1985), *J. Phys. C* **18**, 3037.

[24] M. Mézard and G. Parisi (2001), *Eur. Phys. J. B* **20**, 217.

[25] M. Mézard and G. Parisi (2003), *J. Stat. Phys.* **111**, 1.

[26] S. Franz and M. Leone (2003), *J. Stat. Phys.* **111**, 535.

[27] S. Edwards and P.J. Anderson (1975), *J. Phys.* **F5**, 965.

[28] D.J. Thouless, P.A. Anderson and R.G. Palmer (1977), *Phil. Mag.* **35**, 593.

[29] R.L. de Almeida and D.J. Thouless (1978), *J. Phys. A* **11**, 983.

[30] D. Kastler and D.W. Roberts (1965), *Comm. Math. Phys.* **3**, 151.

[31] D. Ruelle (1988), *Commun. Math. Phys.* **48**, 351.

[32] E. Marinari and G. Parisi (2001), *Phys. Rev. Lett.* **86**, 3887–3890.

[33] P. Contucci, C. Giardinà, C. Giberti and C. Vernia (2006), *Phys. Rev. Lett.* **96**, 217204.

[34] P. Contucci (2003), *J. Phys. A: Math. Gen.* **36**, 10961; P. Contucci, C. Giardinà (2005), *Jour. Stat. Phys.* to appear *math-ph/05050*.

[35] G. Parisi (2006), in *Les Houches Summer School – Session LXXXIII: Mathematical Statistical Physics*, ed. by A. Bovier, F. Dunlop, F. den Hollander and A. van Enter (Elsevier).

[36] I. Giardina's contribution to this school.

[37] D. Iniguez, G. Parisi and J.J. Ruiz-Lorenzo (1996), *J. Phys. A* **29**, 4337.

[38] G. Parisi (2003), *J. Phys. A* **36**, 10773.

[39] G. Parisi (2004), *Europhys. Lett.* **65**, 103.

[40] J. Kurchan, G. Parisi and M.A. Virasoro (1993), *J. Phys. I France* **3**, 1819.

[41] M.E. Ferrero and M.A. Virasoro (1994), *J. Phys. I France* **4**, 1819.

[42] G. Parisi (1994), in *Perspectives on Biological Complexity*, eds. O.T. Solbrig and G. Nicolis (IUBS, Paris), pp. 87–121.

[43] G. Parisi (1993), *J. Stat. Phys.* **72**, 857–872.

[44] G. Parisi and F. Ricci-Tersenghi (2000), *J. Phys. A* **33** 113.

[45] G. Parisi and M. Talagrand (2004), *C.R.A.S.* **339**, 306–306.

[46] S. Franz and G. Parisi (2000), *Europ J. Phys.* **18**, 1434.

[47] D. Ruelle (1987), *Commun. Math. Phys.* **108**, 225.

[48] S. Franz, G. Parisi and M.A. Virasoro (1993), *Europhys. Lett.* **22**, 405.

[49] P. Erdös and A. Rènyi (1959), *Publ. Math. (Debrecen)* **6**, 290.

[50] S. Katsura, S. Inawashiro and S. Fujiki (1979), *Physica* **99A**, 193.

[51] G. Parisi (2002), *On local equilibrium equations for clustering states*, cs.CC/0212047.

[52] M. Mézard, G. Parisi and R. Zecchina (2002), *Science* **297**, 812.

[53] G. Parisi (2002), *On the survey-propagation equations for the random K-satisfiability problem*, cs.CC/0212009.

[54] A. Montanari, G. Parisi and F. Ricci-Tersenghi (2004), *J. Phys. A* **37**, 2073.

[55] T. Temesvari, *Replica symmetric spin glass field theory*, cond-mat/0612523 and reference therein.

[56] G. Parisi, P. Ranieri, F. Ricci-Tersenghi, and J.J. Ruiz-Lorenzo (1997), *J. Phys. A* **30**, 7115.

[57] S. Boettcher, *Stiffness of the Edwards-Anderson Model in all Dimensions*, cond/mat 0508061.

[58] E. Marinari, G. Parisi and J.J. Ruiz-Lorenzo (1998), in *Spin Glasses and Random Fields*, ed. P. Young (World Scientific, Singapore), pp. 58.

[59] L. Gnesi, R. Petronzio and F. Rosati, *Evidence for Frustration Universality Classes in 3D Spin Glass Models*, cond-mat/0208015.

[60] E. Marinari and F. Zuliani (1999), *J. Phys. A* **32**, 7447–7461.

[61] C. Djurberg, K. Jonason and P. Nordblad (1998), *Eur. Phys. J. B* **10**, 15.

[62] E. Marinari, G. Parisi, J. Ruiz-Lorenzo and F. Ritort (1996), *Phys. Rev. Lett.* **76**, 843.

[63] Y.G. Joh, R. Orbach, G.G. Wood, J. Hammann and E. Vincent (1999), *Phys. Rev. Lett.* **82**, 438.

[64] E. Marinari, G. Parisi, F. Ricci-Tersenghi and J.J. Ruiz-Lorenzo (1998), *J. Phys. A* **31**, 2611.

[65] G. Parisi, F. Ricci-Tersenghi and J.J. Ruiz-Lorenzo (1999), *Eur. Phys. J. B* **11**, 317–325.

[66] S. Franz, G. Parisi and M.A. Virasoro (1994), *J. Phys. France* **4**, 1657.

[67] A. Cacciuto, E. Marinari and G. Parisi (1997), *J. Phys. A: Math. Gen.* **30**, L263–L269.

[68] P. Contucci, C. Giardinà, C. Giberti, G. Parisi and C. Vernia, *Ultrametricity in the Edwards-Anderson Model*, cond-mat/0607376.

[69] D. Hérisson and M. Ocio (2002), *Phys. Rev. Lett.* **88**, 257202.

[70] G. Parisi (2006), *PNAS* **103** 7948.

[71] E. Marinari, G. Parisi and F. Ritort (1994), *J. Phys. A* **27**, 7647.

[72] J.-P. Bouchaud and M. Mézard (1994), *J. Physique* **4**, 1109.

[73] M. Degli Esposti, C. Giardinà, S. Graffi and S. Isola (2001), *J. Stat. Phys.* **102**, 1285.

[74] See for example D.J. Amit (1989), *Modeling Brain Functions* (Cambridge University Press, Cambridge).

Course 4

RANDOM MATRICES, THE ULAM PROBLEM, DIRECTED POLYMERS & GROWTH MODELS, AND SEQUENCE MATCHING

Satya N. Majumdar

Laboratoire de Physique Théorique et Modèles Statistiques (UMR 8626 du CNRS), Université Paris-Sud, Bât. 100, 91405 Orsay Cedex, France

J.-P. Bouchaud, M. Mézard and J. Dalibard, eds.
Les Houches, Session LXXXV, 2006
Complex Systems
© *2007 Published by Elsevier B.V.*

Contents

Preamble

In these lecture notes I will give a pedagogical introduction to some common aspects of 4 different problems: (i) random matrices (ii) the longest increasing subsequence problem (also known as the Ulam problem) (iii) directed polymers in random medium and growth models in $(1 + 1)$ dimensions and (iv) a problem on the alignment of a pair of random sequences. Each of these problems is almost entirely a sub-field by itself and here I will discuss only some specific aspects of each of them. These 4 problems have been studied almost independently for the past few decades, but only over the last few years a common thread was found to link all of them. In particular all of them share one common limiting probability distribution known as the Tracy-Widom distribution that describes the asymptotic probability distribution of the largest eigenvalue of a random matrix. I will mention here, without mathematical derivation, some of the beautiful results discovered in the past few years. Then, I will consider two specific models (a) a ballistic deposition growth model and (b) a model of sequence alignment known as the Bernoulli matching model and discuss, in some detail, how one derives exactly the Tracy-Widom law in these models. The emphasis of these lectures would be on how to map one model to another. Some open problems will be discussed at the end.

1. Introduction

In these lectures I will discuss 4 seemingly unrelated problems: (i) random matrices (ii) the longest increasing subsequence (LIS) problem (also known as the Ulam problem after its discoverer) (iii) directed polymers in random environment in $(1 + 1)$ dimensions and related random growth models and (iv) the longest common subsequence (LCS) problem arising in matching of a pair of random sequences (see Fig. 1). These 4 problems have been studied extensively, but almost independently, over the past few decades. For example, random matrices have been extensively studied by nuclear physicists, mathematicians and statisticians. The LIS problem has been studied extensively by probabilists. The models of directed polymers in random medium and the related growth models have been a very popular subject among statistical physicists. Similarly, the LCS problem

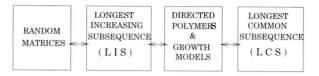

FOUR A PRIORI UNRELATED PROBLEMS

Fig. 1. All 4 problems share the Tracy-Widom distribution.

has been very popular among biologists and computer scientists. Only, in the last 10 years or so, it became progressively evident that there are profound links between these 4 problems. All of them share one common probability distribution function which is called the Tracy-Widom distribution.

This distribution was first discovered in the context of random matrices by Tracy and Widom [1]. They calculated exactly the probability distribution of the *typical* fluctuations of the largest eigenvalue of a random matrix around its mean. This distribution, suitably scaled, is known as the Tracy-Widom (TW) distribution (see later for details). Later in 1999, in a landmark paper [2], Baik, Deift and Johansson (BDJ) showed that the same TW distribution describes the scaled distributions of the length of the longest increasing subsequence in the LIS problem. Immediately after, Johansson [3], Baik and Rains [4] showed that the same distribution also appears in a class of directed polymer problems. Around the same time, Prähofer and Spohn showed [5] that the TW distribution also appears in a class of random growth models known as the polynuclear growth (PNG) models. Following this, it was discovered that the TW distribution also occurred in several other growth models, such as the 'oriented digital boiling' model [6], a ballistic deposition model [7], in PNG type of growth models with varying initial conditions and in various geometries [8, 9] and also in the single-step growth model arising from the totally asymmetric exclusion process [10]. Also, a somewhat direct connection between the stochastic growth models and the random matrix models via the so called 'determinantal point processes' was found in a series of work by Spohn and collaborators [11], which I will not discuss here (see Ref. [11] for a recent review). Finally, the TW distribution was also shown to appear in the LCS problem [12], which is also related to these growth models. Apart from these 4 problems that we will focus here, the TW distribution has also appeared in many other problems, e.g., in the mesoscopic fluctuations of excitation gaps in a dirty metal grain or a semiconductor quantum dot induced by a nearby superconductor [13]. The TW distribution also appears in problems related to finance [14].

The appearance of the TW distribution in so many different problems is really interesting, suggesting an underlying universality that links all these different systems. The purpose of my lectures would be to explore and elucidate the links between the 4 problems stated above. The literature on this subject is huge. I will not try to provide any detailed derivation of the mathematical results here. Instead, I will state precisely the known results that we will need to use and put more emphasis on how one maps one problem to the other. In particular, I will discuss two problems in some detail and show how the TW distribution appears in them. These two problems are: (i) a random growth model in $(1 + 1)$ dimensions that we call the anisotropic ballistic deposition model and (ii) a particular variant of the LCS problem known as the Bernoulli matching (BM) model. In the former case, I will show how to the map the ballistic deposition model to the LIS problem and subsequently use the BDJ results. In the second case, I will show that the BM model can be mapped to a particular directed polymer model that was studied by Johansson. The mappings are often geometric in nature, are nontrivial and they mainly serve two purposes: (a) to elucidate how the TW distribution appears in somewhat unrelated problems and (b) to derive exact analytical results in problems such as the sequence matching models, where precise analytical results were missing so far.

The lecture notes are organized as follows. In Section 3, I will describe some basic results of the random matrix theory and define the TW distribution precisely. In Section 4, the LIS problem will be described along with the main results of BDJ. Section 5 contains a discussion of the directed polymer problems, and in particular the main results of Johansson will be mentioned. In Section 5.1, I will describe how one maps the anisotropic ballistic deposition model to the LIS problem. Section 6 contains a discussion of the LCS problem. Finally, I will conclude in Section 7 with a discussion and open problems.

2. Random matrices: the Tracy-Widom distribution for the largest eigenvalue

Studies of the statistics of the eigenvalues of random matrices have a long history going back to the seminal work of Wigner [15]. Since then, random matrices have found applications in multiple fields including nuclear physics, quantum chaos, disordered systems, string theory and number theory [16]. Three classes of matrices with Gaussian entries have played important roles [16]: $(N \times N)$ real symmetric (Gaussian Orthogonal Ensemble (GOE)), $(N \times N)$ complex Hermitian (Gaussian Unitary Ensemble (GUE)) and $(2N \times 2N)$ self-dual Hermitian matrices (Gaussian Symplectic Ensemble (GSE)). For example, in GOE, one considers an $(N \times N)$ real symmetric matrix X whose elements x_{ij}'s are drawn independently from a Gaussian distribution: $P(x_{ii}) = \frac{1}{\sqrt{2\pi}} \exp[-x_{ii}^2/2]$

and $P(x_{ij}) = \frac{1}{\sqrt{\pi}} \exp[-x_{ij}^2]$ for $i < j$. Thus the joint distribution of all the $N(N + 1)/2$ independent elements is just the product of the individual distributions and can be written in a compact form as $P[X] = A_N \exp[-\text{tr}(X^2)/2]$, where A_N is a normalization constant. One can similarly write down the joint distribution for the other two ensembles [16].

One of the key results in the random matrix theory is due to Wigner who derived, starting from the joint distribution of the matrix elements $P(X)$, a rather compact expression for the joint probability density function (PDF) of the eigenvalues of a random $(N \times N)$ matrix from all ensembles [15]

$$P(\lambda_1, \lambda_2, \ldots, \lambda_N) = B_N \exp\left[-\frac{\beta}{2}\left(\sum_{i=1}^{N} \lambda_i^2 - \sum_{i \neq j} \ln(|\lambda_i - \lambda_j|)\right)\right], \qquad (2.1)$$

where B_N normalizes the pdf and $\beta = 1, 2$ and 4 correspond respectively to the GOE, GUE and GSE. Note that in writing Eq. (2.1), starting from the joint distribution of entries $P[X] = A_N \exp[-\text{tr}(X^2)/2]$, we have rescaled $\lambda \to \sqrt{\beta}\lambda$ for convenience. The joint law in Eq. (2.1) allows one to interpret the eigenvalues as the positions of charged particles, repelling each other via a 2-d Coulomb potential (logarithmic); they are confined on a 1-d line and each is subject to an external harmonic potential. The parameter β that characterizes the type of ensemble can be interpreted as the inverse temperature.

Once the joint pdf is known explicitly, other statistical properties of a random matrix can, in principle, be derived from this joint pdf. In practice, however this is often a technically daunting task. For example, suppose we want to compute the average density of states of the eigenvalues defined as $\rho(\lambda, N) = \sum_{i=1}^{N}\langle\delta(\lambda - \lambda_i)\rangle/N$, which counts the average number of eigenvalues between λ and $\lambda + d\lambda$ per unit length. The angled bracket $\langle\rangle$ denotes an average over the joint pdf. It then follows that $\rho(\lambda, N)$ is simply the marginal of the joint pdf, i.e, we fix one of the eigenvalues (say the first one) at λ and integrate the joint pdf over the rest of the $(N - 1)$ variables.

$$\rho(\lambda, N) = \frac{1}{N}\sum_{i=1}^{N}\langle\delta(\lambda - \lambda_i)\rangle = \int_{-\infty}^{\infty}\prod_{i=2}^{N}d\lambda_i \, P(\lambda, \lambda_2, \ldots, \lambda_N). \qquad (2.2)$$

Wigner was able to compute this marginal and this is one of the central results in the random matrix theory, known as the celebrated Wigner semi-circular law. For large N and for any β,

$$\rho(\lambda, N) = \sqrt{\frac{2}{N\pi^2}}\left[1 - \frac{\lambda^2}{2N}\right]^{1/2}. \qquad (2.3)$$

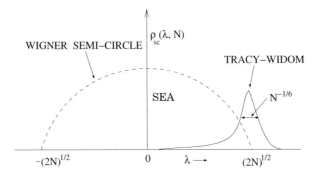

Fig. 2. The dashed line shows the semi-circular form of the average density of states. The largest eigenvalue is centered around its mean $\sqrt{2N}$ and fluctuates over a scale of width $N^{-1/6}$. The probability of fluctuations on this scale is described by the Tracy-Widom distribution (shown schematically).

Thus, on an average, the N eigenvalues lie within a finite interval $\left[-\sqrt{2N}, \sqrt{2N}\right]$, often referred to as the Wigner 'sea'. Within this sea, the average density of states has a semi-circular form (see Fig. 2) that vanishes at the two edges $-\sqrt{2N}$ and $\sqrt{2N}$. Note that since there are N eigenvalues distributed over the interval $\left[-\sqrt{2N}, \sqrt{2N}\right]$, the average spacing between adjacent eigenvalues scales as $N^{-1/2}$.

From the semi-circular law, it is clear that the average of the maximum (or minimum) eigenvalue is $\sqrt{2N}$ $\left(-\sqrt{2N}\right)$. However, for finite but large N, the maximum eigenvalue fluctuates, around its mean $\sqrt{2N}$, from one sample to another. A natural question is: what is the full probability distribution of the largest eigenvalue $\lambda_{\max}$? Once again, this distribution can, in principle, be computed from the joint pdf in Eq. (2.1). To see this, it is useful to consider the cumulative distribution of $\lambda_{\max}$. Clearly, if $\lambda_{\max} \leq t$, it necessarily means that all the eigenvalues are less than or equal to t. Thus,

$$\text{Prob}\,[\lambda_{\max} \leq t, N] = \int_{-\infty}^{t} \prod_{i=1}^{N} d\lambda_i \, P(\lambda_1, \lambda_2, \ldots, \lambda_N), \qquad (2.4)$$

where the joint pdf is given in Eq. (2.1). In practice, however, carrying out this multiple integration in closed form is very difficult. Relatively recently, Tracy and Widom [1] were able to find the limiting form of $\text{Prob}\,[\lambda_{\max} \leq t, N]$ for large N. They showed that the fluctuations of $\lambda_{\max}$ *typically* occur over a very narrow scale of width $\sim N^{-1/6}$ around its mean $\sqrt{2N}$ at the upper edge of the Wigner sea. It is useful to note that this scale $\sim N^{-1/6}$ of typical fluctuations of the largest eigenvalue is much bigger than the average spacing $\sim N^{-1/2}$ between adjacent eigenvalues in the limit of large N.

More precisely, Tracy and Widom showed [1] that asymptotically for large N, the scaling variable $\xi = \sqrt{2}\, N^{1/6} \left[\lambda_{\max} - \sqrt{2N} \right]$ has a limiting N-independent probability distribution, $\text{Prob}[\xi \leq x] = F_\beta(x)$ whose form depends on the value of the parameter $\beta = 1$, 2 and 4 characterizing respectively the GOE, GUE and GSE. The function $F_\beta(x)$ is called the Tracy-Widom (TW) distribution function. The function $F_\beta(x)$, computed as a solution of a nonlinear Painleve differential equation [1], approaches to 1 as $x \to \infty$ and decays rapidly to zero as $x \to -\infty$. For example, for $\beta = 2$, $F_2(x)$ has the following tails [1],

$$F_2(x) \to 1 - O\!\left(\exp[-4x^{3/2}/3] \right) \quad \text{as } x \to \infty$$
$$\to \exp\!\left[-|x|^3/12 \right] \quad \text{as } x \to -\infty. \tag{2.5}$$

The probability density function $f_\beta(x) = dF_\beta/dx$ thus has highly asymmetric tails. A graph of these functions for $\beta = 1$, 2 and 4 is shown in Fig. 3. A convenient way to express these typical fluctuations of $\lambda_{\max}$ around its mean $\sqrt{2N}$ is to write, for large N,

$$\lambda_{\max} = \sqrt{2N} + \frac{N^{-1/6}}{\sqrt{2}}\, \chi \tag{2.6}$$

where the random variable χ has the limiting N-independent distribution, $\text{Prob}[\chi \leq x] = F_\beta(x)$. As mentioned in the introduction, amazingly this TW

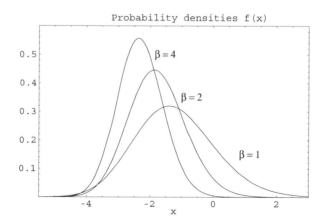

Fig. 3. The probability density function $f_\beta(x)$ plotted as a function of x for $\beta = 1$, 2 and 4 (reproduced from Ref. [1]).

distribution function has since emerged in a growing variety of seemingly unrelated problems, some of which I will discuss in the next sections.

Large Deviations of $\lambda_{\max}$**:** Before we end this section and proceed to the other problems, it is worth making the following remark. The Tracy-Widom distribution describes the probability of *typical and small* fluctuations of $\lambda_{\max}$ over a very narrow region of width $\sim O(N^{-1/6})$ around the mean $\langle \lambda_{\max} \rangle \approx \sqrt{2N}$. A natural question is how to describe the probability of *atypical and large* fluctuations of λ_{max} around its mean, say over a wider region of width $\sim O(N^{1/2})$? For example, what is the probability that all the eigenvalues of a random matrix are negative (or equivalently all are positive)? This is the same as the probability that $\lambda_{\max} \leq 0$ (or equivalently $\lambda_{\min} \geq 0$). Since $\langle \lambda_{\max} \rangle \approx \sqrt{2N}$, this requires the computation of the probability of an extremely rare event characterizing a large deviation of $\sim - O(N^{1/2})$ to the left of the mean. This question naturally arises in any physical system where one is interested in the statistics of stationary points of a random landscape. For example, in disordered systems such as spin glasses one is interested in the stationary points (metastable states) of the free energy landscape. On the other hand, in structural glasses or supercooled liquids, one is interested in the stationary points of the potential energy landscape. In order to have a local minimum of the random landscape one needs to ensure that the eigenvalues of the associated Hessian matrix are all positive [17, 18]. A similar question recently came up in the context of random landscape models of anthropic principle based string theory [19, 20] as well as in quantum cosmology [21]. Here one is interested in the statistical properties of vacua associated with a random multifield potential, e.g., how many minima are there in a random string landscape? These large deviations are also important in characterizing the large sample to sample fluctuations of the excitation gap in quantum dots connected to a superconductor [13].

The issue of large deviations of $\lambda_{\max}$ was addressed in Ref. [3] for a special class of matrices drawn from the Laguerre ensemble that corresponds to the eigenvalues of product matrices of the form $W = X^{\dagger} X$ where X itself is a Gaussian matrix (real or complex). Adopting similar methods as in Ref. [3] one can prove that for Gaussian ensembles, the probability of *large* fluctuations to the left of the mean $\sqrt{2N}$ behaves for large N as,

$$\text{Prob}\,[\lambda_{\max} \leq t, N] \sim \exp\left[-\beta N^2 \Phi_-\left(\frac{\sqrt{2N}-t}{\sqrt{N}}\right)\right] \qquad (2.7)$$

where $t \sim O(N^{1/2}) \leq \sqrt{2N}$ is located deep inside the Wigner sea and $\Phi_-(y)$ is a certain *left* large deviation function. On the other hand, for *large* fluctuations to the right of the mean $\sqrt{2N}$,

$$1 - \text{Prob}\,[\lambda_{\max} \le t,\, N] \sim \exp\left[-\beta N\,\Phi_+\left(\frac{t - \sqrt{2N}}{\sqrt{N}}\right)\right] \tag{2.8}$$

for $t \sim O(N^{1/2}) \ge \sqrt{2N}$ located outside the Wigner sea to its right and $\Phi_+(y)$ is the *right* large deviation function. The problem then is to evaluate explicitly the left and the right large deviation functions $\Phi_\mp(y)$ explicitly. While, for the Laguerre ensemble, an explicit expression of $\Phi_+(y)$ was obtained in Ref. [3] and that of $\Phi_(y)$ recently in Ref. [22], similar expressions for the Gaussian ensemble were missing so far.

Indeed, to calculate the probability that all eigenvalues are negative (or positive) for Gaussian matrices, we need an explicit expression of $\Phi_-(y)$ for the Gaussian ensemble. This is because, the probability that all eigenvalues are negative is precisely the probability that $\lambda_{\max} \le 0$, and hence, from Eq. (2.7)

$$\text{Prob}\,[\lambda_{\max} \le 0,\, N] \sim \exp\left[-\beta N^2 \Phi_-\left(\sqrt{2}\right)\right]. \tag{2.9}$$

The coefficient $\theta = \beta \Phi_-(\sqrt{2})$ of the N^2 term inside the exponential term in Eq. (2.9) is of interest in string theory, and in Ref. [20], the authors provided an approximate estimate (for $\beta = 1$) of $\theta \approx 1/4$, along with numerical simulations. Recently, in collaboration with D.S. Dean, we were able to compute exactly an explicit expression [23] for the full *left* large deviation function $\Phi_-(y)$. I will not provide the derivation here, but the calculation of *large* deviations turns out to be somewhat simpler [23] than the calculation of the *small* deviations 'a la TW. One simply has to minimize the effective free energy of a Coulomb gas using the method of steepest descents and then analyze the resulting saddle point equation (which is an integral equation) [23]. This technique is quite useful, as it can be applied to other problems as well, such as the calculation of the average number of stationary points for a Gaussian random fields with N components in the large N limit [24, 25] and also the large deviation function associated with the largest eigenvalue of other types of matrices, such as the Wishart matrices [22]. In terms of the variable $z = y - \sqrt{2}$, the *left* large deviation function has the following explicit expression [23]

$$\Phi_-(y = z + \sqrt{2}) = -\frac{1}{8}(3 + 2\ln 2) + \frac{1}{216}\left[72z^2 - 2z^4(30z + 2z^3)\sqrt{6 + z^2}\right.$$
$$\left. + 27\left(3 + \ln(1296) - 4\ln(-z + \sqrt{6 + z^2})\right)\right]. \tag{2.10}$$

In particular, the constant θ is given exactly by

$$\theta = \beta\,\Phi(\sqrt{2}) = \beta\,\frac{\ln 3}{4} = (0.274653\ldots)\,\beta. \tag{2.11}$$

Another interesting point about the left large deviation function $\Phi_-(y)$ is the following. It describes the probability of large $\sim O(\sqrt{N})$ fluctuations to the left of the mean, i.e., when $y = (\sqrt{2N} - \lambda_{\max})/\sqrt{N} \sim O(1)$. Now, if we take the $y \to 0$ limit, then $\Phi_-(y)$ should describe the *small* fluctuations to the left of the mean $\sqrt{2N}$. In other words, we expect to recover the left tail of the TW distribution by taking the $y \to 0$ limit in the left large deviation function. Indeed, as $y \to 0$, one finds from Eq. (2.10), that $\Phi_-(y) \approx y^3/6\sqrt{2}$. Putting this expression back in Eq. (2.7) one gets

$$\text{Prob}[\lambda_{\max} \leq t, N] \approx \exp\left[-\frac{\beta}{24}\left|\sqrt{2}\,N^{1/6}\,(t - \sqrt{2N})\right|^3\right] \tag{2.12}$$

Given that $\chi = \sqrt{2}\,N^{1/6}\left(t - \sqrt{2N}\right)$ is the Tracy-Widom scaling variable, we find that the result in Eq. (2.12) matches exactly with the left tail of the Tracy-Widom distribution for all β. For example, for $\beta = 2$ one can easily verify this by comparing Eqs. (2.12) and (2.5). This approach not only serves as a useful check that one has obtained the correct large deviation function $\Phi_-(y)$, but also provides an alternative and simpler way to derive the asymptotics of the left tail of the TW distribution. A similar expression for the right large deviation function $\Phi_+(y)$ for the Gaussian ensemble is still missing and its computation remains an open problem.

Although the Tracy-Widom distribution was originally derived as the limiting distribution of the largest eigenvalue of matrices whose elements are drawn from Gaussian distributions, it is now believed that the same limiting distribution also holds for matrices drawn from a larger class of ensembles, e.g., when the entries are independent and identically distributed random variables drawn from an arbitrary distribution with all moments finite [26, 27]. Recently, Biroli, Bouchaud and Potters [14] extended this result to power-law ensembles, where each entry of a random matrix is drawn independently from a power-law distribution [28, 29]. They showed that as long as the fourth moment of this power-law distribution is finite, the suitably scaled $\lambda_{\max}$ is again TW distributed, but when the fourth moment is infinite, $\lambda_{\max}$ has Fréchet fluctuations [14]. It would be interesting to compute the probability of *large* deviations of $\lambda_{\max}$ for this power-law ensemble, as in the Gaussian case mentioned above. For example, what is the probability that all the eigenvalues of such random matrices (drawn from the power-law ensemble) are negative (or positive), i.e. $\lambda_{\max} \leq 0$? This is an open question.

3. The longest common subsequence problem (or the Ulam problem)

The longest common subsequence (LIS) problem was first stated by Ulam [30] in 1961, hence it is also called the Ulam's problem. Since then, a lot of research, mostly by probabilists, has been done on this problem (for a brief history of the problem, see the introduction in Ref. [2]). The problem can be stated very simply as follows. Consider a set of N distinct integers $\{1, 2, 3, \ldots, N\}$. Consider all $N!$ possible permutations of this sequence. For any given permutation, let us find all possible increasing subsequences (terms of a subsequence need not necessarily be consecutive elements) and from them find out the longest one. For example, take $N = 10$ and consider a particular permutation $\{8, 2, 7, \underline{1}, \underline{3}, \underline{4}, 10, \underline{6}, \underline{9}, 5\}$. From this sequence, one can form several increasing subsequences such as $\{8, 10\}$, $\{2, 3, 4, 10\}$, $\{1, 3, 4, 10\}$ etc. The longest one of all such subsequences is either $\{1, 3, 4, 6, 9\}$ as shown by the underscores or $\{2, 3, 4, 6, 9\}$. The length l_N of the LIS (in our example $l_N = 5$) is a random variable as it varies from one permutation to another. In the Ulam problem one considers all the $N!$ permutations to be equally likely. Given this uniform measure over the space of permutations, what is the statistics of the random variable l_N?

Ulam found numerically that the average length $\langle l_N \rangle$ behaves asymptotically $\langle l_N \rangle \sim c\sqrt{N}$ for large N. Later this result was established rigorously by Hammersley [31] and the constant $c = 2$ was found by Vershik and Kerov [32]. Recently, in a seminal paper, Baik, Deift and Johansson (BDJ) [2] derived the full distribution of l_N for large N. In particular, they showed that asymptotically for large N

$$l_N \to 2\sqrt{N} + N^{1/6}\chi \tag{3.1}$$

where the random variable χ has a limiting N-independent distribution,

$$\text{Prob}(\chi \leq x) = F_2(x) \tag{3.2}$$

where $F_2(x)$ is precisely the TW distribution for the largest eigenvalue of a random matrix drawn from the GUE ($\beta = 2$), as defined in Section 3. Note that the power of N in the correction term in Eq. (3.1) is $+1/6$ as opposed to the asymptotic law in Eq. (2.6) where the power of N in the correction term is $-1/6$. This means that while for random matrices of size ($N \times N$), the typical fluctuation of $\lambda_{\max}$ around its mean value $\sqrt{2N}$ *decreases* with N as $N^{-1/6}$ as $N \to \infty$ (i.e., the distribution gets narrower ans narrower around the mean as N increases), the opposite happens in the Ulam problem: the typical fluctuation in l_N around its mean $2\sqrt{N}$ *increases* as $N^{1/6}$ with increasing N, i.e., the distribution around the mean gets broader and broader with increasing N.

BDJ also showed that when the sequence length N itself is a random variable drawn from a Poisson distribution with mean $\langle N \rangle = \lambda$, the length of the LIS converges for large λ to

$$l_\lambda \to 2\sqrt{\lambda} + \lambda^{1/6}\chi, \tag{3.3}$$

where χ has the Tracy-Widom distribution $F_2(x)$. The fixed N and the fixed λ ensembles are like the canonical and the grand canonical ensembles in statistical mechanics. The BDJ results led to an avalanche of subsequent mathematical works [33].

I will not provide here the derivation of the BDJ results, but I will assume this result to be known and use it later for other problems. As we will see later, in many problems such as in several growth models, the strategy is to map those models into the LIS problem and subsequently use the BDJ results. In these mappings, typically the height of a growing surface in the $(1 + 1)$ dimensional growth models gets mapped to the length of the LIS, i.e., schematically, $H \rightarrow l_N$. Subsequently, using the BDJ results for the distribution of l_N, one shows that the height in growth models is distributed according to the Tracy-Widom law. I will show explicitly how this strategy works for one specific ballistic deposition model in Section 5.1. But to understand the mapping, we need to know one additional fact about the LIS, which I discuss below.

Suppose we are given a specific permutation of N integers. What is a simple algorithm to find the length of the LIS of this permutation? The most famous algorithm goes by the name of Robinson-Schensted-Knuth (RSK) algorithm [34], which makes a correspondence between the permutation and a Young tableaux, and has played a very important role in the development of the LIS problem. But let me not discuss this here, the reader can find a nice readable account in Ref. [33]. Instead, I will discuss another related algorithm known as the 'patience-sorting' algorithm which will be more useful for our purposes. This algorithm was developed first by Mallows [35] who showed its connection to the Young tableaux. I will discuss here the version that was discussed recently by Aldous and Diaconis [33]. This algorithm is best explained in terms of an example. Let us take $N = 8$ and consider a specific permutation, say $\{8, 3, 5, 1, 2, 6, 4, 7\}$. The 'patience sorting' is a greedy algorithm that will easily find the length of the LIS of this sequence. It is like a simple card game of 'patience'. This game goes as follows: start forming piles with the numbers in the permuted sequence starting with the first element which is 8 in our example. So, the number 8 forms the base of the first pile (see Fig. 4). The next element, if less than 8, goes on top of 8. If not, it forms the base of a new pile. One follows a greedy algorithm: for any new element of the sequence, check all the top numbers on the existing piles starting from the first pile and if the new number is less than the top number of an already existing pile, it goes on top of that pile. If the new number is larger than all the top numbers of the existing piles, this new number forms the base of a new pile. Thus in our example, we form 4 distinct piles: $[\{8, 3, 1\}, \{5, 2\}, \{6, 4\}, \{7\}]$. Thus the number of piles is 4. On the other hand, for this particular example, it is easy to check that there are 3 LIS's namely, $\{3, 5, 6, 7\}$, $\{1, 2, 6, 7\}$ and $\{1, 2, 4, 7\}$, all of the same length $l = 4$. So,

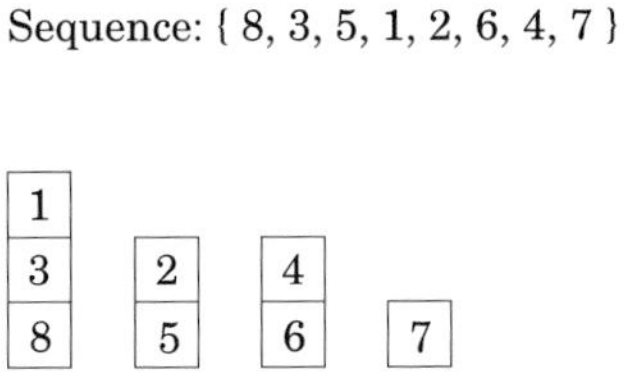

Sequence: { 8, 3, 5, 1, 2, 6, 4, 7 }

Fig. 4. The construction of piles according to the patience sorting game. The number of piles corresponding to the sequence {8, 3, 5, 1, 2, 6, 4, 7} is 4, which is also the length of the LIS of this sequence.

we see that the length of the LIS is 4, same as the number of piles in the patience sorting game. But this is not an accident. One can easily prove [33] that for any given permutation of N integers, the length of the LIS l_N is exactly the same as the number of piles in the corresponding 'patience sorting' algorithm. We will see later that this fact does indeed play a crucial role in our mapping of growth models to the LIS problem.

4. Directed polymers and growth models

The problem of directed polymers in random medium has been an active area of research in statistical physics for the past three decades. Apart from the fact that it is a simple 'toy' model of disordered systems, the directed polymer problem has important links to a wide variety of other problems in physics, such as interface fluctuations and pinning [36], growing interface models of the Kardar-Parisi-Zhang (KPZ) variety [37], randomly forced Burger's equation in fluid dynamics [38], spin glasses [39–41], and also to a single-particle quantum mechanics problem in a time-dependent random potential [42]. There are many interesting issues associated with the directed polymer problem, such as the phase-transition at a finite temperature in $(d + 1)$-dimensional directer polymer when $d > 2$ [43], the nature of the low temperature phase [40, 41], the nature of the transverse fluctuations [36, 44] etc. The literature on the subject is huge (for a review see Ref. [45]).

Here we will focus simply at zero-temperature and a lattice version of the directed polymer problem. This version can be stated as in Fig. 5. Consider a square lattice with O denoting the origin. On each site with coordinates (i, j) of this lattice, there is a random energy $\epsilon_{i,j}$, drawn independently from site to site, but from the identical distribution $\rho(\epsilon)$. For simplicity, we will consider that $\epsilon_{i,j}$'s are all negative, i.e., $\rho(\epsilon)$ has support only over $\epsilon \in [0, -\infty]$. The energy variables $\epsilon_{i,j}$'s are quenched random variables.

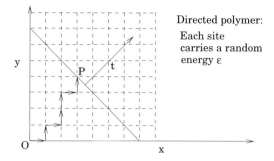

Fig. 5. Directed polymer in $(1+1)$ dimensions with random site energies.

We are interested here only in directed walks for simplicity. Consider all possible directed walk configurations (a walk that can move only north or eastward as shown in Fig. 5) that start from the origin O and end up at a fixed point, say P with co-ordinates (x, y). An example of such a walk is shown in Fig. 5. The total energy $E(W)$ of any given walk W from O to P is just the sum of site energies along the path W, $E(W) = \sum_{i \in W} \epsilon_i$. Thus, for fixed O and P (the endpoints), the energy of a path varies from one path to another (all having the same endpoints O and P). The path having the minimum energy (optimal path) among these will correspond to the ground state configuration, i.e., the polymer will prefer to choose this optimal path at zero temperature. Let $E_0(x, y)$ denote this minimum energy amongst all directed paths that start at O and finish at $P : (x, y)$. Now, this minimum energy $E_0(x, y)$ is, of course, a random variable since it fluctuates from one configuration of quenched disorder to another. One is interested in the statistics of $E_0(x, y)$ for a given fixed (x, y). For example, what is the probability distribution of $E_0(x, y)$ given that $\epsilon_{x,y}$'s are independent and identically distributed random variables each drawn from $\rho(\epsilon)$?

Mathematically, one can write an 'evolution' equation or recursion relation for the variable $E_0(x, y)$. Indeed, the path that ends up at say (x, y), must have visited either the site $(x-1, y)$ or the site $(x, y-1)$ at the previous step. Then clearly,

$$E_0(x, y) = \min\left[E_0(x-1, y), E_0(x, y-1)\right] + \epsilon_{x,y} \qquad (4.1)$$

where $\epsilon_{x,y}$ denotes the random energy associated with the site (x, y). Alternately, we can define $H(x, y) = -E_0(x, y)$ which are all positive variables that satisfy the recursion relation

$$H(x, y) = \max\left[H(x-1, y), H(x, y-1)\right] + \xi_{x,y} \qquad (4.2)$$

where $\xi_{x,y} = -\epsilon_{x,y}$ are positive random variables. The recursion relation in Eq. (4.2) is non-linear and hence is difficult to find the distribution of $H(x, y)$, knowing the distribution of the $\xi_{x,y}$'s. Note that, by interpreting $t = x + y$ as a time-like variable, and denoting by i the transverse coordinate at a fixed t, this recursion relation can also be interpreted as a stochastic evolution equation,

$$H(i, t) = \max\big[H(i + 1, t - 1), H(i - 1, t - 1)\big] + \xi_{i,t} \tag{4.3}$$

where the site energy $\xi_{i,t}$ can now be interpreted as a stochastic noise. In this interpretation, one can think of the directed polymer as a growing model of $(1+1)$ dimensional interface where $H(i, t)$ denotes the height of the interface at the site i of a one dimensional lattice at time t. Only, in this version, the length of one dimensional lattice or the substrate keeps increasing linearly with time t. In this respect, it corresponds to a special version of a polynuclear growth model where growth occurs on top of a single droplet whose linear size keeps increasing uniformly with time. There are, of course, several other variations of this simple directed polymer model [45]. For example, one can consider a version where the random energies are associated with bonds, rather than the sites. Similarly, one can consider a finite temperature version of the model. In the corresponding analogy to the interface model, at finite temperature, the free energy (as opposed to the ground state energy) of the polymer corresponds to the height variable of the interface. This is most easily seen in the continuum formulation of the model by writing down the partition function as a path integral and then showing directly that $H = \ln Z$ satisfies the KPZ equation [46].

A lot is known about the first and the second moment of $H(x, y)$ (or alternatively for $H(i, t)$ in the height language) and the associated universality properties [40, 41, 47]. For example, from simple extensivity properties, one would expect that average ground state energy of the path will increase linearly with the size (number of steps t) of the path. In terms of height, this means $\langle H(i, t) \rangle \to v(i)t$ for large t where $v(i)$ is velocity of the interface at site i of the one dimensional lattice [48]. Also, the standard deviation of height, say of $H(x, x)$ (along the diagonal), is known to grow universally, for large x as $x^{1/3}$ [45]. For the interface, this means that the typical height fluctuation grows as $t^{1/3}$ for large t, a result that is known from the KPZ problem in 1-dimension (via a mapping to the noisy Burgers equation). However, much less was known about the full distribution of $H(x, y)$, till only recently.

Johansson [3] was able to derive the full asymptotic distribution of $H(x, y)$ evolving via Eq. (4.2) for a specific disorder distribution, where the noise $\xi_{x,y}$'s in Eq. (4.2) are i.i.d variables taking nonnegative integer values according to the distribution: Prob($\xi_{x,y} = k) = (1 - p)\, p^k$ for $k = 0, 1, 2, \ldots$, where $0 \le p \le 1$ is a fraction. Interestingly, exactly the same recursion relation as in Eq. (4.2) and also with the same disorder distribution as in Johansson's model also appeared independently around the same time in an anisotropic directed percolation problem studied by Rajesh and Dhar [49], a problem to which we will come back later when we discuss the sequence matching problem. The authors in Ref. [49] were able to compute exactly the first moment, but Johansson computed the full asymptotic distribution. He showed that for large x and y [3]

$$H(x, y) \rightarrow \frac{2\sqrt{pxy} + p(x+y)}{q} +$$
$$+ \frac{(pxy)^{1/6}}{q} \left[(1+p) + \sqrt{\frac{p}{xy}} (x+y) \right]^{2/3} \chi \qquad (4.4)$$

where $q = 1 - p$, χ is a random variable with the Tracy-Widom distribution, $\text{Prob}(\chi \leq x) = F_2(x)$ as in Eq. (3.2). If one sets $x = y = t/2$, then for the growing droplet interpretation, it would mean that the height $H(i = 0, t)$ has a mean that grows linearly with t and a standard deviation that grows as $t^{1/3}$ and when properly centered and scaled, the distribution of $H(0, t)$ tends to the GUE Tracy-Widom distribution. Around the same time, Prähofer and Spohn derived a similar result for a class of PNG models [5]. Moreover, they were able to show that not just the $F_2(x)$, but other Tracy-Widom distributions such as the $F_1(x)$ (corresponding to the GOE ensemble) also arises in the PNG model when one starts from different initial conditions [5].

4.1. Exact height distribution in a ballistic deposition model

In this subsection, we will show explicitly how one can derive the exact height distribution in a specific $(1 + 1)$ dimensional growth model and show that it has a limiting Tracy-Widom distribution. This example will illustrate explicitly how one maps a growth model to the LIS problem [7]. A similar mapping was used by Prähofer and Spohn for the PNG model [5]. But before we illustrate the mapping, it is useful to remark (i) why one studies such growth models and (ii) what does this mapping and subsequent calculation of the height distribution achieve?

The answer to these two questions are as follows. We know that growth processes are ubiquitous in nature. The past few decades have seen extensive research on a wide variety of both discrete and continuous growth models [45, 50, 51]. A large class of these growth models in $(1 + 1)$ dimensions such as the Eden model [52], restricted solid on solid (RSOS) models [53], directed polymers as mentioned before [45], polynuclear growth models (PNG) [54] and ballistic deposition models (BD) [55] are believed to belong to the same universality class as that of the Kardar-Parisi-Zhang (KPZ) equation describing the growth of interface fluctuations [37]. This universality is, however, somewhat restricted in the sense that it refers only to the width or the second moment of the height fluctuations characterized by two independent exponents (the growth exponent β and the dynamical exponent z) and the associated scaling function. Moreover, even this restricted universality is established mostly numerically. Only in very few special discrete models in $(1 + 1)$ dimensions, the exponents $\beta = 1/3$ and $z = 3/2$ can be computed exactly via the Bethe ansatz technique [57]. A natural and important question is whether this universality can be extended beyond the second moment of height fluctuations. For example, is the full distribution of the height fluctuations (suitably scaled) universal, i.e. is the same for different growth models belonging to the KPZ class? Moreover, the KPZ-type equations are usually attributed to models with small gradients in the height profile and the question whether the models with large gradients (such as the BD models) belong to the KPZ universality class is still open. The connection between the discrete BD models and the continuum KPZ equation has recently been elucidated upon [58].

To test whether this more stringent test of universality (going beyond the second moment) of the full distribution is true or not, one needs to calculate the full height distribution in different models which are known to belong to the KPZ universality class as far as only the second moment is concerned. In fact, as mentioned earlier, Prähofer and Spohn were able to calculate the asymptotic height distribution in a class of PNG models and showed that it has the Tracy-Widom distribution [5]. Similarly, we mentioned earlier that Johansson [3] established rigorously that the height distribution, in a specific version of the directed polymer model, is of the Tracy-Widom form. Subsequently, there have been several other works [6] recently, including the ballistic deposition model [7] that we will discuss below, that showed that indeed all these $(1 + 1)$ dimensional growth models share the same common scaled height distribution (Tracy-Widom), thus putting the universality on a much stronger footing going beyond just the second moment.

We now focus on a specific ballistic deposition model. Ballistic deposition models typically try to mimic columnar growth that occur in many natural systems and have been studied extensively in the past with a variety of microscopic rules [55, 56], though an exact calculation of the height distribution remained elusive in any of these microscopic models. In collaboration with S. Nechaev, we found a particular ballistic deposition model which can be explicitly mapped to the LIS problem and hence the full asymptotic height distribution can be computed exactly [7]. In our $(1 + 1)$-D (here D stands for 'dimensional') BD model columnar growth occurs sequentially on a linear substrate consisting of L columns with free boundary conditions. The time t is discrete and is increased by 1 with every deposition event. We first consider the flat initial condition, i.e., an empty substrate at $t = 0$. Other initial conditions will be treated later. At any stage of the growth, a column (say the k-th column) is chosen at random with

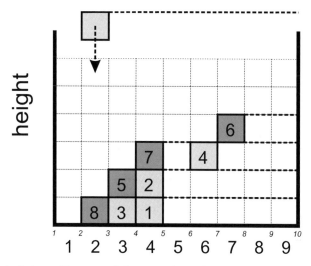

Fig. 6. Growth of a heap with asymmetric long-range interaction. The numbers inside cells show the times at which the blocks are added to the heap.

probability $p = \frac{1}{L}$ and a "brick" is deposited there which increases the height of this column by one unit, $H_k \to H_k + 1$. Once this "brick" is deposited, it screens all the sites at the same level in all the columns to its right from future deposition, i.e. the heights at all the columns to the right of the k-th column must be strictly greater than or equal to $H_k + 1$ at all subsequent times. For example, in Fig. 6, the first brick (denoted by 1) gets deposited at $t = 1$ in the 4-th column and it immediately screens all the sites to its right. Then the second brick (denoted by 2) gets deposited at $t = 2$ again in the same 4-th column whose height now becomes 2 and thus the heights of all the columns to the right of the 4-th column must be ≥ 2 at all subsequent times and so on. Formally such growth is implemented by the following update rule. If the k-th site is chosen at time t for deposition, then

$$H_k(t + 1) = \max\{H_k(t), H_{k-1}(t), \ldots, H_1(t)\} + 1. \tag{4.5}$$

The model is anisotropic and evidently even the average height profile $\langle H_k(t) \rangle$ depends non-trivially on both the column number k and time t. Our goal is to compute the asymptotic height distribution $P_k(H, t)$ for large t.

It is easy to find the height distribution $P_1(H, t)$ of the first column, since the height there does not depend on any other column. At any stage, the height in the first column either increases by one unit with probability $p = \frac{1}{L}$ (if this column is selected for deposit) or stays the same with probability $1 - p$. Thus $P_1(H, t)$ is simply the binomial distribution, $P_1(H, t) = \binom{t}{H} p^H (1 - p)^{t-H}$ with

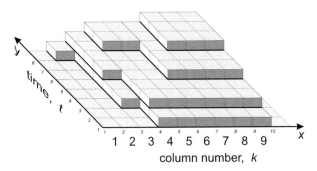

column number, *k*

Fig. 7. $(2 + 1)$ dimensional "terraces" corresponding to the growth of a heap in Fig. 6.

$H \leq t$. The average height of the first column thus increases as $\langle H_1(t) \rangle = pt$ for all t and its variance is given by $\sigma_1^2(t) = tp(1 - p)$. While the first column is thus trivial, the dynamics of heights in other columns is nontrivial due to the right-handed infinite range interactions between the columns. For convenience, we subsequently measure the height of any other column with respect to the first one. Namely, by height $h_k(t)$ we mean the height difference between the $(k+1)$-th column and the first one, $h_k(t) = H_{k+1}(t) - H_1(t)$, so that $h_0(t) = 0$ for all t.

To make progress for columns $k > 0$, we first consider a (2+1)-D construction of the heap as shown in Fig. 7, by adding an extra dimension indicating the time t. In Fig. 7, the x axis denotes the column number, the y axis stands for the time t and the z axis is the height h. In this figure, every time a new block is added, it "wets" all the sites at the same level to its "east" (along the x axis) and to its "north" (along the time axis). Here "wetting" means "screening" from further deposition at those sites at the same level. This $(2 + 1)$-D system of "terraces" is in one-to-one correspondence with the $(1+1)$-D heap in Fig. 6. This construction is reminiscent of the 3D anisotropic directed percolation (ADP) problem studied by Rajesh and Dhar [49]. Note however, that unlike the ADP problem, in our case each row labeled by t can contain only one deposition event.

The next step is to consider the projection onto the 2D (x, y)-plane of the level lines separating the adjacent terraces whose heights differ by 1. In this projection, some of the level lines may overlap partially on the plane. To avoid the overlap for better visual purposes, we make a shift $(x, y) \rightarrow (x + h(x, y), y)$ and represent these shifted directed lines on the 2D plane in Fig. 8. The black dots in Fig. 8 denote the points where the deposition events took place and the integer next to a dot denotes the time of this event. Note that each row in Fig. 8 contains a single black dot, i.e., only one deposition per unit of time can occur. In Fig. 8, there are 8 such events whose deposition times form the sequence $\{1, 2, 3, 4, 5, 6, 7, 8\}$ of

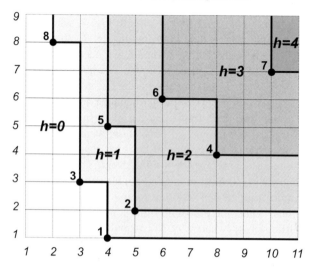

Fig. 8. The directed lines are the level lines separating adjacent terraces with height difference 1 in Fig. 2, projected onto the (x, y) plane and shifted by $(x, y) \rightarrow (x + h(x, y), y)$ to avoid partial overlap. The black dots denote the deposition events. The numbers next to the dots denote the times of those deposition events.

length $N = 8$. Now let us read the deposition times of the dots sequentially, but now column by column and vertically from top to bottom in each column, starting from the leftmost one. Then this sequence reads $\{8, 3, 5, 1, 2, 6, 4, 7\}$ which is just a permutation of the original sequence $\{1, 2, 3, 4, 5, 6, 7, 8\}$. In the permuted sequence $\{8, 3, 5, 1, 2, 6, 4, 7\}$ there are 3 LIS's: $\{3, 5, 6, 7\}$, $\{1, 2, 6, 7\}$ and $\{1, 2, 4, 7\}$, all of the same length $l_N = 4$. As mentioned before (see Fig. 4), this is precisely the number of piles in the patience sorting of the permutation $\{8, 3, 5, 1, 2, 6, 4, 7\}$.

Let us note one immediate fact from Fig. 8. The numbers belonging to the different level lines in Fig. 8 are in one-to-one correspondence with the piles $[\{8, 3, 1\}, \{5, 2\}, \{6, 4\}, \{7\}]$ in Aldous–Diaconis patience sorting game. Hence, each pile can be identified with an unique level line. Now, the height $h(x, t)$ at any given point (x, t) in Fig. 8 is equal to the number of level lines inside the rectangle bounded by the corners: $[0, 0]$, $[x, 0]$, $[0, t]$, $[x, t]$. Thus, we have the correspondence: height $\equiv$ number of level lines $\equiv$ number of piles $\equiv$ length l_n of the LIS. However, to compute l_n, we need to know the value of n which is precisely the number of black dots inside this rectangle.

Once the problem is reduced to finding the number of black dots or deposition events, we no longer need the Fig. 8 (as it may confuse due to the visual shift

$(x, y) \rightarrow (x + h(x, y), y))$ and can go back to Fig. 7, where the north-to-east corners play the same role as the black dots in Fig. 7. In Fig. 7, to determine the height $h_k(t)$ of the k-th column at time t, we need to know the number of deposition events inside the 2D plane rectangle $R_{k,t}$ bounded by the four corners $[0, 0], [k, 0], [0, t], [k, t]$. Let us begin with the last column $k = L$. For $k = L$ the number of deposition events N in the rectangle $R_{L,t}$ is equal to the time t because there is only one deposition event per time. In our example $N = t = 8$. For a general $k < L$ the number of deposition events N inside the rectangle $R_{k,t}$ is a random variable, since some of the rows inside the rectangle may not contain a north-to-east corner or a deposition event. The probability distribution $P_{k,t}(N)$ (for a given $[k, t]$) of this random variable can, however, be easily found as follows. At each step of deposition, a column is chosen at random from any of the L columns. Thus, the probability that a north-to-east corner will fall on the segment of line $[0, k]$ (where $k \leq L$) is equal to k/L. The deposition events are completely independent of each other, indicating the absence of correlations between different rows labeled by t in Fig. 7. So, we are asking the question: given t rows, what is the probability that N of them will contain a north-to-east corner? This is simply given by the binomial distribution

$$P_{k,t}(N) = \binom{t}{N} \left(\frac{k}{L}\right)^N \left(1 - \frac{k}{L}\right)^{t-N}, \tag{4.6}$$

where $N \leq t$. Now we are reduced to the following problem: given a sequence of integers of length N (where N itself is random and is taken from the distribution in Eq.(4.6)), what is the length of the LIS? Recall that this length is precisely the height $h_k(t)$ of the k-th column at time t in our model. In the thermodynamic limit $L \rightarrow \infty$ for $t \gg 1$ and any fixed k such that the quotient $\lambda = \frac{tk}{L}$ remains fixed but is arbitrary, the distribution in Eq.(4.6) becomes a Poisson distribution $P(N) \rightarrow e^{-\lambda} \frac{\lambda^N}{N!}$, with the mean $\lambda = \frac{tk}{L}$. We can then directly use the BDJ result in Eq.(3.3) to predict our main result for the height in the BD model,

$$h_k(t) \rightarrow 2\sqrt{\frac{tk}{L}} + \left(\frac{tk}{L}\right)^{1/6} \chi, \tag{4.7}$$

for large $\lambda = tk/L$, where the random variable χ has the Tracy-Widom distribution $F_2(\chi)$ as in Eq. (3.2). Using the known exact value $\langle \chi \rangle = -1.7711 \ldots$ from the Tracy-Widom distribution [1], we find exactly the asymptotic average height profile in the BD model,

$$\langle h_k(t) \rangle \rightarrow 2\sqrt{\frac{tk}{L}} - 1.7711 \ldots \left(\frac{tk}{L}\right)^{1/6}. \tag{4.8}$$

The leading square root dependence of the profile on the column number k has been seen numerically. Eq. (4.8) also predicts an exact sub-leading term with $k^{1/6}$ dependence. Similarly, for the variance, $\sigma_k^2(t) = \langle [h_k(t) - \langle h_k(t) \rangle]^2 \rangle$, we find asymptotically: $\sigma_k^2(t) \to c_0 \left(\frac{tk}{L} \right)^{1/3}$, where $c_0 = \langle [\chi - \langle \chi \rangle]^2 \rangle = 0.8132 \ldots [1]$. Eliminating the t dependence for large t between the average and the variance, we get, $\sigma_k^2(t) \approx a \langle h_k(t) \rangle^{2\beta}$ where the constant $a = c_0/2^{2/3} = 0.51228 \ldots$ and $\beta = 1/3$, thus recovering the KPZ scaling exponent. In addition to the BD model with infinite range right-handed interaction reported here, we have also analyzed the model (analytically within a mean field theory and numerically) when the right-handed interaction is short ranged. Somewhat surprisingly and pleasantly, we found that the asymptotic average height profile is independent of the range of interaction. A recent analysis of the short range BD model sheds light on this fact [59].

So far, we have demonstrated that for a flat initial condition, the height fluctuations in the BD model follow the Tracy-Widom distribution $F_{\text{GUE}}(x)$ which corresponds to the distribution of the largest eigenvalue of a random matrix drawn from a Gaussian unitary ensemble. In the context of the PNG model, Prähofer and Spohn [5] have shown that while the height fluctuations of a single PNG droplet follow the distribution $F_{\text{GUE}}(x)$, it is possible to obtain other types of universal distributions as well. For example, the height fluctuations in the PNG model growing over a flat substrate follow the distribution $F_{\text{GOE}}(x)$ where $F_{\text{GOE}}(x)$ is the distribution of the largest eigenvalue of a random matrix drawn from the Gaussian orthogonal ensemble. Besides, in a PNG droplet with two external sources at its edges which nucleate with rates ρ_+ and ρ_-, the height fluctuations have different distributions depending on the values of ρ_+ and ρ_-. For $\rho_+ < 1$ and $\rho_- < 1$, one gets back the distribution $F_{\text{GUE}}(x)$. If however $\rho_+ = 1$ and $\rho_- < 1$ (or alternatively $\rho_- = 1$ and $\rho_+ < 1$), one gets the distribution $F_{\text{GOE}}^2(x)$ which corresponds to the distribution of the largest of the superimposed eigenvalues of two independent GOE matrices. In the critical case $\rho_+ = 1$ and $\rho_- = 1$, one gets a new distribution $F_0(x)$ which does not have any random matrix analogy. For $\rho_+ > 1$ and $\rho_- > 1$, one gets Gaussian distribution. These results for the PNG model were obtained in Ref. [5] using a powerful theorem of Baik and Rains [4].

The question naturally arises as to whether these other distributions, apart from the $F_{\text{GUE}}(x)$, can also appear in the BD model considered in this paper. Indeed, they do. For example, if one starts with a staircase initial condition $h_k(0) = k$ for the heights in the BD model, one gets the distribution $F_{\text{GOE}}^2(x)$ for the scaled variable χ. This follows from the fact that for the staircase initial condition, in Fig. 2 there will be a black dot (or a north-to-east corner) at every value of k on the k axis at $t = 0$. Thus the black dots appear on the k axis with unit density. This corresponds to the case $\rho_+ = 1$ and $\rho_- = 0$ of the general results of Baik and Rains which leads to a $F_{GOE}^2(x)$ distribution. Of course, the density ρ_+ can be tuned between 0 and 1, by tuning the average slope of the staircase. For a generic $0 < \rho_+ \leq 1$, one can also vary ρ_- by putting an external source at the first column. Thus one can obtain, in principle, most of the distributions discussed in Ref. [4] by varying ρ_+ and ρ_-. Note that the case $\rho_- = 1$ (external source which drops one particle at the first column at every time step) and $\rho_+ = 0$ (flat substrate) is, however, trivial since the surface then remains flat at all times and the height just increases by one unit at every time step. The distribution $F_{GOE}(x)$ is, however, not naturally accessible within the rules of our model.

5. Sequence matching problem

In this section, I will discuss a different problem namely that of the alignment of two random sequences and will illustrate how the Tracy-Widom distribution appears in this problem. This is based on a joint work with S. Nechaev [12].

Sequence alignment is one of the most useful quantitative methods used in evolutionary molecular biology [60–62]. The goal of an alignment algorithm is to search for similarities in patterns in different sequences. A classic and much studied alignment problem is the so called 'longest common subsequence' (LCS) problem. The input to this problem is a pair of sequences $\alpha = \{\alpha_1, \alpha_2, \ldots, \alpha_i\}$ (of length i) and $\beta = \{\beta_1, \beta_2, \ldots, \beta_j\}$ (of length j). For example, α and β can be two random sequences of the 4 base pairs A, C, G, T of a DNA molecule, e.g., $\alpha = \{A, C, G, C, T, A, C\}$ and $\beta = \{C, T, G, A, C\}$. A subsequence of α is an ordered sublist of α (entries of which need not be consecutive in α), e.g, $\{C, G, T, C\}$, but not $\{T, G, C\}$. A common subsequence of two sequences α and β is a subsequence of both of them. For example, the subsequence $\{C, G, A, C\}$ is a common subsequence of both α and β. There can be many possible common subsequences of a pair of sequences. For example, another common subsequence of α and β is $\{A, C\}$. One simple way to construct different common subsequences (for two fixed sequences α and β) is by drawing lines from one member of the set α to another member of the set β such that the lines can not cross. For example, the common subsequence $\{C, G, A, C\}$ is shown by solid lines in Fig. 9. On the other hand the common subsequence $\{A, C\}$ is shown by the dashed lines in Fig. 9. The aim of the LCS problem is to find the longest of such common subsequences between two fixed sequences α and β.

This problem and its variants have been widely studied in biology [63–66], computer science [61, 67–69], probability theory [70–75] and more recently in

α: A   C   G   C   T   A   C

β: C   T   G   A   C

No. of matches in matching with solid lines = 4
No. of matches in matching with dashed lines = 2

Fig. 9. Two fixed sequences α : $\{A, C, G, C, T, A, C\}$ and β : $\{C, T, G, A, C\}$. The solid lines show the common subsequence $\{C, G, A, C\}$ and the dashed lines denote another common subsequence $\{A, C\}$.

statistical physics [76–78]. A particularly important application of the LCS problem is to quantify the closeness between two DNA sequences. In evolutionary biology, the genes responsible for building specific proteins evolve with time and by finding the LCS of the same gene in different species, one can learn what has been conserved in time. Also, when a new DNA molecule is sequenced *in vitro*, it is important to know whether it is really new or it already exists. This is achieved quantitatively by measuring the LCS of the new molecule with another existing already in the database.

For a pair of fixed sequences of length i and j respectively, the length $L_{i,j}$ of their LCS is just a number. However, in the stochastic version of the LCS problem one compares two random sequences drawn from c alphabets and hence the length $L_{i,j}$ is a random variable. A major challenge over the last three decades has been to determine the statistics of $L_{i,j}$ [70–74]. For equally long sequences $(i = j = n)$, it has been proved that $\langle L_{n,n} \rangle \approx \gamma_c n$ for $n \gg 1$, where the averaging is performed over all realizations of the random sequences. The constant γ_c is known as the Chvátal-Sankoff constant which, to date, remains undetermined though there exists several bounds [71, 73, 74], a conjecture due to Steele [72] that $\gamma_c = 2/(1 + \sqrt{c})$ and a recent proof [75] that $\gamma_c \to 2/\sqrt{c}$ as $c \to \infty$. Unfortunately, no exact results are available for the finite size corrections to the leading behavior of the average $\langle L_{n,n} \rangle$, for the variance, and also for the full probability distribution of $L_{n,n}$. Thus, despite tremendous analytical and numerical efforts, exact solution of the random LCS problem has, so far, remained elusive. Therefore it is important to find other variants of this LCS problem that may be analytically tractable.

Computationally, the easiest way to determine the length $L_{i,j}$ of the LCS of two arbitrary sequences of lengths i and j (in polynomial time $\sim O(ij)$) is via using the recursive algorithm [61, 78]

$$L_{ij} = \max[L_{i-1,j}, L_{i,j-1}, L_{i-1,j-1} + \eta_{i,j}], \tag{5.1}$$

subject to the initial conditions $L_{i,0} = L_{0,j} = L_{0,0} = 0$. The variable $\eta_{i,j}$ is either 1 when the characters at the positions i (in the sequence α) and j (in the sequence β) match each other, or 0 if they do not. Note that the variables $\eta_{i,j}$'s are not independent of each other. To see this consider the simple example – matching of two strings $\alpha =$ AB and $\beta =$ AA. One has by definition: $\eta_{1,1} = \eta_{1,2} = 1$ and $\eta_{2,1} = 0$. The knowledge of these three variables is sufficient to predict that the last two letters will not match, i.e., $\eta_{2,2} = 0$. Thus, $\eta_{2,2}$ can not take its value independently of $\eta_{1,1}$, $\eta_{1,2}$, $\eta_{2,1}$. These residual correlations between the $\eta_{i,j}$ variables make the LCS problem rather complicated. Note however that for two random sequences drawn from c alphabets, these correlations between the $\eta_{i,j}$ variables vanish in the $c \to \infty$ limit.

A natural question is how important are these correlations between the $\eta_{i,j}$ variables, e.g., do they affect the asymptotic statistics of $L_{i,j}$'s for large i and j? Is the problem solvable if one ignores these correlations? These questions naturally lead to the Bernoulli matching (BM) model which is a simpler variant of the original LCS problem where one ignores the correlations between $\eta_{i,j}$'s for all c [78]. The length $L_{i,j}^{BM}$ of the BM model satisfies the same recursion relation in Eq. (5.1) except that $\eta_{i,j}$'s are now independent and each drawn from the bimodal distribution: $p(\eta) = (1/c)\delta_{\eta,1}+(1-1/c)\delta_{\eta,0}$. This approximation is expected to be exact only in the appropriately taken $c \to \infty$ limit. Nevertheless, for finite c, the results on the BM model can serve as a useful benchmark for the original LCS model to decide if indeed the correlations between $\eta_{i,j}$'s are important or not. Unfortunately, even in the absence of correlations, the exact asymptotic distribution of $L_{i,j}^{BM}$ in the BM model has so far remained elusive, mainly due to the nonlinear nature of the recursion relation in Eq. (5.1). The purpose of this Rapid Communication is to present an exact asymptotic formula for the distribution of the length $L_{n,n}^{BM}$ in the BM model for all c. So far, only the leading asymptotic behavior of the average length in the BM model is known [78] using the 'cavity' method of spin glass physics [79],

$$\langle L_{n,n}^{BM} \rangle \approx \gamma_c^{BM} n \tag{5.2}$$

where $\gamma_c^{BM} = 2/(1 + \sqrt{c})$, same as the conjectured value of the Chvátal-Sankoff constant γ_c for the original LCS model. However, other properties such as the variance or the distribution of $L_{n,n}^{BM}$ remained intractable even in the BM model. We have shown [12], as illustrated below, that for large n,

$$L_{n,n}^{BM} \to \gamma_c^{BM} n + f(c) n^{1/3} \chi \tag{5.3}$$

where χ is a random variable with a n-independent distribution, Prob$(\chi \le x) = F_2(x)$ which is precisely the Tracy-Widom distribution in Eq. (3.2). Indeed, we were also able to compute the functional form of the scale factor $f(c)$ exactly for all c [12],

$$f(c) = \frac{c^{1/6}(\sqrt{c}-1)^{1/3}}{\sqrt{c}+1}. \tag{5.4}$$

This allows us to calculate the average including the subleading finite size correction term and the variance of $L_{n,n}^{BM}$ for large n,

$$\langle L_{n,n}^{BM} \rangle \approx \gamma_c^{BM} n + \langle \chi \rangle f(c) n^{1/3}$$

$$\mathrm{Var}\, L_{n,n}^{BM} \approx \left(\langle \chi^2 \rangle - \langle \chi \rangle^2 \right) f^2(c)\, n^{2/3}, \tag{5.5}$$

where one can use the known exact values [1], $\langle \chi \rangle = -1.7711\ldots$ and $\langle \chi^2 \rangle - \langle \chi \rangle^2 = 0.8132\ldots$. These exact results thus invalidate the previous attempt [78] to fit the subleading correction to the mean in the BM model with a $n^{1/2}/\ln(n)$ behavior and also to fit the scaled distribution with a Gaussian form. Note that the recursion relation in Eq. (5.1) can also be viewed as a $(1+1)$ dimensional directed polymer problem [77, 78] and some asymptotic results (such as the $O(n^{2/3})$ behavior of the variance of $L_{n,n}$ for large n) can be obtained using the arguments of universality [77]. However, this does not provide precise results for the full distribution along with the correct scale factors that are obtained here.

It is useful to provide a synopsis of our method in deriving these results. First, we prove the results in the $c \to \infty$ limit, by using mappings to other models. To make progress for finite c, we first map the BM model exactly to a 3-d anisotropic directed percolation (ADP) model first studied by Rajesh and Dhar [49]. This ADP model is also precisely the same as the directed polymer model studied by Johansson [3], as discussed in the previous section and for which the exact results are known as in Eq. (4.4). To extract the results for the BM model from those of Johansson's model, we use a simple symmetry argument which then allows us to derive our main results in Eqs. (5.3)–(5.5) for all c. As a check, we recover the $c \to \infty$ limit result obtained independently by the first method.

In the BM model, the length $L_{i,j}^{BM}$ can be interpreted as the height of a surface over the 2 dimensional (i, j) plane constructed via the recursion relation in Eq. (5.1). A typical surface, shown in Fig. 10 (a), has terrace-like structures.

It is useful to consider the projection of the level lines separating the adjacent terraces whose heights differ by 1 (see Fig.11) onto the 2-D (i, j) plane. Note that, by the rule Eq. (5.1), these level lines never overlap each other, i.e., no two paths have any common edge. The statistical weight of such a projected 2-D configuration is the product of weights associated with the vertices of the 2-D plane. There are five types of possible vertices with nonzero weights as shown in Fig. 11, where $p = 1/c$ and $q = 1 - p$. Since the level lines never cross each other, the weight of the first vertex in Fig. 11 is 0.

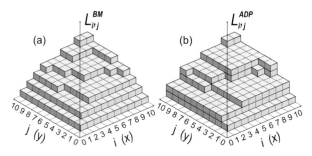

Fig. 10. Examples of (a) BM surface $L_{i,j}^{BM} \equiv \tilde{h}(x, y)$ and (b) ADP surface $L_{i,j}^{ADP} \equiv h(x, y)$.

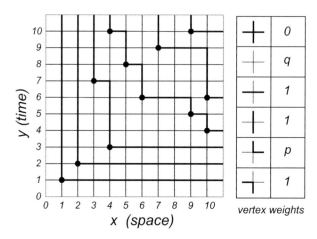

Fig. 11. Projected 2-d level lines separating adjacent terraces of unit height difference in the BM surface in Fig.10 (a). The adjacent table shows the weights of all vertices on the 2-d plane.

Consider first the limit $c \to \infty$ (i.e., $p \to 0$). The weights of all allowed vertices are 1, except the ones shown by black dots in Fig. 11, whose associated weights are $p \to 0$. The number N of these black dots inside a rectangle of area $A = ij$ can be easily estimated. For large A and $p \to 0$, this number is clearly Poisson distributed with the mean $\overline{N} = pA$. The height $L_{i,j}^{BM}$ is just the number of level lines $\mathcal{N}$ inside this rectangle of area $A = ij$. One can easily estimate $\mathcal{N}$ by following precisely the method outlined in the previous subsection in the context of the ballistic deposition model. Following the same analysis as in the ballistic deposition model, it is easy to see that the number of level lines $\mathcal{N}$ inside the rectangle (for large A), appropriately scaled, has a limiting behavior, $\mathcal{N} \to 2\sqrt{\overline{N}} + \overline{N}^{1/6} \chi$, where χ is a random variable with the Tracy-Widom distribution. Using $\overline{N} = pA = ij/c$, one then obtains in the limit $p \to 0$,

$$L_{i,j}^{BM} = \mathcal{N} \to \frac{2}{\sqrt{c}}\sqrt{ij} + \left(\frac{ij}{c}\right)^{1/6}\chi. \tag{5.6}$$

In particular, for large equal length sequences $i = j = n$, we get for $c \to \infty$

$$L_{n,n}^{BM} \to \frac{2}{\sqrt{c}}n + c^{-1/6}n^{1/3}\chi. \tag{5.7}$$

For finite c, while the above mapping to the LIS problem still works, the corresponding permutations of the LIS problem are not generated with equal probability and hence one can no longer use the BDJ results.

For any finite c, we can however map the BM model to the ADP model studied by Rajesh and Dhar [49]. In the ADP model on a simple cubic lattice the bonds are occupied with probabilities p_x, p_y, and p_z along the x, y and z axes and are all directed towards increasing coordinates. Imagine a source of fluid at the origin which spreads along the occupied directed bonds. The sites that get wet by the fluid form a 3-d cluster. In the ADP problem, the bond occupation probabilities are anisotropic, $p_x = p_y = 1$ (all bonds aligned along the x and y axes are occupied) and $p_z = p$. Hence, if the point (x, y, z) gets wet by the fluid then all the points (x', y', z) on the same plane with $x' \geq x$ and $y' \geq y$ also get wet. Such a wet cluster is compact and can be characterized by its bounding surface height $H(x, y)$ as shown in Fig.(1b). It is not difficult to see [49] that the height $H(x, y)$ satisfies exactly the same recursion relation of the directed polymer as in Eq. (4.2) where $\xi_{x,y}$'s are i.i.d. random variables taking nonnegative integer values with $\text{Prob}(\xi_{x,y} = k) = (1 - p)\,p^k$ for $k = 0, 1, 2, \ldots$. Thus the ADP model of Rajesh and Dhar is precisely identical to the directed polymer model studied by Johansson with exactly the same distribution of the noise $\xi(x, y)$.

While the terrace-like structures of the ADP surface look similar to the BM surfaces (compare Figs. (10 a) and (10 b)), there is an important difference between the two. In the ADP model, the level lines separating two adjacent terraces can overlap with each other [49], which does not happen in the BM model. However, by making the following change of coordinates in the ADP model [49]

$$\zeta = x + h(x, y); \quad \eta = y + h(x, y) \tag{5.8}$$

one gets a configuration of the surface where the level lines no longer overlap. Moreover, it is not difficult to show that the projected 2-D configuration of level lines of this shifted ADP surface has exactly the same statistical weight as the projected 2-D configuration of the BM surface. Denoting the BM height by

$\tilde{h}(x, y) = L^{BM}_{x,y}$, one then has the identity, $\tilde{h}(\zeta, \eta) = h(x, y)$, which holds for each configuration. Using Eq. (5.8), one can rewrite this identity as

$$\tilde{h}(\zeta, \eta) = h\big(\zeta - \tilde{h}(\zeta, \eta), \eta - \tilde{h}(\zeta, \eta)\big). \tag{5.9}$$

Thus, for any given height function $h(x, y)$ of the ADP model, one can, in principle, obtain the corresponding height function $\tilde{h}(x, y)$ for all (x, y) of the BM model by solving the nonlinear equation (5.9). This is however very difficult in practice. Fortunately, one can make progress for large (x, y) where one can replace the integer valued discrete heights by continuous functions $h(x, y)$ and $\tilde{h}(x, y)$. Using the notation $\partial_x \equiv \partial/\partial x$ it is easy to derive from Eq. (5.8) the following pair of identities,

$$\partial_x h = \frac{\partial_\zeta \tilde{h}}{1 - \partial_\zeta \tilde{h} - \partial_\eta \tilde{h}}; \quad \partial_y h = \frac{\partial_\eta \tilde{h}}{1 - \partial_\zeta \tilde{h} - \partial_\eta \tilde{h}}. \tag{5.10}$$

In a similar way, one can show that

$$\partial_\zeta \tilde{h} = \frac{\partial_x h}{1 + \partial_x h + \partial_y h}; \quad \partial_\eta \tilde{h} = \frac{\partial_y h}{1 + \partial_x h + \partial_y h}. \tag{5.11}$$

We then observe that Eqs. (5.10) and (5.11) are invariant under the simultaneous transformations

$$\zeta \to -x; \quad \eta \to -y; \quad \tilde{h} \to h. \tag{5.12}$$

Since the height is built up by integrating the derivatives, this leads to a simple result for large ζ and η,

$$\tilde{h}(\zeta, \eta) = h(-\zeta, -\eta). \tag{5.13}$$

Thus, if we know exactly the functional form of the ADP surface $h(x, y)$, then the functional form of the BM surface $\tilde{h}(x, y)$ for large x and y is simply obtained by $\tilde{h}(x, y) = h(-x, -y)$. Changing $x \to -x$ and $y \to -y$ in Johansson's expression for the ADP surface in Eq. (4.4) we thus arrive at our main asymptotic result for the BM model

$$L^{BM}_{x,y} = \tilde{h}(x, y) \to \frac{2\sqrt{pxy} - p(x + y)}{q} +$$

$$+ \frac{(pxy)^{1/6}}{q} \left[(1 + p) - \sqrt{\frac{p}{xy}}\,(x + y)\right]^{2/3} \chi, \tag{5.14}$$

where $p = 1/c$ and $q = 1 - 1/c$. For equal length sequences $x = y = n$, Eq. (5.14) then reduces to Eq. (5.3).

To check the consistency of our asymptotic results, we further computed the difference between the left- and the right-hand sides of Eq. (5.9),

$$\Delta h(\zeta, \eta) = \tilde{h}(\zeta, \eta) - h\big(\zeta - \tilde{h}(\zeta, \eta), \eta - \tilde{h}(\zeta, \eta)\big), \tag{5.15}$$

with the functions $h(x, y)$ and $\tilde{h}(x, y)$ given respectively by Eqs. (4.4) and (5.14). For large $\zeta = \eta$ one gets

$$\Delta h(\zeta, \zeta) \to \big[p^{1/3}\chi^2/3(1 - \sqrt{p})^{4/3}\big]\zeta^{-1/3}. \tag{5.16}$$

Thus the discrepancy falls off as a power law for large ζ, indicating that indeed our solution is asymptotically exact. We have also performed numerical simulations of the BM model using the recursion relation in Eq. (5.1) for $c = 2, 4, 9, 16, 100$. Our preliminary results [12] for relatively small system sizes (up to $n = 5000$) are consistent with our exact results in Eqs. (5.3)–(5.5).

Thus, the Tracy-Widom distribution also describes the asymptotic distribution of the optimal matching length in the BM model, for all c. Given that the correlations in the original LCS model become negligible in the $c \to \infty$ limit, it is likely that the BM asymptotics in Eq. (5.7) would also hold for the original LCS model in the $c \to \infty$ limit. An important open problem is to determine whether the Tracy-Widom distribution also appears in the LCS problem for finite c. The precise distribution obtained here (including exact prefactors) for all c in the BM model will serve as a useful benchmark to which future simulations of the LCS problem can be compared.

6. Conclusion

In these lectures I have discussed 4 a priori unrelated problems and tried to give a flavour of the recent developments that have found a deep connection between these problems. These connections have now established the fact that they all share one common limiting distribution, namely the Tracy-Widom distribution that describes the asymptotic distribution law of the largest eigenvalue of a random matrix. I have also discussed the probabilities of large deviations of the largest eigenvalue, in the range outside the validity of the Tracy-Widom law. As examples, I have demonstrated in detail, in two specific models a ballistic deposition model and a sequence alignment problem, how they can be mapped on to the longest increasing subsequence problem and consequently proving the existence of the Tracy-Widom distribution in these models.

There have been many other interesting recent developments in this rather broad area encompassing different fields that I did not have the scope to discuss in these lectures. There are, of course, plenty of open questions that need to be addressed, some of which I mention below.

Finite size effects in growth models: We have discussed how the Tracy-Widom distribution appears as the limiting scaled height distribution in several $(1 + 1)$ dimensional growth models that belong to the KPZ universality class of fluctuating interfaces. Indeed, for a fluctuating surface with height $H(x, t)$ growing over a substrate of infinite size one now believes that at long times $t \gg 1$

$$H(x, t) = vt + bt^{1/3}\chi \tag{6.1}$$

where χ is a time-independent random variable with the Tracy-Widom distribution. The prefactors v (the velocity of the interface) and b are model dependent, but the distribution of the scaled variable $\chi = (H - vt)/bt^{1/3}$ is universal for large t. The non-universal prefactors are often very hard to compute. We have shown two examples in these lectures where these prefactors can be computed exactly. Note, however, that the result in Eq. (6.1) holds only in an infinite system. In any real system with a finite substrate size L, the result in Eq. (6.1) will hold only in the growing regime of the surface, i.e., when $1 \ll t \ll L^z$, where z is the dynamical exponent characterizing the surface evolution. For example, for the KPZ type of interfaces in $(1 + 1)$ dimensions, $z = 3/2$. However, when $t \gg L^z$, the probability distribution of the height fluctuation $H - \langle H \rangle$ will become time-independent. For example, for $(1 + 1)$ dimensional KPZ surfaces with periodic boundary conditions, it is well known [45] that the stationary distribution of the height fluctuation is a simple Gaussian, Prob$[H - \langle H \rangle = x] \propto \exp[-x^2/a_0 L]$ where a_0 is a non-universal constant and the typical fluctuation scales with the system size as $L^{1/2}$. An important open question is how does the distribution of the height fluctuation crosses over from the Tracy-Widom form to a simple Gaussian form as t becomes bigger than the crossover time L^z. It would be nice to show this explicitly in any of the simple models discussed above.

A direct connection between the growth models and random matrices: The existence of the Tracy-Widom distribution in many of the growth models discussed here, such as the polynuclear growth model or the ballistic deposition model, rely on the mapping to the LIS problem and subsequently using the BDJ results that connect the LIS problem to random matrices. It is certainly desirable to find to a direct mapping between the growth models and the largest eigenvalue of a random matrix. Recent work by Spohn and collaborators [11] linking the top edge of a PNG growth model to Dyson's Brownian motion of the eigenvalues of a random matrix perhaps provides a clue to this missing link.

Largest Lyapunov exponent in population dynamics: The Tracy-Widom distribution and the associated large-deviation function discussed in Section 3 conceivably have important applications in several systems where the largest eigenvalue controls the spectral properties of the system. Some examples were discussed in Section 3. Recently, it has been shown that the statistics of largest eigenvalue (the largest Lyapunov exponent) is also of importance in population growth of organisms in fluctuating environments [80]. It would be interesting to see if Tracy-Widom type distribution functions also appear in these biological problems.

Sequence matching, directed polymer and vertex models: In the context of the sequence matching problem discussed in Section 6, we have demonstrated how the statistical weights of the surface generated in the Bernoulli matching model of the sequence alignment are exactly identical to that of a 5-vertex model on a square lattice (see Fig. 11). This is a useful connection because there are many quantities in the 5-vertex models that can be computed exactly by employing the Bethe ansatz techniques and subsequently one can use those results for the sequence alignment or equivalently for the directed polymer model. Recently, in collaboration with K. Mallick and S. Nechaev, we have made some progress in these directions [81]. A very interesting open issue is if one can derive the Tracy-Widom distribution by using the Bethe ansatz techniques.

Other issues related to the sequence matching problem: There are also many other interesting open questions associated with the sequence matching problem. We have shown that the length of the longest matching is Tracy-Widom distributed only in the Bernoulli matching model which is a simpler version of the original LCS problem. In the BM model one has ignored certain correlations, as we discussed in detail. This approximation is exact in the $c \to \infty$ limit, where c is the number of different types of alphabets, e.g. for DNA, $c = 4$. Is this approximation good even for finite c? In other words, is the optimal matching length in the original LCS problem also Tracy-Widom distributed? It would also be interesting if one can make a systematic $1/c$ expansion of the LCS model, i.e., keeping the correlations up to $O(1/c)$. Numerical simulations the LCS problem [82] for binary sequence $c = 2$ indeed indicates that the standard deviation of the optimal matching length scales as $n^{1/3}$ where n is the sequence size, as in the BM model, the question is if the scaled distribution is also Tracy-Widom or not. For the original LCS problem, there is also a curious result due to Bonetto and Matzinger [82] that claims that if the value of c for the two sequences are not the same (for example, the first sequence may be drawn randomly from 3 alphabets and the second may be a binary sequence), then the standard deviation of the optimal matching length scales as $n^{1/2}$ for large n, which is rather surprising! It would be interesting to study the statistics of optimal matches between more than two sequences. Finally, here we have just mentioned the matching of random sequences. It would be interesting and important to study the statistics of optimal matching lengths between non-random sequences, e.g., when there are some correlations between the members of any given sequence.

Acknowledgments

My own contribution to this field that is presented here was developed partly in collaboration with D.S. Dean and partly with S. Nechaev. It is a pleasure to thank them. I also thank O. Bohigas, K. Mallick and P. Vivo for collaborations on related topics. esides, I acknowledge useful discussions with G. Biroli, J.-P. Bouchaud, A.J. Bray, A. Comtet, D. Dhar, S. Leibler, O.C. Martin, M. Mézard, R. Rajesh and C. Tracy. I also thank the organizers J.-P. Bouchaud and M. Mézard and all other participants of this summer school for physics, for fun, and for making the school a memorable one.

References

[1] C. Tracy and H. Widom, Comm. Math. Phys. **159**, 151 (1994); **177**, 727 (1996); For a review see *Proceedings of the International Congress of Mathematicians*, Beijing 2002, Vol. I, ed. LI Tatsien, Higher Education Press, Beijing 2002, pp. 587–596.

[2] J. Baik, P. Deift, and K. Johansson, J. Amer. Math. Soc. **12**, 1119 (1999).

[3] K. Johansson, Comm. Math. Phys. **209**, 437 (2000).

[4] J. Baik and E.M. Rains, J. Stat. Phys. **100**, 523 (2000).

[5] M. Prähofer and H. Spohn, Phys. Rev. Lett. **84**, 4882 (2000); Physica A **279**, 342 (2000).

[6] J. Gravner, C.A. Tracy, and H. Widom, J. Stat. Phys. **102**, 1085 (2001).

[7] S.N. Majumdar and S. Nechaev, Phys. Rev. E **69**, 011103 (2004).

[8] T. Imamura and T. Sasamoto, Nucl. Phys. **B699**, 503 (2004); J. Stat. Phys. **115**, 749 (2004).

[9] P.L. Ferrari, Commun. Math. Phys. **252**, 77 (2004).

[10] T. Sasamoto, J. Phys. A.: Math. Gen. **38**, L549 (2005).

[11] H. Spohn, Physica A **369**, 71 (2006) and references therein.

[12] S.N. Majumdar and S. Nechaev, Phys. Rev. E **72**, 020901(R) (2005).

[13] M.G. Vavilov, P.W. Brouwer, V. Ambegaokar, and C.W.J. Beenaker, Phys. Rev. Lett. **86**, 874 (2001); A. Lamacraft and B.D. Simons, Phys. Rev. B **64** 014514 (2001); P.M. Ostrovsky, M.A. Skvortsov, and M.V. Feigel'man, Phys. Rev. Lett. **87**, 027002 (2001); J.S. Meyer and B.D. Simons, Phys. Rev. B **64**, 134516 (2001); A. Silva and L.B. Ioffe, Phys. Rev. B **71**, 104502 (2005); A. Silva, Phys. Rev. B **72**, 224505 (2005).

[14] G. Biroli, J-P. Bouchaud, and M. Potters, cond-mat/0609070, Europhys. Lett. **78**, 10001 (2007).

[15] E.P. Wigner, Proc. Cambridge Philos. Soc. **47**, 790 (1951).

[16] M.L. Mehta, Random Matrices, 2nd Edition, (Academic Press) (1991).

[17] A. Cavagna, J.P. Garrahan, and I. Giardina, Phys. Rev. B. **61**, 3960 (2000).

[18] Y.V. Fyodorov Phys. Rev. Lett. **92**, 240601 (2004); *ibid* Acta Physica Polonica B **36**, 2699 (2005).

[19] L. Susskind, arXiv:hep-th/0302219; M.R. Douglas, B. Shiffman, and S. Zelditch, Commu. Math. Phys. **252**, 325 (2004).

[20] A. Aazami and R. Easther, J. Cosmol. Astropart. Phys. JCAP03 013 (2006).

[21] L. Mersini-Houghton, Class. Quant. Grav. **22**, 3481 (2005).

[22] P. Vivo, S.N. Majumdar, and O. Bohigas, in preparation.

[23] D.S. Dean and S.N. Majumdar, Phys. Rev. Lett. **97**, 160201 (2006).

[24] A.J. Bray and D.S. Dean, cond-mat/0611023.

[25] Y.V. Fyodorov, H-J. Sommers, and I. Williams, cond-mat/0611585.

[26] A. Soshnikov, Commu. Math. Phys. **207**, 697 (1999).

[27] J. Baik, G. Ben Arous, and S. Péché, Ann. Proab. **33**, 1643 (2005).

[28] P. Cizeau and J.-P. Bouchaud, Phys. Rev. E **50**, 1810 (1994).

[29] Z. Burda et al., cond-mat/0602087.

[30] S.M. Ulam, *Modern Mathematics for the Engineers*, ed. by E.F. Beckenbach (McGraw-Hill, New York, 1961), p. 261.

[31] J.M. Hammersley, *Proc. VI-th Berkeley Symp. on Math. Stat. and Probability*, (University of California, Berkeley, 1972), Vol. 1, p. 345.

[32] A.M. Vershik and S.V. Kerov, Sov. Math. Dokl. **18**, 527 (1977).

[33] For a review, see D. Aldous and P. Diaconis, Bull. Amer. Math. Soc. **36**, 413 (1999).

[34] C. Schensted, Canad. J. Math. **13**, 179 (1961).

[35] C.M. Mallows, Bull. Inst. Math. Appl. **9**, 216 (1973).

[36] D.A. Huse and C.L. Henley, Phys. Rev. Lett. **54**, 2708 (1985).

[37] M. Kardar, G. Parisi, and Y.C. Zhang, Phys. Rev. Lett. **56**, 889 (1986).

[38] D. Forster, D.R. Nelson, and M.J. Stephen, Phys. Rev. A **16**, 732 (1977).

[39] B. Derrida and H. Spohn, J. Stat. Phys. **51**, 817 (1988).

[40] M. Mezard, J. Phys. Fr. **51**, 1831 (1990).

[41] D.S. Fisher and D.A. Huse, Phys. Rev. B **43**, 10728 (1991).

[42] M. Kardar, Nucl. Phys. **B290**, 582 (1987).

[43] J.Z. Imbrie and T. Spencer, J. Stat. Phys. **52**, 609 (1988); J. Cook and B. Derrida, J. Stat. Phys. **57**, 89 (1989).

[44] M. Kardar and Y.C. Zhang, Phys. Rev. Lett. **58**, 2087 (1987); M. Kardar, Phys. Rev. Lett. **55**, 2923 (1989).

[45] T. Halpin-Healy and Y.C. Zhang, Phys. Rep. **254**, 215 (1995).

[46] D.A. Huse, C.L. Henley, and D.S. Fisher, Phys. Rev. Lett. **55**, 2924 (1985).

[47] J. Krug, P. Meakin, and T. Halpin-Healy, Phys. Rev. A **45**, 638 (1992).

[48] J. Krug and T. Halpin-Healy, J. Phys. A **31**, 5939 (1998).

[49] R. Rajesh and D. Dhar, Phys. Rev. Lett. **81**, 1646 (1998).

[50] P. Meakin, *Fractals, Scaling, and Growth Far From Equilibrium* (Cambridge University Press, Cambridge, 1998).

[51] J. Krug and H. Spohn, in *Solids Far From Equilibrium* (ed. by C. Godrèche) (Cambridge University Press, New York, 1991).

[52] M. Eden, in *Proc. IV-th Berkeley Symp. on Math. Sciences and Probability*, ed. by F. Neyman (University of California, Berkeley, 1961), Vol. 4, p. 223.

[53] J.M. Kim and J.M. Kosterlitz, Phys. Rev. Lett. **62**, 2289 (1989).

[54] F.C. Frank, J. Cryst. Growth **22**, 233 (1974); J. Krug and H. Spohn, Europhys. Lett. **8**, 219 (1989). J. Kertész and D.E. Wolf, Phys. Rev. Lett. **62**, 2571 (1989).

[55] M.J. Vold, J. Colloid Sci. **14**, 168 (1959); P. Meakin, P. Ramanlal, L.M. Sander, and R.C. Ball, Phys. Rev. A **34**, 5091 (1986); J. Krug and H. Spohn, Phys. Rev. A **38**, 4271 (1988).

[56] J. Krug and P. Meakin, Phys. Rev. A **40**, 2064 (1989); *ibid*, **43**, 900 (1991).

[57] D. Dhar, Phase Transitions **9**, 51 (1987); L.-H. Gwa and H. Spohn, Phys. Rev. Lett. **68**, 725 (1992); D. Kim, Phys. Rev. E **52**, 3512 (1995).

[58] E. Katzav and M. Schwartz, Phys. Rev. E **70**, 061608 (2004).

[59] E. Katzav, S. Nechaev, and O. Vasilyev, cond-mat/0611537.

[60] M.S. Waterman, *Introduction to Computational Biology* (Chapman & Hall, London, 1994).

[61] D. Gusfield, *Algorithms on Strings, Trees, and Sequences* (Cambridge University Press, Cambridge, 1997).

[62] R. Dubrin, S. Eddy, A. Krogh, and G. Mitchison, *Biological Sequence Analysis* (Cambridge University Press, Cambridge, 1998).

[63] S.B. Needleman and C.D. Wunsch, J. Mol. Biol. **48**, 443 (1970).

[64] T.F. Smith and M.S. Waterman, J. Mol. Biol. **147**, 195 (1981); Adv. Appl. math. **2**, 482 (1981).

[65] M.S. Waterman, L. Gordon, and R. Arratia, Proc. Natl. Acad. Sci. USA **84**, 1239 (1987).

[66] S.F. Altschul et al., J. Mol. Biol. **215**, 403 (1990).

[67] D. Sankoff and J. Kruskal, *Time Warps, String Edits, and Macromolecules: The theory and practice of sequence comparison* (Addison Wesley, Reading, Massachussets, 1983).

[68] A. Apostolico and C. Guerra, Alogorithmica **2**, 315 (1987).

[69] R. Wagner and M. Fisher, J. Assoc. Comput. Mach. **21**, 168 (1974);

[70] V. Chvátal and D. Sankoff, J. Appl. Probab. **12**, 306 (1975).

[71] J. Deken, Discrete Math. **26**, 17 (1979).

[72] J.M. Steele, SIAM J. Appl. Math. **42**, 731 (1982).

[73] V. Dancik and M. Paterson, in STACS94, Lecture Notes in Computer Science **775**, 306 (Springer, New York, 1994).

[74] K.S. Alexander, Ann. Appl. Probab. **4**, 1074 (1994).

[75] M. Kiwi, M. Loebl, and J. Matousek, math.CO/0308234.

[76] M. Zhang and T. Marr, J. Theor. Biol. **174**, 119 (1995).

[77] T. Hwa and M. Lassig, Phys. Rev. Lett. **76**, 2591 (1996); R. Bundschuh and T. Hwa, Discrete Appl. Math. **104**, 113 (2000).

[78] J. Boutet de Monvel, European Phys. J. B **7**, 293 (1999); Phys. Rev. E **62**, 204 (2000).

[79] M. Mézard, G. Parisi, and M.A. Virasoro, eds., *Spin Glass Theory and Beyond* (World Scientific, Singapore, 1987).

[80] E. Kussell and S. Leibler, Science **309**, 2075 (2005).

[81] S.N. Majumdar, K. Mallick, and S. Nechaev, in preparation.

[82] F. Bonetto and H. Matzinger, arXiv:math.CO/0410404.

Course 5

ECONOMIES WITH INTERACTING AGENTS

Alan Kirman

GREQAM, Université d'Aix Marseille III, Ecole des Hautes Etudes en Sciences Sociales, Institut Universitaire de France

J.-P. Bouchaud, M. Mézard and J. Dalibard, eds.
Les Houches, Session LXXXV, 2006
Complex Systems

Contents

1. Introduction

Why is economic activity so fascinating? Most people when they first come to economics find it extraordinary that the myriad of disparate individual economic activities come to be coordinated. A modern economy is composed of a mass of networks of interacting agents and these in turn are linked. Like a biological organism, the closer one looks at the structure, the more complicated its organisation appears. The natural question is, how is it that from the highly complex networks of interaction between agents reasonably coherent aggregate behaviour emerges? These individuals are essentially aware of what is happening around them but are not necessarily conscious of the more global picture. One reaction to this is to say that, if it is aggregate behaviour that interests us, we should be concerned only with looking at the relationship among aggregate variables and not be concerned with the, possibly complicated, interrelated micro-behaviour that generates those relationships. Whilst any economist knows that individual economic agents constantly interact with each other in different ways and for different purposes, it could be argued that to analyse this is unnecessary for the explanation of macroeconomic phenomena. For example, it is far from obvious that, in the neurosciences, it is necessary to revert to the study of the behaviour of neurons to explain thought processes. Yet the practice of analysing macro-relationships without considering their micro-foundations is now, in economics, almost universally considered as unscientific. Furthermore, the microfoundations in question are particularly restrictive. Thus economics is criticised from two points of view, firstly there is a lack of satisfaction with the basic model of the rational individual and this is the attack led by behavioural economists. Secondly there is the problem highlighted in this contribution, that of aggregation, the passage from the micro to the macro level.

Although one has to be careful about what one means by "explain", economists are not alone in claiming that macro phenomena should be explained by an analysis of the underlying micro-behaviour of a system. Where they are more alone is in suggesting that the behaviour of the aggregate can be captured by considering that aggregate as a typical or "representative" individual. To see this, it is enough to take a look at the standard economic model. This is one in which agents make their decisions independently of each other as a function of some generally available market signals such as prices. Thus the only way in which agents interact is through the price system. Yet direct interaction between agents is an intrinsic part of economic organization. Agents trade with each other, communicate with each other, give to and obtain information from other agents, and furthermore they infer things from the actions of others. Taking such interaction into account involves changing the notion of the relationship between micro and macro behaviour. The usual way to "derive" macro behaviour is to assume that one can represent aggregate behaviour as that corresponding to some average or "representative" agent. This might seem to be a plausible approach when agents act in isolation responding only to market signals but even in this case it is problematic and may lead to erroneous analysis (see Kirman (1992), Summers (1991), and Jerison (2006)). However, in models in which individuals interact directly and possibly randomly, as is well known from statistical mechanics for example, macro regularities may appear which cannot be thought of as corresponding to average individual behaviour.

One important problem arises immediately. The notion of equilibrium in the normal economic model is simple and well defined. Agents make choices as a function of market signals. An equilibrium signal is one which engenders choices which satisfy some rule of consistency. In the market framework the signal is the price vector and the consistency condition is that markets clear. In a system where there is interaction the idea must be modified.

Consider the opposite extreme to the model with isolated decision makers, that in which every individual reacts to every other one. This is the situation in a non-cooperative game, every individual takes account of the strategies of every other player and what is more, knows that the other does so and knows that he knows and so forth. Despite the complexity of the reasoning imputed to the individuals, the basic equilibrium notion, that referred to as the Nash equilibrium is also clear and well defined. It is given by a strategy for each player which cannot be improved upon, given the strategies of the others. It is clear that market equilibrium can be recast in this way, no agent given the market rules could make a better choice (see Debreu (1952)).

Nash equilibrium and the full-blown game theoretic model share fundamental defects with the market model. Such equilibria are not, in general, unique and there is no obvious way in which they would be arrived at. Furthermore the complexity of these games and the reasoning involved means that they pose problems of logical consistency (see Binmore (1992)) and that only very simple examples are analytically tractable.

The purpose of these notes is to describe models in which the interaction between agents is between the two extremes I have just described. Agents may, for example, interact with any other agents (global interaction) even though they may actually only do so with some randomly drawn sample or they may meet only with their neighbours (local interaction) and thus the consequences of their behaviour may take some time to have an effect on agents with whom they are not directly linked. Although the static equilibrium notion is worth examining, perhaps more interesting is the dynamic evolution of the systems I have in mind. Here there are a number of possibilities. One can think of a repeated series of markets or games, each of which is the same as its predecessor and in which the payoffs in the current game are unaffected by the players' actions in previous rounds. In such a case one might look, for example, at a stationary equilibrium which would be just a sequence of identical equilibria for each round. Of course, even if current payoffs only reflect current strategies the latter will and should take into account histories of play to date. This simple consideration leads to the "folk theorem" sort of results with a large class of equilibria. Here an equilibrium would be the choice of a strategy which defined an action at each point in time as a function of the past history of the game. This would define the outcome in each of the sequences of identical games. An interesting problem here and one which has been widely investigated is under what circumstances will players, by learning, converge to some particular equilibrium? This will depend crucially on what information they have at each point and in the context here who they play against in each period.

In general economic models, current actions influence future outcomes and, in particular, the payoff from future actions. When agents take account of this and the consequences of the actions of other agents for current and future payoffs the situation becomes highly complicated. Either one can think of the individuals as trying to solve the full blown equilibrium problem *ab initio* or alternatively one might ask whether players would learn or adjust from one period to the other and whether such behaviour would converge to any specific outcome.

Economic agents will not necessarily learn from the experience of <u>all</u> other agents but might, for example, learn from the experience of a limited randomly drawn subset. Another way of looking at this problem is to deprive the players of any rationality to identify them with strategies and allow Darwinian selection to operate through the "birth" of new agents and the attribution of more successful strategies to those agents. This is the approach adopted by evolutionary game theory and which represents a rather radical simplification of the over-complex general repeated game model. In local interaction models one is interested in knowing how the configuration of states of the individuals who are related to each other through a specific network behaves over time, whether this settles to any stationary situation and whether this is characterised by local "clusters" of individuals in a particular state.

Perhaps the most interesting problem of all in this context it to model how the links between individuals develop and hence how market structure itself evolves. This can be thought of in two ways. Firstly one can think of some physical, say spatial, framework in which locations could be thought of as a lattice structure for example. In this case agents might choose where to situate themselves in lattice and one could observe the evolution of activity and of resources over time (see e.g. Durlauf (1990)). Alternatively, one could think of individuals as interacting with each other in different ways, trading, communicating for example, and reinforcing their links with those people with whom interaction has proved to be more profitable in the past. For example in repeated prisoners' dilemmas in an evolutionary context players are typically matched with each other with a uniform probability. Yet, in reality it is unlikely that one would keep interacting with players who chose strategies which lead to poor pay-offs for their partners, or rather opponents. It is therefore of interest to see what happens when players can refuse to play with partners with whom they have had bad experiences in the past (see Stanley et al. (1994)).

A final point is in order. Each of the models described below captures some aspect of economic reality and each should be thought of as giving some insight into a particular phenomenon. We are far from having a general model which captures all the different types of economic interaction one can observe. For the moment we must content ourselves with looking at simple lattice-like models when trying to get insights into spatial problems, random matching models for the labour market and some goods markets, temporary equilibrium models for looking at some aspects of financial markets and so forth.

When we allow for the full feedback from observed outcomes to current behaviour and for the evolution of new types of behaviour even the examples I shall present display considerable analytical problems. Thinking of the economy as a whole, as a complex, adaptive, interactive system is still an exercise for the future but the examples presented here represent small steps in this direction.

2. Models of segregation: a physical analogy

At the end of the 60's Tom Schelling (a good summary of the variants of his model is given in Schelling (1978), introduced a model of segregation which showed essentially that even if people only have a very mild preference for living with neighbours of their own colour, as they move to satisfy their preferences complete segregation will occur.

This result was greeted with surprise and has provoked a large number of reactions. Before asking why this was it is worth looking at the model that Schelling introduced.

Take a large chess board, and place a certain number of black and white counters on the board, leaving some free places. Then take a counter at random. The counter prefers to be on a square where at most four of his eight neighbours are of a different colour than his own. This "utility function" is illustrated in Figure 1.

Schelling's individual preferences

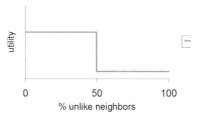

Fig. 1.

If an individual is not happy he moves to the nearest unoccupied space where he is happy. A sequence of moves illustrating this is shown in Figure 2 and it is clear that segregation develops rapidly.

In a recent paper. Kirman and Vinkovic (2006), we have developed an argument to suggest that this result is not, in fact, surprising and that some simple physical theory can explain the segregation phenomenon. Numerous papers have been written using Schelling's original model as a starting point and they have used many variants. Typically, the form of the utility function used by Schelling has been questioned as have the number of neighbours, the rules for moving, the amount of unoccupied space and all of these have been claimed to be relevant for the results.

Using simulations in general, a number of authors have argued that modifying the parameters just mentioned yields different patterns of segregation. What has been lacking so far is a theoretical structure which can incorporate all of these considerations and then can produce analytic results for each variation. It is precisely this that we propose in the recent paper mentioned above, Perhaps the nearest approach to ours is that adopted by Pollicot and Weiss (2005). They however, examine the limit of a Laplacian process in which individuals' preferences are strictly increasing in the number of like neighbours. In this situation it is intuitively clear that there is a strong tendency to segregation. Yet, Schelling's result has become famous precisely because the preferences of individuals for segregatin were not particularly strong whereas in the paper by Pollicot and Weiss there is a strong preference for neighbours of one's own colour.

Indeed this observation leads us directly to the interesting question as to why people have paid so much attention to this model. Pancs and Vriend (2006) give three explanations. Firstly, it is the "micro motives" of the agents that lead to the aggregate result, in other words a macro phenomenon "emerges" from individual behaviour. Furthermore, as we have said, the result at the aggregate level is more extreme than what would have been necessary to make the individuals happy. Secondly, the model is remarkably simple and its workings can be understood

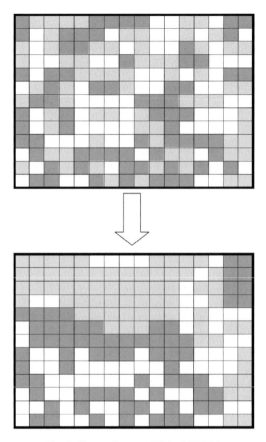

Fig. 2. (Source Pancs and Vriend (2006).)

immediately. Lastly the model concerns a topic of great concern in the developed world, that of segregation whether it be on racial or any other grounds.

Three things are of particular interest to economists. The first is the organisation of the system into "regions" or clusters, each containing individuals of only one colour and second the shape of the frontier between the regions.

A second question concerns the importance of the number of empty spaces. How does this affect the final configuration? In some analyses, agents are simply swapped with each other if this makes them both happier and there are no free spaces. The question here is as to the role of the free space and will it inevitably wind up as a "no-man's land" between two clusters?

A third question concerns the distance which an unhappy individual may traverse in order to reach a position in which he is happier. If only local movement is allowed, as opposed to movement over any distance, does this affect the final outcome? A natural economic interpretation would be that there is a cost to moving and that this is an increasing function of distance.

In our article we develop a physical analogy to the Schelling model. The agent's happiness or satisfaction in this interpretation corresponds to the energy stored in the agent. An increase in utility would correspond to a decrease in the internal energy. An agent, therefore, wants to minimize his energy, and this may be achieved, either by taking some action or through the interaction with his environment. In the Schelling model the utility of an agent depends on her local environment and the agent moves if the utility declines below a certain value. The interpretation of this situation as that of a particle system immediately poses problem for philosophers such as Schabas (2005) who insist on the importance of intention and reason in economic models and these attributes cannot be given to particles. She also argues that a clear analogy between energy and utility has yet to be found even though many references to such an analogy have been made in the literature. As to the latter, it seems that the case here provides a counter-example. As to the argument about reason and intent, this argument is not relevant here. If agents act in a certain way which corresponds to the behaviour of particles in a system, then the reason for their behaviour does not prevent us using models of inanimate particles that act in the same way. To analyse the behaviour of a stone thrown by a human does not require one to know why he threw it. In the particle analogue the internal energy depends on the local concentration (number density) of like or unlike particles. This analogue is a typical model description of microphysical interactions in dynamical physical systems of gases, liquids, solids, colloids, solutions, etc. Interactions between particles are described with potential energies, which result in inter-particle forces driving particles' dynamics.

The goal of such models is to study the collective behaviour of a large number of particles. In the Schelling model the number of particles is conserved and the total volume in which they move around is constant (that is, the underlying lattice is fixed). The pressure can be also considered constant. The system is not closed, however, because the energy lost by a particle is not transferred to other particles, but transmitted out of the system (for an economist this is not a zero sum game). Similarly, a particle can gain energy from outside the system when an unlike particle moves into the neighborhood and lowers the particle's utility (once again one would say in economics terms that agents generate externalities by their actions). Hence, in physical terms, the system can change its energy only by emitting or absorbing radiation and not by changing its volume or pressure or number of particles.

Since the basic tendency of such a physical system is to minimize its total energy, it can do that only by arranging particles into structures (clusters) that reduce the individual internal energy of as many particles as possible. In other words, interparticle forces attract particles into clustering and the stability of a cluster is determined by this force. Hence, all we need to do is to look at the behaviour of this force on the surface of a cluster to see if the surface will be stable or if it will undergo deformations and ripping. If we think of the Schelling model and the utility function that he chose it is clear that the only individuals who do not have high utility are those on the boundary and it is for this reason that the evolution and final form of the boundary are of particular interest.

The initial configuration of the system may well be out of equilibrium. A typical initial state of the Schelling model involves randomized positions of particles in the system. After that we let the system evolve and reach its equilibrium condition. Since we can consider different types of utility functions, the equivalent physical systems can also differ in their underlying definition of interparticle forces. This, in turn will lead to different types of « equilibria ». Three types of equilibria can be immediately predicted based on physical analogues:

i) Those which are *dynamically unstable*: particles cannot create any stable structure and the whole system is constantly changing, with clusters forming and breaking apart all the time. Water clouds are an example of this equilibrium, where water droplets are constantly forming and reevaporating. In a society this would correspond to constantly changing neighbourhoods with no propensity to settle to any clear spatial structure.

ii) Those which are *dynamically stable*: well defined clusters are formed. They maintain their size despite constant change of its surface due to rearrangement of the surface particles. A wobbling droplet is such an example. A special case of this equilibrium is a cluster that seemingly does not change its shape because moving particles do not accumulate at any specific point on the surface. A small raindrop maintaining its spherical shape is an example of this. This, in the context of spatial arrangements of society, corresponds to the idea that areas on the border between two segregated zones will see movement since it is precisely in these areas that individuals wish to change. The boundary will shift slightly all the time but will not be fundamentally retraced.

iii) Those which « *freeze* »: all particles stop their movement because all of them reach the energy minimum or do not have any available space to move into. The frozen cluster structure can be either completely irregular (amorphous) or well structured. Glass is an example of the former and crystals of the latter. This sort of frozen structure can happen when there is relatively little free space available and when particles can only move locally.

As we have seen, we can study the formation and stability of a cluster by considering the energy of particles on its surface. The interparticle forces derived from the utility function will determine the optimal change of the cluster's surface as it tries to minimise the energy of surface particles. The first step toward this analysis is a transition from the discretized lattice of the Schelling model into a continuous medium by refining the lattice and taking the limit. Hence instead of counting the number of agents ΔN in a discrete area ΔA we can write dA/dN. Next we transform the utility function from counting the individuals in a neighborhood around an agent into the measurement of the total solid angle covered by different particles around the differential area dA (Figure 3). Utility is replaced with energy $\varepsilon(\theta)$, with high utility corresponding to low energy and vice versa. This gives the total energy $dE = n\varepsilon(\theta)\,dA$ stored in dA *or* the energy per unit area, $dE/dA = n\varepsilon(\theta)$. Since we are interested in the cluster surface, we take a differential length dL of the surface and write its energy per unit length as $dE = n\varepsilon(\theta)\,dL$.

The total surface energy of a cluster is an integral of dE over the whole cluster surface. It is clear that the energy at a contact point depends on the local curvature θ at that point. We thus have an expression for the surface tension which is given by taking a local gradient along dL and the *tension* force at a point r is given by

$$F(r) = -L\nabla_r E(r) = n\varepsilon\big(\theta(r)\big)$$

where L is a unit vector along dL. The energy thus depends on the utility function chosen, since this is what determines $n(\theta)$. In Schelling's model with the basic step function then energy is a constant if $\theta > 180$ and 0 otherwise. This corresponds to the discrete model in which the utility of an individual is 1 if at least half of the neighbours are similar and 0 otherwise.

A cluster tries to minimize this surface energy and this is what drives its dynamics. In physics, this force is usually called *surface tension*. By analysing this we can predict the behaviour of a cluster for a given utility function for the individuals. In the continuous version we show that any convex cluster surface experiences a tangential force pushing the surface toward flattening. The outcome of this surface reshaping are clusters that "prefer" flat surfaces, except for parts covered by a boundary layer made of empty space. In Figure 3 we can see the approximation of the discrete model by a continuum.

In Figure 4 we can see the forces at work on the frontier trying to "flatten" the surface.

When clusters start to grow they compete for particles they can collect. A bigger cluster has a bigger surface for collecting particles, thus also a higher probability of collecting them. Eventually, one cluster grows bigger than others and grows until it collects all available particles. Smaller clusters can exist if they manage to « freeze » their particles, not letting them « evaporate » and travel to the big cluster. If the volume of the big cluster is not much smaller than the total

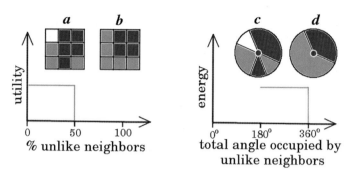

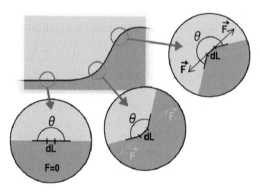

Fig. 3. (Source Vinkovic and Kirman (2006).)

Fig. 4.

available volume then particles segregate into two distinct clusters separated by a flat surface. Alternatively, a cluster at the centre of the available volume forms. Since particles have the same probability to reach the cluster's surface from any direction, such a cluster will stay close to a spherical shape. In Vinkovic and Kirman we give detailed illustrations of the sort of cluster patterns that can form under different parameter values.

The growth of one big cluster out of a mixture of particles is a common natural phenomenon and is well known in the physics of metal surfaces, where the evolution of fewer but larger nano-structures out of many smaller nano-structures is an actively studied phenomenon generally known as « coarsening ». The Schelling model is, therefore, just a discretized model of such physical phenomena.

Thus this analysis allows us not only to obtain the well-known « surprising » result of Schelling but also can show that other configurations can appear depending on the distance that individuals can move and how much space is available and the particular utility function that is attributed to them. A rather simple physical analysis allows us to consider a number of interesting phenomena which may provide a richer panoply of possibilities than the pure segregation found in the original model.

3. Market relations

The second example concerns the network of buyer-seller relations on markets. Markets are frequented by agents, buyers and sellers, who establish relationships over time. In a study of the Marseille wholesale fish market, for which we have full details of every transaction over three years, we have examined the network of relations between buyers and sellers. We have developed (Weisbuch et al. (2000)), a theoretical framework within which to analyse the emergence of links between agents in such a market.

Consider the simplest market in which there are n buyers i and m sellers j. Buyers visit one seller per day and the quantity that they buy determines the profit they will obtain outside the market. Sellers offer a quantity which is fixed by them each day before the market opens. Thus when a buyer chooses a seller he will either obtain a unit or he will be unlucky because the seller that he has chosen is out of stock. The cumulated profit that a buyer i obtains from a seller j is given by the following expression,

$$J_{ij}(t) = (1 - \gamma)J_{ij}(t - 1) + \pi_{ij}(t)$$

where obviously the current profit is zero if the seller has already sold all of his stock.

Now we have to say how buyers choose sellers. We will assume that the rule which determines the probability that buyer i will purchase from seller j is made on the basis of the profit that they obtained in the past from those sellers. In particular we will assume that the probability p_{ij} that i will visit j in that period is given by:

$$p_{ij}(t) = \frac{\exp \beta J_{ij}(t)}{\sum_k \exp \beta J_{ik}(t)}$$

where β is a reinforcement parameter which describes how sensitive the individual is to past profits. This non-linear updating rule will be familiar from discrete choice problems and is known as the "quantal response function" in game theory. It is widely used in statistical physics.

This rule can be derived as resulting from a trade-off between "exploration" and "exploitation". Agents in many economic situations have to face a choice between using their past experience as to actions which have been profitable and trying out new actions. Exploration can be thought of as trying to choose those actions which will provide the maximum information about what is profitable and in this case, with no prior information, one would choose probabilities to maximise the entropy of the system. This amounts to sampling the possible actions with uniform probability. Exploitation would involve maximising expected profit given the performance of the rules in the past. Maximising a linear combination of the two,

$$\underset{p_j}{Max}\,\beta \sum_j p_{ij} J_{ij}(t) + \sum_j p_j \log p_j$$

yields the rule given above.

Given this rule, we can now derive the equilibrium probabilities for this situation. We shall use a technique adopted from physics called the "mean field" approach. It is clear that if we use our basic approach directly we have to model a stochastic process in discrete time. At each point buyers will visit sellers. Depending on whether they are successful or not they will then sell the quantity, in the simplest case one unit, on the outside market and make a profit. This profit will modify the probability of their choosing different sellers in the next period. Once they have made their choice in that period the process will recommence.

Now the mean field approach makes two approximations. Firstly, it replaces the random variables by their expected values and secondly it approximates a discrete process by a continuous one. To see how this works consider the following expression where a continuous representation is given and where the profit is now replaced by the expected profit.

$$d J_{ij}/dt = -\gamma J_{ij} + \langle \pi_{ij} \rangle$$

where the expected profit is given by the following:

$$\langle \pi_{ij} \rangle = \Pr(q_j > 0) \pi_{ij} \frac{\exp(\beta J_{ij})}{\sum_k \exp(\beta J_{ik})}$$

where $\Pr(q_j > 0)$ represents the probability that seller j still has a unit to sell when buyer i arrives. Now let us consider the very simplest case in which there are two sellers who both charge the same price and therefore who generate the same profit for the buyer and furthermore they always have sufficient fish to satisfy all customers, that is we will assume that

$$\Pr(q_j > 0) = 1 \quad \text{and} \quad \pi_{ij} = \pi \quad \text{for all } i \text{ and } j$$

Clearly the cumulated profits will evolve over time and the equilibrium will be reached when the probabilities no longer change that is when,

$$d J_{ij}/dt = 0$$

The equilibrium conditions in this case for seller 1 and symmetrically for seller 2 are given by,

$$\gamma J_1 = \pi \frac{\exp(\beta J_1)}{\exp(\beta J_1) + \exp(\beta J_2)}$$

$$\gamma J_2 = \pi \frac{\exp(\beta J_2)}{\exp(\beta J_1) + \exp(\beta J_2)}$$

Consider now the difference, that is, $\Delta = J_1 - J_2$ and we have,

$$\Delta = \frac{\pi \exp(\beta \Delta - 1)}{\gamma \exp(\beta \Delta + 1)}$$

What are the solutions? There are two cases to examine.

Firstly, if $\beta < \beta_c = \frac{2\gamma}{\pi}$ then there is one solution which is given by $\Delta = 0$ and $J_1 = J_2 = \frac{\pi}{2\gamma}$.

Secondly, if $\beta > \beta_c$ there are three solutions and the equiprobability solution $\Delta = 0$ is now unstable, furthermore there is a rapid transition to two distinct solutions at $\beta = \beta_c$ as shown below in Figure 5.

What this says is that if individuals have a β above the critical level then they will have a much higher probability of going to one seller than another even though visiting each seller is equally profitable. Thus, in the market we would expect to see such individuals being very loyal to a particular seller. In the Weisbuch et al. model, as I have said, this sort of "phase transition" is derived using the "mean field" approach. The latter is open to the objection that random variables are replaced by their means and, in consequence, the process derived is only an approximation. The alternative is to consider the fully stochastic process but this is often not tractable, and one can then resort to simulations to see whether the theoretical results from the approximation capture the features of the simulated stochastic process.

Consider, for example, the case of 3 sellers, and 30 buyers. The simplex in figure 6 below can be used to represent this situation. Each of the 30 buyers has certain probabilities of visiting each of the sellers and thus can be thought of as a point in the simplex. If he is equally likely to visit each of the three sellers then he can be represented as a point in the centre of the triangle. If, on the other hand, he visits one of the sellers with probability one then he can be shown as a point at one of the apexes of the triangle.

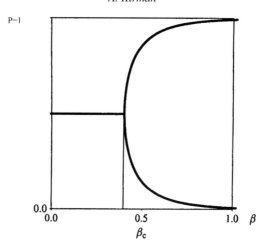

Plot of both equilibrium preference coefficients versus the discrimonation rate β. Below transition rate β_c, preference coefficients are equal, but they rise or plummet sharply when the discrimonation rate β increases above the transition. When profits from both sellers are equal (as in this figure), loyalty to one describes the upper branch, while loyalty to the other describes the lower branch.

Fig. 5. (Source Weisbuch et al. (2000).)

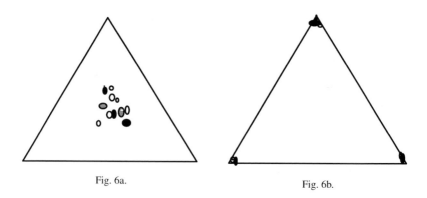

Fig. 6a.

Fig. 6b.

Thus, at any one point in time, the market is described by a cloud of points in the triangle and the question is how will this cloud evolve? If buyers all become loyal to particular sellers then the result will be that all the points, corresponding to the buyers will be at the apexes of the triangle as in figure 6a, this might be thought of as a situation in which the market is "ordered". On the other hand, if buyers learn to search randomly amongst the sellers then the result will be a cluster of points at the centre of the triangle, as in Figure 6b. In network terms the first case corresponds to a situation in which every buyer has a link with every

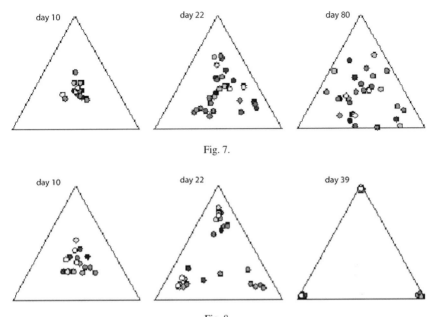

Fig. 7.

Fig. 8.

seller and in the second case each buyer has a link to only one seller. Which of these situations will develop depends crucially on the parameters β, γ and the profit π per transaction. The stronger the reinforcement, the slower the individual forgets and the higher the profit the more likely is it that order will emerge. In particular, as mentioned, the transition from disorder to order is very sharp with a change in β.

Why is it important to understand the nature of the trading relationships that emerge? The answer is that the aggregate efficiency of the market depends on them. A market where agents are predominantly searchers is less efficient than one with many loyal traders. When searching is preeminent, sellers' supply will often not be equal to the demand they face. Some buyers will be unsatisfied and some sellers will be left with stock on their hands. This is particularly important in markets for perishable goods.

To illustrate the evolution of the networks involved in this model, it is worth looking at Figures 7 and 8. As before we are looking at a situation with three sellers and thirty buyers and the evolution of the loyalty of the individuals and the group can be seen as time passes.

What was done to produce these figures was to simulate the actual stochastic process to see whether the results obtained in the deterministic approximation still held up in the actual process. Thus at each step the market participants decide which rule to use according to its success and then form their offer prices, for the sellers, and reservation prices for the buyers. This, in turn determines the profit the buyers and sellers make in that period and this conditions the buyers' new probability of visiting each seller. They draw a seller according to these probabilities and the process starts again. The only way in which the two figures above differ is in the value of β. Figure 7 shows the situation in which beta is small and below the critical value which is determined by the profit and the rate of discount. The graph corresponding to this situation is one in which each buyer visits several sellers with a positive probability. Although each agent changes his probability of visiting the sellers they are essentially still in the centre of the simplex.

In Figure 8 however, β is above the critical value and each of the buyers rapidly become loyal to a single seller. Thus the graph is essentially deterministic and is characterised for each individual by a single link with one seller. It is clear that the graph structure is very different in the two cases.

As has already been observed the deterministic graph is more efficient in terms of resource utilisation. Buyers know where to find what they want and sellers know how much to supply. Now, what predictions could we make about an empirical situation. Recall that we have always talked about high and low values of β itself. Yet, since we know nothing about the βs of each individual we must assume that they are all the same. However, the critical value of β is different for each individual and depends on the two parameters γ and π. In the Marseille wholesale fish market where the model was tested those buyers who have a low critical value of β will be those who have a low γ, or a high π. That is they will be those who come often to the market and who make large profits.

Yet, for the results of the simulations to be convincing we have to deal with one unsatisfactory feature of the way in which the model was simulated. Why do buyers with high betas always find what they want in the simulated model? It may be because there are no random buyers to interfere with their learning process. Yet what we are claiming is that in a market where both types of buyer are present those with high betas develop deterministic links while the low betas continue to shop randomly. Our simulations are of markets with all traders with high betas or all traders with low betas. Might it not be the case that when mixed our results no longer hold?

To test this we ran a simulation with a mixed population. Half of them had low critical values of β and half high values. The result can be seen in Figure 9.

Here the circles at the corners represent buyers with a low critical value while the individuals at the centre have high critical value. There is clearly a separation and if the situation were represented as a stochastic graph one would have a mixture of links with probability $p_{ij} = 1$ and others with almost $p_{ij} = \frac{1}{3}$ but none with intermediate values.

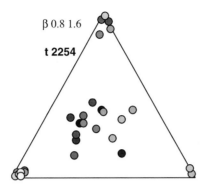

Fig. 9.

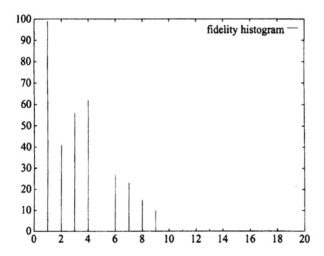

The sample of buyers include those visiting more than once per month, and present in the market for more than six months. The distribution is clearly bimodal, with one peak corresponding to fidelity to one seller, and another peak centred on visting 4 sellers on average.

Fig. 10. (Source Weisbuch et al. (2000).)

Now what about the evidence? The empirical data, which as I have said, cover every single transaction over three years, reveals that this is precisely the case, frequent buyers and those who buy large quantities are, indeed, extremely loyal whilst the others "shop around" and do not become loyal at all. The distribution of the number of sellers visited per month is bimodal with a peak at 1 as shown in Figure 10.

Thus the use of a simple model from physics enabled us to explain the data from an empirical situation and to test a well defined hypothesis as to what distinguishes loyal and disloyal buyers in a market, and furthermore to suggest that the buyers should be sharply divided into two groups.

4. Financial markets

The standard model of financial markets, is based on a number of very simple hypotheses. All information is contained in asset prices. As a result the only movements that can occur cannot be anticipated, otherwise there would be a possible arbitrage to be made. Thus, in essence the prices must follow essentially a random walk, which may have a drift. This observation was first made by Bachelier (1900) and the basic idea of prices following Brownian motion is now widely used. The problem with models based on the standard assumptions is that they do not reproduce certain well established stylised facts about empirical price series.

In standard models where there is uncertainty the usual way of achieving consistency is to assume that agents have common and "rational" expectations. Yet, if agents have such common expectations how can there be trade? Indeed there are many "no trade" theorems for such markets. How then do we deal with the fact that agents do, in fact, differ in their opinions and forecasts and that this is one of the main sources of trade? There is also an old problem of "excess volatility", that is prices have a higher variance than the returns on the assets on which they are based. One answer is to allow for direct interaction between agents other than through the market mechanism. Models reminiscent of the Ising model from physics have been used to do this. For example, one might suggest that individuals may change their opinion or forecast as a function of those of other agents. In simple models of financial markets such changes may be self reinforcing. If agents forecast an increase in the price of an asset and others are persuaded by their view the resultant demand will drive the price up thereby confirming the prediction. However, the market will not necessarily "lock on" to one view forever. Indeed if agents make stochastic rather than deterministic choices, then it is sure that the system will swing back to a situation in which another opinion dominates. The stochastic choices are not irrational, however. The better the results obtained when following one opinion, the higher the probability of continuing to hold that opinion.

Such models will generate swings in opinions, regime changes and "long memory" all of which are hard to explain with standard analysis. An essential feature of these models is that agents are wrong for some of the time, but, whenever they are in the majority they are essentially right. Thus they are not systematically irrational (for examples of this sort of model see, Lux and Marchesi (2000), Brock and Hommes (1999) and Kirman and Teyssiere (2002) and for a recent survey De Grauwe and Grimaldi (2005)). Thus the behaviour of the agents in the market cannot be described, to use Greenspan's well known phrase, as "irrational exuberance".

Economists faced with this sort of model are often troubled by the lack of any equilibrium notion. The process is always moving, agents are neither fully rational nor systematically mistaken. Worse the process never settles down to a particular price even without exogenous shocks. There remains one final question. Suppose that we accept this kind of model, can we say anything about the time series that result? If we consider some of these models, for certain configurations of parameters they could become explosive. There are two possible reactions to this. Since we will never observe more than a finite sample it could well be that the underlying stochastic process is actually explosive but this will not prevent us from trying to infer something about the data that we observe. Supposing however, that we are interested in being able, from a theoretical point of view, to characterise the long run behaviour of the system. Furthermore if we treat the process as being stochastic and do not make a deterministic approximation, then we have to decide what constitutes an appropriate long run equilibrium notion.

With Hans Foellmer and Ulrich Horst (Foellmer et al. (2005)), we have analysed formally the sort of price process discussed here and we have been able to produce some analytical results characterising the process. Furthermore, we can give a long run equilibrium notion which is not the convergence to a particular price vector. If prices change all the time how may we speak of "equilibrium"? The idea is to look at the evolving distribution of prices and to see if we can characterise its long-run behaviour. We have to examine the process governing the evolution of prices and profits and to see under what conditions we can show that it is ergodic i.e that the proportion of time that the price takes on each possible value converges over time and that the limit distribution is unique. This means that, unlike the "anything" can happen often associated with deterministic chaos we can say that in the long run the price and profits process does have a well-defined structure.

If we think about the sort of model which involve chartists and fundamentalists interacting on a market, it is clear that what destabilises the price process is the presence of chartists. Indeed, we know that if there were only chartists the process would explode. What we can show in the model that I will present here, is that the distribution of the time averages of prices converges, if the probability of any individual becoming a chartist is not too high. Thus although the presence of chartists prevents the price process from being bounded, if they do not dominate for too long, which is what is guaranteed by putting a bound on the probability of becoming a chartist, then the distribution of the time averages of prices is stable. In other words, although prices are always changing and cannot be bounded, structure emerges in the time series and this is probably the appropriate notion of equilibrium for such models.

The model presented here is from Foellmer et al. (2005). Consider a finite set A of agents who trade a single risky financial asset on a market.

At each point in time agent $a \in A$ has a log-linear excess demand function given by:

$$e_t^a(p, \omega) = c_t^a\left(\hat{S}_t^a(\omega) - \log p\right) + \eta_t^a(\omega)$$

where $\hat{S}_t^a$ and η_t^a represent the agents current "reference level" for the price in question, and their liquidity demand. This is already a departure from the previous models in that the excess demand is not derived from any underlying utility function but such demand or excess demand functions have been widely used in the mathematical finance literature, see Foellmer and Schweizer (1990) for example. The advantage of working in logs is evident since it avoids the problem of the underlying prices becoming negative. The excess demand function can be envisaged as having two components, an excess demand based on the difference between the current price and a reference level price and a liquidity demand which is random. The latter convention is often used, but can be dispensed with in two ways. Either one can introduce a second market, if for example one thinks of the foreign exchange market this would be the « rest of the world » or one introduces an aggregate excess demand for the asset or for foreign denominated liabilities, which arises not from the financial market itself but from simple trade requirements. At each point in time the temporary equilibrium price is that which makes the *excess demand* for the asset equal to zero.

$$S_t = \frac{1}{c_t} \sum_{a \in A} c_t^a \hat{S}_t^a(\omega) + \eta_t$$

Thus, equilibrium prices are a weighted average of individual reference prices and the liquidity demand. Where do the reference prices of the individuals come from? A simple idea is that they are provided by m experts or « gurus » among whom agents choose. An alternative interpretation is that these reference prices are forecasts obtained by using one of m forecasting rules. This can be represented as

$$\hat{S}_t^a \in \{R_t^1, \ldots, R_t^m\}$$

The proportion of agents who choose rule or guru I at time t is then given by

$$\pi_t^i = \frac{1}{c_t} \sum_{a \in A} c_t^a 1_{\{\hat{S}_t^a = R_t^i\}}$$

where the last term is the indicator function for the choice of guru, that is, it takes on value 1 if guru i is chosen and 0 otherwise. Given this, we can rewrite the expression for the log equilibrium prices as

$$S_t = \sum_{i=1}^{m} \pi_t^i R_t^i + \eta_t$$

Now the question arises as to how the reference prices are determined. The idea is that each guru, $i \in \{1, \ldots, m\}$ has a fundamental value F^i but also takes account of the price trend. Thus the reference value for guru i is given by

$$R_t^i = S_{t-1} + \alpha^i \left[F^i - S_{t-1} \right] + \beta^i [S_{t-1} - S_{t-2}]$$

Now we can look at the evolution of the price of the asset over time which is given by

$$S_t = F(S_{t-1}, S_{t-2}, \tau_t)$$
$$= \left[1 - \alpha(\pi_t) + \beta(\pi_t) \right] S_{t-1} - \beta(\pi_t) S_{t-2} + \gamma(\pi_t, \eta_t)$$

and in these expressions there are two random elements one of which, the choice probabilities of the agents is endogenous and the other, the liquidity demand is exogenous. Thus the random environment in which the process evolves is given by,

$$\{\tau_t\} = \left\{ (\pi_t, \eta_t) \right\}.$$

Now let me look a little closer at the characteristics of the gurus and see how they can be assimilated to the « chartists » and « fundamentalists » used widely in many models of financial markets. If we think of several « gurus » or fundamentalist rules, what identifies rule i is a fundamental value F_i to which those people who follow this rule believe that the exchange rate or asset price will return. So the forecast will have the form

$$R_t^i = S_{t-1} + \alpha^i \left[F^i - S_{t-2} \right]_t, \quad \alpha^i \in (0, 1)$$

Suppose, for the moment that there were only fundamentalist rules or gurus then the evolution of prices would behave as

$$S_t = \left[1 - \alpha(\pi_t) \right] S_{t-1} + \gamma \left((\pi_t, \eta_t) \right), \quad \alpha(\pi_t) = \sum_{i=1}^{m} \alpha^i \pi_t^i$$

since $\alpha_i \in (0, 1)$, the sequence of temporary price equilibria will be mean-reverting. The process can be considered as an Ornstein–Uhlenbeck process in a random environment. In a sense fundamentalists stabilise the market dynamics.

This is clear since the fundamental values which are finite in number bound the range of possible prices.

Now, consider the chartists who are simply extrapolating from previous prices. They may do so in any more or less sophisticated way, but what is important is that their forecast should only be based on previous prices. Consider the simplest case in which chartists forecast that the change in the future will be based on the last change. Forecasts will differ in the proportion of that change that is expected next time. Formally the recommendations take the form

$$R_t^i = S_{t-1} + \beta^i [S_{t-1} - S_{t-2}] + \eta_t, \quad \beta^i \in (0, 1)$$

Now, we can ask what would happen if there were only chartists on the market. In this case with m chartist rules or gurus the process would be given by

$$S_t - S_{t-1} = \beta(\pi_t)[S_{t-1} - S_{t-2}] + \eta_t, \quad \beta(\pi_t) = \sum_{i=1}^{m} \beta^i \pi_t^i$$

What is interesting here is that now, returns become mean reverting but prices will explode. Thus, in contrast to fundamentalists, chartists have a destabilising effect on prices.

Now put the two pieces together and the process becomes

$$S_t = \left[1 - \alpha\big((\pi_t)\big) + \beta(\pi_t)\right]S_{t-1} - \beta(\pi_t)S_{t-2} + \gamma(\pi_t, \eta_t)$$

It should be clear, from the above, that when there are few chartists the process will be stable but that the price process will become explosive if the chartists dominate. There will be bubbles and crashes but these will be temporary if the probability of the chartists continuing to dominate is not, « too high ».

We now need to know how the probability that an agent i will follow guru j is determined. In this model we will assume that the probability of using a rule is dependent on the past experience the agent had, using that rule. To evaluate the agent's experience define the profits obtained at time t associated with the rule i as

$$P_t^i = \left(R_t^i - S_{t-1}\right)\left(e^{S_t} - e^{S_{t-1}}\right)$$

and then, by discounting past profits, we can define cumulated profits as

$$U_t^i = \alpha U_{t-1}^i + P_t^i = \sum_{j=0}^{t} \alpha^{t-j} P_j^i$$

It will be these profits that determine the choice probabilities, i.e.

$$\pi_{t+1} \sim Q(U_t : \cdot) \quad \text{where} \quad U_t = (U_t^1, \ldots, U_t^m)$$

We make the assumption that the probabilities of choosing a guru depend on the experience with that guru in the past and will assume, in particular, the better the performance, the higher the probability of choosing that guru.

Two features then emerge from the model:

1. The more agents adopt a guru's recommendation, the more impact that guru will have on the evolution of prices.

2. The stronger a guru's impact on prices the better will be his performance.

Thus, the simple fact that agents prefer successful gurus will lead to a self reinforcing move to the guru who is currently doing best.

Now that we have established all the building blocks we can see the overall structure

The asset price evolves according to the linear stochastic difference equation defined earlier i.e.

$$S_t = \big[1 - \alpha(\pi_t) + \beta(\pi_t)\big]S_{t-1} - \beta(\pi_t)S_{t-2} + \gamma(\pi_t, \eta_t)$$

and this in a random environment $\{(\pi_t, \eta_t)\}$. The latter has an endogenous component π_t which depends on past prices and performances and a further component, the liquidity demand which is exogenously given. Note that the latter may be made endogenous by introducing a second population of agents, who can be simply thought of as the inhabitants of the foreign country, in the exchange rate context. We assume here that the aggregate liquidity demand is an i.i.d. process but this would not be relevant if it were made endogenous. The past prices have an influence on the environment because the choices made by the agents depend on profits, which, in turn are dependent on prices.

The process governing the evolution of prices and of profits is a Markov chain which we can write as

$$\xi_t = (S_t, S_{t-1}, U_t)$$

The dynamics of this stochastic process ξ_t are given by

$$\xi_{t+1} = V(\xi_t, \tau_t) = \begin{bmatrix} F(S_t, S_{t-1}, \tau_t) \\ S_t \\ \alpha U_t + P(S_t, S_{t-1}, \tau_t) \end{bmatrix}, \quad \tau_t \sim Z(U_t : \cdot)$$

The analysis of this process is not simple and, in particular, it should be noted that the map

$$(S_t, S_{t-1}) \rightarrow P\big((S_t, S_{t-1}, \tau_t)\big)$$

is non-linear, and that this means that some of the standard methods for analysing such problems cannot be applied.

Where is the fundamental problem with this sort of model which should not be unfamiliar to physicists? From the economics point of view, it is quite simply that prices might explode and that, as a consequence, no long run properties of the system could be established. To show that this is not the case we need to have a property known as a « mean contraction » condition. The details of this are spelled out in Foellmer et al. (2005) from which this model is taken. The condition can be translated into putting a bound on the probability that an individual becomes a chartist. For example, if only fundamentalists are present on the market the condition holds automatically.

However, what is interesting is that it is enough to bound the *probability* that individuals switch to chartism, since one might intuitively think that at some point, even though it is a low probability event, almost all agents *could* become chartists. However, this situation will not persist, if the condition holds, since the probability that the system will stay in that configuration is very low. Thus, although a chartist bubble may start, it inevitably collapses. The advantage of this approach is that it allows for temporary swings to the dominance of chartism but guarantees that such periods will not last. Thus, without having to bound prices or the proportion of chartists, we can still get the results that we are looking for.

There are two main results. Firstly, the price and profit process will not explode. This is important from the economics point of view since many models of bubbles have the property that once a bubble starts it will expand indefinitely. The following result shows that this cannot happen.

Theorem 1 (Foellmer et al. (2005)). *Under the mean contraction condition the Markov chain is tight i.e.*

$$\lim_{c \to \infty} \sup_{t} P\big[|\xi_t| \geq c\big] = 0$$

Now, the second result says that not only do prices not explode, they actually have a stable structure in the long run.

Theorem 2 (Foellmer et al. (2005)). *Under the mean contraction condition, the Markov chain has a unique stationary distribution μ and the time averages of prices and profits converge to their expected value under μ.*

The important thing to emphasise here is that we place no a priori bounds on prices nor on the capital gains that can be made with the rules, but the limitation on the probability that an agent will become a chartist is sufficient to guarantee that radical departures from fundamentals will be relatively short in duration.

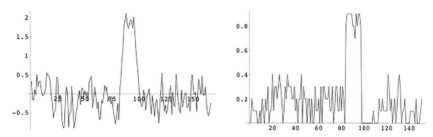

Fig. 11. (Source Foellmer et al. (2005).)

Although prices may, in principle "explode", they only exhibit this sort of transient behaviour for limited periods of time. Nevertheless, the influence of these episodes on long run average behaviour is sufficient to modify the distribution which would have been obtained had there been only fundamentalists and, what is more, it is this modification that produces the features such as long memory. It is also this sort of behaviour that makes the sort of statement that one hears from traders, such as, "only in the long run do fundamentals matter" meaningful.

The system will never settle to some steady state but we can see a certain long run regularity in the behaviour of prices. However, in addition to showing that, with both fundamentalists and chartists present, we can guarantee the existence of a unique limit distribution it is also interesting to see how the characteristics of that distribution are affected by the presence of chartists. We can see, using simulations of a simple example, that chartists add to the noise in the system and that bubbles and crashes appear. The translation of this in terms of the limit distribution is increased kurtosis and variance. Thus, loosely speaking, chartists increase the volatility of the price process. The switching of agents' expectations between fundamentalist and chartist provides an answer to the "excess volatility" puzzle mentioned earlier. The following figures enable us to see this clearly.

The idea is then, that changes in the distribution of expectations as a result of the mutual influences of agents, play a key role in explaining the evolution of prices. Of course, in the standard "representative agent" model there is no place for any such interaction and resultant heterogeneity of expectations. Indeed, many authors have suggested that this is the reason for the poor predictive power of such models. This argument has been made in particular for the foreign exchange market that we have modelled here and empirical evidence for this is given by Meese and Rogoff (1988), Frankel and Rose (1995) and Cheung, Chinn, and Pascual (2002). It is worth emphasising again, in the sort of model proposed here, heterogeneity is largely a temporary phenomenon, and for much of the time people have similar expectations. Furthermore, the expectations are essentially self fulfilling and as a consequence there are few periods in which people are sim-

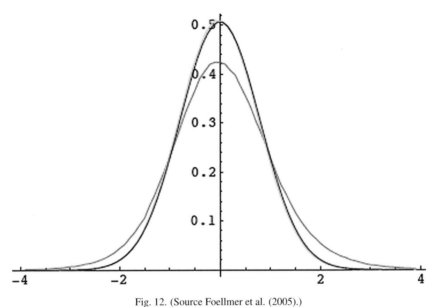

Fig. 12. (Source Foellmer et al. (2005).)

ply forecasting badly. In general they forecast well but not as a result of clever individual calculations.

5. Contributions to public goods

An enormous literature has developed on simple experiments involving contributions to public goods games. In the most basic version of these games the Nash equilibrium involves nobody contributing whereas in the social optimum everyone would give everything. The results of these experiments reveal that some cooperation emerges though, in general, the social optimum is not obtained (see Ledyard (1997)). The standard explanation is that the players are "learning to play Nash" and that, if the game went on long enough, all the players would be free-riding and all heterogeneity of behaviour would have disappeared. Yet, careful examination of the individuals' behaviour (see Hichri and Kirman (2006)) shows that heterogeneity persists and that some players play in a rather altruistic way, some are non-cooperative all the time and others seem to signal their intentions to other players. This mixture of behaviour does not disappear. The explanation for this phenomenon is far from clear and the puzzle remains. The question becomes whether people have more or less altruistic preferences and what is the origin of this. One suggestion is that altruism is selected for but this selection cannot again be uniform. A large literature on the emergence and persistence of fairness and altruism has sprung up (see Fehr and Gächter (2000)) and the debate is far from closed.

5.1. The basic public goods game

In the basic game of private contribution to a public good, each subject i ($i = 1, \ldots, N$) has to split an initial endowment E into two parts: the first part ($E - c_i$) represents his private share and the other part c_i represents his contribution to the public good. The payoff of each share depends on and varies with the experimental design, but in most experiments is taken to be linear (Andreoni (1995)). The total payoff π_i of individual i, in that case is given by the following expression:

$$\pi_i = E - c_i + \theta \sum_{j=1}^{N} c_j$$

This linear case gives rise to a corner solution. In fact, assuming that it is common knowledge that players are rational payoff maximisers, such a function gives a Nash equilibrium (NE) for the one shot game at zero and full contribution as social optimum. The dominant strategy for the finite repeated game, is to contribute zero at each step. Nevertheless, experimental studies show that there is generally over-contribution (30 to 70% of the initial endowments) in comparison to the NE.

Attempts to explain this difference between the theoretical and the experimental results are the main subject of the literature on private contribution to public goods. To do so, several pay-off functions with different parameters have been tested in various contexts to try to see the effect of their variation on subjects' contributions (for surveys, see Davis and Holt (1993), Ledyard (1995) and Keser (2000)).

A standard explanation for the observed contributions is that players simply make errors. In the linear case, given that the NE is at zero, and giving that subjects could not contribute negative amounts to the public good, error can only be an over-contribution. This, of course, undermines the idea of purely random errors. To test the error hypothesis experimentally, Keser (1996) performed a new experiment. She proposed a new design in which the payoff function is quadratic and the equilibrium is a dominant strategy in the *interior of the strategy space*. With such a design, under-contribution becomes possible and error on average could be expected to be null. The results of Keser's experiment show that in each period, contributions are above the dominant solution, which leads to the rejection of this error hypothesis.

5.2. Our model

The theoretical model and design used for the experiments concerns a public goods game with a "Nash equilibrium treatment" which does not involve zero contributions from each of the players. The individual payoff function is

$$\pi_i = E - c_i + \theta \left(\sum_{j=1}^{N} c_j \right)^{1/2}$$

The Nash equilibrium and the social optimum corresponding to this payoff structure are not trivial solutions but are in the interior of the set of the possible choices. The Nash equilibrium for individuals is not a dominant strategy for the finite repeated game. Indeed the solution for that game poses problems for a simple reason. There is a unique Nash equilibrium in the sense that for any Nash equilibrium the group contribution is the same. However, that contribution can be obtained by several combinations of individual contributions. Since there are many Nash equilibria for the one-shot game, precisely what constitutes an equilibrium for the repeated game is unclear. For a group of N subjects, at the CO the total contribution is given by the following expression:

$$\overline{Y} = \sum_{i=1}^{N} y_i = N^2 \frac{\theta^2}{4}$$

and at the NE is equal to:

$$Y^* = N.y^*$$
$$\text{where } y^* = \frac{\theta^2}{4}$$

and y^* is the symmetric individual Nash equilibrium.

With such a design, the Nash equilibrium and the social optimum vary with the value of θ. We shall see whether there is any difference between the evolution of individual and aggregate contributions under the different treatments and how well learning models explain this evolution.

We gave θ four different values, which give four levels for the CO and the NE. The following tables summarize the four treatments (Low, Medium, High and Very High) the different levels of interior solutions for each group of four ($N = 4$) persons (table 1) and for the individual subjects in each group (table 2).

---

[1] These are approximate values. The exact values are respectively: 4, 5.6568542, 6.9282032 and 8.9442719. We choose these values such that the CO corresponds respectively to 64, 128, 192 and 280.

Table 1

The NE and the CO values for the four treatments for one group

Value of θ^1	Treatment	Endowment	Symmetric NE	CO
4	L	280	4	64
5.66	M	280	8	128
6.93	H	280	12	192
8.94	VH	280	20	280

Table 2

The symmetric NE and the CO values for the four treatments for one subject

Value of θ	Treatment	Endowment	Symmetric NE	CO
4	L	70	1	16
5.66	M	70	2	32
6.93	H	70	3	48
8.94	VH*	70	5	70

The set of possible group contributions in our model is very large. In fact, given that each one of the four individuals of a group is endowed with 70 tokens, each group can contribute an amount that varies between zero and 280.

5.3. The experimental results

Our initial analysis will be at the aggregate level where we have for each treatment the average contribution of the six groups compared to the aggregate NE and to the aggregate CO. These results are reported in Figure 2 for H treatment and the others are similar and can be found in Hichri and Kirman (2006).

Basic results

The first thing we observe when analysing the experimental results is the fact that the average group contribution $(\overline{Y})$ decreases over time. In the particular case of the H treatment, the average group contribution $(\overline{Y})$ decreases during the 10 first periods and stays at a steady level during the rest of the periods of the game. Generally, if we do not consider the first five periods that could be assimilated to "*learning periods,*" contributions are almost steady over the twenty last periods for all the treatments.

Our results show that contributions vary with the CO level. There is over-contribution in comparison to the NE. This over-contribution increases with the CO level in absolute value. Nevertheless, average contributions remain proportionately as far from the CO as the level of the latter gets high. Computing these contributions in relative values by calculating an over-contribution index that takes into account the NE and the CO, shows that, except in the VH treatment, this ratio is constant.

That means that an increase the CO level leads to, a proportional increase in individual and hence, total contributions.

The group and the individual level
Although at the aggregate level behaviour seems rather consistent, in all treatments, there are at the group level different behaviours in the same group and between groups. The latter are not regular and we can clearly identify different attitudes to contribution. Figure 13 shows the contribution of the six groups in treatment H and gives us an idea about this variation between groups.

Within groups, at the individual level, there is also a difference between the individual and the aggregate behaviour. In fact, individuals behave differently. Moreover, the individual behaviour is more difficult to classify because of the

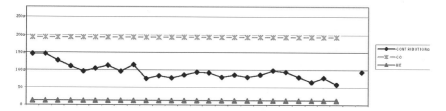

Fig. 13. Average total contribution in treatment H (source Hichri and Kirman (2006)).

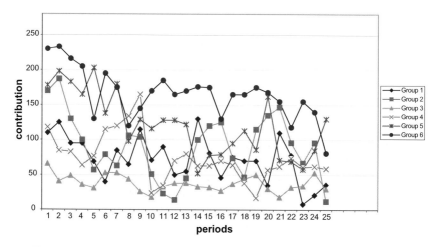

Fig. 14. Contributions of the six groups in treatment H (source Hichri and Kirman (2006)).

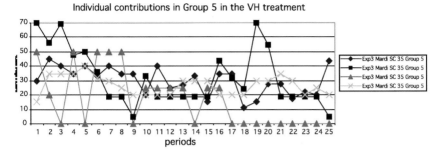

Fig. 15. (Source Hichri and Kirman (2006).)

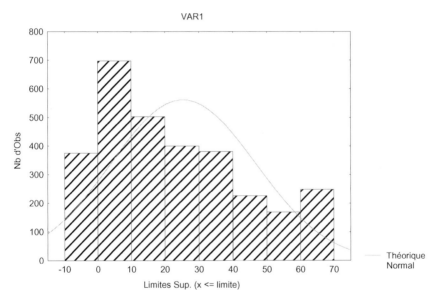

Fig. 16. Number of individual contributions for each interval for the 5 treatments (3000 decisions) (source Hichri and Kirman (2006)).

considerable volatility of the contributions of each subject during the 25 periods of the game.

To have an idea about the different levels of contributions of individuals, we classify contributions in 8 intervals of ten each and we present in Figure 15 the number of times individuals make a contribution belonging to each interval. As in each of the 5 treatments there are 24 persons playing 25 periods, we have then 3000 observations or decision of contribution. As we can see in Figure 16 almost all the intervals are significant.

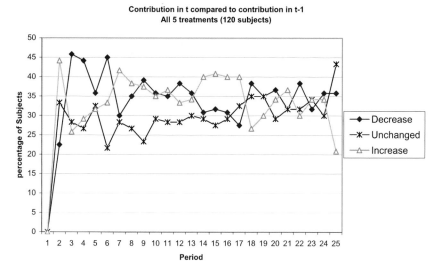

Fig. 17. Percentage of increasing, decreasing and unchanged individual contributions in t compared to $t-1$ for the 5 treatments (3000 observations) (source Hichri and Kirman (2006)).

Also, to isolate the different strategies that could explain the differences between the behaviour of different individuals, we compare for the 5 treatments and for the 25 decisions of each individual her contribution in period (t) to her contribution in ($t-1$). This allows us to know whether individuals react in response to past contributions. We classify this comparison into three possibilities: contribution in t increases, decreases and remains unchanged in comparison to contributions in $t-1$. Figure 17 shows that all these strategies are significant.

6. Conclusion

The basic idea of these notes is to show how viewing the economy as a complex interactive system can explain phenomena that are difficult to explain with standard economic models and that also paint a more realistic picture of how the economy functions. A number of the examples use the methods and models of physics. This is not new in economics since a good part of our modelling heritage has been based on classical mechanics but the latter approach has revealed its limits over time and the introduction of more recent physical models may help us to advance.

Taking the examples in turn: The analysis of the Schelling model based on a physical analogue enables better to understand the nature of the segregation that follows from individuals using apparently innocuous rules based on a rather weak aversion to the presence of others. The nature of the segregated groups and how they develop becomes much clearer once one adopts this approach and the results from the original Schelling model no longer seem surprising.

The study of the Marseille fish market provides an analysis of a curious but sharp division of buyers on a real market into those who are completely loyal, and those who always "shop around". This can be understood by studying the stochastic process generated by the reinforcement learning of the market participants. Furthermore the learning process can be justified as the result of a trade-off between "exploitation and exploration".

The model of a financial market offers an explanation of how the interaction between individuals and the consequences of that interaction for their choice of forecasting rules can generate bubbles and crashes. The model is based on a self reinforcing switching process through which one forecasting rule comes to dominate in a market and then is succeeded by another. This generates periods of stability interspersed with bubbles which inevitably burst. This model reproduces standard features of market price data such as "excess volatility" and "long memory". One can also have an equilibrium notion which does not involve a steady state or convergence to an "equilibrium price" but still, in contrast with deterministic chaos, shows that there is well defined long term structure in the time series.

The experiments on contributions to public goods reveal how misleading it is to draw behavioural conclusions from aggregate data. Although one has the impression that collectively the population "learns to play Nash", this does not stand up to an examination of the individual data. There we see a considerable variety of behaviours and no systematic convergence to a particular contribution. The learning models which provide good fits at the aggregate level do badly at the group and individual level. Furthermore, the "convergence" to a Nash equilibrium in which players contribute different amounts and how players achieve this coordination remains to be explained.

Yet, in all of these examples we see the presence of self-organisation and the emergence of coordination without any need for imposing sophisticated analytical capacities on the individuals in the models. *Homo economicus* is surely not an ant but he may behave more like one than economists have been prepared to admit.

References

Andreoni, James (1995), "Cooperation in Public-Goods Experiments: Kindness or Confusion?," American Economic Review, American Economic Association, vol. 85(4), pages 891–904, September.

Bachelier, L. (1900), "Théorie de la spéculation", *Annales Scientifiques de l'Ecole Normale Supérieure*, troisième série 17, pp. 21–88.

Binmore, Ken (1992), Fun and Games: A Text on Game Theory. Lexington, MA: D. C. Heath.

Brock, W.A., and Hommes, C.H. (1998), Heterogeneous beliefs and routes to chaos asset pricing model, *Journal of Economic Dynamics and Control* vol. 22, pp. 1235–1274.

Cheung Yin-Wong, Menzie D. Chinn and Antonio G. Pascual (2002), "Empirical Exchange Rate Models of the Nineties: Are they Fit to Survive?" NBER WP 9393.

Davis, D.D. and Holt, C.A. (1993), Experimental economics (Princeton University Press, Princeton, NJ).

De Grauwe, P. and Grimaldi, M. (2006), *The Exchange Rate in a Behavioral Finance Framework*, Princeton University Press, Princeton N.J.

Debreu, G. (1952), Definite and Semi-definite Quadratic Forms, "Econometrica", n. 20, pp. 295–300.

Durlauf, S. (1990), "Locally interacting systems, coordination failure, and the behaviour of aggregate activity", Working Paper, Stanford University.

Fehr Ernst and Simon Gächter (2000), "Fairness and Retaliation: The Economics of Reciprocity" Journal of Economic Perspectives, vol. 14, pp. 159–181.

Frankel, Jeffrey A. and Rose, A.K. (1995), "Empirical Research on Nominal Exchange Rates", in G.M. Grossman and K. Rogoff (eds), Handbook of International Economics, Vol. III, North Holland, Amsterdam.

Hichri, W. and Kirman, A. (2007), "The Emergence of Coordination in Public Goods Games", *European Physics Journal B*, Vol. 55, pp. 149–159.

Jerison, M. (2006), "Nonrepresentative Representative Consumers" Working paper, Department of Economics, SUNY, Albany.

Keser, C. (1996), "Voluntary Contributions to a Public Good when Partial Contribution is a Dominant Strategy", *Economics Letters*, 50, pp. 359–366.

Keser, C. (2000), "Cooperation in Public Goods Experiments", Working paper, Centre interuniversitaire de recherche en analyse des organisations (CIRANO) Scientific series no. 2000s-04.

Kirman, A. (1992), "What or whom does the representative individual represent?", *Journal of Economic Perspectives*, vol. 6, pp. 117–36.

Kirman, A.P. and Teyssière, G. (2002), "Micro-economic Models for Long Memory in the Volatility of Financial Time Series" in P.J.J. Herings, G. Van der Laan and A.J.J. Talman (eds), The Theory of Markets, pp. 109–137, North Holland, Amsterdam.

Ledyard, J.O. (1995), "Public Goods: A Survey of Experimental Research," in Kagel, J. and Roth, A. (eds.), The Handbook of Experimental Economics, Princeton University Press.

Ledyard, J. (1997), "Public Goods: A Survey of Experimental Research" in Kagel John H. and Alvin E. Roth, eds, *Handbook of Experimental Economics*, Princeton University Press, Princeton.

Lux T. and Marchesi, M. (2000), "Volatility clustering in financial markets: a microsimulation of interacting agents," International Journal of Theoretical and Applied Finance.

Meese, R. & Rogoff, K. (1988). "Was It Real? The Exchange Rate-Interest Differential Ralation Over The Modern Floating-Rate Period," Working papers 368, Wisconsin Madison – Social Systems.

Pancs, R., & Vriend, N.J. (2006) Schelling's Spatial Proximity Model of Segregation Revisited. Journal of Public Economics, 91, 1–24.

Pollicot, M. and Weiss, H. (2005), "The dynamics of Schelling-type segregation models and a non-linear graph Laplacian variational problems" Advances in Applied Mathematics 27, pp. 17–40.

Schabas, M. (2005), *The Natural Origins of Economics*, Chicago, Chicago University Press.

Schelling, T.S. (1978), Micromotives and Macrobehavior; W.W. Norton and Co, N.Y. (Trad. fr. : La tyrannie des petites décisions, PUF, Paris, 1980).

Stanley, E.A., Ashlock, D. and Tesfatsion, L. (1994), "Iterated prisoner's dilemma with choice and refusal of partners" in Artificial Life III, C.G. Langton (ed.), Santa Fe Institute Studies in the Sciences of Complexity, proc. vol. XVII, Addison-Wesley. Summers.

Vinkovic, D. and Kirman, A. (2006), A physical analogue of the Schelling model, *Proceedings of the National Academy of Sciences*, vol. 103, pp. 19261–19265.

Weisbuch, G., Kirman, A., & Herreiner, D. (2000), "Market Organisation and Trading Relationships, Economic Journal, 110, pp. 411–436.

Course 6

CRACKLING NOISE AND AVALANCHES: SCALING, CRITICAL PHENOMENA, AND THE RENORMALIZATION GROUP

James P. Sethna

Laboratory of Atomic and Solid State Physics, Cornell University, Ithaca, NY, USA

J.-P. Bouchaud, M. Mézard and J. Dalibard, eds.
Les Houches, Session LXXXV, 2006
Complex Systems
© *2007 Published by Elsevier B.V.*

Contents

1. Preamble

In the past two decades or so, we have learned how to understand crackling noise in a wide variety of systems. We review here the basic ideas and methods we use to understand crackling noise—critical phenomena, universality, the renormalization group, power laws, and universal scaling functions. These methods and tools were originally developed to understand continuous phase transitions in thermal and disordered systems, and we also introduce these more traditional applications as illustrations of the basic ideas and phenomena.

We focus largely on crackling noise in magnetic hysteresis, called *Barkhausen noise*. These lecture notes are distilled from a review article written with Karin Dahmen and Christopher Myers [1], from a book chapter written with Karin Dahmen and Olga Perkovi c [2], and from a chapter in my textbook [3].

2. What is crackling noise?

Many systems, when stressed or deformed slowly, respond with discrete events spanning a broad range of sizes. We call this *crackling noise*. The Earth crackles, as the tectonic plates rub past one another. The plates move in discrete earthquakes (Fig. 1a), with many small earthquakes and only a few large ones. If the earthquake series is played as an audio clip, sped up by a factor of ten million, it sounds like crackling [4]. A piece of paper [6] or a candy wrapper [10] will crackle as it is crumpled (try it!), emitting sharp sound pulses as new creases form or the crease pattern reconfigures (Fig. 2(a)). Paper also tears in a series of avalanches [7]; fracture in many other systems exhibit avalanche distributions and jerky motion [11, 12]. Foams (a head of beer) move in jerky avalanches as they are sheared [13], and as the bubbles pop [14]. Avalanches arise when fluids invade porous media in avalanches [15, 16] (such as water soaking into a sponge). I used to say that metals were an example of something that did not crackle when bent (permanently plastically deformed), but there is now excellent data that ice crackles when it is deformed [8], and recent data on micron-scale metal deformation also shows crackling noise [17, 18]. We will focus here on *Barkhausen noise*, the magnetic pulses emitted from (say) a piece of iron as it is placed inside an increasing external field (see Fig. 3(a)).

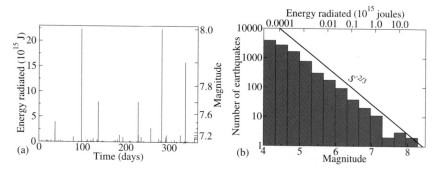

Fig. 1. **Earthquake sizes.** (a) Earthquake energy release versus time in 1995. There are only a few large earthquakes, and many small ones. This time series, when sped up, sounds like crackling noise [4]. (b) Histogram of the number of earthquakes in 1995 as a function of their magnitude M. Notice the logarithmic scales; the smallest earthquakes shown are a million times smaller and a thousand times more probable than the largest earthquakes. The fact that the earthquake size distribution is well described by a power is the Gutenberg–Richter law [5].

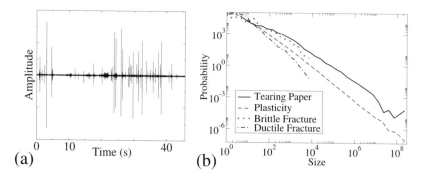

Fig. 2. (a) **Crackling noise in paper** $V(t)$ for multiple pulses, from [6]. Again, there are many small pulses, and only a few large events. (b) **Power laws** of event sizes for paper tearing [7], plastic flow in ice [8], and brittle and ductile fracture in steel [9].

All of these systems share certain common features. They all have many more tiny events than large ones, typically with a *power-law* probability distribution (Figs 1(a), 2(a)). For example, the histogram of the number of earthquakes of a given size yields a straight line on a log–log plot (Figure 1b). This implies that the probability of a large earthquake goes as its size (energy radiated) to a power. The tearing avalanches in paper, dislocation avalanches in deformed ice, and rupture avalanches in steel all have power-law avalanche size distributions (Fig. 2(b)). Because crackling noise exhibits simple, emergent features (like these power law size distributions), it encourages us to expect that a theoretical understanding might be possible.

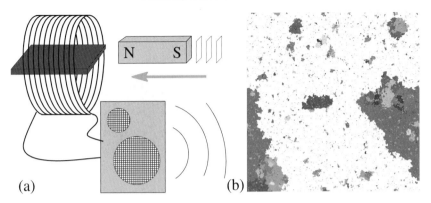

Fig. 3. (a) **Barkhausen noise experiment.** By increasing an external magnetic field $H(t)$ (bar magnet approaching), the magnetic domains in a slab of iron flip over to align with the external field. The resulting magnetic field jumps can be turned into an electrical signal with an inductive coil, and then listened to with an ordinary loudspeaker. Barkhausen noise from our computer experiments can be heard on the Internet [4]. (b) **Cross section of all avalanches** in a billion–domain simulation of our model for Barkhausen noise at the critical disorder [1]. The white background is the infinite, spanning avalanche.

What features of, say, an earthquake fault do we expect to be important for such a theoretical model? If earthquake faults slipped only in large, snapping events we would expect to need to know the shape of the tectonic plates in order to describe them. If they slid more-or-less smoothly we anticipate that the nature of the internal rubble and dirt (fault gouge) would be important. But since earthquakes come in all sizes, we expect that neither the microscopic rock-scale nor the macroscopic continental-scale details can be crucial. What, then, is important to get right in a model?

An important hint is provided by looking at the dynamics of an individual avalanche. In many (but not all) systems exhibiting crackling noise, the avalanches themselves have complex, internal structures. Fig. 4 shows that the avalanches producing individual Barkhausen pulses in magnets proceed in an irregular, jerky fashion, almost stopping several times in between sub-avalanches. Fig. 5(a) shows that the spatial structure of an avalanche in a model magnet is also irregular. Both the time and spatial structures are *self-similar*; if we take one of the sub-avalanches and blow it up, it looks statistically much like the original avalanche. Large avalanches are built up from multiple, similar pieces, with each sub-avalanche triggering the next. It is the way in which one sub-avalanche triggers the next that is crucial for a theoretical model to get right.

Our focus in these lecture notes will be on understanding the emergent behavior in crackling noise (power laws and scaling functions) by exploring the

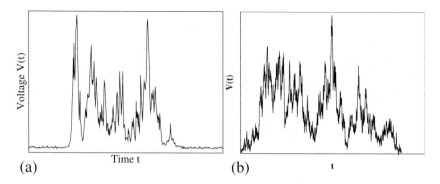

Fig. 4. Internal avalanche time series (a) during a single experimental Barkhausen noise pulse [19], and (b) during a single avalanche in a simulation of magnetic Barkhausen noise [1]. The experiment measures the voltage $V(t)$, which in this experiment measures the volume swept out per unit time by the moving magnetic domain wall; the simulation measures the number of domains flipped per unit time. In these two cases, the total area under the curve gives the size S of the avalanche. Notice how the avalanches almost stop several times; if the forcing were slightly smaller, the large avalanche would have broken up into two or three smaller ones. The fact that the forcing is just large enough to on average keep the avalanche growing is the cause of the self-similarity; a partial avalanche of size S will on average trigger one other of size S.

consequences of this self-similar structure. In section 3, we illustrate the process by which models are designed and tested by experiments using Barkhausen noise. In section 4 we introduce the *renormalization group*, and use it to explain *universality* and self-similarity in these (and other) systems. In section 5 we use the renormalization group to explain the power laws characteristic of crackling noise, and also use it to derive the far more powerful *universal scaling functions*.

3. Hysteresis and Barkhausen noise in magnets

Microscopically, iron at room temperature is always magnetized; non-magnetic bulk iron is composed of tiny magnetic domains, whose north poles point in different directions giving a net zero magnetization. An external magnetic field, as in Fig. 3(a), will attract a piece of iron by temporarily moving the domain walls to enlarge the regions whose north poles are aligned towards the south pole of the external field. When the external field is removed, these domain walls do not completely return to their original positions, and the iron will end up partially magnetized; this history dependence is called *hysteresis*. The dependence of the magnetization of the iron M on the history of the external field H (the *hysteresis loop*) can have an interesting hierarchical structure of sub-loops (Fig. 6).

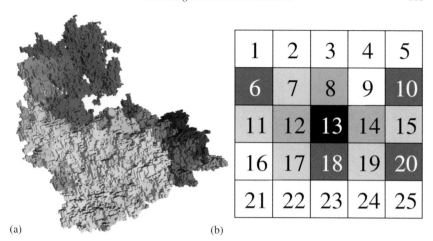

(a)   (b)

Fig. 5. (a) **Fractal spatial structure of an avalanche.** This moderate–sized avalanche contains 282,785 domains (or spins) [1]. The shading depicts the time evolution: the avalanche started in the dark region in the back, and the last domains to flip are in the upper, front region. The sharp changes in shading are real, and represent sub-avalanches separated by times where the avalanche almost stops (see Fig. 4). (b) **Avalanche propagation.** The avalanches in many models of crackling noise are first nucleated when the external stress induces a single site to transform. In this two-dimensional model, site #13 is the nucleating site which is first pushed over. The coupling between site #13 and its neighbors triggers some of them to flip (#4, 8, and 12), which in turn trigger another shell of neighbors (#15, 19, 7, 11, and 17), ending eventually in a final shell which happens not to trigger further sites (#6, 10, 18, and 20). The time series $V(t)$ plotted in figure 4(b) is just the number of domains flipped in shell #t for that avalanche.

These hysteresis loops may look smooth, but when Fig. 6 is examined in detail we see that the magnetization grows in discrete jumps, or avalanches. These avalanches are the origin of magnetic Barkhausen noise. In section 3.1 we will develop a simple model for this Barkhausen noise. In section 3.2 we shall observe that our model, while capturing some of the right behavior, is not the correct model for real magnets, and introduce briefly more realistic models that do appear to capture rather precisely the correct behavior.[1]

3.1. Dynamical random-field Ising model

We introduce here a caricature of a magnetic material. We model the iron as a cubic grid of magnetic domains S_i, whose north pole is either pointing upward ($S_i = +1$) or downward ($S_i = -1$). The external field pushes on our domain with a force $H(t)$, which starts pointing down ($H(t = 0) \ll 0$) and will increase

---

[1] That is, we believe they are in the right *universality class*, section 4.1.

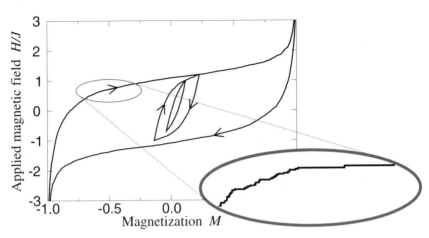

Fig. 6. Hysteresis loop and sub-loops: The magnetization in our model [20], as the external field H is ramped up and down. Our focus will primarily be on the upper, outer loop as the external field is ramped from $-\infty$ to ∞. **Barkhausen jumps** (exploded region): The hysteresis loop appears smooth, but when examined in detail is composed of discrete, abrupt jumps in magnetization, corresponding to avalanches in the positions of the walls of the magnetic domains. This jerky magnetization is what emits *Barkhausen noise*.

with time. Iron is magnetic because a domain has lower energy when it is magnetized in the same direction as its neighbors; the force on site i from the six neighboring domains S_j in our model is of strength $J \sum_j S_j$. Finally, we model the effects of impurities, randomness in the domain shapes, and other kinds of disorder by introducing a random field[2] h_i, different for each domain and chosen from a normal distribution with standard deviation $R \times J$. The net force on S_i is thus

$$F_i = H(t) + \sum_j J S_j + h_i, \tag{3.1}$$

corresponding to the energy function or Hamiltonian

$$\mathcal{H} = -J \sum_{\langle ij \rangle} S_i S_j - \sum_i \left(H(t) + h_i \right) S_i. \tag{3.2}$$

---

[2]Most disorder in magnets is not microscopically well described by random fields, but better modeled using random anisotropy or random bonds which do not break time-reversal invariance. For models of hysteresis, time-reversal symmetry is already broken by the external field, and all three types of randomness are probably in the same universality class. That is, random field models will have the same statistical behavior as more realistic models, at least for large avalanches and long times (see section 4.1).

This model is called the *random-field Ising model* (RFIM), and its thermal equilibrium properties have historically been studied as an archetypal disordered system with glassy dynamics. Here we ignore temperature (specializing to magnets where the thermal fluctuations are too small to de-pin the domain walls), and assume that each domain S_i reorients whenever its local field F_i changes sign, to minimize its energy.

In our model, there are two situations where a domain will flip. It may be induced to flip directly by a change in the external field $H(t)$, or it may be triggered to flip by the flipping of one of its neighboring domains. The first case corresponds to nucleating a new avalanche; the second propagates the avalanche outward from the triggering domain, see Fig. 5(b). We assume that the external field $H(t)$ changes slowly enough that each avalanche finishes (even ones which sweep over the entire magnet) before the field changes appreciably.

These avalanches can become enormous, as in Fig. 5(a). How can an avalanche grow to over 10^5 domains, but then halt? In our model, it happens only when the disorder R and the field $H(t)$ are near a *critical point* (Fig. 7). For large disorder compared to the coupling between domains $R \gg J \equiv 1$, each domain turns over almost independently (roughly when the external field $H(t)$ cancels the local random field h_i, with the alignment of the neighbors shifting the transition only slightly); all avalanches will be tiny. For small disorder compared to the coupling $R \ll J$, there will typically be one large avalanche that flips most of the domains in the sample. (The external field $H(t)$ needed to nucleate the first domain flip will be large, to counteract the field $-6J$ from the un-flipped neighbors without substantial assistance from the random field; at this large external field most of the neighboring domains will be triggered, ... eventually flipping most of the domains in the entire system). There is a critical disorder R_c separating two qualitative regimes of behavior—a phase $R > R_c$ where all avalanches are

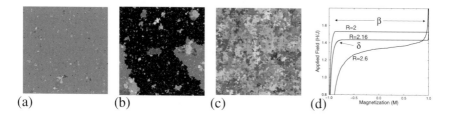

(a)   (b)   (c)   (d)

Fig. 7. Phase transition in hysteresis and avalanche model; (a) one enormous avalanche (background) for small disorder $R < R_c$, (b) avalanches at all scales at $R = R_c$ (c) many small avalanches for large disorder $R > R_c$. (d) The upper branch of the hysteresis loop develops a macroscopic jump in magnetization below R_c, whose size diverges as $\Delta M \propto (R - R_c)^\beta$; at $R = R_c$ the magnetization has a power-law singularity $M - M_c \sim (H - H_c)^{1/\delta}$.

small and a phase $R < R_c$ where one 'infinite' avalanche flips a finite fraction of all domains in the system (Fig. 7). Near[3] R_c, a growing avalanche doesn't know whether it should grow forever or halt while small, so a distribution of avalanches of all sizes might be expected— giving us crackling noise.

3.2. Real Barkhausen noise

Is our model a correct description of Barkhausen noise in real magnets? Our model does capture the qualitative physics rather well; near R_c it exhibits crackling noise with a broad distribution of avalanche sizes (Fig. 3(b)), and the individual avalanches have the same kind of internal irregular dynamical structure as seen in real Barkhausen avalanches (Fig. 4).

However, when examined in detail we find that there are problems with the model. A typical problem might be that the theoretically predicted power-law is wrong (Fig. 8). The probability density $\mathcal{D}(S)$ of an avalanche of size S experimentally [21] decays as a power law $D(S) \sim S^{-\tau}$, with values of the exponent that cluster around either $\tau = 1.27$ or $\tau = 1.5$. In our model at R_c and near H_c we find $\tau = 1.6 \pm 0.06$, which is marginally compatible with the data for the second group of samples (see Fig. 15(a)). If the power law were significantly

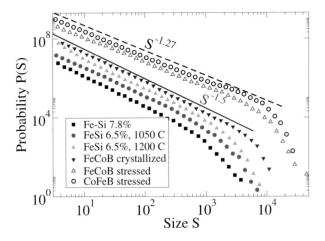

Fig. 8. **Avalanche size distributions** for magnetic Barkhausen noise in various materials, from [21]. The materials fall into two families, with power law exponent $D(S) \sim S^{-\tau}$ with $\tau \sim 1.5$ and $\tau \sim 1.27$. The cutoffs in the avalanche size distribution at large S are due to demagnetization effects, and scale as $S_{\max} k^{-1/\sigma_k}$ (section 5.1, Fig. 17).

---

[3]One must also be near the field $H(t) = H_c$ where the infinite avalanche line ends.

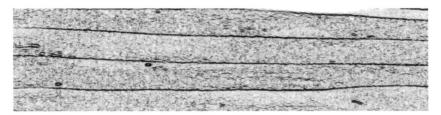

Fig. 9. **Domain structure in a real magnet**, measured with a scanning electron microscope, digitally processed to detect edges. The horizontal lines are magnetic domain walls. Courtesy of K. Zaveta, from [29].

different, our model would have been ruled out. We shall see in section 5 that the power laws are key *universal* predictions of the theory, and if they do not agree with the experiment then the theory is missing something crucial.

However, there are two other closely related avalanche models that describe the two groups of experiments well. One is a front propagation or fluid invasion model (which preceded ours [15, 22–24]), also usually written as a random-field Ising model but here starting with an existing front and allowing new avalanches only contiguous to previously flipped domains (water can invade new pores only next to currently wet pores). The front-propagation model does well in describing the dynamics and size distribution of the materials with $\tau \approx 1.27$. The other model can either be viewed as the motion of a flat domain wall [25, 26] or the mean-field limit of the random-field Ising model [27, 28]; it has a value $\tau = 3/2$ nicely describing the other class of materials. These models incorporate the long-range magnetic interactions between flipped domains, which experimentally produce demagnetizing fields that induce the magnetic domain wall to move rather rigidly.

It is easy to measure the exponents of a power-law decay, but it is difficult to measure them well. We could argue that $\tau = 1.6$ is close enough to some of the experimental measurements that our model is not ruled out. But even a brief investigation of the qualitative behavior in these systems leads us to abandon hope. Fig. 9 shows the domain-wall structure in a real magnet. The domain walls are indeed flat, and the way in which they advance (looking at animated versions of this figure, or observing directly through a microscope) is not by nucleating new walls, but by motion of existing boundaries.

4. Why crackling noise?

In this section we introduce *universality*, the *renormalization group*, and *self-similarity*. These three notions grew out of the theory of continuous phase transitions in thermal statistical mechanics, and are the basis of our understanding of a broad variety of systems, from quantum transitions (insulator to superconductor) to dynamical systems (the onset of chaos) [3, chapter 12]. We will illustrate and explain these three topics not only with our model for crackling noise in magnets, but also using classic problems in statistical mechanics—percolation and the liquid-gas transition. The renormalization group is our tool for understanding crackling noise.

4.1. Universality

Consider holding a sheet of paper by one edge, and sequentially punching holes of a fixed size at random positions. If we punch out only a few holes, the paper will remain intact, but if the density of holes approaches one the paper will fall apart into small shreds. There is a phase transition somewhere in between where the paper first ceases to hold together. This transition is called *percolation*.

Fig. 10 shows two somewhat different microscopic realizations of this problem. On the top we see a square mesh of bonds, where we remove all but a fraction p of the bonds. On the bottom we see a lattice of hexagonal regions punched out at random. Microscopically, these two processes seem completely different (left figures). But if we observe large simulations and examine them at long length scales, the two microscopic realizations yield statistically identical types of percolation clusters (right Figs 10).

Universality arises when the statistical morphology of a system at a phase transition is largely independent of the microscopic details of the system, depending only on the type (or *universality class*) of the transition. This should not come as a complete surprise; we describe most liquids with the same continuum laws (the Navier-Stokes equations) despite their different molecular makeups, with only the mass density and the viscosity depending on microscopic details. Similarly most solids obey elasticity theory on scales large compared to their constituent particles. For these phases we understand the material independence of the constituent equations by taking a continuum limit, assuming all behavior is slowly varying on the scale of the particles. At critical points, because our system is rugged and complicated all the way down to the lattice scale, we will need a more sophisticated way of understanding universality (section 4.2).

Most thermal phase transitions are not continuous, and do not have structure on all scales. Water is water until it is heated past the boiling point, after which it abruptly turns to vapor (with a big drop in density ρ). Even boiling water, though, can show big fluctuations at high enough pressure. As we raise the pressure the boiling temperature gets larger and the density drop gets smaller, until at a certain pressure and temperature (P_c, T_c) the two densities become equal (to ρ_c). At this

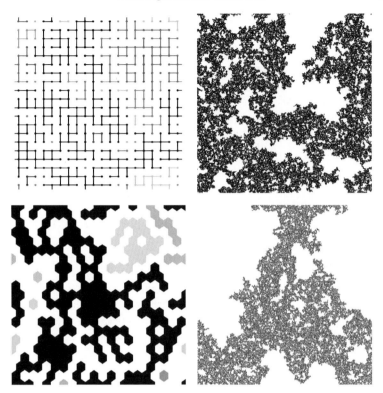

Fig. 10. **Universality in percolation**. Universality suggests that the entire morphology of the percolation cluster at p_c should be independent of microscopic details. On the top, we have bond percolation, where the bonds connecting nodes on a square lattice are occupied at random with probability p; the top right shows the infinite cluster on a 1024 × 1024 lattice at the critical point. On the bottom, we have site percolation on a triangular lattice, where it is the hexagonal sites that are occupied. Even though the microscopic lattices and occupation rules are completely different, the resulting clusters look statistically identical at their critical points. (One should note that the site percolation cluster is slightly less dark. Universality holds up to overall scale changes, here up to a change in the density.)

critical point the H_2O molecules do not know which phase they want to be in, and they show fluctuations on all scales.

Figure 11(a) shows that the approach to this critical point has universal features. If we divide the density by ρ_c and the temperature by T_c, many different liquids share the same ρ, T phase diagram,

$$\rho^{Ar}(T) = A\rho^{CO}(BT). \tag{4.1}$$

This is partly for mundane reasons; most of the molecules are roughly spherical, and have weak interactions.

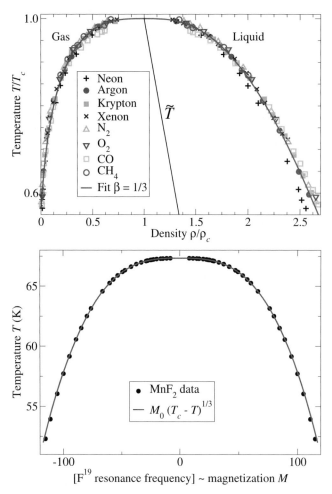

Fig. 11. **Universality.** (a) Universality at the liquid–gas critical point. The liquid–gas coexistence lines ($\rho(T)/\rho_c$ versus T/T_c) for a variety of atoms and small molecules, near their critical points (T_c, ρ_c) [30]. The gas phases lie on the upper left; the liquid phase region is to the upper right; densities in between will separate into a portion each of coexisting liquid and gas. The curve is a fit to the argon data, $\rho/\rho_c = 1 + s(1 - T/T_c) \pm \rho_0 (1 - T/T_c)^\beta$ with $s = 0.75$, $\rho_0 = 1.75$, and $\beta = 1/3$ [30]. (b) Universality: ferromagnetic–paramagnetic critical point. Magnetization versus temperature for a uniaxial antiferromagnet MnF_2 [31]. The tilted line in (a) corresponds to the vertical axis in (b).

But figure 11(b) shows the phase diagram of a completely different system, a magnet whose magnetization $M(T)$ vanishes as the temperature is raised past its T_c. The curve here does not agree with the curves in the liquid-gas collapse, but it can agree if we shear the axes slightly:

$$\rho^{\mathrm{Ar}}(T) = A_1 M(BT) + A_2 + A_3 T. \tag{4.2}$$

Nature has decided that the 'true' natural vertical coordinate for the liquid-gas transition is not T but the line $T = (\rho - A_2)/A_3$. Careful experiments, measuring a wide variety of quantities, show that the liquid-gas transition and this type of magnet share many properties, except for smooth changes of coordinates like that in eq 4.2. Indeed, they share these properties also with a theoretical model—the (thermal, non-random, three-dimensional) Ising model. Universality not only tied disparate experiments together, it also allows our theories to work.

The agreement between Figs 11(a) and (b) may not seem so exciting; both are pretty smooth curves. But notice that they do not have parabolic tops (as one would expect from the maxima of a typical function). Careful measurements show that the magnet, the liquid–gas critical point, and the Ising model all vary as $(1 - T/T_c)^\beta$ with the same, probably irrational exponent β; the best theoretical estimates have $\beta = 0.325 \pm 0.005$ [32, chapter 28]. This characteristic power law represents the effects of the large fluctuations at the critical points (the peaks of these graphs). Like the avalanche size distribution exponent τ, β is a *universal critical exponent* (section 5.1).

4.2. Renormalization group

Our explanation for universality, and our theoretical framework for studying crackling noise, is the *renormalization group*. The renormalization group starts with a remarkable abstraction: it works in an enormous 'system space'. Different points in system space represent different materials under different experimental conditions, or different theoretical models with different interactions and evolution rules. For example, in Fig. 12(a) we consider the space of all possible models and experiments on hysteresis and avalanches, with a different dimension for each possible parameter (disorder R, coupling J, next-neighbor coupling, ...) and for each parameter in an experiment (chemical composition, annealing time, ...). Our theoretical model will traverse a line in this infinite-dimensional space as the disorder R is varied.

The renormalization group studies the way in which system space maps into itself under *coarse-graining*. The coarse-graining operation shrinks the system and removes microscopic degrees of freedom. Ignoring the microscopic degrees of freedom yields a new physical system with the same properties at long length scales, but with different (renormalized) values of the parameters. As

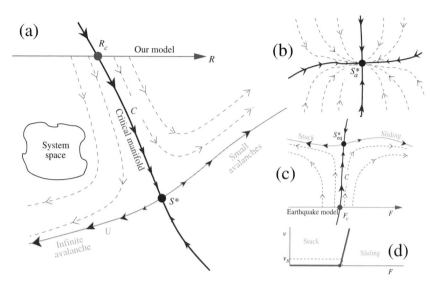

Fig. 12. **Renormalization–group flows** [1]. (a) **Flows describing a phase transition.** The renormalization–group uses coarse–graining to longer length scales to produce a mapping from the space of physical systems into itself. Consider the space of all possible systems exhibiting magnetic hysteresis (including, both real models and experimental systems). Each model can be coarse–grained, removing some fraction of the microscopic domains and introducing more complex dynamical rules so that the remaining domains still flip over at the same external fields. This defines a mapping of our space of models into itself. A fixed point **S*** in this space will be self–similar: because it maps into itself upon coarse–graining, it must have the same behavior on different length scales. Points that flow into **S*** under coarse–graining share this self–similar behavior on sufficiently long length scales: they all share the same *universality class*. (b) **Attracting fixed point** [1]. Often there will be fixed points that attract in all directions. These fixed points describe typical behavior: phases rather than phase transitions. Most phases are rather boring on long length scales. In more interesting cases, like random walks, systems can exhibit self-similarity and power laws without special tuning of parameters. This is called *generic scale invariance*. (c,d) **Flows for a front–propagation model** [1]. The front propagation model has a critical field F_c at which the front changes from a pinned to sliding state. (c) Coarse-graining defines a flow on the space of earthquake models. The critical manifold **C**, consisting of models which flow into $\mathbf{S}^*_{fp}$, separating stuck faults from faults which slide forward with an average velocity $v(F)$. (d) The velocity varies with the external force F across the fault as a power law $v(F) \sim (F - F_c)^\beta$. Clever experiments, or long–range fields, can act to control not the external field, but the position: changing the front displacement slowly sets $v \approx 0$, thus self–tuning $F \approx F_c$. This is one example of *self-organized criticality*.

an example, figure 13 shows a real-space renormalization-group 'majority rule' coarse-graining procedure applied to the Ising model. Many approximate methods have been devised to implement this coarse-graining operation (real-space, ϵ-expansions, Monte Carlo...) which we will not discuss here.

Fig. 13. **Ising model at T$_c$: coarse-graining**. Coarse-graining of a snapshot of the two-dimensional Ising model at its critical point. Each coarse-graining operation changes the length scale by a factor $B = 3$. Each coarse-grained spin points in the direction given by the majority of the nine fine-grained spins it replaces. This type of coarse-graining is the basic operation of the real-space renormalization group.

Under coarse-graining, we often find a fixed point S^* in system space. All of the systems that flow into this fixed point under coarse-graining will share the same long-wavelength properties, and will hence be in the same universality class.

Fig. 12(a) shows the case of a fixed point S^* with one unstable direction. Points deviating from S^* in this direction will flow away from it under coarse-graining. There is a surface C of points which do flow into the fixed point, which separates system space into two different phases (say, one with all small avalanches and one with one system-spanning, 'infinite' avalanche). The set C represents a universality class of systems at their critical points. Thus, fixed points with one unstable direction represent phase transitions.

Cases like the liquid-gas transition with two tuning parameters (T_c, ρ_c) determining the critical point will have two unstable directions. What happens when we have no unstable directions? The fixed-point S_a in Fig. 12(b) represents an entire region in system space sharing the same long-wavelength properties; it represents a *phase* of the system. Usually phases do not show structure on all scales, but some cases (like random walks) show this *generic scale invariance*.

Sometimes the external conditions acting on a system naturally drive it to stay at or near a critical point, allowing one to spontaneously observe fluctuations on all scales. A good example is provided by certain models of earthquake fault dynamics. Fig. 12(c) schematically shows the renormalization-group flow for an earthquake model. It has a *depinning* fixed point representing the external force F across the fault at which the fault begins to slip; for $F > F_c$ the fault will slide with a velocity $v(F) \sim (F - F_c)^\beta$. Only near F_c will the system exhibit earthquakes of all sizes. Continental plates, however, do not impose constant forces on the faults between them; they impose a rather small constant velocity (on the order of centimeters per year, much slower than the typical fault speed during an earthquake). As illustrated in Fig. 12(d), this naturally sets the fault almost precisely at its critical point, tuning the system to its phase transition. This is called *self-organized criticality*.

In broad, the existence of fixed points under renormalization group coarse-graining is our fundamental explanation for universality. Real magnets, Ising models, and liquid-gas systems all apparently flow to the same fixed points under coarse graining, and hence show the same characteristic fluctuations and behaviors at long length scales.

5. Self-similarity and its consequences

The most striking feature of crackling noise and other critical systems is self-similarity, or scale invariance. We can see this vividly in the patterns observed in the Ising model (upper left, Fig. 13), percolation (Fig. 10), and our RFIM for crackling noise (Fig. 5a). Each shows roughness, irregularities, and holes on all scales at the critical point. This roughness and fractal-looking structure stems at root from a hidden symmetry in the problem: these systems are (statistically) invariant under a change in *length scale*.

Consider Fig. 14, depicting the self-similarity of the avalanches in our RFIM simulation at the critical disorder. The upper-right figure shows the entire system, and each succeeding picture zooms in by another factor of two. Each zoomed-in picture has a black 'background' showing the largest avalanche spanning the entire system, and a variety of smaller avalanches of various sizes. If you blur your eyes a bit, the figures should look roughly alike. This rescaling and blurring process is the renormalization-group coarse-graining transformation.

How does the renormalization group explain self-similarity? The fixed point S^* under the renormalization group is the same after coarse-graining (that's what it means to be a fixed point). Any other system that flows to S^* under coarse graining will also look self-similar (except on the microscopic scales that are removed in the first few steps of coarse-graining, during the flow to S^*). Hence systems at their critical points naturally exhibit self-similarity.

This scale invariance can be thought of as an emergent symmetry, invariance under changes of length scale. In a system invariant under a translational symmetry, the expectation of any function of two positions x_1 and x_2 can be written in terms of the separation between the two points, $\langle g(x_1, x_2) \rangle = \mathcal{G}(x_2 - x_1)$. In just the same way, scale invariance will allow us to write functions of N variables in terms of *scaling functions* of $N - 1$ variables—except that these scaling functions are typically multiplied by power laws in one of these variables.

5.1. Power laws

Let us begin with the case of functions of one variable. Consider the avalanche size distribution $D(S)$ for a model, say the real earthquakes in Fig. 1(a), or our model for hysteresis at its critical point. Imagine taking the same system, but increasing the units of length with which we measure the system—stepping back, blurring our eyes, and looking at the system on a coarse-grained level. Let us multiply the spacing between markings on our rulers by a small amount $B = (1 + \epsilon)$. After coarsening, any length scales in the problem (like the spatial extent L of a particular avalanche) will be divided by B. The avalanche size (volume) S after coarse-graining will also be smaller by some factor[4] $C = (1 + c\epsilon)$. Finally, the overall scale of $D(S)$ after coarse-graining will be changed by some factor

---

[4]If the size of the avalanche were the cube of its length, then c would equal three since $(1+\epsilon)^3 = 1 + 3\epsilon + O(\epsilon^2)$. In general, c is the *fractal dimension* of the avalanche.

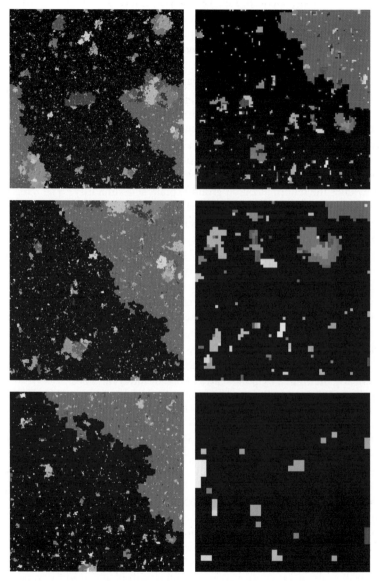

Fig. 14. **Avalanches: scale invariance**. Magnifications of a cross-section of all the avalanches in a run of our hysteresis model each one the lower right-hand quarter of the previous. The system started with a billion domains (1000^3). Each avalanche is shown in a different shade. Again, the larger scales look statistically the same.

$A = 1 + a\epsilon$.[5] Hence under the coarse-graining we have

$$
\begin{aligned}
L' &= L/B = L/(1+\epsilon), \\
S' &= S/C = S/(1+c\epsilon), \\
D' &= AD = D(1+a\epsilon).
\end{aligned}
\tag{5.1}
$$

Now the probability that the coarse-grained system has an avalanche of size S' is given by the rescaled probability that the original system had an avalanche of size $S = (1+c\epsilon)S'$:

$$
D'(S') = AD(CS') = (1+a\epsilon)D\big((1+c\epsilon)S'\big).
\tag{5.2}
$$

Here $D'(S')$ is the distribution measured with the new ruler: a smaller avalanche with a larger probability density. Because we are at a self-similar critical point, the coarse-grained distribution $D'(S')$ should equal $D(S')$. Making ϵ infinitesimal leads us to a differential equation:

$$
\begin{aligned}
D(S') = D'(S') &= (1+a\epsilon)D\big((1+c\epsilon)S'\big), \\
0 &= a\epsilon D + c\epsilon S' \frac{dD}{dS}, \\
\frac{dD}{dS} &= -\frac{aD}{cS},
\end{aligned}
\tag{5.3}
$$

which has the general solution

$$
D = D_0 S^{-a/c}.
\tag{5.4}
$$

Because the properties shared in a universality class only hold up to overall scales, the constant D_0 is system dependent. However, the exponents a, c, and a/c are *universal*—independent of experiment (within the universality class). Some of these exponents have standard names: the exponent c giving the fractal dimension of the avalanche is usually called d_f or $1/\sigma\nu$. The exponent a/c giving the size distribution law is called τ in percolation and in most models of avalanches in magnets[6] and is related to the Gutenberg–Richter exponent for earthquakes[7] (Fig. 1(b)). Most measured quantities depending on one variable will have similar power-law singularities at the critical point. For example, the distribution of avalanche durations and peak heights also have power-law forms.

---

[5]The same avalanches occur independent of your measuring instruments! But the probability $D(S)$ changes, because the *fraction* of large avalanches depends upon how many small avalanches you measure, and because the fraction per unit S changes as the scale of S changes.

[6] In our RFIM for hysteresis, we use τ to denote the avalanche size law at the critical field and disorder ($D(S, R_c, H_c) \sim S^{-\tau}$); integrated over the hysteresis loop $D_{\text{int}} \propto S^{-\bar{\tau}}$ with $\bar{\tau} = \tau + \sigma\beta\delta$.

[7]We must not pretend that we have found the final explanation for the Gutenberg–Richter law. There are many different models that give exponents $\approx 2/3$, but it remains controversial which of these, if any, are correct for real-world earthquakes.

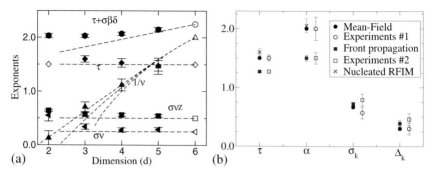

Fig. 15. (a) **Universal Critical Exponents in Various Spatial Dimensions.** [1] We test our ϵ-expansion predictions [34] by measuring the various critical exponents numerically in up to five spatial dimensions [35, 36]. The various exponents are described in the text. All of the exponents are calculated only to linear order in ϵ, except for the correlation length exponent ν, where we use results from other models. The agreement even in three dimensions is remarkably good, considering that we're expanding in $6 - D$ where $D = 3$! (b) **Universal Critical Exponents vs. Experiment** [21]. The exponent τ gives the probability density $\sim S^{-\tau}$ of having an avalanche of size S, α gives the probability density $\sim T^{-\alpha}$ of having an avalanche of duration T, and the exponents σ_k and Δ_k describe the cutoff in the avalanche sizes and durations, respectively, due to the demagnetizing field (see [21]). The experimental samples group naturally into two groups, described by the mean-field and front-propagation universality classes; none appear to be described well by our nucleated RFIM.

This is because power laws are the only self-similar functions. If $f(x) = x^{-\alpha}$, then on a new scale multiplying x by B, $f(Bx) = B^{-\alpha}x^{-\alpha} \propto f(x)$.

Crackling noise involves several power laws. We've seen that the probability of having an avalanche of size S goes as $S^{-\tau}$. In our RFIM model for hysteresis, if one is at a distance $R - R_c$ from the critical point, there will be a cutoff in the avalanche size distribution. The typical largest spatial extent L of an avalanche is called the *correlation length* ξ, which scales as $\xi \sim (R - R_c)^{-\nu}$. The cutoff in the avalanche size S scales as $(R - R_c)^{-\sigma}$ (Fig 17). In other models, demagnetization effects (parameterized by k) lead to cutoffs in the avalanche size distribution [21], with analogous critical exponents ν_k and σ_k. The size and spatial extent of the avalanches are related to one another by a power law $S \sim L^{d_f}$, where $d_f = 1/\sigma\nu$ is called the *fractal dimension*. The duration of an avalanche goes as L^z. The probability of having an avalanche of duration T goes as $T^{-\alpha}$, where $\alpha = (\tau - 1)/\sigma\nu z + 1$.[8] In our RFIM, the jump in magnetization goes as $(R - R_c)^{\beta}$, and at R_c the magnetization $(M - M_c) \sim (H - H_c)^{1/\delta}$ (Fig. 7(d)).

To specialists in critical phenomena, these exponents are central; whole conversations often rotate around various combinations of Greek letters. We know

---

[8]Notice that we can write d_f and α in terms of the other exponents. These are *exponent relations*; all of the exponents can typically be written in terms of two or three basic ones. We shall derive some exponent relations in section 5.2.

how to calculate critical exponents from various analytical approaches; given an implementation of the renormalization group they can be derived from the eigenvalues of the linearization of the renormalization-group flow around the fixed-point S^* in Fig. 12. Figure 15(a) shows our numerical estimates for several critical exponents of the RFIM model for Barkhausen noise [35, 36], together with our $6 - \epsilon$ expansions results [34, 37]. Of course, the challenge is not to get analytical work to agree with numerics: it is to get theory to agree with experiment. Figure 15(b) compares recent Barkhausen experimental measurements of the critical exponents to values from the three theoretical models.

5.2. Scaling functions

Critical exponents are not everything, however. Many other scaling predictions are easily extracted from numerical simulations, even if they are inconvenient to calculate analytically. (Universality should extend even to those properties that we have not been able to write formulæ for.) In particular, there are an abundance of functions of two or more variables that one can measure, which are predicted to take universal *scaling forms*.

5.2.1. Average pulse shape

For example, let us consider the time history $V(t)$ of the avalanches (Fig. 4). Each avalanche has large fluctuations, but one can average over many avalanches to get a typical shape. Averaging $V(t)$ together for large and small avalanches would seem silly, since only the large avalanches will last long enough to contribute at late times. The experimentalists originally addressed this issue [40] by rescaling the voltages of each avalanche by the peak voltage and the time by the duration, and then averaging these rescaled curves (dark circles in Fig. 16b). But there are so many more small avalanches than large ones that it seems more sensible to study them separately.

Consider the average voltage $\bar{V}(T, t)$ over all avalanches of duration T. Universality suggests that this average should be the same for all experiments and (correct) theories, apart from an overall shift in time and voltage scales:

$$\bar{V}_{\exp}(T, t) = A \bar{V}_{\text{th}}(T/B, t/B). \tag{5.5}$$

Comparing systems with a shifted time scale becomes simpler if we change variables; let $v(T, t/T) = \bar{V}(T, t)$. Now, as we did for the avalanche size distribution in section 5.1, let us compare a system with itself, after coarse-graining by a small factor $B = 1/(1 - \epsilon)$:

$$v(T, t/T) = Av(T/B, t/T) = (1 + a)v\big((1 - \epsilon)T, t/T\big). \tag{5.6}$$

Again, making ϵ small we find $av = T\, \partial v/\partial T$, with solution $v(T, t/T) = v_0(T)T^a$. Here the integration constant v_0 will now depend on t/T, so we arrive

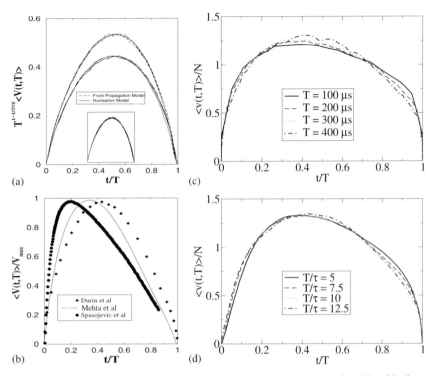

Fig. 16. (a) **Theoretical average pulse shape scaling functions** for our nucleated model and the front propagation model [38]. The overall height is non-universal; the two curves are otherwise extremely similar. The front propagation model has $1/\sigma v z = 1.72 \pm 0.03$ in this collapse; our nucleation model has $1/\sigma v z = 1.75 \pm 0.03$ (there is no reason to believe these two should agree). The inset shows the two curves rescaled to the same height (the overall height is a non–universal feature): they are quantitatively different, but far more similar to one another than either is to the experimental curves in part (b). (b) **Comparison of experimental average pulse shapes** for fixed pulse duration, as measured by three different groups [21, 38–40]. Notice that both theory curves are much more symmetric than those of the experiments. Notice also that the three experiments do not agree. This was a serious challenge to our ideas about the universality of the dynamics of crackling noise [1]. (c) **Pulse shape asymmetry experiment** [19]. Careful experiments show a weak but systematic duration dependence in the collapse of the average Barkhausen pulse shape. The longer pulses (larger avalanches) are systematically more symmetric (approaching the theoretical prediction). (d) **Pulse shape asymmetry theory** [19]. Incorporating the time-retardation effects of eddy currents into the theoretical model produces a similar systematic effect. The non-universal effects of eddy currents are in principle irrelevant for extremely large avalanches.

at the scaling form

$$\bar{V}(T, t) = T^{a} \mathcal{V}(t/T) \tag{5.7}$$

where the scaling function $\mathcal{V} \equiv v_0$ (up to an overall constant factor) is a universal prediction of the theory.

Can we write the exponent a in terms of the exponents we already know? Since the size of an avalanche is defined as the integral of $V(t)$, we can use the scaling relation (eq 5.7) to write an expression for the average size of an avalanche of duration T,

$$\bar{S}(T) = \int \bar{V}(t, T)\, dt = \int T^a \mathcal{V}(t/T)\, dt \sim T^{a+1}. \tag{5.8}$$

We also know that avalanches have fractal dimension $d_f = 1/\sigma\nu$, so $S \sim L^{1/\sigma\nu}$, and that the duration of an avalanche of size L goes as $T \sim L^z$. Hence $\bar{S}(T) \sim T^{1/\sigma\nu z} \sim T^{a+1}$, and $a = 1/\sigma\nu z - 1$. This is an example of an *exponent relation*.

Can we use the experimental pulse shape to figure out which theory is correct? Fig. 16(a) shows a *scaling collapse* of the pulse shapes for our RFIM and the front propagation model. The scaling collapse tests the scaling form eq 5.7 by attempting to plot the scaling function $\mathcal{V}(t/T) = T^{-a} V(t, T)$ for multiple durations T on the same plot. The two theoretical pulse shapes look remarkably similar to one another, and almost perfectly time-reversal symmetric. The mean-field model also is time-reversal symmetric.[9] Thus none of the theories describe the strongly skewed experimental data (Fig. 16). Indeed, the experiments did not even agree with one another, calling into question whether universality holds for the dynamical properties.

This was recognized as a serious challenge to our whole theoretical picture [1, 38, 43]. An elegant, convincing physical explanation was developed by Colaiori *et al.* [44], who attribute the asymmetry to eddy currents, whose slow decay lead to a time-dependent damping of the domain wall mobility. Incorporating these eddy current effects into the model leads to a clear correspondence between their eddy-current theory and experiment, see Fig. 16(c,d). In those figures, notice that the scaling 'collapses' are imperfect, becoming more symmetric for avalanches of longer durations. These eddy-current effects are theoretically *irrelevant* perturbations; under coarse-graining, they disappear. So, the original models are in principle correct,[10] but only for avalanches far larger and longer-lasting than those actually seen in the experiments.

5.2.2. *Avalanche size distribution*
Finally, let us conclude by analyzing the avalanche size distribution in our RFIM for hysteresis. The avalanche size distribution not only connects directly to many

---

[9]The mean–field model apparently has a scaling function which is a perfect inverted parabola [41]. The (otherwise similar) rigid-domain-wall model, interestingly, has a different average pulse shape, that of one lobe of a sinusoid [42, eq. 3.102].

[10]One might wonder how the original models do so well for the avalanche size distributions and other properties, when they have such problems with the pulse shape. In models which obey *no passing* [45, 46], the domains flipped in an avalanche are independent of the dynamics; eddy currents in these models won't change the shapes and sizes of the avalanches.

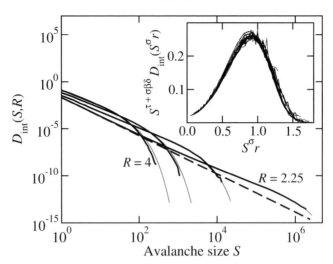

Fig. 17. **Avalanche size distribution**. The distribution of avalanche sizes in our model for hysteresis. Notice the logarithmic scales.[12] (i) Although only at $R_c \approx 2.16$ do we get a pure power law (dashed line, $D(S) \propto S^{-\bar\tau}$), we have large avalanches with hundreds of domains even a factor of two away from the critical point, and six orders of magnitude of scaling at 5% above R_c. (ii) The curves have the wrong slope except very close to the critical point. Be warned that a power law over two decades (although often publishable [47]) may not yield a reliable exponent. (iii) The scaling curves (thin lines) work well even far from R_c. Inset: We plot $D(S)/S^{-\bar\tau}$ versus $S^\sigma (R - R_c)/R$ to extract the universal scaling curve $\mathcal{D}(X)$ (eqn 5.15). (We use the exponent relation $\bar\tau = \tau + \sigma\beta\delta$ without deriving it; see footnote 6 on page 279.) Varying the critical exponents and R_c to get a good collapse allows us to measure the exponents far from R_c, where power-law fits are still unreliable.

of the experiments (Fig. 8), it also involves a non-obvious example of a *scaling variable*). Fig. 17 shows the distribution of avalanche sizes $D_{\text{int}}(S, R)$ for different disorders $R > R_c$, integrated over the entire hysteresis loop (Fig. 6). We observe a power-law distribution near $R = R_c \sim 2.16$, that is cut off at smaller and smaller avalanches as we move away from R_c.[13]

Let us derive the scaling form for $D_{\text{int}}(S, R)$. By using scale invariance, we will be able to write this function of two variables as a power of one of the variables times a universal, one-variable function of a combined scaling variable. From our treatment at R_c (eqns 5.1) we know that

$$S' = S/(1 + c\epsilon),$$
$$D' = D(1 + a\epsilon).$$

(5.9)

---

[13]This cutoff is the one described in section 5.1, scaling as $(R - R_c)^{-\sigma}$.

A system at $R = R_c + r$ after coarse-graining will have all of its avalanches reduced in size, and hence will appear similar to a system further from the critical disorder (where the cutoff in the avalanche size distribution is smaller, Fig. 7), say at $R = R_c + Er = R_c + (1 + e\epsilon)r$. Hence

$$D(S', R_c + Er) = D'(S', R_c + r) = AD(CS', R_c + r),$$
$$D\big(S', R_c + (1 + e\epsilon)r\big) = (1 + a\epsilon) D\big((1 + c\epsilon) S', R_c + r\big). \tag{5.10}$$

To facilitate deriving the scaling form for multiparameter functions, it is helpful to change coordinates so that all but one variable remains unchanged under coarse-graining (the scaling variables). In the average pulse shape of section 5.2.1, the time t and the duration T change in the same way under coarse-graining, so the ratio was a scaling variable. For the avalanche size distribution, consider the combination $X = S^{e/c}r$. After coarse-graining $S' = S/C$ and shifting to the higher disorder $r' = Er$ this combination is unchanged:

$$X' = S'^{e/c}r' = (S/C)^{e/c}(Er) = \big(S/(1 + c\epsilon)\big)^{e/c}\big((1 + e\epsilon)r\big)$$
$$= S^{e/c}r\left(\frac{1 + e\epsilon}{(1 + c\epsilon)^{e/c}}\right) = S^{e/c}r + O(\epsilon^2) = X + O(\epsilon^2). \tag{5.11}$$

Let $\bar{D}(S, S^{e/c}R) = D(S, R)$ be the size distribution in terms of the variables S and X. Then $\bar{D}$ coarse-grains much like a function of one variable, since X stays fixed. Equation 5.10 now becomes

$$\bar{D}(S', X') = \bar{D}(S', X) = (1 + a\epsilon) \bar{D}\big((1 + c\epsilon) S', X\big), \tag{5.12}$$

so

$$a\bar{D} = -cS'\frac{\partial \bar{D}}{\partial S'} \tag{5.13}$$

and hence

$$\bar{D}(S, X) = S^{-a/c} = S^{-\bar{\tau}}\mathcal{D}(X) \tag{5.14}$$

with the universal *scaling function* $\mathcal{D}(X)$. Rewriting things in terms of the original variables and the traditional Greek names for the scaling exponents ($c = 1/\sigma\nu$, $a = \bar{\tau}/\sigma\nu$, and $e = 1/\nu$), we find the scaling form for the avalanche size distribution:

$$D(S, R) \propto S^{-\bar{\tau}}\mathcal{D}(S^{\sigma}(R - R_c)). \tag{5.15}$$

We can use a *scaling collapse* of the experimental or numerical data to extract this universal function, by plotting $D/S^{-\bar{\tau}}$ against $X = S^{\sigma}(R - R_c)$; the inset of Fig. 17 shows this scaling collapse.

In broad terms, most properties that involve large scales of length and time at a critical point will have universal scaling forms; any N variable function will be writable in terms of a power law times a universal function of $N - 1$ variables, $F(x, y, z) \sim z^{-\alpha} \mathcal{F}(x/z^{\beta}, y/z^{\gamma})$. The deep significance of the renormalization-group predictions are only feebly illustrated by the power-laws most commonly studied. The universal scaling functions, and other morphological self-similar features are the best ways to measure the critical exponents, the sharpest tests for the correctness of theoretical models, and provide the richest and most complete description of the complex behavior observed in these systems.

References

[1] J.P. Sethna, K.A. Dahmen, and C.R. Myers. Crackling noise. *Nature*, 410:242, 2001.

[2] James P. Sethna, Karin A. Dahmen, and Olga Perković. Random-field models of hysteresis, http://arxiv.org/abs/cond-mat/0406320. In *The Science of Hysteresis, Vol. II*, pages 107–179. Academic Press, 2006.

[3] James P. Sethna. *Statistical Mechanics: Entropy, Order Parameters, and Complexity*, http://www.physics.cornell.edu/sethna/StatMech/. Oxford University Press, Oxford, 2006.

[4] M.C. Kuntz, P. Houle, and J.P. Sethna. Crackling noise. simscience.org/, 1998.

[5] B. Gutenberg and C.F. Richter. *Seismicity of the Earth and Associated Phenomena*. Princeton University, Princeton, NJ, 1954.

[6] P.A. Houle and J.P. Sethna. Acoustic emission from crumpling paper. *Physical Review E*, 54:278, 1996.

[7] L.I. Salminen, A.I. Tolvanen, and M.J. Alava. Acoustic emission from paper fracture. *Phys. Rev. Lett.*, 89:185503, 2002.

[8] M. Carmen Miguel, A. Vespignani, S. Zapperi, J. Weiss, and J.-R. Grasso. Intermittent dislocation flow in viscoplastic deformation. *Nature*, 410:667, 2001.

[9] F. Kun, Gy.B. Lenkey, N. Takács, and D.L. Beke. Structure of magnetic noise in dynamic fracture. *Phys. Rev. Lett.*, 93:227204, 2004.

[10] E.M. Kramer and A.E. Lobkovsky. Universal power law in the noise from a crumpled elastic sheet. *Phys. Rev. E*, 53:1465–1469, 1996.

[11] L.C. Krysac and J.D. Maynard. Evidence for the role of propagating stress waves during fracture. *Phys. Rev. Lett.*, 81:4428–31, 1998.

[12] Knut Jørgen Måløy, Stéphane Santucci, Jean Schmittbuhl, and Renaud Toussaint. Local waiting time fluctuations along a randomly pinned crack front. *Phys. Rev. Lett.*, 96:045501, 2006.

[13] S. Tewari, D. Schiemann, D.J. Durian, C.M. Knobler, S.A. Langer, and A.J. Liu. Statistics of shear-induced rearrangements in a two-dimensional model foam. *Phys. Rev. E*, 60:4385–4396, 1999.

[14] N. Vandewalle, J.F. Lentz, S. Dorbolo, and F. Brisbois. Avalanches of popping bubbles in collapsing foams. *Phys. Rev. Lett.*, 86:179, 2001.

[15] M. Cieplak and M.O. Robbins. Dynamical transition in quasistatic fluid invasion in porous media. *Phys. Rev. Lett.*, 60:2042–2045, 1988.

[16] O. Narayan and D.S. Fisher. Threshold critical dynamics of driven interfaces in random media. *Phys. Rev. B*, 48:7030–42, 1993.

[17] Michael D. Uchic, Dennis M. Dimiduk, Jeffrey N. Florando, and William D. Nix. Sample dimensions influence strength and crystal plasticity. *Science*, 305:986–9, 2004.

[18] Dennis M. Dimiduk, Chris Woodward, Richard LeSar, and Michael D. Uchic. Scale-free intermittent flow in crystal plasticity. *Science*, 312:1188–90, 2006.

[19] Stefano Zapperi, Claudio Castellano, Francesca Colaiori, and Gianfranco Durin. Signature of effective mass in crackling-noise asymmetry. *Nature Physics*, 1:46–9, 2005.

[20] James P. Sethna, Karin Dahmen, Sivan Kartha, James A. Krumhansl, Bruce W. Roberts, and Joel D. Shore. Hysteresis and hierarchies: Dynamics of disorder-driven first-order phase transformations. *Physical Review Letters*, 70:3347–50, 1993.

[21] G. Durin and S. Zapperi. Scaling exponents for barkhausen avalanches in polycrystalline and amorphous ferromagnets. *Phys. Rev. Lett.*, 84:4705–4708, 2000.

[22] H. Ji and M.O. Robbins. Percolative, self–affine, and faceted domain growth in random 3-dimensional magnets. *Phys. Rev. B*, 46:14519, 1992.

[23] O. Narayan. Self-similar barkhausen noise in magnetic domain wall motion. *Phys. Rev. Lett.*, 77:3855–3857, 1996.

[24] B. Koiller and M.O. Robbins. Morphology transitions in three-dimensional domain growth with gaussian random fields. *Phys. Rev. B*, 62:5771–5778, 2000. and references therein.

[25] B. Alessandro, C. Beatrice, G. Bertotti, and A. Montorsi. Domain–wall dynamics and barkhausen effect in metallic ferromagnetic materials. 2. experiments. *J. Appl. Phys.*, 68:2908–2915, 1990.

[26] B. Alessandro, C. Beatrice, G. Bertotti, and A. Montorsi. Domain–wall dynamics and barkhausen effect in metallic ferromagnetic materials. 1. theory. *J. Appl. Phys.*, 68:2901–2907, 1990.

[27] P. Cizeau, S. Zapperi, G. Durin, and H.E. Stanley. Dynamics of a ferromagnetic domain wall and the barkhausen effect. *Phys. Rev. Lett.*, 79:4669, 1997.

[28] S. Zapperi, P. Cizeau, G. Durin, and E.H. Stanley. Dynamics of a ferromagnetic domain wall: Avalanches, depinning transition, and the barkhausen effect. *Phys. Rev. B*, 58:6353–6366, 1998.

[29] Gianfranco Durin. The barkhausen effect. http://www.ien.it/~durin/barkh.html, 2001.

[30] E.A. Guggenheim. The principle of corresponding states. *Journal of Chemical Physics*, 13:253–61, 1945.

[31] P. Heller and G.B. Benedek. Nuclear magnetic resonance in MnF_2 near the critical point. *Physical Review Letters*, 8:428–32, 1962.

[32] J. Zinn-Justin. *Quantum field theory and critical phenomena (3rd edition)*. Oxford University Press, 1996.

[33] M.E.J. Newman. Power laws, Pareto distributions and Zipf's law. *Contemporary Physics*, 46:323–51, 2005. arXiv.org/abs/cond-mat/0412004/.

[34] K. Dahmen and J.P. Sethna. Hysteresis, avalanches, and disorder induced critical scaling: A renormalization group approach. *Phys. Rev. B*, 53:14872, 1996.

[35] O. Perković, K. Dahmen, and J.P. Sethna. Avalanches, barkhausen noise, and plain old criticality. *Phys. Rev. Lett.*, 75:4528, 1995.

[36] Olga Perković, Karin A. Dahmen, and James P. Sethna. Disorder-induced critical phenomena in hysteresis: Numerical scaling in three and higher dimensions. *Phys. Rev. B*, 59:6106–19, 1999.

[37] K. Dahmen and J.P. Sethna. Hysteresis loop critical exponents in 6-epsilon dimensions. *Phys. Rev. Lett.*, 71:3222–5, 1993.

[38] A. Mehta, A.C. Mills, K.A. Dahmen, and J.P. Sethna. Universal pulse shape scaling function and exponents: Critical test for avalanche models applied to barkhausen noise. *Phys. Rev. E*, 65:046139/1–6, 2002.

[39] L. Dante, G. Durin, A. Magni, and S. Zapperi. Low–field hysteresis in disordered ferromagnets. *Phys. Rev. B*, 65:144441, 2002.

[40] D. Spasojevic, S. Bukvic, S. Milosevic, and H.E. Stanley. Barkhausen noise: Elementary signals, power laws, and scaling relations. *Phys. Rev. E*, 54:2531–2546, 1996.

[41] M. Kuntz. *Barkhausen noise: simulations, experiments, power spectra, and two dimensional scaling*. PhD thesis, Cornell University, 1999.

[42] G. Durin and S. Zapperi. The barkhausen effect. In G. Bertotti and I. Mayergoyz, editors, *Science of Hysteresis, Vol. II*. Elsevier, London, 2004.

[43] M. Kuntz and J.P. Sethna. Noise in disordered systems: The power spectrum and dynamics exponents in avalanche models. *Phys. Rev. B*, 62:11699, 2000.

[44] F. Colaiori, M.J. Alava, G. Durin, A. Magni, and S. Zapperi. Phase transitions in a disordered system in and out of equilibrium. *Phys. Rev. Lett.*, 92:257203, 2004.

[45] Alan A. Middleton. Asymptotic uniqueness of the sliding state for charge-density waves. *Physical Review Letters*, 68:670, 1992.

[46] A.A. Middleton and D.S. Fisher. Critical behavior of charge-density waves below threshold: numerical and scaling analysis. *Phys. Rev. B*, 47:3530–52, 1993.

[47] O. Malcai, D.A. Lidar, O. Biham, and D. Avnir. Scaling range and cutoffs in empirical fractals. *Physical Review E*, 56:2817–28, 1997.

Course 7

BOOTSTRAP AND JAMMING PERCOLATION

C. Toninelli

*Laboratoire de Probabilités et Modèles Aléatoires CNRS-UMR 7599, Univ. Paris VI-VII,
4 Pl. Jussieu, Paris, France*

J.-P. Bouchaud, M. Mézard and J. Dalibard, eds.
Les Houches, Session LXXXV, 2006
Complex Systems
© *2007 Published by Elsevier B.V.*

Contents

1. Introduction

The aim of these notes is to review some results for Bootstrap Percolation (BP) [1–8] and the more recently introduced Jamming Percolation (JP) models [9–13].

Let us start by presenting two examples to explain what is (and what is not) bootstrap and jamming percolation. Take a square box and fill it with insulating and metal balls, then shake it in order to reach a random mixture. A natural question arises: Does the overall box behave like a conductor or an insulator? In other words: Does there exist a group of metal balls touching one another and reaching the opposite sides of the box? the answer depends on the size of the box and on the fraction of metal vs insulating balls: the higher this fraction the higher the probability of an overall conductive behavior. The simplest model for this is *Site Percolation* (SP): a lattice model in which each site is independently occupied or empty with probability p and $1 - p$ (metal and conducting balls) and the focus is on the properties of occupied clusters, i.e. group of nearest neighbour occupied sites. SP tells us that, in the limit of large box size, a cluster of metal balls which spans or percolates the box occurs if and only if p exceeds a finite threshold, $p > p_c$. This means that a sharp change in the global behavior of the system occurs upon a small variation of the density: a *phase transition* occurs at p_c. This is a second-order transition: the continuously changing order parameter, $\theta(p)$, is the probability that the origin belongs to an infinite cluster or, which is the same thanks to translation invariance, the density of the infinite cluster itself. Indeed, the main appeal of SP relies on being the simplest model which displays a second-order phase transition. Thus, the natural issues of estimating the value of p_c and determining the behavior of $\theta(p)$ as well as the other cluster properties around criticality can be analyzed by using standard tools for second-order transitions, in particular renormalization group theory. From this we know that $\theta(p)$ and the other cluster properties scale as *power laws*, $\theta(p) = (p - p_c)^\beta$, and their *finite size scaling* has the form $\theta(p, L) = L^{-\beta/\nu} F[(p - p_c)L^{1/\nu}]$ where β is a quantity dependent *critical exponent* and ν is the exponent of the power law divergence of the incipient cluster size. Thanks to the knowledge of finite size effects, one can extrapolate via numerical simulations the value of p_c and of all *critical exponents*. Finally it is important to recall that there is a large degree of *universality* for SP models: p_c is model dependent but the critical exponents depend only on the spatial dimension of the lattice. This means that the occurrence and the character of the transition is not influenced by microscopic details.

Now, instead of the mixture of conducting and isolating balls, put inside the square box just one kind of balls without filling it completely. Then put the box on a horizontal plane, shake it in order to start again from a random configuration and let the system evolve with a dynamics in which a given ball can move only if "there is enough empty space around". You could either imagine a deterministic Newtonian dynamics, e.g. balls interacting via a hard sphere potential, or a stochastic dynamics, e.g. the motion of a suspension of colloidal particles which is driven by the thermal agitation of the solvent. The natural question which now arises is the following: Is the whole system mobile or does there exist a *frozen backbone* which does not evolve during dynamics? In other words: if we focus on a specific ball will it always move (provided we wait long enough) or could it belong to a finite fraction of balls which are forever blocked? Of course, the answer depends again on the density of particles and on the size of the box. But now, as we will see, it also crucially depends on the microscopic details, namely on the specific meaning of the words "enough empty space" when defining the dynamics. Turning to real systems, numerical simulations for sphere packings [14] and experiments for continuum particle systems [15] show that there is a packing density where a transition occurs. This can roughly be regarded as the transition from a fluid to an amorphous solid and should correspond to the formation of a frozen backbone for the system. The onset of this transition is sharp and displays a peculiar mixture of first-order and critical character. The average coordination number of a particle is discontinuous but displays a singularity (as for a continuous transition) when the critical density is approached from above. Also, a singularity in the typical relaxation time occurs when the transition is approached from below. Similar mixed first-order and continuous properties are observed in other systems of very different microscopic nature where upon tuning a proper external parameter dynamics is slowed down leading ultimately to the formation of an amorphous solid. These include supercooled liquids when temperature is lowered until the formation of a glass phase [16] and non-thermal systems such as vibrated granular materials [17]. The understanding of these phenomena, which are usually referred to as *glass or jamming transitions*, still leaves many unsettled and fascinating questions for condensed matter physicists. From the theoretical point of view, the first natural step is to devise a microscopic model simple enough to be analyzed and displaying a dynamical arrest with this mixture of first order and critical features. Another property one would like to reproduce is the unconventional scaling of relaxation times which for molecular fragile liquids diverge faster than power law, a very popular fit being the Vogel Fulcher form: $\log \tau \simeq 1/T - T_c$.

In the course of this lecture we will see that in a sense BP is the easiest discrete model one could imagine to describe the constrained dynamics of the balls in the box. However BP fails to provide a jamming transition: a frozen cluster in the thermodynamic limit never occurs on any finite dimensional hyper-cubic lattice (square lattice in $d = 2$, cubic lattice in $d = 3, \ldots$). On the other hand, if one keeps the lattice size finite and raises the density, a sharp onset which is very reminiscent of a first-order phase transition occurs at a (size dependent) density threshold. This metastability phenomenon is due to the fact that proper large and rare critical droplets of empty sites are necessary in order to make the system mobile. In turn, this induces time scales which diverge very rapidly when the density is raised towards one.

We will explain how, by a different choice of the microscopic dynamics, it is possible to construct models which instead display a jamming transition on finite dimensional lattices [9–13]: a frozen cluster occurs in the thermodynamic limit for $p \geq p_c$ with $0 < p_c < 1$. These are what we call Jamming Percolation (JP) models. Here we define the simplest JP model, the so called Spiral Model (SM) [11, 12], and we provide the tools which allow to identify and analyze its transition. This has a character which is different from SP and all standard percolation transitions: the density of the frozen cluster is discontinuous at p_c, namely the clusters are compact rather than fractal at criticality, and the size of the incipient cluster diverges when $p \nearrow p_c$ faster than any power law. Furthermore, since the transition is not second order, we do not have the large degree of universality which holds for SP: the existence and character of the transition (and not only its location) depends on the specific choice of the microscopic details. However, we will see that it is possible to identify a class of models which give rise to a jamming transition and belong to the same universality class.

2. Bootstrap Percolation (BP)

2.1. *Definition and main results*

Consider a d-dimensional lattice and fix a parameter m, $1 < m < 2d$. Let each site be independently occupied or empty with probability p and $1 - p$, as for SP. The objects of interest for BP are *blocked (or frozen) clusters* with respect to the dynamics in which the "enough empty space" required to move a particle corresponds to (at least) m empty nearest neighbours. More precisely, blocked clusters are sets of occupied sites each of which has less than m empty neighbours (more than $2d - m$ occupied neighbours). Denoting by $R(L, p)$ the probability of finding the cluster on a lattice of linear size L with filled boundary conditions,[1] it is

---

[1] Unless otherwise stated, we always refer to filled boundary conditions for finite volume lattices (since with empty b.c. blocked clusters do not occur).

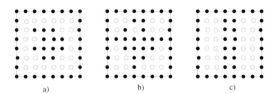

Fig. 1. a) and b): configurations without blocked clusters; c) configuration with a blocked cluster.

immediate to check that $R(L, 0) = 0$, $R(L, p)$ increases with p and $R(L, 1) = 1$. Does there exists a non trivial critical density, $0 < p_c < 1$, such that in the thermodynamic limit a blocked cluster never occurs ($\lim_{L \to \infty} R(L, p) = 0$) for $p < p_c$, while it always occurs ($\lim_{L \to \infty} R(L, p) = 1$) for $p > p_c$? And, if such transition takes place, is it second-order (as for SP) or first-order? namely, is the density of the frozen cluster, $\Phi(p)$, zero or greater than zero at p_c? How do $\Phi(p)$ and the other cluster properties scale around criticality?

Before answering to these questions let us explain more in detail what are blocked clusters and give a receipt which allows to identify them. Focus for simplicity on the case of a two-dimensional lattice with $m = 2$. It is immediate to check that the configuration of Fig. 1a) does not contain a blocked cluster. This is also true for any configuration which does not have occupied clusters spanning the lattice: the external particles are not blocked and starting from them one can subsequently unblock the whole cluster. Consider instead the configurations in Fig. 1b) and c) which contain a spanning cluster: do they also contain a blocked cluster? As one can directly check the answer is no for b) and yes for c): the existence of a spanning cluster is a necessary but not sufficient condition for a blocked cluster. How can we decide whether a blocked cluster occurs or not? Should we check all subsets of nearest neighbouring particles and verify whether there exists (at least) one in which all particles are blocked? after some thought you can realize that the following *pruning or leaf removal algorithm* does the job in a polynomial time. Take the initial configuration and remove all particles that have at least two empty neighbours. Then repeat the procedure on the new configuration and continue this culling of particles until reaching a stable configuration: either one which is completely empty or one that contains only groups of particles each of which have less than two empty neighbors. The first result means that the initial configuration had no blocked cluster, the second one that a blocked cluster was present. Furthermore the blocked cluster coincides with the set of particles which remain in the final configuration. The search for a receipt to identify blocked clusters has thus lead us to an alternative definition for BP (which is indeed the more common one). BP is a cellular automaton with a discrete time deterministic dynamics in which at each time unit the configuration is updated according to a local and translation invariant rule: an empty site remains empty and a filled one gets emptied if and only if it has at least m empty nearest neighbors.[2] The questions on the existence and character of the transition can be rephrased by asking which is the threshold, p_c, below which the final configuration is completely empty and whether the final density at criticality is zero or positive. Note that now we can also ask dynamical questions, for example which is the velocity of convergence to the empty lattice for $p < p_c$.

At first sight you could think that the occurrence of a transition at a finite p_c follows by a simple extension of the arguments that work for SP. Let us explain why this is not the case focusing again on the case $d = 2, m = 2$. Since the existence of blocked cluster requires a spanning cluster, as for SP we can conclude that $p_c > 0$. Indeed, if p is too low, the cost for a given path being occupied overcomes the entropic factor given by the number of all possible paths going from the origin to infinity, thus $\Phi(p)$ equals zero. On the other hand, since for BP a spanning cluster is not necessarily a blocked one, Peierls type contour arguments leading to $p_c < 1$ for standard percolations, here cannot be applied. Indeed the contrary can be proved: for $m = 2, d = 2$ $p_c = 1$ [3]. Moreover, on any d-dimensional lattice and for any choice of m, BP transition never occurs at a non trivial p_c: either $p_c = 0$ (for $m > d$) or $p_c = 1$ (for $m \leq d$) [5]. The first result is an easy consequence of the fact that, if $m > d$, it is possible to construct frozen clusters with a finite number of particles. These will always be present in the thermodynamic limit $\forall p > 0$, therefore $p_c = 0$. Before explaining the result $p_c = 1$ for the more interesting cases $m \leq d$, we wish to recall that prior to above rigorous results the phase transition scenario for BP was not at all clear. The technique that had been used to measure p_c relied (as for SP) on finite size scaling. From numerical simulations the value of the concentration at which a system of size L contains with probability 0.5 a blocked cluster, $p_{0.5}(L)$, was evaluated. Then p_c was extrapolated by using $p_{0.5}(L) - p_c \propto f(L)$ with $f(L) = L^{-1/\nu}$, getting $p_c \simeq 0.965$ for $d = 2, m = 2$ and $p_c \simeq 0.896$ for $d = 3, m = 3$ [2]. These incorrect numerical estimates of p_c were due to the fact that the correct finite size scaling have an unconventional slow decay, $f(L) \propto \ln L$ for $d = 2, m = 2$ [4] and $f(L) \propto \ln \ln L$ for $d = 3, m = 3$ [7].

2.2. Unstable voids: $p_c = 1$ and finite size effects for $d = 2, m = 2$

The first proof of $p_c = 1$ for $d = 2, m = 2$ was given by van Enter [3] following an argument of Straley. Here we will instead present a modified proof [4,5]

---

[2]The convention we choose (empty sites being stable) is the one that is usually adopted in physical literature, while in mathematical literature the inverse one is considered. Of course the two models can easily be mapped one into the other. Another warn concerns the meaning of the parameter m: here m is the minimal number of vacancies required for not being blocked, in other works it is the minimal number of nearest neighbour occupied site which do block (our $2d - m$).

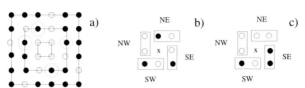

Fig. 2. a) The empty nucleus plus the minimal vacancies required for emptying; b) and c) Site x and its NE,SE,NW,SW neighbours. For the Spiral Model x is blocked in b) and unblocked in c).

which, as we shall explain in next section, can be readily extended to higher dimensions. The argument is based on the following observation: *large voids are unstable*. What does it mean? Focus on a configuration which has a 2×2 empty square centered at the origin and, on any subsequent shell, at least one empty site on each side as in Fig. 2a). If this construction is continued up to the boundary there cannot be a blocked cluster. This can be checked starting from the first internal shell (of linear size $2 + 2$) and removing the particles which are surrounded by the dotted circle in Fig. 2a), i.e. those that are adjacent to the vacancy and to the empty 2×2 nucleus. Then we can subsequently erase all sites in the first shell, thus expanding the empty square of size 2×2 to one of size 4×4. The procedure can then be performed analogously in the next shell and so on, until emptying the whole lattice. The probability of the empty square centered in the origin plus (at least) the minimal vacancies on all shells, $E(p, L) = (1 - p)^4 \prod_j^L (1 - p^j)^4$, converges to a positive value when $L \to \infty$ for any $p < 1$. Thus, by the ergodic theorem, in the thermodynamic limit BP reaches an empty configuration with unit probability, namely $p_c = 1$. This roughly corresponds to the fact that, since there are $O(L^2)$ possible positions for the empty nucleus plus vacancies and a finite (though small) probability that this occurs on a given site, in the limit $L \to \infty$ we will always find (at least) one such structure from which we can perform the emptying procedure, i.e. $\lim_{L \to \infty} R(L, p) = 0 \ \forall p, 1$. On the other hand, if we fix L and raise the density towards one, all lattice will be occupied: $\lim_{p \nearrow 1} R(L, p) = 1$. What does it happen if we let simultaneously $L \to \infty$ and $p \nearrow 1$? there exists a crossover length, $L_c \simeq \exp(c/(1 - p))$ with $c = \pi^2/9$ [4,6], such that $R(L, p)$ goes to one (to zero) if we take $L \to \infty$ and $p \nearrow 1$ with $L < L_c(p)$ ($L > L_c(p)$). In other words, L_c is the threshold value such that for lattice size $L > L_c$ a critical droplet from which we can empty the whole lattice typically occurs, while for $L < L_c$ blocked clusters typically occur. Therefore L_c can be regarded as the size of the incipient blocked cluster and corresponds to the typical time needed to converge to the empty configuration (see [5] for a nice renormalization argument which connects L_c to the first time at which the origin gets emptied in the BP procedure). We stress that large corrections to the asymptotic value of L_c occur. It has indeed been estimated that a linear size of the order 10^{47} [18] should be considered to see the true asymptotic limit. Therefore, while on the one hand numerical simulations should be taken with utmost care when extrapolating asymptotic results (recall the initial incorrect prediction of $p_c < 1$), the same should be done when one uses the rigorous asymptotic results for applications to real systems.

2.3. *Generalisation to $d > 2$, $m \leq d$*

The result $p_c = 1$ can immediately be extended to the cases $d > 2$, $m = 2$ by "monotonicity" arguments, i.e. by using the fact that a site has more neighbours then in the 2-dimensional case and yet the condition for being emptiable is the same. On the other hand, extensions to the cases $d > 2$, $m \neq 2$ are not immediate and rely on the following observation [5]. Consider the case $d = 3$, $m = 3$ and focus on a configuration which contains a $\ell \times \ell \times \ell$ empty cube around the origin. What is the minimal requirement we should now impose to expand this cube of one step? After some thought you should realize that it is sufficient to require that each two-dimensional face of the cube is "emptiable" with respect to BP dynamics with parameters $d = 2$, $m = 2$. The probability of the one step expansion for the cube is therefore $O(E(p, \ell)^4)$, with $E(p, \ell)$ converging to one and defined in previous section. Since $\prod_{i=\ell}^{\infty} E(p, i)$ converges to a finite value of order $1/\exp\exp(c/(1-p))$, we conclude again that $p_c = 1$. The crossover length now scales as [7,8] $L_c \simeq \exp\exp(c/1-p)$, i.e. roughly as exponential of the one for $d = 2$, $m = 2$ (the exact value of c has not been established). All other cases can be treated analogously iterating from models in lower dimension, obtaining the results $p_c = 1$ [5] and $L_c \propto \exp^{o(m-1)}(c/1 - p^{d-m+1})$ [7,8], where $\exp^{o(s)}$ stands for the exponential iterated s times.

2.4. *Other applications and general graphs*

The results in above sections imply that, as far as the search of a finite dimensional model displaying a jamming transition is concerned , BP is unsuccessful. However, BP has been successfully used to model other systems (see references in [1]) both in physics and in different fields such as biology (infection models), geology (crack formation models) and more recently in computer science. For the latter application occupied sites can be regarded as the functional units of memory arrays [19] or computer networks [20] and the study of BP on the corresponding graphs is relevant to analyze stability against random damage, namely to find the minimal value of connectivity which is necessary to maintain a proper level of inner communication and data mirroring. In this context the problem is usually referred to as *k-core percolation* [21], where the k-core of a graph is the maximal subgraph with degree at least k for each edge. Note that the graph which are relevant for these applications are not necessarily the hyper-cubic d-dimensional lattices discussed so far. Thus it is important to know that for BP on a mean field lattice a transition can occur at a non trivial critical density. Consider for example a Bethe lattice with connectivity $k + 1$, namely a graph taken at random among those with connectivity $k + 1$ on each site. In this case an analytical solution can be obtained thanks to the local tree-like structure and it is found [10, 20–22] that a transition occurs at $p_c < 1$ for $1 < m \leq k$. This (except for the choice $m = k$) has mixed first order and critical features: the giant frozen cluster appears discontinuously and the critical length diverges as power law.

3. Jamming Percolation (JP)

As explained in the introduction, Jamming Percolation models [9–13] have been introduced with the aim of finding a *jamming transition*, namely a first-order critical transition on a finite dimensional lattice. Here we will review the easiest example of a JP model, namely the two-dimensional model which has been introduced in [11, 12] and dubbed *Spiral Model* (SM). For SM the existence of a jamming transition has been rigorously proved [11, 12] and the exact value of p_c has been identified: p_c coincides with the critical threshold of directed site percolation (DP) in two dimensions, $p_c^{DP} \simeq 0.705$. The discontinuity of the density of the frozen cluster, $\Phi(p)$, and the faster then power law divergence of the crossover length, $L_c(p)$, have also been proved [12] modulo the standard conjecture on the existence of two different correlation lengths for DP [23].

In the following we will sketch the arguments which lead to the above results providing the tools one should use to analyze this transition and explaining the underlying mechanism: it is the consequence of two perpendicular directed percolation processes which together can form a compact network of frozen directed paths at criticality. From this discussion it will emerge how one can modify the microscopic rules without loosing the jamming transition. This is relevant since the transition is not second order, thus we do not have the large degree of universality with respect to microscopic details which holds for Site Percolation. This should not come as a surprise: we already know that for BP (which can be regarded as a microscopic modification of SM) a transition does not occur. The extension and universality of the jamming percolation transition of SM remain fundamental questions to be investigated. However, as it has been discussed in [11], it is possible to identify a class of rules which give rise to a jamming transition and belong to the same universality class of SM: as $p \nearrow p_c$ the divergence of the incipient frozen cluster follows the same scaling. A model that belongs to this class is for example the Knight model defined in [9]. Note that in general, at variance with SM, it will not be possible to determine analytically the exact value of p_c. It is therefore important to analyze finite size effects and give a proper receipt which allows to obtain a reliable estimate of p_c from numerical simulations (since, as for BP, convergence to the asymptotic results can be extremely slow). For an extended discussion on this we refer to [11], where for example the value of the critical density for the Knight model has been derived, $p_c^{Knight} \simeq 0.635$. Note that this differs from the original conjecture $p_c^{Knight} = p_c^{DP}$ [9] which was due to the overlooking of some blocked structures [13].

3.1. The simplest example: the Spiral Model (SM)

When defining the new rules we should keep in mind the lessons we learned from BP. One the one hand, in order to have $p_c < 1$, blocked clusters should occur with higher probability than for BP. This means that we should allow a larger variety of shapes: blocked clusters should "bend more" than BP ones. At the same time we should not let them "bend too much": if they can close on themselves forming finite blocked clusters we would get $p_c = 0$, as for BP with $m > d$. Is it possible to realize such a compromise?

Consider a square lattice and, for each site x, define among its first and second neighbours the couples of its North-East (NE), South-West (SW), North-West (NW) and South-East(SE) neighbours as in Fig. 2b) and c), namely NE $= (x + e_1, x + e_1 + e_2)$, SW $= (x - e_2, x - e_1 - e_2)$, NW $= (x - e_1, x - e_1 + e_2)$ and SE $= (x + e_1, x + e_1 - e_2)$, where e_1 and e_2 are the coordinate unit vectors. The update rule for the Spiral Model is the following: empty sites remain empty (as for BP), occupied sites get emptied if both its NE *and/or* both its SW neighbours are empty *and* both its SE *and/or* both its NW neighbours are empty too (see Fig. 2b) and c) for examples in which the constraint for x is satisfied and not satisfied, respectively). These rules can be rephrased by saying that at least one among the four sets $NE \cup SE$, $SE \cup SW$, $SW \cup NW$ and $NW \cup NE$ should be completely empty. Frozen or blocked clusters are, as for BP, the groups of particles that can never be erased under the iteration of the update rule. Again: Which is the critical density p_c above which a frozen cluster occurs? What is the value of the density of the frozen cluster, $\Phi(p)$, at p_c? How does the size of the the incipient blocked cluster scales when $p \nearrow p_c$?

3.2. SM: occurrence of blocked clusters for $p > p_c^{DP}$

As for BP, arguments analogous to those for SP can be applied only to conclude that $p_c > 0$. In order to establish $p_c < 1$ we will identify a set of configurations which contain a blocked cluster and prove that they cover the configuration space for $p > p_c^{DP}$, therefore $p_c \leq p_c^{DP} < 1$.

Let us start by recalling the definition and a few basic results on DP (see e.g. [23]). Take a square lattice and put two arrows going out from each site x

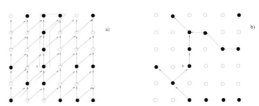

Fig. 3. a) The directed lattice obtained drawing arrows from each site towards its NE neighbours. Particles inside the continuous line belong to a spanning cluster on the directed lattice and form a blocked cluster. b) A non spanning NE-SW cluster blocked by T-junctions with NW-SE clusters.

towards its neighbours in the positive coordinate directions, $x + e_1$ and $x + e_2$. On this directed lattice a continuous percolation transition occurs at a non trivial critical density $p_c^{DP} \simeq 0.705$ (a percolating cluster is now one which spans the lattice following the direction of the arrows). This transition is second order, as for SP, but belongs to a different universality class. In particular, due the anisotropy of the lattice, the typical sizes of the incipient percolating cluster in the parallel ($e_1 + e_2$) and transverse ($e_1 - e_2$) directions diverge with different exponents, $\xi_\parallel \simeq (p_c^{DP} - p)^{-\nu_\parallel}$ and $\xi_\parallel \simeq \xi_\perp^z$ with $\nu_\parallel \simeq 1.74$ and $z \simeq 1.58$.

Back to the Spiral Model, let us consider the directed lattice that is obtained from the square lattice putting two arrows from each site towards its NE neighbours, as in Fig. 3a). This lattice is equivalent to the one of DP, simply tilted and squeezed. Therefore, for $p > p_c^{DP}$, there exists a cluster of occupied sites which spans the lattice following the direction of the arrows (cluster inside the continuous line in Fig. 3a)). We denote by *NE-SW clusters* the occupied sets which follow the arrows of such lattice and *NW-SW clusters* those that follow instead the arrows drawn starting from each site towards its NW neighbours. Consider now a site in the interior of a spanning NE-SW cluster, e.g.site x in the Fig. 3a): by definition there is at least one occupied site in both its NE and SW neighbouring couples, therefore x is blocked with respect to the update rule of SM. Thus, the presence of the DP cluster implies a blocked cluster and $p_c \leq p_c^{DP}$ follows. Note that these results would remain true also for a different updating rule with the milder requirement that only at least one among the two couples of NE and SW sites is completely empty (and no requirement on the NW-SE direction). As we will see, the coexistence of the constraint in the NE-SW and NW-SE directions will be crucial to find a discontinuous transition for SM, otherwise we would have a standard DP-like continuous transition.

3.3. SM: absence of blocked clusters for $p < p_c^{DP}$

Before proving that below p_c^{DP} blocked clusters do not occur, a few remarks are in order. If instead of SM we were considering the milder rules described at the

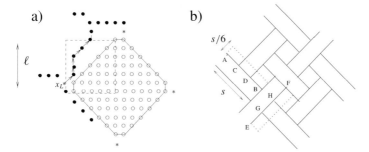

Fig. 4. a) Q_ℓ (continuous line), S_ℓ (dashed line) and the four sites (∗) which should be emptied to expand Q_ℓ. We draw the necessary condition for x to be frozen: it should belong to a NE-SW cluster spanning S_ℓ and supported by NW-SE cluster from the exterior. b)The frozen structure described in the text: continuous lines stand for occupied NE-SW or SW-NE clusters. Each of these clusters is blocked since it ends in a T-junction with a cluster along the transverse diagonal. The dotted rectangle adjacent to cluster AB (EF) are the regions in which this cluster can be displaced and yet a frozen backbone is preserved: the T-junctions in C and D (G and H) will be displaced but not disrupted.

end of previous section, the result would follow immediately since the presence of a blocked cluster would imply the existence of a DP one. On the other hand for SM rules, since blocking can occur along either the NE-SW or the NW-SE direction (or both), a directed path implies a blocked cluster but the converse is not true. This is because a NE-SW non spanning cluster can be blocked if both its ends are blocked by a T-junction with NW-SE paths, as shown in Fig. 3b). By using such T-junctions it is also possible to construct frozen clusters which do not contain a percolating cluster neither in the NE-SW direction nor in the NW-SE one: all NE-SW (NW-SE) clusters are finite and are blocked at both ends by T junctions by finite NW-SE (NE-SW) ones (see figure 3b)). As we will show in section 3.4 these T-junctions are crucial to make the behavior of the transition for SM very different from DP transition, although they share the same critical density. This also means that the fact that spanning DP clusters do not occur for $p < p_c^{DP}$ is not sufficient to conclude that also blocked clusters do not occur. What strategy could we use? Recalling BP results, a possible idea is to search for proper unstable voids from which we can iteratively empty the whole lattice. Of course, since we already know that blocked clusters occur when $p \geq p_c^{DP}$, here something should prevent this unstable voids to expand at high density.

Consider the region Q_ℓ inside the continuous line in Fig. 4, namely a "square" of linear side $\sqrt{2}\ell$ tilted of 45 degrees with respect to the coordinate axis and with each of the four vertexes composed by two sites. If Q_ℓ is empty and the four sites external and adjacent to each vertex denoted by ∗ in Fig. 4 are also empty, then it is possible to enlarge the empty region Q_ℓ to $Q_{\ell+1}$. Indeed, as can be directly checked, all the sites external to the top right side can be subsequently emptied starting from the top one and going downward. For the sites external to the other three sides of Q_ℓ we can proceed analogously, some care is only required in deciding whether to start from top sites and go downward or bottom ones and go upward. Therefore we can expand Q_ℓ of one step provided all the four $*$ sites are empty or can be emptied after some iterations of the dynamics. Let us focus one of these $*$ sites, e.g. the left one, x_L in Fig. 4. As it can be proved by an iterative procedure (see [12]), in order for x_L not to be emptiable there should exist a NE-SW cluster which spans the square S_ℓ of size ℓ containing the top left part of Q_ℓ (region inside the dashed line in Fig. 4). This is due to the fact that any directed path in the NW-SE direction can be unblocked starting from the empty part of S_ℓ below the diagonal in the $e_1 + e_2$ direction. Therefore in order to block x_L we need a NE-SW cluster which is at least of length ℓ since it can be supported from NW-SE clusters only outside S_ℓ (see Fig. 4). Therefore, for large ℓ, the cost for a one step expansion of Q_ℓ is proportional to the probability of not finding such a DP path, $1 - 4\exp(-c\ell/\xi_\parallel)$. The probability that the emptying procedure can be continued up to infinity is bounded from below by the product of these single step probabilities[3] which goes to a strictly positive value for $p < p_c^{DP}$ since $\xi_\parallel < \infty$. Note that, as we already knew from the results of section 3.2, this is not true for $p > p_c^{DP}$: the presence of long DP paths prevents the expansion of voids. As for BP, by the ergodic theorem we conclude that in the thermodynamic limit with probability one the final configuration is completely empty, therefore $p_c \geq p_c^{DP}$. This, together with the result of section 3.2, yields $p_c = p_c^{DP}$.

3.4. SM: discontinuity of the transition

In the two previous sections we have shown that the percolation transition due to the occurrence of a frozen backbone for the Spiral Model occurs at p_c^{DP}. We will now explain why it is qualitatively different from DP and any conventional percolation transition: the density of the frozen cluster is discontinuous, $\Phi(p_c^{DP}) > 0$, namely the frozen structures are compact rather than fractal at criticality, and their typical size increases faster than any power law for $p \nearrow p_c^{DP}$.

In order to prove discontinuity we construct a set of configurations, $\mathcal{B}$, for which the origin is occupied and that cannot be unblocked under the iteration of the update rule. The probability of this set, $P_{\mathcal{B}}$, gives a lower bound for $\Phi(p)$, thus it is sufficient to show that $P_{\mathcal{B}}(p_c^{DP}) > 0$ to conclude $\Phi(p_c^{DP}) > 0$. To construct $\mathcal{B}$ we make use of the T-junctions which have been introduced in Sec-

---

[3]The event that a directed spanning cluster does not occur when expanding from ℓ to $\ell + 1$ and from $\ell + 1$ and $\ell + 2$ are not independent since S_ℓ and $S_{\ell+1}$ do intersect (and the same is true on the following steps). However these events are positively correlated, therefore the joint probability is bounded from below by the product of single step probabilities.

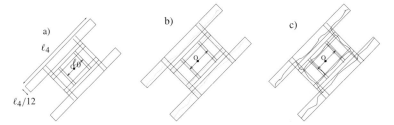

Fig. 5. a) The sequence of intersecting rectangles. b) Dotted non straight line stand for NE-SW (NW-SE) clusters spanning the rectangles c) Frozen structure containing the origin.

tion 3.3. Consider a configuration in which the origin belongs to a NE-SW path of length $\ell_0/2$: this occurs with probability $q_0 > 0$. Now focus on the infinite sequence of pairs of rectangles of increasing size $\ell_i \times \ell_i/12$ with $\ell_1 = \ell_0$, $\ell_i = 2\ell_{i-2}$ and intersecting as in Fig. 5a). A configuration belongs to $\mathcal{B}$ if *each* of these rectangles with long side along the NE-SW (NW-SE) diagonal contains a NE-SW (NW-SE) percolating path (dotted lines in Fig. 5b)). This implies that an infinite backbone of particles containing the origin (cluster inside the continuous line in Fig. 5c)) survives thanks to the T-junctions among paths in intersecting rectangles. Therefore $\Phi(p) > q_0 \prod_{i=1,\infty} P(\ell_i)^2$, where $P(\ell_i)$ is the probability that a rectangle of size $\ell_i \times 1/12\ell_i$ with short side in the transverse direction is spanned by a DP cluster.[4] Recall that there is a parallel and a transverse length for DP with different exponents, i.e. a cluster of parallel length ℓ has typically transverse length ℓ^z. Let us divide the $\ell_i \times 1/12\ell_i$ rectangle into ℓ_i^{1-z} slices of size $\ell_i \times 1/12\ell_i^z$. For each slice the probability of having a DP cluster along the parallel direction at ρ_c^{DP} is order unity. Thus, the probability of *not* having a DP cluster in each of the slice is $1 - P(\ell_i) = O[\exp(-c\ell_i^{1-z})]$. From this result and above inequality we get $\Phi(p_c^{DP}) > 0$. Therefore the infinite cluster of jamming percolation is "compact", i.e. of dimension $d = 2$ at the transition. Note that to obtain discontinuity two ingredients of the SM rules are crucial: the existence of two transverse blocking directions each with an underlying percolation transition and the anisotropy of these transitions. Indeed, anisotropy is necessary to obtain that the probability that the above rectangles are spanned converges to one exponentially fast as their size is increased. In turn, this is necessary to get a finite probability for the construction which freezes the origin.

---

[4]Again, since the rectangles do intersect the events that they are spanned are not independent. However these are positively correlated and the lower bound follows.

3.5. *SM: dynamical correlation length*

In the previous sections we have analyzed the behavior of SM in the thermodynamic limit. We now turn to the finite volume behavior and establish that the typical size, L_c, below which frozen clusters occur on finite lattices, namely the size of the incipient blocked cluster, diverges faster than any power law when $p \nearrow p_c^{DP}$: $\log L_c(p) \simeq k(p - p_c^{DP})^{-\mu}$, where $\mu = \nu_{\parallel}(1 - 1/z) \simeq 0.64$. We sketch separately the arguments leading to lower and upper bounds for L_c.

Consider the set of NE-SW and NW-SE paths of length s intersecting as in Fig. 4b). As can be directly checked, this structure can be emptied only starting from its border since each finite directed path terminates on both ends with T-junctions with a path in the transverse direction. Therefore, if the structure is continued up to the border of the lattice it is completely frozen. Furthermore, a similar frozen backbone exists also if one (or more) of the finite paths is displaced inside an adjacent rectangular region of size $s \times s/6$, as shown in Fig. 4b). Therefore the probability that there exists a frozen cluster, $1 - R(L, \rho)$, is bounded from below by the probability that *each* of the $O(L/s)^2$ rectangles contains at least one path connecting its short sides. This leads to $R(L, \rho) \leq (L/s)^2 \exp(-cs^{1-z})$ since the probability of not having a DP path in a region $s \times s/6$ is $\simeq \exp(-cs^{1-z})$ as long as $s \leq \xi_{\parallel}$. Thus $\lim_{L \to \infty, p \nearrow p_c^{DP}} R(L, \rho) = 0$ for $L/\xi_{\parallel} \exp(c\xi_{\parallel}^{1-z}) \to 0$, therefore $\log L_c \geq k_l(p - p_c^{DP})^{-\mu}$.

On the other hand, in order to establish an upper bound on L_c, we determine the size L above which unstable voids which can be expanded until emptying the whole lattice typically occur. The results in Section 3.3 imply that the probability of expanding an empty nucleus to infinity is dominated by the probability of expanding it up to $\ell = \xi_{\parallel}$. Indeed, above this size the probability of an event which prevents expansion is exponentially suppressed. Therefore, considering the $O(L/\xi_{\parallel})^2$ possible positions for a region that it is guaranteed to be emptyable up to size $\xi_{\parallel}$, we can bound the probability that a lattice of linear size L is emptiable as $R(L, \rho) \geq L^2 \delta$, where δ is the probability that a small empty nucleus can be expanded until size $\xi_{\parallel}$. In the emptying procedure described in Section 3.3 we evaluated the cost for expanding of one step the empty region Q_ℓ. Analogously, the cost of expanding directly from Q_ℓ to $Q_{2\ell}$ can be bounded from below by $C^{\ell^{1-z}}$, with C a positive constant independent from ℓ. This can be done by dividing the region contained in $Q_{2\ell}$ and not in Q_ℓ into ℓ^{1-z} strips with parallel and transverse length of order ℓ and ℓ^z, requiring that none of them contains a DP path which percolates in the transverse direction and using for this event the scaling hypothesis of directed percolation when $p \nearrow p_c^{DP}$. Thus for the expansion up to size $\xi_{\parallel}$ we get $\delta \geq \prod_{i=1}^{\log_2 \xi_{\parallel}} C^{2^{i(1-z)}} = \exp(-C'\xi_{\parallel}^{1-z})$, with $C' > 0$. This, together with above inequality, yields $\log L_c \leq k_u(p - p_c^{DP})^{-\mu}$.

4. Related stochastic models

In the above sections we studied the Spiral Model and showed that this, at variance with BP, is a JP model, namely a cellular automaton with a jamming transition at a non trivial p_c. Starting from this result it is possible [9] to define a stochastic lattice gas with Glauber dynamics (i.e. one in which elementary moves are birth and death of particles) which displays a purely dynamical transition with the same mixed first-order and critical character. This is done by considering the correspondent *kinetically constrained model* (KCM), namely a stochastic lattice gas in which a site is filled at rate p and emptied at rate $1 - p$ provided the surrounding configuration satisfies the constraint that is required in the SM model in order to empty the same site if it is occupied. Otherwise, if this constraint is not satisfied, both the birth and death rates are zero. As an immediate consequence of SM percolation transition, an ergodicity breaking transition occurs for this stochastic model at p_c^{DP}. Furthermore, the discontinuous character of SM transition implies a discontinuous jump of the order parameter $q_{EA} = < \eta_x(t)\eta_x(0) >_c$. Indeed, when raising the density towards p_c^{DP} the correlation curves clearly display a two step relaxation with this developing (discontinuous) plateau. The length of the plateau, namely the relaxation time $\tau(p)$ to equilibrium, diverges faster than any power law as $p \nearrow p_c^{DP}$ with a form similarly to the Vogel Fulcher law for supercooled liquids. In particular in [24] it has been proved that τ diverges at least as L_c. One can also construct a kinetically constrained model with Kawasaki dynamics (evolution is now a sequence of particle jumps) which displays the same behavior: the rate of a given jump is different from zero if SM constraints is satisfied both in the initial and final position of the particle.[5] So, back to example discussed in in the introduction, this models provide a possible choice of the "enough empty space" a particle should find in order to move which gives a jamming transition at a finite critical density. Of course much work remains to be done to investigate the connection of this and similar models with real systems undergoing jamming and glass transition, such as hard spheres or colloidal suspensions. This would require on the one hand analysing the extension and universality of this jamming percolation transition. A partial step in this direction has been undertaken in [11], where a class of rules which belong to the same universality class of SM has been identified. It remains to be understood if one can devise different (possibly more isotropic) rules displaying a jamming transition with a different scaling for the size of the incipient

---

[5]The KCM corresponding to BP constraints with Glauber or Kawasaki dynamics have also been analyzed. They are the so called Fredrickson Andersen [25] and Kob Andersen [26], respectively. In both cases a dynamical transition does not occur. The result for Glauber dynamics follows immediately from the fact that a jamming transition does not occur for BP. For Kawasaki dynamics additional arguments are required [27].

cluster as $p \nearrow p_c$. Also, for the connection with real systems it would be important to analyze the effect of constraint-violating processes and to compare the geometrical and statistical properties of the clusters of JP models with those that can be measured for the real systems.

I am very happy to thank G.Biroli and D.Fisher for a longstanding collaboration on the subject of Jamming Percolation and G.Biroli for a careful reading of this manuscript.

References

[1] J. Adler, *Physica A* **171**, 435 (1991).

[2] J. Adler and A. Aharony, *J. Phys. A* **21**, 1387 (1988).

[3] A.C.D. van Enter, *J. Stat. Phys.* **48**, 943 (1987).

[4] M. Aizenmann, J.L. Lebowitz, *J. Phys. A* **21**, 3801 (1988).

[5] R.H. Schonmann, *Ann. of Probab.* **20**, 174 (1992).

[6] A. Holroyd, *Probab. Th. and Related Fields* **125**, 195 (2003).

[7] R. Cerf, E. Cirillo, *Ann. of Probab.* **27**, 1837 (1999).

[8] R. Cerf, F. Manzo, *Stoch. Proc. Appl.* **101**, 69 (2002).

[9] C. Toninelli, G. Biroli, D.S. Fisher, *Phys. Rev. Lett* **96**, 035702 (2006).

[10] J. Schwarz, A. Liu, L.Q. Chayes, *Europ. Letters* **73**, 570 (2006).

[11] C. Toninelli, G. Biroli, D.S. Fisher, cond-mat/0612485.

[12] G. Biroli, C. Toninelli, *J. Stat. Phys* in press.

[13] M. Jeng, J. Schwartz, cond-mat/0612484.

[14] C.S. O'Hern et al., *Phys. Rev. Lett* **88** 075507 (2002), *Phys. Rev. E* **68**, 011306 (2003).

[15] E.R. Weeks et al., *Science* **287**, 627 (2000); V. Trappe et al., *Nature* **411**, 722 (2001).

[16] P.G. De Benedetti, F.H. Stillinger, *Nature* **410**, 267 (2001).

[17] G. Marty, O. Dauchot, *Phys. Rev. Lett* **94**, 015701 (2005).

[18] P. De Gregorio, A. Lawlor, P. Bradley, K. Dawson, *Proc. Natl. Acad. Sci.* **102**, 5669 (2005).

[19] S. Kirkpatrick, W. Wilcke, R. Garner, H. Huels, *Phyica A* **314**, 220 (2002).

[20] N. Dorogotsev, A.V. Goltsev, J.F.F. Mendes, *Phys. Rev. Lett* **96**, 040601 (2006).

[21] B. Pittel, J. Spencer, N. Wormald, *J. Comb. Th. B* **67**, 111 (1996).

[22] J. Chalupa, P.L. Leath, R. Reich, *J. Phys. C: Solid State Phys.* **12**, L31 (1979).

[23] H. Hinrichsen, *Adv. in Phys* **49**, 815 (2000).

[24] N. Cancrini, C. Roberto, F. Martinelli, C. Toninelli, cond-mat/0603745 and math.PR/0610106.

[25] G.H. Fredrickson, H.C. Andersen, *Phys. Rev. Lett.* **53**, 1244 (1984); *J. Chem. Phys.* **84**, 5822 (1985).

[26] W. Kob and H.C. Andersen, *Phys. Rev. E* **48**, 4364 (1993).

[27] C. Toninelli, G. Biroli, D.S. Fisher, *Phys. Rev. Lett.* **92** 185504 (2004); *J. Stat. Phys.* **120**, 167 (2005).

Course 8

COMPLEX NETWORKS

M.E.J. Newman

Department of Physics, University of Michigan, Randall Laboratory, 450 Church Street, Ann Arbor, MI 48109-1040, USA

J.-P. Bouchaud, M. Mézard and J. Dalibard, eds.
Les Houches, Session LXXXV, 2006
Complex Systems
© *2007 Published by Elsevier B.V.*

Contents

Preamble

Networks provide a useful mathematical abstraction of the structure of complex systems. The theory of networks is well developed and embraces empirical methods from the social, biological and information sciences, traditional methods of mathematical graph theory and a host of more recent developments from physics, computer science and statistics. Here we give an introduction to the field, describing both empirical results and theory and reviewing the current areas of research interest.

1. Introduction

A *network* or *graph* is, in its simplest form, a set of points joined in pairs by lines—see Fig. 1. In the jargon of the field the points are known as *nodes* or *vertices* (singular: *vertex*) and the lines are known as *edges*. Networks have found wide use in the scientific literature as a convenient mathematical representation of the structure of many different complex systems. For instance, Fig. 2 shows a network representation of the Internet. In this figure the nodes of the network represent "autonomous systems"—groups of computers on the network that are under single administrative control, such as the computers at a single university or company. The edges represent data connections between autonomous systems, principally optical fibre.[1] The image is certainly very complicated, and one could not hope to take in all its details just by eye, but some features are nonetheless clear. In particular, note that while the network is by no means regular it is also not completely random either. There is clearly some structure present, visible for instance in the star-like formations scattered about the picture and the greater density of edges towards the centre. This is a crucial insight: there are significant structural patterns and deviations from randomness in many real-world networks. These patterns turn out to have profound implications for the function of networked systems and the discovery and analysis of structural features is one of the primary goals of research in the field.

---

[1] For clarity, given the level of detail present in this network, the nodes are not drawn as actual dots; they are present only as the meeting points of edges. Also, for technical correctness, note that the edges really represent "direct peering relations" under the Border Gateway Protocol, which usually correspond to data connections, but not always.

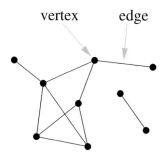

Fig. 1. A small example network with 9 vertices and 10 edges.

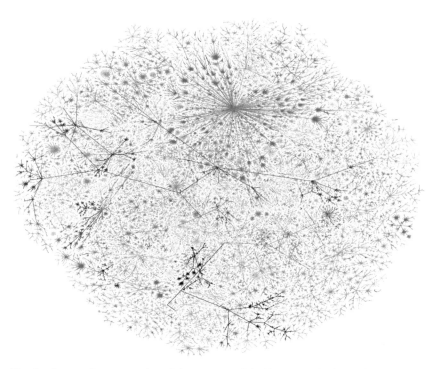

Fig. 2. A network representation of the structure of the Internet at the level of "autonomous systems"—groups of computers under single administrative control. The edges in the network represent data ("direct peering") connections between autonomous systems.

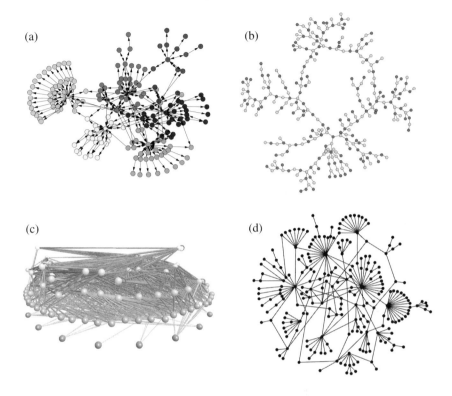

Fig. 3. Four example networks as described in the text: (a) a web site, (b) a social network, (c) a food web and (d) a protein interaction network.

Networks appear in many branches of science. Fig. 3 shows an assortment of examples as follows:

a) A portion of the web site of a large corporation (Newman and Girvan 2004). In this network the vertices represent web pages and the edges represent hyperlinks from page to page. Note that the edges have arrows on them—they are directed. Hyperlinks run *from* one web page *to* another, in a particular direction.

b) A network representing dating patterns in a US high school, created using data from Bearman *et al.* (2004). The vertices represent students in the school and the edges indicate who dated whom during a specified time period. The vertices are colour coded blue and pink to represent boys and girls and the figure shows only the largest "component" of the network, i.e., the largest connected group of individuals—separate components that are not connected to the main group are not shown.

c) The food web of species in a freshwater lake in Wisconsin, USA called Little Rock Lake (Martinez 1991). The vertices represent species and the edges represent predator-prey interactions between species. Technically, this is a directed network—an edge in this network has a direction indicating which of the two species it connects is the predator and which the prey. In this particular image, however, the directions are not indicated.

d) A protein interaction network representing interactions between proteins in the yeast *S. cerevisiae* (Maslov and Sneppen 2002). As in network (b) above, only the largest component of this network is shown.

In the following sections I describe some of the basic structural features observed in these and other networks and some of the mathematical tools developed for understanding the behaviour of networked systems.

2. Network expansion and the small-world effect

One of the most remarkable properties of networks is the *expansion property*. When we examine most real-world networks we find that the "surface area" enclosing a "volume" in the network is *proportional to that volume*. I will explain precisely what this means in a moment, but first let's examine a more familiar situation to get some bearings.

In ordinary low-dimensional space, such as in three dimensions, we are accustomed to objects having a surface area that is, in some sense, "smaller" than their volume. Take an ordinary sphere, for instance, which has volume $V = \frac{4}{3}\pi r^3$, where r is the sphere's radius. The surface area is $A = 4\pi r^2$ and, eliminating r between these two expressions, we find that

$$A = \sqrt[3]{36\pi}\, V^{2/3}. \tag{2.1}$$

Thus the area increases more slowly than the volume as the sphere gets larger. More generally, in dimension d we have $A \sim r^{d-1}$ and $V \sim r^d$, so that

$$A \sim V^{(d-1)/d}, \tag{2.2}$$

and for any finite d the exponent on the right-hand side is always less than one. In networks, however, it is possible for "area" to increase simply in proportion to "volume", an odd and counterintuitive state of affairs that gives rise to some interesting network phenomena.

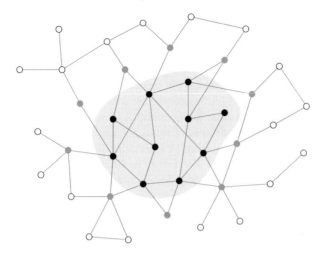

Fig. 4. A "volume" in a network (vertices in the shaded region) and its "surface" (vertices immediately adjacent to the shaded region).

Networks are not Euclidean objects so we need to make clear exactly what we mean by area and volume for a network. In the discussion that follows, the network plays the role not of the sphere in the example above but of the *space* in which the sphere lives. The role of the sphere will be played by a connected set S of nodes in the network, as shown in Fig. 4. The volume of this set of nodes is defined as the number of nodes $|S|$ in the set, while the surface area is defined as the number of nodes immediately adjacent to the set (i.e., nodes not belonging to the set but connected by one or more edges to at least one member of the set).

An *expander graph* is a network of n vertices in which the surface area of every set S of size $|S| < \alpha n$ is at least $k|S|$ for some constants $\alpha, k > 0$. Put more simply, the surface area of every small volume in the graph is at least some fraction k of that volume: surface area and volume are proportional. Looking back at Eq. (2.2), we can think of this as meaning that the graph is an infinite-dimensional object $d \to \infty$, and indeed there are many cases in which it appears sensible to view networks as having infinite dimension.

Perfect expander graphs are rare in the real world. There are some mathematical classes of networks, such as some k-regular graphs, that are perfect expanders, and a few real-world examples too, but most networks are not perfect expanders. (For a start, every vertex in a network has to have at least k neighbours for the network to be an expander with expansion constant k.) Nonetheless, real networks are often sufficiently good approximations to expander graphs that we can gain insight about their behaviour by studying perfect expanders. (Typically, we find that most sets S of vertices in a graph satisfy the expander condition, even if a few do not.)

Why is the expansion property interesting? Because it leads to something called the *small-world effect*, which has profound implications for the operation of networks in the real world. Consider the set consisting of just a single vertex i in an expander graph with expansion constant k. Call this set S_0, with $|S_0| = 1$. By definition S_0 has a surface area of at least k, i.e., i has at least k immediate neighbours in the network. Let us call the set composed of i plus all these immediate neighbours S_1, with $|S_1| \geq k + 1 > k$. Again by definition S_1 has a surface area of at least $k|S_1| > k^2$, implying that i has more than k^2 neighbours at distance 2 or less, so $|S_2| > k^2$. Repeating this argument, it's easy to show that i must always have more than k^m neighbours out to distance m, and this argument goes on working until the size of our set S_m becomes greater than or equal to αn for some α, where n again is the total number of vertices in the network; after this, according to the definition, the expansion property no longer holds.

Now let us start this same process from two different vertices i and j, and keep on adding vertices to the sets centred on i and j until each is at least $\sqrt{n}$ in size. Since the expansion property holds for sets of size up to αn then, no matter what the value of α, the expansion property will hold up to $\sqrt{n}$ provided n is large enough, which we will assume it is. But once we have our two sets of size $\sqrt{n}$ they will on average intersect at one vertex[2] and once they intersect then there must exist a *path* from i to j along edges of the network, via the vertices in the two sets—see Fig. 5.

Now here is the interesting result. Our set of size $\sqrt{n}$ around vertex i is composed of i and all its network neighbours out to distance no more than m, where $k^m < \sqrt{n}$, and similarly for the set around vertex j. This means that the maximum number of steps we have to take along the path from i to j is $2m$ and hence that there must exist a path from i to j of length at most ℓ where $k^{\ell/2} = \sqrt{n}$ or equivalently

$$\ell = \frac{\ln n}{\ln k}. \tag{2.3}$$

In other words, the typical length of the shortest path we have to traverse to pass between two randomly chosen vertices in an expander graph—and hence, roughly speaking, between vertices in real-world networks as well—is (at most) of order $\ln n$. Since the logarithm is a very slowing increasing function of its

---

[2]If I choose two sets of r vertices from a network of n vertices total then the probability that a particular vertex belongs to both sets is r^2/n^2, and the expected number of vertices that belong to both sets is $n \times r^2/n^2 = r^2/n$. Thus if $r = \sqrt{n}$ the expected number of vertices in both sets is 1.

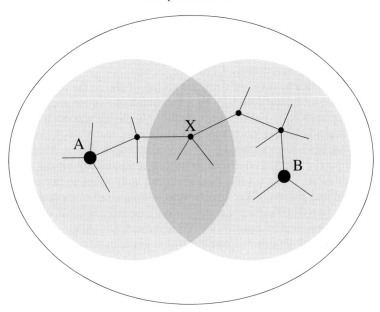

Fig. 5. Two "volumes" centred on vertices A and B, as described in the text, are expected to overlap at one or more vertices X if the volumes contain around $\sqrt{n}$ vertices each. This then means that there is a path through the network from A to B via X.

argument, this means that even for very large networks the typical length of the path through the network from one vertex to another can still be remarkably short.

These arguments to not constitute a rigorous proof, but the same basic ideas can be used to form a rigorous proof. See for instance Chung and Lu (2002b) and references therein.

2.1. Milgram's small-world experiment

The small-world effect has been of particular interest in social networks. If one considers people to be the nodes in a network and one places edges between pairs of nodes whenever the corresponding individuals are acquainted (where acquaintance should be defined suitably precisely if we wish to be scientific about things), then the argument above implies that there will be a short path of acquaintances joining most pairs of individuals in the world, provided the social network is, approximately at least, an expander graph. This result was suspected on empirical grounds long before any mathematical arguments were written down. The Hungarian novelist Frigyes Karinthy wrote a remarkably prescient short story in 1929 in which one of his characters claimed to be able to connect any two individuals in the world by a chain of acquaintances no more than five people long (Karinthy 1929). (Karinthy's story is reproduced in an English translation in the collection by Newman *et al.* (2006).)

In 1959, a mathematical argument somewhat similar to the one given above was written down by Ithiel de Sola Pool and Manfred Kochen and circulated in preprint form within the academic community, although it was not actually published until nearly twenty years later (Pool and Kochen 1978). A decade later, Pool and Kochen's ideas inspired the social psychologist Stanley Milgram, then a professor at Harvard University, to conduct an ingenious experiment to test the small-world effect. Milgram sent packages containing letters to 296 individuals in Nebraska and Boston.[3] The letters were all addressed to the same "target" individual, a friend of Milgram's, a stockbroker who lived in the Boston suburbs, and were accompanied by instructions asking the recipients of the packages to try to get the letters to the target. Recipients were not allowed to simply mail their letter directly to the target, however. Instead, they were allowed to send the letter only to someone they knew on a first-name basis. (This is where the definition of "acquaintance" enters the experiment.) Since most recipients probably would not know the target individual, their best bet would be to forward the letter to someone they did know whom they judged to be closer to the target than they were. "Closer" in this context could mean actually geographically closer, but it could also mean closer in some social sense—recipients might for instance forward the letter to someone they knew who also worked in the financial industry. The individuals to whom the letters were forwarded were then asked to repeat the process and forward the letters again to someone *they* knew, and so forth until, with luck, the letters found their way to the intended target.

What is perhaps most remarkable about this experiment is that a significant fraction of the letters did reach the Boston stockbroker: 64 out of the 296 made it to the target, or about 22%. I think this indicates that people must have been a lot less jaded about junk mail in 1969 than they are today. I imagine most people receiving such a package in the mail today would throw it in the rubbish unopened. For the packages that did arrive, Milgram found that they took on average about six steps to travel from their initial recipient to the target, a result that has since been immortalized in pop culture in the expression "six degrees of separation".[4] (In the light of this result, the receipt by the target of 22% of the letters is all the more remarkable, since each letter had to be passed on by six

---

[3] Actually, the "letters" were small booklets, which Milgram called "passports". It's possible they were designed to look official in the hope that this would encourage people to participate in the experiment.

[4] The expression was not coined by Milgram. It comes from the title of a 1990 Broadway play by John Guare in which the protagonist discusses Milgram's experiment.

other people. If any one of those people had discarded the letter it would never have reached the end of the line.)

The small-world effect is today one of the most well-known results in experimental psychology. It also forms the basis for a popular parlour game, "The Six Degrees of Kevin Bacon", an American television drama and many web sites and on-line pursuits, it is mentioned often in the press and it appears widely in movies, advertisements and other forms of popular culture. However, it has a serious side to it as well. The small-world effect is the fundamental reason why many networked systems work at all. For instance, it ensures that data packets transmitted across the Internet will never have to make too many "hops" between nodes of the network to get from their source to their destination. The Internet would certainly work much more poorly than it does if the average number of hops a packet makes were a thousand or a million, rather than the 20 or 30 we observe in practice. The small-world effect also has profound implications for epidemiology. Since diseases are mostly communicated by physical contact between individuals, the pattern and rate of their spread is dictated in large part by the structure of the network of such contacts and in recent years networks have as a result played a significant role in our understanding of the spread of disease. Clearly it is the case that a disease will spread through a population much faster if the typical number of steps or hops from an initial carrier to others is six, as opposed to a hundred or a thousand.

3. Degree distributions

The *degree* of a vertex in a network is the number of edges attached to that vertex. In the acquaintance networks of the previous section, for instance, the degree of an individual would be the number of acquaintances they have.

One of the most important properties of a network is the frequency distribution of the degrees, the so-called *degree distribution*. The degree distribution can be visualized by making a histogram of the degrees. For instance, the left panel of Fig. 6 shows a histogram of the degrees of the vertices in a network representation of the Internet at the autonomous system level (similar to the network shown in Fig. 2, although the particular snapshot used to calculate the histogram was not the same as that of Fig. 2). This histogram reveals an interesting pattern: the degree distribution is heavily *right skewed*, meaning that most nodes have low degree but there is a "fat tail" to the distribution consisting of a small number of nodes with degree much higher than the average. These *hub nodes* are a feature common to many networks and play an important role in network phenomena.

A clearer picture of the degree distribution can be derived by re-plotting it on logarithmic scales, as in the right panel of Fig. 6. When plotted in this way

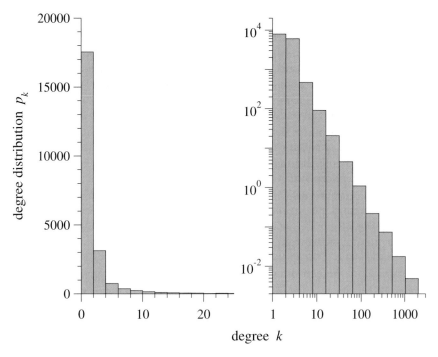

Fig. 6. Left: a histogram of the degree distribution of the Internet at the autonomous system level, made from a snapshot of the network from July 2006. Right: the same data re-binned and replotted on logarithmic scales.

we see a further interesting pattern: the distribution appears to fall roughly on a straight line, at least for the large-k part of its range, implying that the tail of the distribution obeys a *power law*; if p_k is the fraction of vertices having degree k, then

$$p_k \sim k^{-\alpha}, \tag{3.1}$$

where α is a constant known as the *exponent* or *scaling parameter* of the power law. Many degree distributions are found to possess power-law tails of this kind (Barabási and Albert 1999; Dorogovtsev and Mendes 2003), with exponents typically falling in the range $2 \le \alpha \le 3$. Networks with power-law degree distributions are sometimes called *scale-free networks*, although this name can be confusing: keep in mind that it is not the network itself that is scale-free, only its degree distribution. There are also many other networks that are not strictly scale-free in the sense of following a power law, but nonetheless possess right-skewed degree distributions with a small number of high-degree hubs.

A word is perhaps in order here about the statistical analysis of data that appear to follow power laws. In much of the physics literature it is common to present plots such as that in Fig. 6, which appear approximately to follow a straight line, and on the basis of a visual inspection assert that the data have a power-law distribution. The exponent of the power law, which is a parameter of interest in many theories of network behaviour, can be extracted by performing a least-squares fit of a straight line to the tail of the observed distribution. The region corresponding to the tail is usually also chosen by inspection.

These methods, however, present problems in many situations. Extracting exponents by least-squares fitting, for example, is known to produce systematically biased answers (Goldstein *et al.* 2004). A much better technique is to use maximum likelihood fitting, which leads to a direct formula for the exponent. Suppose the degree of vertex i is denoted k_i and the lower limit of the power-law tail in the distribution is k_{min}. And suppose also that there are n_{tail} vertices in the tail. Then the exponent α is the solution of the equation

$$\frac{\zeta'(\alpha, k_{min})}{\zeta(\alpha, k_{min})} = -\frac{1}{n_{tail}} \sum_{i \mid k_i \geq k_{min}} \ln k_i, \tag{3.2}$$

where $\zeta(\alpha, x)$ is the generalized or Hurwitz zeta function $\zeta(\alpha, x) = \sum_{k=0}^{\infty}(k + x)^{-\alpha}$ and the prime indicates differentiation with respect to the first argument. Unfortunately, this equation cannot be rearranged to give α in closed form and is quite tedious to solve numerically: the solution normally involves making numerical approximations to the zeta functions and then finding α by, for instance, binary search. A good approximation to the solution can, however, be derived and is given by

$$\alpha = 1 + n_{tail} \left[\sum_{i \mid k_i \geq k_{min}} \ln \frac{k_i}{k_{min} - \frac{1}{2}} \right]^{-1} + O(k_{min}^{-2}). \tag{3.3}$$

In practice, this expression seems to give results in agreement with (3.2) to a percentage point or so for $k_{min} \gtrsim 4$ and typical network sizes.

The value of k_{min} can also be estimated from data in a rigorous fashion, for instance by using a Kolmogorov–Smirnov test, and one can make statements about whether the degree distribution really does follow a power law or not by using goodness-of-fit indicators (such as the Kolmogorov–Smirnov test again) or likelihood ratio tests. For a detailed discussion, see Clauset *et al.* (2007).

3.1. An explanation for skewed degree distributions

Degree distributions have been a topic of study in the networks literature for a long time. In his early work on the social networks of school children, Anatol Rapoport and his collaborators noted in qualitative fashion that the degree distribution appeared to be right-skewed (Rapoport and Horvath 1961). However, the first detailed quantitative observations were probably those of Derek de Solla Price, a physicist-turned-historian-of-science who in the 1960s was one of the first to perform serious studies of academic citation networks—networks in which the nodes are academic papers and the edges between nodes represent citation of one paper by another. In an influential paper in the journal *Science*, Price presented evidence that the degree distribution of the citation network followed a power law with an exponent around $\alpha = 3$ (Price 1965).

A citation network is a directed network, meaning its edges have direction—citations run *from* one paper *to* another. (We saw another example of a directed network in the web graph of Fig. 3a.) In such a network each vertex has two degrees, its *in-degree* (the number of papers citing it) and its *out-degree* (the number of papers it cites). Price was interested primarily in the citations papers receive and hence in the in-degree. A power-law distribution of in-degree would imply that there are a small number of "hub" papers that receive many more citations than the average, which perhaps implies in turn that these papers are more influential than most. (Indeed, a substantial majority of papers never get cited at all, so having even just a single citation already puts a paper in an elite club.) Price's results have been duplicated many times since his work (Seglen 1992; Redner 1998), enough so that his 1965 paper has itself become something of a citation classic, with about 500 citations at the time of the writing of this article.

Price was curious about the power-law degree distribution he observed and in a follow-up paper he proposed a possible mechanism for its generation (Price 1976). He proposed that papers receive citations at a rate roughly proportional to the number that they already have. If people come across papers because they are cited by others, then the more times a paper is cited the more likely people are to find it, and hence perhaps to cite it themselves. Thus it seems not unreasonable to posit a "rich get richer" mechanism whereby the papers with the most citations receive the most citations. Price called this process *cumulative advantage*, although it is today known more often as *preferential attachment*, a term coined by Barabási and Albert (1999). Technically it is not possible for the rate of new citations to be exactly proportional to the current number of citations a paper has, since then no paper would ever receive any citations, given that all papers have zero citations to begin with. To get around this problem, Price proposed that a paper with k citations receives new citations at a rate proportional to $k+a$, where a is a constant. (In fact, Price only considered the case $a = 1$, but it is a trivial generalization to extend his model to $a > 1$.) The parameter a is a measure of the rate at which previously uncited papers get cited: it is as if each paper gets a citations for free when first published, to get it started in the citation race.

Price then assumed that, among papers with the appropriate degree, citations are awarded completely at random. This results in a model of the citation process that can be solved exactly for its in-degree distribution. (The solution given here is not identical to Price's, although it is very similar. Price based his method of solution on previous work by Simon (1955).)

Let $p_k(n)$ be the fraction of vertices (i.e., papers) having in-degree k in a citation network of n vertices. The network size n is not a constant, since new papers are assumed to be published regularly and hence n increases. Suppose that each newly appearing paper cites c previously existing ones on average. A particular citation goes to another paper i with probability proportional to the degree k_i of that paper plus a, or with normalized probability

$$\frac{k_i + a}{\sum_{i=1}^{n}(k_i + a)} = \frac{k_i + a}{n(c + a)}, \tag{3.4}$$

where I have made use of the fact that there are by definition nc outgoing edges from all vertices, and hence also nc ingoing ones (since every edge that starts somewhere must also end somewhere), and thus $\sum_i k_i = nc$.

Now, given that each new paper cites c others, and noting that there are $np_k(n)$ vertices of in-degree k in the network, the expected number of new citations to vertices of in-degree k upon publication of a new paper is

$$np_k(n)c\frac{k + a}{n(c + a)} = p_k(n)\frac{c}{c + a}(k + a). \tag{3.5}$$

Now we can write a *rate equation* governing $p_k(n)$ as follows. When a new paper is published, the number of vertices increases by 1, and the new number with in-degree k is given by

$$(n + 1)p_k(n + 1) = np_k(n) + \frac{c}{c + a}\big[(k - 1 + a)p_{k-1}(n) - (k + a)p_k(n)\big], \tag{3.6}$$

where the terms in the square brackets $[\ldots]$ represent respectively the increase from vertices previously of in-degree $k - 1$ that receive a new citation and thereby achieve degree k and the decrease from vertices previously of in-degree k that receive a new citation and therefore no longer have degree k.

Equation (3.6) applies so long as $k > 0$. If $k = 0$ then there are no vertices of in-degree $k - 1$, and the first term in the brackets is modified so that the equation reads

$$(n + 1)p_0(n + 1) = np_0(n) + 1 - \frac{ca}{c + a}p_0(n). \tag{3.7}$$

The "+1" on the right-hand side represents the fact that one new vertex of degree zero is added when n increases by 1.

We are interested in the degree distribution of the network as n becomes large. Letting $n \to \infty$ and denoting $p_k(\infty) = p_k$, terms cancel in Eqs. (3.6) and (3.7) and we find

$$p_0 = 1 - \frac{ca}{c+a}p_0, \qquad\qquad \text{for } k = 0, \qquad (3.8)$$

$$p_k = \frac{c}{c+a}\big[(k-1+a)p_{k-1} - (k+a)p_k\big], \qquad \text{for } k > 0. \qquad (3.9)$$

Equation (3.8) can be rearranged to give

$$p_0 = \frac{1+a/c}{1+a+a/c}, \qquad (3.10)$$

while Eq. (3.9) can be rearranged to give the recursion

$$p_k = \frac{k-1+a}{k+1+a+a/c}p_{k-1}. \qquad (3.11)$$

Iterating this recursion, we then find the complete solution for the degree distribution thus:

$$\begin{aligned} p_k &= \frac{(k-1+a)\ldots(1+a)a}{(k+1+a+a/c)\ldots(2+a+a/c)}p_0 \\ &= (1+a/c)\frac{\prod_{r=0}^{k-1}(r+a)}{\prod_{r=0}^{k}(r+1+a+a/c)}. \end{aligned} \qquad (3.12)$$

This expression can be simplified further by making use of the gamma function

$$\Gamma(x) = \int_0^\infty t^{x-1}e^{-t}\,dt, \qquad (3.13)$$

which (integrating by parts) satisfies $\Gamma(x+1) = x\Gamma(x)$ and hence

$$\prod_{r=0}^{k-1}(r+x) = \frac{\Gamma(k+x)}{\Gamma(x)}. \qquad (3.14)$$

Thus Eq. (3.12) can be written

$$\begin{aligned} p_k &= (1+a/c)\frac{\Gamma(k+a)\Gamma(1+a+a/c)}{\Gamma(a)\Gamma(k+2+a+a/c)} \\ &= \frac{1+a/c}{1+a+a/c}\frac{B(k+a, 2+a/c)}{B(a, 2+a/c)}, \end{aligned} \qquad (3.15)$$

where $B(x, \alpha)$ is the Legendre beta function:

$$B(x, \alpha) = \frac{\Gamma(x)\Gamma(\alpha)}{\Gamma(x + \alpha)}. \tag{3.16}$$

The distribution (3.15) is known as the *Yule distribution*, after Udny Yule who first derived it (using completely different means) as the limit distribution of a cumulative advantage process. The Yule distribution has a tail that asymptotically follows a power-law distribution, which we can demonstrate as follows.

It can be shown (by repeated integration by parts) that

$$B(x, \alpha) = \int_0^1 u^{x-1}(1 - u)^{\alpha-1}\, du. \tag{3.17}$$

Making a change of variables to $t = x(1 - u)$, we then get

$$B(x, \alpha) = \frac{1}{x}\int_0^x \left(1 - \frac{t}{x}\right)^{x-1}\left(\frac{t}{x}\right)^{\alpha-1}\, dt. \tag{3.18}$$

In the limit of large x this tends to

$$B(x, \alpha) = \frac{1}{x^\alpha}\int_0^\infty t^{\alpha-1}e^{-t}\, dt = \Gamma(\alpha)\, x^{-\alpha}, \tag{3.19}$$

which follows a power law with exponent α.

Comparing with Eq. (3.15), we thus conclude that the degree distribution p_k in Price's model of a citation network follows a power law for large k with exponent

$$\alpha = 2 + \frac{a}{c}, \tag{3.20}$$

where c again is the average number of citations made by a paper and a is the offset parameter introduced by Price to represent the average rate at which previously uncited papers receive citations. Unfortunately, while c can typically be measured quite simply, a usually cannot, which means that we cannot predict the value of the exponent of the power law. Given detailed citation data, one might be able to measure the rate at which previously uncited papers are cited, but to my knowledge no one has done this at present (although it is possible that there are published measurements of other quantities that could be used to extract a figure for a indirectly).

In practice, most authors have approached the problem from the opposite direction, using the observed value of the exponent of the degree distribution to estimate the value of a. For instance, Price observed that his citation networks had an exponent around 3, which would imply that $a \simeq c$ in that case.

Price's work, while interesting and important, was not, at least until recently, well known outside of the literature on citation networks, and in particular was not familiar to most physicists (even though one of Price's two PhDs was in physics). The idea of cumulative advantage was introduced into the physics literature by Barabási and Albert, who independently rediscovered it in 1999 (Barabási and Albert 1999). Unaware of Price's work, they gave it the new name *preferential attachment* by which it is commonly known today. They also proposed a model for the process which, as shown later by Dorogovtsev *et al.* (2000), is equivalent to Price's model in the special case where $a = c$. Using essentially the method presented above, Dorogovtsev *et al.* and independently Krapivsky *et al.* (2000) derived the exact solution for this special case as well as for several more general cases. (Barabási and Albert presented an approximate solution of the model making use of a completely different mean-field-like approach.)

Preferential attachment is not the only possible explanation for skewed degree distributions. Quite a number of others have been proposed, including vertex duplication processes (Kleinberg *et al.* 1999) and optimization processes (Valverde *et al.* 2002). Certainly, however, preferential attachment has received the largest amount of attention and is widely thought to be essentially correct for citation networks as well as some others, although clearly it represents a substantial simplification of true citation processes.

3.2. Configuration model

The models described in the preceding section are *generative models*. That is, they are models of the process by which a network is generated. Their aim is to offer a possible explanation (albeit simplified) of how a network comes to have some observed feature or features. This is, however, not the only reason why one might build models of networks. Another important line of questioning is, given that a network *does* have some observed feature, what we should expect its structure to look like. In the particular case where the feature of interest is the degree distribution, this question has been thoroughly investigated and leads to a beautiful set of results concerning the "configuration model" (Molloy and Reed 1995, 1998; Aiello *et al.* 2000; Callaway *et al.* 2000; Newman *et al.* 2001; Chung and Lu 2002a,b).

Suppose we are told or somehow discern the degree sequence of a network but we are given no other information. That is, we are given only the degrees k_i, $i = 1 \ldots n$ of each of the n vertices in the network. (I will assume it to be an undirected network, although a straightforward generalization exists of the results here to the directed case.) Then our best assumption in the absence of any information to the contrary is to assume that the vertices of the network are connected at random, respecting the required degree sequence. (In fact, this

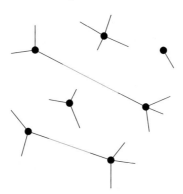

Fig. 7. The construction of the configuration model. The degrees are represented by stubs or spokes emerging from the vertices, and then pairs of stubs are chosen at random and joined together to make a complete network.

assumption turns out to be equivalent to maximizing the entropy of the resulting ensemble of networks, and hence can be justified on quite rigorous grounds. See, for instance, Park and Newman (2004).)

A procedure for generating such a network is the following (Molloy and Reed 1995). Imagine drawing the n vertices on a sheet of paper and adding k_i "stubs" of edges emerging from each vertex i—see Fig. 7. Then we choose pairs of those stubs uniformly at random and connect them together to create whole edges. When all stubs have been connected, we have a random network with the desired degree sequence. This random model of a network is called the *configuration model*.

More rigorously, the configuration model is the *ensemble* of networks generated by this process. That is, it is the set of networks such that any edge stub emerging from a vertex has equal chance to be connected to any other stub anywhere in the network. When people talk about properties of the configuration model, they are typically referring to properties averaged over this ensemble, in a manner similar to the ensemble averages of statistical mechanics, and in fact there are close mathematical similarities between the treatment of ensemble models of networks and ensembles in statistical mechanics.

Sometimes we do not know the precise degree sequence of a network but we do know its degree distribution: that is, the probability distribution over degrees is specified. In this case, we can draw a particular degree sequence from that probability distribution and use it to generate an instance of the configuration model. If we can calculate properties of such an instance, we can then average over many instances to find average properties of networks created from the specified degree distribution.

3.2.1. The excess degree distribution

Networks possess a interesting property that sets them apart from structures such as regular lattices that are perhaps more familiar to physicists. It is principally this property that gives them their sometimes counterintuitive behaviour and the configuration model provides an elegant setting in which to explore it.

As before, let p_k be the degree distribution of our network, as before, i.e., the fraction of vertices having degree k. Looked at another way, p_k is the probability that a randomly chosen vertex in the network will have degree k. But now take a randomly chosen *edge* in the network and consider either of the vertices at the ends of that edge. The distribution of vertices encountered in this way is not in general the same as the distribution p_k of vertices overall. Why not? Because the probability of arriving at a vertex by following an edge depends on the number of edges attached to that vertex. If we choose an edge at random and then move at random to one of the vertices at its ends, a vertex with ten edges attached to it (i.e., with degree $k = 10$) has ten times as many chances to be chosen as a vertex with only one edge (i.e., with degree $k = 1$). And a vertex with no edges attached to it at all will never be chosen.

To be precise, if there are m edges in our network and we choose one of them at random, then further choose one of the two ends of that edge at random, we are in effect choosing uniformly at random among the $2m$ ends of edges in the network. Of those $2m$ ends, k of them are attached to any particular vertex with degree k. And given that there are np_k such vertices overall, the total probability of our landing at a vertex of degree k is

$$np_k \frac{k}{2m} = \frac{kp_k}{\langle k \rangle}, \tag{3.21}$$

where $\langle k \rangle$ indicates the average degree of a vertex in the network and I have made use of the observation that $2m = \sum_{i=1}^{n} k_i$, so that

$$\frac{2m}{n} = \frac{1}{n} \sum_{i=1}^{n} k_i = \langle k \rangle. \tag{3.22}$$

Thus the probability distribution of the degree of a vertex at the end of an edge is proportional not to p_k but to kp_k. This effect is not seen in regular lattices, for example, since every vertex has the same degree in such a lattice (the same *coordination number*, to use the traditional physics jargon), so it makes no difference whether we choose a vertex at random or an edge. But in networks the extra factor of k makes for some interesting phenomena, as we will see.

As a simple demonstration, note that the expected number of "friends" (i.e., network neighbours) a vertex has in the configuration model is simply $\langle k \rangle$. But, strangely, this is not equal to the expected number of friends that one of those friends itself has. The "friend" of a vertex i is, by definition, the vertex at the end of one of the edges attached to i. And the degree k of a vertex at the end of an edge is distributed according to Eq. (3.21). Thus the average degree of a friend is:

$$\sum_{k=0}^{\infty} k \frac{k p_k}{\langle k \rangle} = \frac{\sum_k k^2 p_k}{\langle k \rangle} = \frac{\langle k^2 \rangle}{\langle k \rangle}, \qquad (3.23)$$

which is in general different from $\langle k \rangle$.

In fact, the difference between the expected degree of a vertex and the expected degree of its friends is just

$$\frac{\langle k^2 \rangle}{\langle k \rangle} - \langle k \rangle = \frac{\langle k^2 \rangle - \langle k \rangle^2}{\langle k \rangle} = \frac{\sigma^2}{\langle k \rangle}, \qquad (3.24)$$

where $\sigma^2 = \langle k^2 \rangle - \langle k \rangle^2$ is the variance of the degree distribution. Being the square of a real number, the variance is nonnegative, and furthermore is zero only if every vertex in the network has exactly the same degree. Let's assume this not to be the case, in which case $\sigma > 0$ strictly. Similarly $\langle k \rangle > 0$ unless every vertex has degree zero, which we'll also assume not to be the case. Then the difference in Eq. (3.24) is strictly greater than zero, which implies that

$$\frac{\langle k^2 \rangle}{\langle k \rangle} > \langle k \rangle. \qquad (3.25)$$

In plain English: the average number of friends your friends have is greater than the average number you have. This result was first demonstrated (to my knowledge) by Feld (1991).

Thus, if it appears to you that your friends have more friends than you do, *you're absolutely right.* They do. But it's not your fault. It's a simple mathematical fact about networks. It arises because vertices with high degree are, by definition, the "friends" of more other vertices than vertices of low degree. So in taking the average of the number of friends your friends have, these vertices dominate and push up the average value. You can test this result with real networks and it really is true, even though real networks are not exactly the same as the configuration model. For instance, in a paper in 2001 the present author compiled and studied a network of collaborations of physicists drawn from the physics preprint archive at arxiv.org. In this network the vertices represent physicists who authored one or more papers appearing in the archive and edges connect any pair of authors whose names appeared as coauthors of the same paper or papers. In this network, it turns out that the average number of collaborators a physicist has is 9.3. But the average number of collaborators their *collaborators* have is 34.7. Thus, on average, your collaborators have more collaborators than you do.

This seems counterintuitive, since surely it is exactly the same population of people we are counting in both averages. This is true, but in the second average, the average of the number of collaborators your collaborators have, people with high degree get counted not once in the calculation but repeatedly, because they appear over and over as the collaborators of many other individuals.

In the calculations that follow, we will be interested primarily not in the degree of a neighbour of vertex i, but in the number of edges connected to that neighbour *other* than the edge connecting it to i. This number is just one less than the total degree of the neighbour and is sometimes referred to as the *excess degree*, a term coined by Vazquez and Weigt (2003). Thus if k is the excess degree of a vertex, then $k + 1$ is its total degree, which is distributed according to (3.21), and hence the distribution q_k of the excess degree is

$$q_k = \frac{(k+1)p_{k+1}}{\langle k \rangle}. \tag{3.26}$$

3.2.2. Formation of the giant component

One of the most interesting properties of the configuration model is the "giant component". If the degree sequence is such that most vertices in the network have low degree—say zero or one—then vertices will be connected together in twos or perhaps threes, but there will be no large groups of connected vertices, or *components* as they are called. If the degrees of vertices are increased, however, then vertices will connect to more others and will form into larger components. At some point, if the network becomes dense enough, a large fraction of the vertices will connect together to form a single *giant component* that fills an extensive portion of the network. The transition between the regime in which there are only small components and the regime in which there is a giant component is a geometric phase transition similar in nature to the percolation phase transition. We can shed light on this transition by the following argument.

Consider a randomly chosen vertex i that is not a member of the giant component. A vertex is not a member of the giant component if and only if all of its neighbours are not members of the giant component. And for a neighbour itself not to be a member of the giant component it must be that all of the other edges from that neighbour connect to vertices not in the giant component, and so forth. Let us define u to be the probability of this latter event, i.e., u is the probability that a vertex reached by following an edge is not connected to the giant component via any of its other edges. If there are k such other edges then this happens, by definition, with probability u^k. But k is precisely the excess degree of the vertex in question, and hence is distributed according to Eq. (3.26). Averaging over

k, we then find that the mean value of u within the configuration model ensemble is given by

$$u = \sum_{k=0}^{\infty} q_k u^k. \tag{3.27}$$

The function

$$G_1(u) = \sum_{k=0}^{\infty} q_k u^k \tag{3.28}$$

is called the *probability generating function* of the excess degree distribution and it crops up repeatedly in many situations in the theory of networks. Here we see that the probability u is a fixed point of the generating function:

$$u = G_1(u). \tag{3.29}$$

A few properties of the generating function are worth pointing out. First, note that, since u is a probability, we are only interested in the value of $G_1(u)$ in the range $0 \leq u \leq 1$. Second, G_1 is a polynomial function with coefficients q_k that are probabilities and hence are nonnegative. This means that the y-axis intercept of the function, $G_1(0) = q_0$, is nonnegative, as are the first, second and all higher derivatives of the function. Third, the value at $u = 1$ is

$$G_1(1) = \sum_{k=0}^{\infty} q_k = 1, \tag{3.30}$$

since q_k is a properly normalized probability distribution. Thus, in the range $0 \leq u \leq 1$, $G_1(u)$ is a positive-semidefinite, upward-concave function as sketched in Fig. 8. Equation (3.30) means that (3.29) always has a solution at $u = 1$ and the form of the function then ensures that there is, at most, one other solution in the range of interest, as shown in the figure. The solution at $u = 1$ corresponds to the situation where the probability of a vertex not being connected to the giant component is 1, meaning that no vertex belongs to the giant component and hence there is no giant component. Only if the other solution also exists can we have a giant component.

Inspecting the figure, we see that this implies that there will be a giant component only if the gradient of the generating function at $u = 1$ is greater than unity, i.e., $G_1'(1) > 1$. Making use of Eq. (3.26), we find that

$$G_1'(u) = \sum_{k=0}^{\infty} k q_k u^{k-1}, \tag{3.31}$$

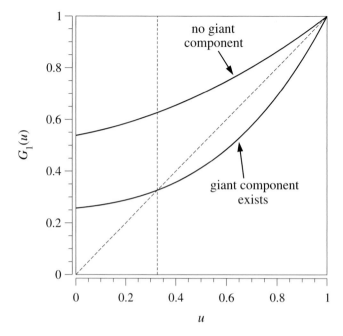

Fig. 8. Two possible forms for the generating function $G_1(u)$ as a function of u (solid lines). Solutions to Eq. (3.29), $u = G_1(u)$, occur where $G_1(u)$ crosses the diagonal dashed line. Since $G_1(1) = 1$ (Eq. (3.30)) there is always a solution at $u = 1$ (top right in the plot). There will be another, nontrivial solution at $u < 1$ if and only if $G_1'(1) > 1$. This occurs for the lower curve in the plot (with the nontrivial solution indicated by the vertical dashed line) but not for the upper curve.

and hence

$$G_1'(1) = \sum_{k=0}^{\infty} k q_k = \frac{\sum_k k(k+1) p_{k+1}}{\langle k \rangle} = \frac{\sum_k (k-1) k p_k}{\langle k \rangle} = \frac{\langle k^2 \rangle - \langle k \rangle}{\langle k \rangle}. \tag{3.32}$$

Then the criterion for the existence of a giant component is $(\langle k^2 \rangle - \langle k \rangle)/\langle k \rangle > 1$ or equivalently

$$\langle k^2 \rangle - 2\langle k \rangle > 0. \tag{3.33}$$

This criterion was first derived, using a different method, by Molloy and Reed (1995).

We can also calculate the size of the giant component when there is one. A vertex of total degree k will not be in the giant component only if all of its k

neighbours are not, which happens with probability u^k. Averaging over the distribution p_k of k, the probability of a randomly chosen vertex not belonging to the giant component is then given by

$$G_0(u) = \sum_{k=0}^{\infty} p_k u^k. \tag{3.34}$$

The function $G_0(u)$ is another probability generating function, in this case for the degree distribution p_k. The probability that a vertex *is* in the giant component is now given by $1 - G_0(u)$, and this also is the fraction of vertices that belong to the giant component, which is usually denoted S:

$$S = 1 - G_0(u). \tag{3.35}$$

Along with Eq. (3.29) for u, this equation tells us the size of the giant component.

As an example of the application of these results, consider the case of a network with a power-law degree distribution of the form

$$p_k = \frac{k^{-\alpha}}{\zeta(\alpha)}, \tag{3.36}$$

for $k \geq 1$, where the Riemann zeta function $\zeta(\alpha) = \sum_{k=1}^{\infty} k^{-\alpha}$ is a normalizing constant that ensures $\sum_k p_k = 1$. Then the criterion for the existence of a giant component, Eq. (3.33), becomes

$$\frac{\zeta(\alpha - 2)}{\zeta(\alpha - 1)} > 2. \tag{3.37}$$

Although this expression cannot be solved in closed form, we can plot the left-hand side as a function of α and find that it is greater than 2 provided $\alpha <$ 3.4788.... This result was first demonstrated, by different means, by Aiello *et al.* (2000).

In practice, most networks with power-law degree distributions have exponents in the range $2 \leq \alpha \leq 3$, so the result of Aiello *et al.* seems to imply that we should expect these networks to have giant components, and—satisfyingly—this is usually found to be the case. However, one should treat this apparent agreement between theory and experiment with caution. It turns out that the existence of the giant component is, in this case, controlled in large part by the vertices of low degree in the network. As discussed in Section 3, however, most real-world scale-free networks show power-law behaviour only above some minimum degree $k_{\min}$ and deviate from the power law below this point. This deviation strongly affects the condition for the existence of a giant component: typically there is still an upper limit on the exponent α above which no giant component exists, but its

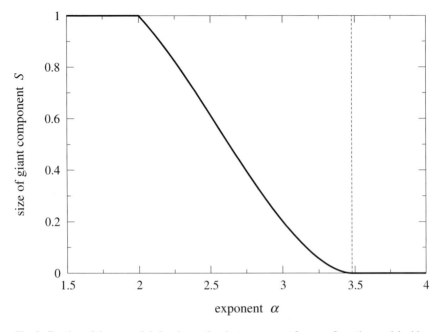

Fig. 9. Fraction of the network belonging to the giant component for a configuration model with a power-law degree distribution of the form given in Eq. (3.36) as a function of the exponent α of the power law. The giant component occupies 100% of the network for $\alpha \leq 2$ and vanishes altogether for $\alpha > 3.4788$—see Eq. (3.37).

value may be quite different from the value given by Eq. (3.37). The result above should therefore be taken only as an example of the type of calculation that can be performed, rather than as a quantitative description of real-world networks.[5]

Figure 9 shows the size of the giant component in a power-law network as a function of α. As the figure makes clear, the giant component does indeed vanish around $\alpha = 3.4788$ (vertical dashed line in the plot). Below this value the giant component grows with decreasing α until, at $\alpha = 2$, it fills 100% of the network—for this value, which is at the lower end of the range seen in real life, the giant component fills the entire network (except, possibly, for a small number of vertices that constitute a zero fraction of the network in the limit of large system size).

---

[5]There are other reasons why one should take this result with a pinch of salt. In particular, it was derived using the configuration model, whereas most real-world networks are only approximately described by the configuration model.

4. Further directions

There is by now a substantial volume of literature describing both theories of and experiments on real-world networks, extending the ideas introduced here in many directions. In this concluding section I discuss briefly a small selection of topics that the interested reader may like to investigate further. There are also a number of reviews and books in the field that the reader may find useful—see Albert and Barabási (2002), Dorogovtsev and Mendes (2002, 2003), Newman (2003b), Bornholdt and Schuster (2003), Boccaletti *et al.* (2006) and Newman *et al.* (2006), and references therein.

There is an interesting aspect of Milgram's small-world experiment that Milgram himself never commented on: the experiment reveals not only that short paths exist between most members of a population, but also that ordinary people are, collectively, good at finding those paths. Although no individual knows the entire social network of a country and it is unlikely that participants in the experiment would know the entire path a letter needed to take to get to the target individual, participants were nonetheless able to find short paths in a cooperative fashion. As Kleinberg (2000) has pointed out, the task of finding short paths would be difficult unless the network is structured in quite specific ways. Models of this structure and of the way in which humans "navigate" their social networks have been proposed and studied by Kleinberg (2000; 2002) and by Watts *et al.* (2002).

Other explanations have been proposed for the fat-tailed degree distributions observed in many networks. In particular, for biochemical networks such as protein interaction and metabolic networks, and possibly for some computer networks, vertex copying may be a plausible mechanism of network generation. In this scenario, networks grow by adding new vertices that are approximate copies of already existing ones—their connections are the same as those of some other vertex, apart from perhaps some small changes. Gene duplication provides a possible example of this kind of copying in biochemical networks and it has been shown that copying can give rise to networks with power-law degree distributions (Kleinberg *et al.* 1999; Solé *et al.* 2002; Vázquez *et al.* 2003).

The generating function techniques of Section 3.2 can be extended to other types of calculations as well. For instance, they have been successfully applied to percolation problems on networks and to epidemiological models. The site percolation problem is interesting as a model of network resilience, the tolerance of a network to removal of its vertices. In numerical work, Albert *et al.* (2000) showed that real-world networks with skewed degree distributions are highly resilient to random deletion of their vertices while at the same time being very sensitive to targeted deletion of their highest-degree vertices. That is, if we delete the hubs in a network it will quickly break apart into disconnected segments. Similar results have been shown analytically using percolation processes on configuration models (Cohen *et al.* 2000, 2001; Callaway *et al.* 2000).

The configuration model of Section 3.2 assumes that vertices are essentially connected at random, allowing for the constraints imposed by the degree sequence. There is strong evidence however that real networks show a number of interesting deviations from randomness in their connection patterns. Watts and Strogatz (1998), for instance, pointed out that short loops of vertices connected by edges, and in particular triangles, occur much more frequently in many networks than one would expect on the basis of chance. They proposed a measure of the density of triangles in a network, called a *clustering coefficient*, and used it to quantify the structure of a variety of different networks. A number of other authors have more recently made studies of the densities of longer loops in networks (Fronczak *et al.* 2002; Bianconi and Capocci 2003; Bianconi *et al.* 2005). Recent calculations also suggest that when degree sequence is properly taken into account some networks in fact have only as many short loops as expected by chance and not more (Newman and Park 2003; Bianconi and Marsili 2006).

The degrees of adjacent vertices in many networks turn out to be correlated. For instance, in some networks it is found that vertices with high degree are connected together more often than one would expect by chance. Networks with this property are said to show *assortative mixing*. Other networks show *disassortative mixing*, meaning that high-degree vertices are preferentially connected to low-degree ones and *vice versa*. For example, Pastor-Satorras *et al.* (2001) showed that the Internet, represented at the level of autonomous systems, is a disassortative network, with high-degree vertices mostly connected to low-degree ones. Maslov and Sneppen (2002) showed a similar result for protein interaction networks. In a study of networks from a number of different fields, Newman (2002b) found that by and large most social networks are assortatively mixed, while most nonsocial networks are disassortatively mixed. Degree correlations are believed to have substantial effects on the operation of many networked systems. For example, they affect the formation of a giant component (Newman 2002b) and the resilience of networks to damage (Newman 2002b, 2003a; Vázquez and Moreno 2003).

The vertices in many networks are found to divide strongly into groups, clusters, or communities, such that vertices tend to connect principally to others in their own group (Girvan and Newman 2002). This phenomenon seems to be particularly prevalent in social networks, biochemical networks and some technological networks, such as transportation networks. A great deal of interest has been directed at methods for detecting communities in networks (Danon *et al.* 2005) and at connections between community structure and the function of networks (Krause *et al.* 2003; Guimerà and Amaral 2005; Palla *et al.* 2005).

Finally, and perhaps most importantly, progress has begun towards developing an understanding of the link between the structure of networks and the dynamics of the systems they describe. In a sense this is the ultimate goal of most recent studies of networks. The main reason for studying the structure of, for instance, the Internet is to understand better how the Internet functions, and similarly for other networks. Two particular lines of research are worth highlighting in this respect. The first is the study of the spread of disease over contact networks between individuals (Ball *et al.* 1997; Keeling 1999; Moore and Newman 2000; Pastor-Satorras and Vespignani 2001; Sander *et al.* 2002; Newman 2002a; Jones and Handcock 2003; Eubank *et al.* 2004). This "network epidemiology" is one of the most successful attempts at establishing a connection between structure and dynamics in networks. The second is the study of synchronization in networks of coupled oscillators. Building on a beautiful set of analytic results by Barahona and Pecora (2002) relating oscillator dynamics to the spectrum of the graph Laplacian, a number of authors have in recent years assembled a comprehensive theory of the synchronization transition in networks of various kinds. Although oscillator synchronization is only one possible type of network dynamics, these studies provide an intriguing hint at the kinds of techniques that could ultimately enable us to understand quantitatively the dynamics of networked systems.

References

Aiello, W., Chung, F., and Lu, L., 2000. A random graph model for massive graphs. In *Proceedings of the 32nd Annual ACM Symposium on Theory of Computing*, pp. 171–180. Association of Computing Machinery, New York.

Albert, R. and Barabási, A.-L., 2002. Statistical mechanics of complex networks. *Rev. Mod. Phys.* **74**, 47–97.

Albert, R., Jeong, H., and Barabási, A.-L., 2000. Attack and error tolerance of complex networks. *Nature* **406**, 378–382.

Ball, F., Mollison, D., and Scalia-Tomba, G., 1997. Epidemics with two levels of mixing. *Annals of Applied Probability* **7**, 46–89.

Barabási, A.-L. and Albert, R., 1999. Emergence of scaling in random networks. *Science* **286**, 509–512.

Barahona, M. and Pecora, L. M., 2002. Synchronization in small-world systems. *Phys. Rev. Lett.* **89**, 054101.

Bearman, P. S., Moody, J., and Stovel, K., 2004. Chains of affection: The structure of adolescent romantic and sexual networks. *Am. J. Sociol.* **110**, 44–91.

Bianconi, G. and Capocci, A., 2003. Number of loops of size h in growing scale-free networks. *Phys. Rev. Lett.* **90**, 078701.

Bianconi, G. and Marsili, M., 2006. Effect of degree correlations on the loop structure of scale-free networks. *Phys. Rev. E* **73**, 066127.

Bianconi, G., Caldarelli, G., and Capocci, A., 2005. Loops structure of the internet at the autonomous system level. *Phys. Rev. E* **71**, 066116.

Boccaletti, S., Latora, V., Moreno, Y., Chavez, M., and Hwang, D.-U., 2006. Complex networks: Structure and dynamics. *Physics Reports* **424**, 175–308.

Bornholdt, S. and Schuster, H. G. (eds.), 2003. *Handbook of Graphs and Networks*. Wiley-VCH, Berlin.

Callaway, D. S., Newman, M. E. J., Strogatz, S. H., and Watts, D. J., 2000. Network robustness and fragility: Percolation on random graphs. *Phys. Rev. Lett.* **85**, 5468–5471.

Chung, F. and Lu, L., 2002a. Connected components in random graphs with given degree sequences. *Annals of Combinatorics* **6**, 125–145.

Chung, F. and Lu, L., 2002b. The average distances in random graphs with given expected degrees. *Proc. Natl. Acad. Sci. USA* **99**, 15879–15882.

Clauset, A., Shalizi, C. R., and Newman, M. E. J., 2007. Statistical tests for power-law behavior. Working paper, Santa Fe Institute.

Cohen, R., Erez, K., ben-Avraham, D., and Havlin, S., 2000. Resilience of the Internet to random breakdowns. *Phys. Rev. Lett.* **85**, 4626–4628.

Cohen, R., Erez, K., ben-Avraham, D., and Havlin, S., 2001. Breakdown of the Internet under intentional attack. *Phys. Rev. Lett.* **86**, 3682–3685.

Danon, L., Duch, J., Diaz-Guilera, A., and Arenas, A., 2005. Comparing community structure identification. *J. Stat. Mech.* p. P09008.

Dorogovtsev, S. N. and Mendes, J. F. F., 2002. Evolution of networks. *Advances in Physics* **51**, 1079–1187.

Dorogovtsev, S. N. and Mendes, J. F. F., 2003. *Evolution of Networks: From Biological Nets to the Internet and WWW*. Oxford University Press, Oxford.

Dorogovtsev, S. N., Mendes, J. F. F., and Samukhin, A. N., 2000. Structure of growing networks with preferential linking. *Phys. Rev. Lett.* **85**, 4633–4636.

Eubank, S., Guclu, H., Kumar, V. S. A., Marathe, M. V., Srinivasan, A., Toroczkai, Z., and Wang, N., 2004. Modelling disease outbreaks in realistic urban social networks. *Nature* **429**, 180–184.

Feld, S., 1991. Why your friends have more friends than you do. *Am. J. Sociol.* **96**, 1464–1477.

Fronczak, A., Hołyst, J. A., Jedynak, M., and Sienkiewicz, J., 2002. Higher order clustering coefficients in Barabasi-Albert networks. *Physica A* **316**, 688–694.

Girvan, M. and Newman, M. E. J., 2002. Community structure in social and biological networks. *Proc. Natl. Acad. Sci. USA* **99**, 7821–7826.

Goldstein, M. L., Morris, S. A., and Yen, G. G., 2004. Problems with fitting to the power-law distribution. *Eur. Phys. J. B* **41**, 255–258.

Guimerà, R. and Amaral, L. A. N., 2005. Functional cartography of complex metabolic networks. *Nature* **433**, 895–900.

Jones, J. H. and Handcock, M. S., 2003. Sexual contacts and epidemic thresholds. *Nature* **423**, 605–606.

Karinthy, F., 1929. Chains. In *Everything is Different*. Budapest.

Keeling, M. J., 1999. The effects of local spatial structure on epidemiological invasion.

Proc. R. Soc. London B **266**, 859–867.

Kleinberg, J. M., 2000. Navigation in a small world. *Nature* **406**, 845.

Kleinberg, J. M., 2002. Small world phenomena and the dynamics of information. In T. G. Dietterich, S. Becker, and Z. Ghahramani (eds.), *Proceedings of the 2001 Neural Information Processing Systems Conference*. MIT Press, Cambridge, MA.

Kleinberg, J. M., Kumar, S. R., Raghavan, P., Rajagopalan, S., and Tomkins, A., 1999. The Web as a graph: Measurements, models and methods. In T. Asano, H. Imai, D. T. Lee, S.-I. Nakano, and T. Tokuyama (eds.), *Proceedings of the 5th Annual International Conference on Combinatorics and Computing*, no. 1627 in Lecture Notes in Computer Science, pp. 1–18. Springer, Berlin.

Krapivsky, P. L., Redner, S., and Leyvraz, F., 2000. Connectivity of growing random networks. *Phys. Rev. Lett.* **85**, 4629–4632.

Krause, A. E., Frank, K. A., Mason, D. M., Ulanowicz, R. E., and Taylor, W. W., 2003. Compartments revealed in food-web structure. *Nature* **426**, 282–285.

Martinez, N. D., 1991. Artifacts or attributes? Effects of resolution on the Little Rock Lake food web. *Ecological Monographs* **61**, 367–392.

Maslov, S. and Sneppen, K., 2002. Specificity and stability in topology of protein networks. *Science* **296**, 910–913.

Molloy, M. and Reed, B., 1995. A critical point for random graphs with a given degree sequence. *Random Structures and Algorithms* **6**, 161–179.

Molloy, M. and Reed, B., 1998. The size of the giant component of a random graph with a given degree sequence. *Combinatorics, Probability and Computing* **7**, 295–305.

Moore, C. and Newman, M. E. J., 2000. Epidemics and percolation in small-world networks. *Phys. Rev. E* **61**, 5678–5682.

Newman, M. E. J., 2002a. Spread of epidemic disease on networks. *Phys. Rev. E* **66**, 016128.

Newman, M. E. J., 2002b. Assortative mixing in networks. *Phys. Rev. Lett.* **89**, 208701.

Newman, M. E. J., 2003a. Mixing patterns in networks. *Phys. Rev. E* **67**, 026126.

Newman, M. E. J., 2003b. The structure and function of complex networks. *SIAM Review* **45**, 167–256.

Newman, M. E. J. and Girvan, M., 2004. Finding and evaluating community structure in networks. *Phys. Rev. E* **69**, 026113.

Newman, M. E. J. and Park, J., 2003. Why social networks are different from other types of networks. *Phys. Rev. E* **68**, 036122.

Newman, M. E. J., Strogatz, S. H., and Watts, D. J., 2001. Random graphs with arbitrary degree distributions and their applications. *Phys. Rev. E* **64**, 026118.

Newman, M. E. J., Barabási, A.-L., and Watts, D. J., 2006. *The Structure and Dynamics of Networks*. Princeton University Press, Princeton.

Palla, G., Derényi, I., Farkas, I., and Vicsek, T., 2005. Uncovering the overlapping community structure of complex networks in nature and society. *Nature* **435**, 814–818.

Park, J. and Newman, M. E. J., 2004. The statistical mechanics of networks. *Phys. Rev. E* **70**, 066117.

Pastor-Satorras, R. and Vespignani, A., 2001. Epidemic spreading in scale-free networks. *Phys. Rev. Lett.* **86**, 3200–3203.

Pastor-Satorras, R., Vázquez, A., and Vespignani, A., 2001. Dynamical and correlation properties of the Internet. *Phys. Rev. Lett.* **87**, 258701.

Pool, I. de S. and Kochen, M., 1978. Contacts and influence. *Social Networks* **1**, 1–48.

Price, D. J. de S., 1965. Networks of scientific papers. *Science* **149**, 510–515.

Price, D. J. de S., 1976. A general theory of bibliometric and other cumulative advantage processes. *J. Amer. Soc. Inform. Sci.* **27**, 292–306.

Rapoport, A. and Horvath, W. J., 1961. A study of a large sociogram. *Behavioral Science* **6**, 279–291.

Redner, S., 1998. How popular is your paper? An empirical study of the citation distribution. *Eur. Phys. J. B* **4**, 131–134.

Sander, L. M., Warren, C. P., Sokolov, I., Simon, C., and Koopman, J., 2002. Percolation on disordered networks as a model for epidemics. *Math. Biosci.* **180**, 293–305.

Seglen, P. O., 1992. The skewness of science. *J. Amer. Soc. Inform. Sci.* **43**, 628–638.

Simon, H. A., 1955. On a class of skew distribution functions. *Biometrika* **42**, 425–440.

Solé, R. V., Pastor-Satorras, R., Smith, E., and Kepler, T. B., 2002. A model of large-scale proteome evolution. *Advances in Complex Systems* **5**, 43–54.

Valverde, S., Cancho, R. F., and Solé, R. V., 2002. Scale-free networks from optimal design. *Europhys. Lett.* **60**, 512–517.

Vázquez, A. and Moreno, Y., 2003. Resilience to damage of graphs with degree correlations. *Phys. Rev. E* **67**, 015101.

Vázquez, A. and Weigt, M., 2003. Computational complexity arising from degree correlations in networks. *Phys. Rev. E* **67**, 027101.

Vázquez, A., Flammini, A., Maritan, A., and Vespignani, A., 2003. Modeling of protein interaction networks. *Complexus* **1**, 38–44.

Watts, D. J. and Strogatz, S. H., 1998. Collective dynamics of 'small-world' networks. *Nature* **393**, 440–442.

Watts, D. J., Dodds, P. S., and Newman, M. E. J., 2002. Identity and search in social networks. *Science* **296**, 1302–1305.

Course 9

MINORITY GAMES

Damien Challet

Nomura Centre for Quantitative Finance,
Mathematical Institute,
St Gile's 24-29, Oxford OX1 3LB, UK

J.-P. Bouchaud, M. Mézard and J. Dalibard, eds.
Les Houches, Session LXXXV, 2006
Complex Systems
© *2007 Published by Elsevier B.V.*

343

Contents

Preamble

The Minority Game is the prototype model of global competition between adaptive heterogeneous agents. The mathematical equivalence between agent heterogeneity and physical disorder makes it exactly solvable. Starting from the intractable El Farol bar problem, this chapter explains how to derive the Minority Game and how to solve it, and finally illustrates the universality of minority mechanisms by building more and more complex and still exactly solvable models.

1. Introduction

The minority game (MG thereafter) is a universal model of collective competition. At the same time it illustrates perfectly what statistical physics can bring to the emergent field of complex systems, in particular its methods for solving exactly systems of heterogeneous agents. The history of the MG is a nice example of how to simplify a complex model first into a workable model which still requires sophisticated analytical methods and long calculus, and then into a very simple Markovian model to which standard stochastic calculus applies.

These notes provide an overview of the historical motivations leading to the introduction of the model, the properties of the model and its solution, and will end with a discussion about its relevance in modelling.

1.1. Limited resources

Any resource is in limited supply. This is of course most obvious in the case of petrol, precious metals and other commodities, but we face many situations on a daily basis where we compete for more down-to-earth limited resources, such as a seat in a restaurant, a quietness in a plane, or being accepted at Les Houches summer school on complex systems. In some cases, as we shall see, the possible actions are very simple, thus ideally suitable for designing simple models.

Limited resources are distributed *via* interactions with other people or entities. The decision of how much one exploits a limited resource lies in one's hands, but the final outcome depends on everybody. Therefore, decision making and synchronization are key issues to be considered in this chapter. This means that one should characterize theoretically what optimal synchronization is. In addition, one must find what behaviour leads to an optimal state.

1.2. Game theory

The most relevant concept in game theory for this chapter is that of Nash equilibrium (NE thereafter) [25]. Simply put, a system is in such an equilibrium if no agent has an incentive to deviate from his current action. In other words, it is a local minimum in the payoff landscape. Assuming that there are only two possible actions $+1$ and -1, agent i ($i = 1, \ldots, N$) takes action a_i, and receives payoff $U_i(\{a_j\}) = U(a_i, \{a_{-i}\})$, the notation $\{a_{-i}\}$ referring to the actions of all the agents except agent i. The set $\{a_j\}$ is a NE if $U(-a_i, \{a_{-i}\}]) \leq U(\{a_j\})$ for all i;

If the possible actions of all the agents are fixed in time, one speaks about a NE in pure strategies and about a NE in mixed strategies if the actions are taken probabilistically according to $P(a_i = 1) = m_i$.

Generally, a NE may exist if all the agents are rational and if each of them believes that all the other ones are rational. For simple situations, it is reasonable to think that these conditions can be met. However, it quite clear that we, human beings, are not good at being rational, for instance when the situation involves too many information sources or too much uncertainty (we over-estimate or simply ignore small probabilities), and when the computing power we have is too scarce.[1]

1.3. The El Farol bar problem

Simon was a well-known proponent of bounded rationality, and advocated satisfiability instead of optimality [45]. Since there is only one way to be perfect, the agents are often assumed to be the same in standard Economics literature, which therefore replaces a whole system by a single, representative agent. On the other hand there of many ways of being imperfect, which opens the way to heterogeneous agents. In its famous El Farol Bar Problem (EFBP thereafter) [4], Brian Arthur proposed a situation where one cannot act rationally and where heterogeneity is not only possible, but also needed. Imagine El Farol, a popular bar with very nice live music every Thursday. Say that a 100 people are potentially interested in attending the bar; unfortunately, there are only 60 seats. The question that the agents must answer each week is whether to stay at home if they think that the bar will be over-crowded or go to the bar otherwise.

---

[1]See Chapter 5 by Kirman.

The NE in pure strategies is straightforward: 60 customers go to the bar and 40 stay at home. The NE in mixed strategies reached when everybody goes to the bar with probability 0.6.[2]. These two types of NE correspond to two benchmarks, one being optimal for the society and the other one a purely random outcome whose average is 60. In real life however, the problem cannot be summarized by a discussion on NE, first because some of the 40 frustrated customers *will* eventually go the bar and break the equilibrium, and because human beings are very poor at drawing numbers at random. A final reason for the breakdown of rationality is that the situation is too complex for the human brain (there are 2^N possible sets of actions), and that in such case even the most clever agent cannot assume that the other ones are rational.

One needs therefore to describe mathematically a reasonable agent dynamics and examine its properties in the stationary state. It makes sense to assume that people take into account the past attendances in order to take their decisions, in which case the representative agent approach breaks down completely, as it leads to either nobody or everybody in the bar. Therefore, heterogeneity and bounded rationality are needed. Arthur proposed a concrete solution, namely induction, and specified precisely how to model it: the agents have a finite number of functions that predict the next attendance from past ones; each agent keeps track on how precise each of his functions (thereafter called strategies) is, and uses the currently best one. In this way, an agent can learn from the past, albeit in an indirect way, and also pickup subtle changes of behaviour of the other agents via the global attendance. There is no a priori best strategy, only a best one at a given moment. Nevertheless, the average attendance converges to about 60, which was perceived at the time as a little miracle, and motivated the many papers discussing the average attendance in EFBP.

The rest of this chapter is mainly devoted to two questions: how to compute analytically global quantities such as fluctuations and average attendance and what dynamics leads to a Nash equilibrium (and which one).

2. The minority game: definition and numerical simulations

In order to make any progress it is best to separate the issues of attendance convergence and fluctuations. It is simpler to investigate first the fluctuations and then the convergence. Indeed, the fluctuations of the attendance N_+ are by definition $\langle (N_+ - \langle N_+ \rangle)^2 \rangle$, which implies that in order to compute them, one first need the be able to compute $\langle N_+ \rangle$, the average attendance in the stationary state.

---

[2]There are many more NEs where some of the agents play pure strategies and the others mixed strategies [39].

The key idea is to simplify the problem by ensuring $\langle N_+ \rangle = L$ by design, where L is the resource level. Our natural taste for symmetry suggests to set $L = N/2$. This bring in the Minority Game: whatever the meaning of the choices $+1$ and -1, the aim is to be in the minority at each time step. In a sense the MG is more abstract and generic than the EFBP because of its additional symmetry.

The best example of a minority game is provided by Zig-Zag-Zoug, a game played by French-speaking Swiss children whenever three of them need to elect a chief or share something unfairly. They place themselves in a triangle, put forward one foot each, say the magic words 'Zig-Zag-Zoug' and at 'Zoug' either leave or remove their feet. The one in the minority wins. It is a frustrating game when repeated, because one usually does not seem to be able to learn much from the past; on the other hand, the losers are in the majority and can sometimes coerce the winner into playing another round of the game.

The notations will be as follows: agent i takes decision $a_i = \pm 1$ at time t; the aggregate outcome is $A = N_+ - N_- = \sum_{i=1}^{N} a_i(t)$, and the individual payoff is

$$-a_i(t)A(t) \tag{2.1}$$

The minus sign is the signature of minority games, and means that one is rewarded for acting against the crowd. As discussed below, any agent-based model whose agent payoff contains such a term share some common properties with the original MG.

2.1. No attendance history

Although the idea of Arthur was to base the strategies of the agents on precise past attendance, the simplest re-reinforcement learning scheme is to remember the bias in past choices of the agents [36, 38]: each agent i has a register Δ_i which evolves according to

$$\Delta_i(t+1) = \Delta_i(t) - \frac{A(t)}{N} \tag{2.2}$$

and takes his decision probabilistically, following

$$P[a_i(t) = 1] = \frac{1 + \tanh[\Gamma_i \Delta_i(t)]}{2} \tag{2.3}$$

The simplest case is $\Gamma_i = \Gamma$ and $\Delta_i(0) = 0$ for all agents. In that case, the index i can be omitted and Eq. (2.2) becomes

$$\Delta(t+1) = \Delta(t) - \frac{A(t)}{N} = \Delta(t) - \tanh[\Gamma \Delta(t)] + \eta(t) \tag{2.4}$$

where $\eta(t)$ is a noise term with zero average and

$$\langle \eta(t)\eta(t')\rangle = \delta_{t,t'}\left[1 - \tanh\left(\Gamma\Delta(t)\right)^2\right]/N. \tag{2.5}$$

It vanishes therefore in the $N \to \infty$ limit, which we will take for a while. It is easy to find by linear stability analysis that the fixed point $\Delta^{(0)} = 0$ is stable if $\Gamma < \Gamma^* = 2$ and unstable otherwise; in the latter case, a period 2 dynamics emerges, with the stable points determined by replacing $\Delta(t + 1)$ with $-\Delta(t)$ in Eq. (2.4) [37]

$$\Delta^{(1)} = \tanh(\Gamma\Delta^{(1)})/2. \tag{2.6}$$

The fluctuations $\sigma^2/N = \langle A^2\rangle/N \propto 1$ when $\Gamma < \Gamma^*$ and $\langle A^2\rangle/N^2 \propto 1$ otherwise. A Taylor expansion to the third order of Eq. (2.6) gives $\Delta_1 \simeq \pm\sqrt{\frac{3(\Gamma-2)}{\Gamma^3}}$ for Γ close to 2; on the other hand, $\frac{A}{2} - \Delta_1 \propto \exp(-\Gamma)$ for large Γ. A way to check numerically the value of Γ^* is to observe the onset of the change of $\langle \Delta^2\rangle$ from $O(N^{-1})$ to $O(1)$ as a function of Γ (which requires to take several system size and sweeping over Γ). Heterogeneous initial conditions $\Delta_i(0) \neq 0$ help to stabilize the fixed point by raising Γ^* [36].

In general, one needs to keep N finite, hence, to take into account the noise term of Eq. (2.4), and obtain the distribution of Δ in the stationary state. For this purpose, one can write the equivalent Fokker-Planck equation

$$\frac{\partial P(\Delta)}{\partial t} = \frac{\partial^2}{2\partial\Delta^2}\left[\frac{1 - (N-1)\tanh(\Gamma\Delta)^2}{N}P(\Delta)\right] + \frac{\partial}{\partial\Delta}\left[\tanh(\Gamma\Delta)P(\Delta)\right] \tag{2.7}$$

Solving it in the stationary state is straightforward and gives

$$P(\Delta) = \frac{1}{Z}\left[2 + N\left(\cosh(2\Gamma\Delta) - 1\right)\right]^{\frac{-(1+\Gamma)}{\Gamma}}\cosh(\Gamma\Delta)^2 \tag{2.8}$$

where Z is a normalization factor. When Γ is small, P can be approximated by a Gaussian with zero average and $\langle \Delta^2\rangle = 1/[2\Gamma[(N(\Gamma+1)-\Gamma]]$. Therefore

$$\frac{\langle A^2\rangle}{N} \simeq 1 + (N-1)\frac{\Gamma}{2[N(\Gamma+1)-\Gamma]} \to 1 + \frac{\Gamma}{2(\Gamma+1)} \simeq 1 + \frac{\Gamma}{2}$$
$$N \to \infty,\ \Gamma \ll 1 \tag{2.9}$$

Figure 1 shows that the Fokker-Planck equation provides a good description of the stationary state for $\Gamma \leq 1$, whereas for larger Γ the hypothesis of small jumps in Δ is clearly wrong; this is due to the fact that with $\Gamma > 1$, the drift term makes Δ change sign on average at each time step. The Gaussian approximation

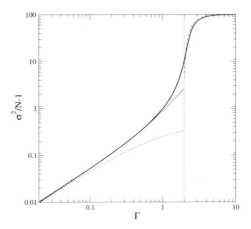

Fig. 1. Fluctuations versus Γ from numerical simulations (thick line; $N = 100$, 10^6 iterations per point), from the Fokker-Plank equation (thin line, $\Gamma < 2$), the Gaussian approximation (dashed red line), and Eq. (2.10) (dash-dotted line, $\Gamma > 2$)

bends in the wrong way as Γ increases and should not be used for $\Gamma > 0.01$. For $\Gamma > 2$, $P(\Delta)$ separates into two symmetric peaks, centered roughly at $\pm\Delta_1$; the fluctuations can therefore be approximated by

$$P(\Delta) \simeq \frac{1}{2Z}\left[2 + N\left(\cosh(2\Gamma(\Delta - \Delta_1)) - 1\right)\right]^{\frac{-(1+\Gamma)}{\Gamma}} \cosh\left(\Gamma(\Delta - \Delta_1)\right)^2$$
$$+ \frac{1}{2Z}\left[2 + N\left(\cosh(2\Gamma(\Delta + \Delta_1)) - 1\right)\right]^{\frac{-(1+\Gamma)}{\Gamma}} \cosh\left(\Gamma(\Delta + \Delta_1)\right)^2.$$

$$(2.10)$$

This equation makes the transition from $\Gamma < 2$ to $\Gamma > 2$ smooth as it is equivalent to Eq. (2.8) for $\Gamma < 2$.

The above discussion can be extended to heterogeneous learning rates and memory in order to investigate how being different provides an advantage over the other players. In the limit of very small Γ_i, one finds that the gain advantage for agent i is proportional to $\langle\Gamma\rangle - \Gamma_i$: one gains more if one has a smaller than average learning rate. This is because it is better not to react to meaningless fluctuations of A. A finite memory about the past is implemented by considering a recursive exponential decay in Δ:

$$\Delta(t + 1) = \Delta(t)(1 - \lambda) - A(t).$$

$$(2.11)$$

One easily checks that the memory length is of order $1/\ln|\lambda| \propto 1/\lambda$ for small λ. In contrast to heterogeneous learning rates, introducing heterogeneous λ_i requires one equation for each λ_i. The simplest case is to have two groups of agents of equal size same Γ, one with λ_1 and the other one with λ_2; the advantage of group 1 over group 2 is proportional to $\lambda_1 - \lambda_2$: forgetting faster gives an edge because it makes the agents more adaptive. In short, learning more slowly and forgetting faster than everybody else is a great idea, culminating in people learning nothing and forgetting it immediately, that is, playing completely at random: it is the only way not to have a predictable behaviour at next time step. One should not be depressed about this seemingly negative result. It simply states that the only way not to be exploited in such a game is to act as randomly as possible, which is obviously a NE. However, human beings are not very good at it because they cannot help taking into account past actions in order to make sure that their next actions are, in their view, as random as possible. This suggests that one should study behavioral models where the actions of the agents are conditional on the past (recent) history of the game thus the strategy space proposed by Arthur is sensible.

So far, one has found a way of reaching the not very efficient NE. What about the NE where the population splits into two halves? Remarkably, a minor modification to the above simple learning dynamics is enough to achieve optimal synchronization. The canonical form of payoff, $-a_i A$, seems reasonable at first sight, but conceals an important assumption from agent i about the nature of A: he considers it as an external process in which he takes no part. This is obviously wrong, as $A = A_{-i} + a_i$. The agent should therefore remove his contribution from A when assessing his performance. It is not an easy task in general, because the payoff could be an unknown non-linear function of A. Therefore, agent i will approximate its impact by removing ηa_i from his payoff. For a homogeneous population, Eq. (2.4) becomes

$$\Delta(t+1) = \Delta(t) - \frac{A - \eta a_i(t)}{N} \tag{2.12}$$

It is easy to see that the dynamics minimizes $H_\eta = (\sum_i m_i)^2 - \eta \sum_i m_i^2$ whose minima are in pure strategies as a soon as $\eta > 0$ [36].

Solving this extremely simple MG does not require any sophisticated tools from statistical mechanics of disordered systems because of the absence of fixed look-up tables: the heterogeneity is scalar instead of vectorial. It is therefore advisable to first study this version of the game, as its behaviour displays many of the properties of more complicated games, such as the influence of learning on fluctuations and the role of game impact.

2.2. From EFB to MG

As stated above, it is very likely that at least one customer will try to extract information from past attendance, and because of this, the next attendances will depend on past attendances. In turn, this will motivate other agents to try and exploit this dependence. Therefore, we cannot content ourselves with the discussion on games with no attendance histories. The strategy space proposed by Arthur amounts to consider auto-regressive functions of the form $f(\vec{A}) = \sum_{\tau=1}^{M} w_\tau A(t - \tau)$ where the weights w are real numbers; it is a space of infinite dimension. The success of the MG is due in a large part to its simple, finite strategy space.

The key observation is that the agents are ultimately interested in binary decisions, not on predicting the attendance. Why then trying to predict the exact attendance? A binary output is sufficient and reduces the dimensionality of the strategy space, but not nearly enough. Predicting a binary output does not really require a vector of exact attendances; therefore, one can simplify the vector of attendances to a vector of bits, each of them encoding a past winning decision. In this way, the dimension of the strategy space is 2^{2^M}. It is a fairly substantial number even for moderately large M but above all does not depend on N. As a consequence, one expects that large fluctuations arise when the number of agents exceeds the number of available strategies.

The full definition of the canonical MG is the following: there are N agents, each with a personal set of S strategies, drawn at random at the beginning of the game and kept fixed afterwards. Each strategy is a look-up table which prescribes an action for each of 2^M possible game history bit-strings, denoted by μ. To each strategy s of agent i, denoted by $a_{i,s}$ is associated a score $U_{i,s}$ that evolves according to

$$U_{i,s}(t + 1) = U_{i,s}(t) - a_{i,s}^{\mu(t)} A(t). \tag{2.13}$$

A score contains the cumulated payoff that this strategy would have obtained, had it been played since the beginning by player i.[3] Agents i plays its best strategy at time t

$$s_i(t) = \arg \max_s U_{i,s}(t) \tag{2.14}$$

and the outcome of the game is

$$A(t) = \sum_{i=1}^{N} a_{i,s_i(t)}. \tag{2.15}$$

The history bit-string $\mu(t)$ is then updated to

$$\mu(t + 1) = \left[2\mu(t)\right]\text{MOD } 2^M + \text{sign}\left[A(t)\right], \tag{2.16}$$

and the next turn can take place.

---

[3]This is of course a wrong assumption because of game impact, as we shall see later.

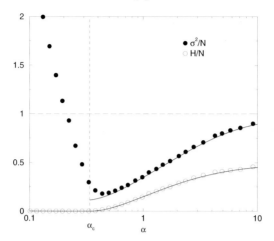

Fig. 2. Fluctuations σ^2/N and predictability H as a function of $\alpha = P/N$ from numerical simulations ($P = 128$) and the exact solution (continuous line). Average over 200 samples.

By construction, $\langle A \rangle = 0$. The next interesting quantity is the fluctuations $\sigma^2/N = \langle A^2 \rangle/N$ whose behaviour is reported in Fig. 2. When the number of agents is small compared with 2^M (right hand side of the figure), the fluctuations tend to the random choice benchmark: the information is too complicated for the agents. Increasing the number of agents decreases the scaled fluctuations, which become smaller than 1: the agents are said to cooperate. The fluctuations reach a minimum, and clearly something happens at $\alpha = 2^M/N \simeq 0.4$, as σ^2 sharply increases when α is lowered further; for smaller α, $\sigma^2 \propto N^2$. This clearly comes from the finiteness of the strategy space. Interestingly, the relevant control parameter is not proportional to 2^{2^M} but to 2^M, which was intuitively explained by a counting argument about the number of strategies that really differ from each other [20, 47]. The exact solution sketched below relates this number to the number of possible histories.

The next relevant quantity is the predictability of next outcome. Since the agent behave conditionally on the history μ, one should measure the conditional average outcomes $\langle A|\mu \rangle$. Any deviation from zero conditional average can be exploited *in principle*.[4] A measure of predictability is given by the scaled squared norm of $\langle A|\mu \rangle$ seen as a vector:

$$H = \frac{1}{P} \sum_{\mu} \langle A|\mu \rangle^2. \tag{2.17}$$

---

[4] See Ref. [16] for a detailed discussion about when predictability can be exploited.

Figure 2 displays H/N as a function of α. The game is predictable when there are few agents. Adding more of them decreases the predictability, which makes sense. Remarkably, H vanishes exactly at the point where σ^2 reaches its minimum, $\alpha_c = 0.3374\ldots$. This is a phase transition of the second order, i.e. H is smooth at α_c. Thus one calls asymmetric the phase with $H > 0$, and symmetric the one with $H = 0$; alternatively, one also speaks about predictable and unpredictable phases. An important difference between the two phases is that the fluctuations are independent from the initial conditions (score valuations) in the asymmetric phase, where as they decrease as the initial score valuation biases towards one strategy increases. In the case of very large bias, $\sigma^2/N \propto \alpha$ ($\sigma^2/P \propto 1$); this is to be compared with $\sigma^2/N \propto 1/\alpha$ ($\sigma^2/P \propto N^2$) in the case of zero initial bias.

2.2.1. Symmetric formalism

We shall restrict the following discussion to the simplest case $S = 2$.[5] The mathematical formalism is made more elegant and powerful by incorporating all the possible symmetries. Instead of denoting the two strategies by 1 and 2, one calls them -1 and $+1$. The strategy played by agent i, still denoted by $s_i = \pm 1$, is therefore very much similar to a spin. It should be noted that the structure of the strategies adds a layer to the decision process: the choice of the agent is between his strategies, not between the actual choices.

When choosing which strategy to play, agent i only takes into account the difference of scores, not their absolute values. It makes sense therefore to focus on the dynamics of $y_i = U_{i,+} - U_{i,-}$. The action of agent i can be decomposed into a constant and a variable parts

$$a_{i,s_i}^{\mu} = \frac{a_{i,+}^{\mu} + a_{i,-}^{\mu}}{2} + s_i \frac{a_{i,+}^{\mu} + a_{i,-}^{\mu}}{2} = \omega_i^{\mu} + s_i \xi_i^{\mu}. \tag{2.18}$$

As a consequence, the evolution of y_i is rewritten as

$$y_i(t+1) = y_i(t) - 2\xi_i^{\mu(t)}\left[\Omega^{\mu(t)} + \sum_j \xi_j^{\mu(t)} s_j(t)\right], \tag{2.19}$$

where $\Omega^{\mu} = \sum_j \omega_j^{\mu}$ is the constant part of A for a given μ.

This formalism allows for explicit formulae of σ^2 and H as a function of averages of the strategy choices: simplifying further the notations by replacing the averages over μ by the shorthand $\sum_{\mu} X^{\mu}/P = \overline{X}$, and setting $\langle s_i \rangle = m_i$, one finally obtains

---

[5]The case $S > 2$ has been partly solved with replica calculus [39]; generating functions have been recently used to find the full solution [1,42].

$$H = \overline{\Omega^2} + 2\sum_i h_i m_i + \sum_{i,j} J_{i,j} m_i m_j \qquad (2.20)$$

where $h_i = \overline{\Omega \xi_i}$ and $J_{i,j} = \overline{\xi_i \xi_j}$, and

$$\sigma^2 = \overline{\Omega^2} + 2\sum_i h_i m_i + \sum_{i,j} J_{i,j} \langle s_i s_j \rangle \qquad (2.21)$$

These two quantities look very much like Hamiltonians ('energy functions') of interacting spin systems. The double sum over i and j does not contain any element of physical distance, therefore the MG falls into the category of mean-field systems. It is good news, as the solution of such systems is usually more easily tractable. Going back to the game payoff, one sees that any repeated game where the players interact synchronously via a global quantity that they all contribute to create may be of mean-field type.

The other important insight given by these formulae is the direct explanation of why any global quantity varies slightly between different runs of the game: this is due to the particular draws of the strategies, which stay constant for the whole duration of the game and is equivalent to fixed physical disorder which occurs in disordered systems such as spin glasses [40]. Therefore, there is a direct equivalence between agent heterogeneity and physical disorder. As a consequence one can use the powerful methods first designed for disordered physical systems to solve models of socio-economic interest containing vectorial heterogeneity. In retrospect, this also shows why Economics could not solve the El Farol Bar problem with its traditional toolbox.

A Hamiltonian is minimized by the dynamics of its system at zero temperature. One must simplify the MG a little more before being able to find out which quantity is minimized by the agents. One might wonder about the consequences of ever simplifying the model on the relationship between the resulting model and the original MG, or worse, the EFBP. As we shall see, the phenomenology of the model is very robust with respect to many modifications. One should also keep in mind the fact that the MG stands exactly at the border between intractable and solvable models, as a whole phase (the symmetric non-ergodic one) still escapes full analytical understanding.

3. Exact solutions

3.1. Further simplifications

3.1.1. Partial linearization
Conceptual simplicity sometimes conflicts with mathematical tractability. The original MG contains many sources of non-linearity. For instance, the "use the

best" strategy rule is equivalent to a sign function

$$s_i(t) = \text{sign}[\Delta_i(t)]. \tag{3.1}$$

Smoothing it can be achieved stochastically, by introducing a probability that $s_i(t) = 1$ as a function of $\Delta_i(t)$:

$$P[s_i(t) = 1] = \frac{1 + f(\Delta_i(t))}{2}, \tag{3.2}$$

where $f(-\infty) = -1$ and $f(+\infty) = 1$ is a continuous function of x, the most common choice being $f(x) = \tanh(\Gamma x)$ [9]; Γ is a learning rate that specifies how much the average behaviour of an agent changes in reaction to a variation of Δ_i.

The second source of non-linearity is more obvious: the agent payoff of the original MG is a sign function, possibly the worst choice from a mathematical point of view. This is why we only considered the payoff $-a_i(t)A(t)$.

3.1.2. Markovian histories

The last and most problematic obstacle to a (not overly complex) mathematical solution of the MG comes from the history process. Bit-strings of length M live on a De Bruijn graph, of which Fig. 3 shows an example. The very structure of these graphs and the fact that the transition probabilities are not constant in time make it tempting to replace the time evolution of μ by a Markovian transition matrix $W_{\mu \to \nu}$, the simplest being uniform and constant $W_{\mu \to \nu} = 1/P$. Cavagna [8] showed that the global behaviour of the game is not changed tremendously when one replaces real histories by random ones. Later work pointed out the differences between the two cases in the asymmetric [13] and symmetric phases [32]. A phenomenological extension of the exact solution of the asymmetric phase was first proposed by Ref. [13], while the full self-consistent solution for real histories was found later by Coolen, with an explicit and systematic expansion near the critical point in the asymmetric phase [22, 23].

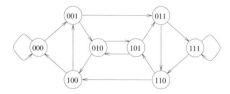

Fig. 3. The de Bruijn graph on which an $M = 3$ bit-string history evolve.

3.2. Stationary state: replica

This method is based on the knowledge of the quantity that the dynamics minimizes. It is a precious knowledge, as it sheds light on the global consequence of the microscopic dynamics, but also because it allows for an intuitive understanding of the dynamics. In the original MG, the quantity minimized by the agents is the predictability H. Without resorting to continuous time transformation to prove it,[6] one can convince oneself of the specific role H by writing down the average increase of y_i:

$$\langle y_i(t+1) - y_i(t) \rangle = \overline{-2\xi_i \left(\Omega + \sum_j \langle s_j \rangle \xi_j \right)} \qquad (3.3)$$

There are two possibilities: either the score difference increases linearly with time or oscillates around 0. In the first case, it is clear that agent i ends up playing only one of his strategies, hence $m_i = \pm 1$; the agents of this type are *frozen*. All the other agents play their two strategies alternatively, generally with a preference for one or the other ($|m_i| < 1$), as their y hovers around 0. Therefore their average score increase is zero:

$$\overline{-2\xi_i \left(\Omega + \sum_j m_j \xi_j \right)} = 0. \qquad (3.4)$$

It is easy to check that this equation is obtained by differentiating H with respect to m_i.

3.2.1. Algebraical interpretation of the phase transition

The above equation raises the question of when the agents actually manage to cancel H. This is done by definition when

$$\sum_j \xi_j^\mu m_j = -\Omega^\mu. \qquad (3.5)$$

This is a system of P linear equations with N degrees of freedom. In principle, P degrees of freedom should be enough to solve this system of equations, but since the m_is are bounded, more than P agents are needed (no MG with $S = 2$ has ever had a critical point at $\alpha_c > 1$).

If there was no Ω^μ term on the right hand side of Eq. (3.5), the system of equations would be homogeneous and the solution $m_i = 0$ for all i would always exist; it corresponds to the case where all the agents play at random either of

---

[6] See Ref. [37] for more details.

their strategies at each time step, which results in $\sigma^2/N \rightarrow 1$ in the limit of infinite systems: random outcome, no more cooperation. Similarly, $\alpha_c = 1$, and for $\alpha < \alpha_c$, the fluctuations can larger or smaller than the random choice benchmark. Since Ω^μ is the aggregate sum of the common parts between every agent's two strategies, canceling all the Ω^μs are obtained by giving two fully opposite strategies to each agent [16,43,44].

3.2.2. Results from replica calculus

The inner workings of the method are explained by Parisi in Chapter 3. Let us only sketch the major steps, and above all the results of the method. The quantity to minimize is $H(\{m_i\}, \{a\})$, where the variables $\{m_i\}$ will be replicated and the strategies (the disorder) $\{a\}$ is kept constant. One wishes to compute

$$Z(\beta, \{a\}) = \text{Tr}_m e^{-\beta H(\{m_i,a\})=} \tag{3.6}$$

where β is an inverse temperature and $\{a\}$ represents a particular realization of the disorder. The minimum of H corresponds to the $\beta \rightarrow \infty$ limit

$$\min_{\{m_i\}} H(\{m_i, a\}) = -\lim_{\beta \rightarrow \infty} \frac{1}{\beta} \ln Z(\beta, a). \tag{3.7}$$

The MG is a self-averaging system. In other words, min H does not depend on the realization of the disorder in the limit $N \rightarrow \infty$. As a consequence, in numerical simulations, one can choose either to average relevant quantities over a few samples of large systems or over a large number of small systems. In practice, one should favor few samples of large systems in order to avoid as much as possible finite-size effects, and at any rate study the latter on their own. Reversely, computing $\langle \min H \rangle_{\{a\}}$ in this limit gives the correct result. The difficulty of computing $\langle \ln Z \rangle_{\{a\}}$ is overcome by using the replica ansatz

$$\langle \ln Z \rangle_{\{a\}} = \lim_{n \rightarrow 0} \frac{1}{n} \ln \langle Z^n \rangle_{\{a\}} \tag{3.8}$$

for which one needs to introduce a set of variables m_i for each replica $c = 1, \ldots, n$, denoted by $\{m_i^c\}$. Since H is positive by definition, one uses the symmetric ansatz, and finally obtain the exact result for $\alpha > \alpha_c$

$$\lim_{N \rightarrow \infty} \frac{1}{N} \left\langle \min_{\{m_i\}} H\{m_i\} \right\rangle_{\{a\}} = \frac{1 + Q}{2(1 + \chi)^2}, \tag{3.9}$$

where the parameters Q and χ are given by

$$Q(\zeta) = 1 - \sqrt{\frac{2}{\pi}} \frac{e^{-\zeta^2/2}}{\zeta} - \left(1 - \frac{1}{\zeta^2}\right)\text{erf}\left(\frac{\zeta}{\sqrt{2}}\right) \tag{3.10}$$

$$\chi(\zeta) = \frac{\text{erf}\left(\frac{\zeta}{\sqrt{2}}\right)}{\alpha - \text{erf}\left(\frac{\zeta}{\sqrt{2}}\right)} \tag{3.11}$$

and ζ is determined by

$$\alpha = \left[1 + Q(\zeta)\right]\zeta^2, \tag{3.12}$$

Since $Q = \sum_i m_i^2/N \geq 0$, $H = 0$ can only happen if $\chi = \infty$, which in turn needs $\alpha = \text{erf}\left(\frac{\zeta}{\sqrt{2}}\right)$, which determines α_c. In order to plot H, one should read backward the above equations, and note that Eq. (3.12) is the only equation to solve numerically. The easiest and by far fastest way to do it is solving iteratively the fixed point equation $\zeta_{n+1} = \sqrt{\alpha/[1 + Q(\zeta_n)]}$ with $\zeta_0 = 1$. From Eqs (2.20) and (2.21) one sees that H and σ^2 are related for infinite systems according to

$$\sigma^2 = H + \frac{1}{2}(1 - Q) + \sum_{i \neq j} J_{i,j}\langle(s_i - m_i)(s_j - m_j)\rangle. \tag{3.13}$$

The best one can do with this approach is to neglect the last term, which is not relevant in the asymmetric phase. In the symmetric phase however, it is responsible for the emergence of large fluctuations. H and σ^2 are reported in Fig 2. Finally, distribution of strategy use $P(m)$ is

$$P(m) = \frac{\phi}{2}\delta(m + 1) + \frac{\phi}{2}\delta(m - 1) + \frac{\zeta}{\sqrt{2\pi}}e^{-\zeta^2 m^2/2}, \tag{3.14}$$

where the fraction of frozen agents $\phi = \text{erfc}(\zeta/\sqrt{2})$.

3.3. Dynamics: generating functionals

Basing an analytical solution on a minimized quantity is static in essence but works remarkably well as long as one can find it[7] and it is not zero. Generating functionals, first used by Coolen and Heimel in the MG context [23,30] allow for a much deeper understanding of the dynamics of the MG, at the price of more intricate calculus. Still, the idea behind generating functionals is wonderfully simple and consists in a Generalisation of the venerable generating function. Assume that one is interested in some quantity $x_i(t)$ in a system whose dynamical variables are $\{y_i(t)\}$. The dynamics of $x_i(t)$ is obviously fully determined by that of $\{y_i(t)\}$, hence, one should restrict the generating functional to the possible paths of $\{y_i(t)\}$. In addition, one wishes to be able to extract information about

---

[7]This is not possible for MG with finite score memory [19].

each time step and each x_i, which requires one auxiliary variable per time step. Therefore the generating functional is given by

$$Z[\{x_i\}] = \left\langle \exp\left[\sum_{i,t} \hat{x}_i(t)x_i(t)\right]\right\rangle_y, \tag{3.15}$$

where $\hat{x}_i(t)$ is the auxiliary variable of $x_i(t)$ and the average $\langle \cdot \rangle$ is over all possible paths of $\{y_i(t)\}$. Therefore, one needs to determine the time evolution of $P(\{y_i(t)\})$, which is easier now that the histories are Markovian: by Chapman-Kolmogorov

$$P(\{y_i(t+1)\}) = \int \prod_j dy_j(t) P(\{y_i(t)\}) W(\{y_i(t)\} \to \{y_i(t+1)\}). \tag{3.16}$$

The transition probability density W is by definition

$$W(\{y_i(t)\} \to \{y_i(t+1)\})$$
$$= \prod_i \delta\left(y_i(t+1) - y_i(t) + 2\xi_i^{\mu(t)}\left[\Omega^{\mu(t)} + \sum_j \xi_j^{\mu(t)} s_j(t)\right]\right). \tag{3.17}$$

One readily sees that it is enough to write the Dirac functions in their exponential representation to transform W into a product of exponentials, whose argument is linear in Ω and ξ_i. Therefore, averaging Z over the strategies is straightforward. As a consequence, the dynamics is also disorder-averaged, and becomes the one of all the games in the limit $N \to \infty$. In this limit, the system evolution is represented by a single representative equation

$$y(t+1) = y(t) + \alpha \sum_{t' \leq t}(1+G)_{t,t'}^{-1}\text{sign}[y(t')] + \sqrt{\alpha}\eta(t), \tag{3.18}$$

where G is the response function of the system at time t to a perturbation occurring at time t'.[8] The denomination 'representative' should not mislead the reader into thinking that this mathematical method brings us back to the Economics representative agent approach. Instead, this equation contains the dynamics of *all* the possible agents. This is because the noise η has long memory; each of its realization gives a different set of properties to $y(t)$. For instance, a frozen agent will have $y(t)/t > 0$ in the limit $t \to \infty$. Of course, there will be ϕ such agents. In the stationary state of the asymmetric phase, one recovers understandably the same equations for H, Q and ζ as the ones found with the replica method. This method enlightens us about the dynamics of the MG: first it shows

---

[8]Eq. (3.18) is valid for the batch MG, where the agents sample over all possible histories before possible changing strategies [30].

that χ is the integrated response function, that is, the cumulative change in the system behaviour caused by a small perturbation. The fact that it is infinite in the symmetric phase is related to the sensitivity of the system to initial conditions. The second dynamical insight is about σ^2: it is tricky to compute, and one must resort to reasonable approximations in most cases. As usual, the asymmetric phase yields more easily to mathematical pressure. The fluctuations in the stationary state depend in principle on every single time step in the past, but if one assumes that the non-persistent correlations of the systems decay very fast, one recovers the result of the replica calculus. It is difficult to find a discrepancy between numerical simulations and the theory with the last assumption for large systems and provided that the system is given enough time to equilibrate (of order $200P/(\alpha - \alpha_c)$). Generating functionals also explain rigorously why continuous time equations [37] give the correct prediction for σ^2. However, there is no reasonable approximation in the symmetric phase; this is mainly due to the excessive non-linearity.[9]

3.4. The role of game impact

Remarkably, the only reason why the non-frozen agents keep changing strategies is because they do not take into account their own impact on A. If one modifies the strategy scores update equations in order to discount the self-impact from A, as in the case with no information, one finds that large fluctuations cannot arise anymore and that a Nash Equilibrium is reached: instead of minimizing the predictability H, the agents minimize the fluctuations, or equivalently, their losses [15, 39]. Because of the discrete nature of strategy choice, there are many NE ($O(2^N)$). The one reached by the dynamics is determined by the initial conditions. Replica calculus and generating functionals can be used to solve the resulting model [15, 31].

4. Application and extensions

Now that the MG is well understood thanks to its exact solution, it is possible to try and extend the realm of application of the methods applied to the MG. The first candidate is obviously a simplified version of the El Farol bar problem; one aims at giving answers regarding the converge of the attendance to the resource level.

---

[9]The so-called spherical minority games solve this problem by linearizing $s_i = \text{sign}(\Delta)$ and imposing a constraint on $\sum_i s_i^2$ [26, 27].

4.1. El Farol redux

In El Farol bar, the resource level is $L = 60$ seats for $N = 100$ customers. Using symmetric mathematical framework of the MG, the payoff of action a is $-a(A - K)$ where $K = 2L - 1$. Since this payoff is still linear, the extension of the exact solution is easy. The only technical point of interest is the rescaling of the discrepancy between the a priori (i.e. random strategy choice) average agent choice $\bar{a}N$ and the resource level K: since $A - \bar{a}N = O(\sqrt{N})$ and $K = O(N)$, the thermodynamical limit is problematic unless one rescales properly $\bar{a} - K$. This is done by introducing

$$\gamma = (\bar{a} - K)\sqrt{N}. \tag{4.1}$$

γ is the speed at which the discrepancy $\bar{a} - K$ must decrease in the thermo-dynamical limit. Comparing the exact solution to direct numerical simulations (inevitably with finite N) with resource level K is done by setting γ to the value given by Eq. (4.1).

After some routine calculations, we find that the predictability H/N is given in the asymmetric phase by

$$H = N \frac{\sigma_a^2(1 + Q(\zeta)) + \gamma^2/\alpha}{[1 + \chi(\zeta)]^2}, \tag{4.2}$$

and the fluctuations are equal to

$$\sigma^2 = H + N\sigma_\xi^2[1 - Q(\zeta)]. \tag{4.3}$$

In these two equations, Q and χ are given by

$$Q(\zeta) = 1 - \sqrt{\frac{2}{\pi}} \frac{e^{-\zeta^2/2}}{\zeta} - \left(1 - \frac{1}{\zeta^2}\right) \mathrm{erf}\left(\frac{\zeta}{\sqrt{2}}\right)$$

$$\chi(\zeta) = \left[\frac{\alpha}{\mathrm{erf}(\zeta/\sqrt{2})} - 1\right]^{-1} \tag{4.4}$$

whereas the parameter ζ is uniquely determined by the transcendental equation

$$\frac{\alpha}{\zeta^2} - Q(\zeta) - 1 - 2\frac{\gamma^2}{\alpha} = 0 \tag{4.5}$$

as a function of α and γ. σ_X^2 is the variance of $P(X)$ for $X = a, \xi$.

Therefore, the El Farol bar problem with the strategy ensemble studied here and the MG have the same behavior of fluctuations when $\gamma = 0$, that is, when the average decision over the whole strategy space is equal to the resource level per agent ($N\bar{a} = L$); such a strategy space is *consistent*. The case $\gamma \neq 0$ reduces

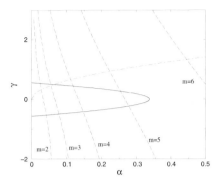

Fig. 4. Phase diagram of the El Farol bar problem. The dashed lines correspond to the trajectories of systems with $L = 60$, $\bar{a} = 1/2$ and $m = 2, \ldots, 6$ as the number of agents increases (from bottom to top). The dot-dashed line corresponds to a typical trajectory of a system with fixed L, N and $\bar{a} > L/N$ as the agents' memory changes.

to $\gamma > 0$ since all quantities depend on γ^2. When γ is small, there are still two phases as illustrated by the phase diagram in Figure 4. The critical line separates the asymmetric phase ($H > 0$) from the symmetric phase ($H = 0$). It crosses $\gamma = 0$ at $\alpha_c(0) = 0.3374\ldots$, the critical point of the standard MG [15,30]; when γ increases, α_c decreases (one needs more and more agents to counterbalance the increasing strategy bias) and $\alpha_c = 0$ for $\gamma = 1/\sqrt{\pi}$.

The meaning of the phase diagram is easy to interpret: $H = 0$ implies $\langle A \rangle = L$. The symmetric phase is the region of parameters where the average attendance converges to the comfort level. This region is also characterized by large collective fluctuations σ^2 for non-biased initial conditions and by a dependence on initial conditions; in particular, the fluctuations decrease if the difference of strategy *a priori* valuation increases as discussed in the MG literature [28, 30, 37, 38]. On the contrary, there is no equality between $\langle A \rangle$ and L in the asymmetric phase if $\gamma > 0$. Since $H = \sigma^2$ when the agents take into account their impact on the game, there is never equality of average attendance and resource level in a pure strategy Nash Equilibrium.

It is therefore questionable to use very inconsistent binary strategies in an El Farol bar problem context because the dynamics might be dominated by spurious binary effects. When the strategies are consistent, every single result on the MG about fluctuations is directly relevant for the El Farol Bar problem.

It did not take much work to extend the solution of the MG to the El Farol bar problem. But this ended the search for new mathematical methods in Economics that would be able to solve such models [7]. El Farol has always been thought as the flagship of broader class of models; to what extend these mathematical methods are still applicable and how universal the fluctuations/predictability properties of these models are is a central issue to be explored in the rest of this chapter.

4.2. Global ultimatum game

So far, we have seen how to generalize the calculus from the $-aA$ payoff, to a $-a(A - L)$ payoff. The next level of difficulty is to have a time-evolving resource level. Instead of designing an *ad hoc* mechanism that changes L according to some rule, be it deterministic or stochastic, but arbitrary nonetheless, let me consider commodity markets, or equivalently global ultimatum games [10] that perfectly show how minority mechanisms emerge when groups of players learn collectively an implicit resource level. In a standard ultimatum game [29], some generous but perverse donor proposes 1\$ to player 1, to be shared with player 2, provided that player 2 accepts the share offered by player 1. Rationality dictates that player 2 should accept any proposal, because he does not receive anything at all otherwise. But experiences show that a human player is not likely to propose nor to accept arbitrarily low proposals, as they are perceived as unfair. Mathematically, player 1 proposes a and player 2 expects at least b. Then player 1's payoff is

$$(1 - a)\Theta(a - b) \tag{4.6}$$

and the one of player 2 is

$$a\Theta(a - b) \tag{4.7}$$

where $\Theta(x) = 1$ if $x > 0$ and 0 otherwise. Suppose now that there is a group of N_A people who have to share $N_A\$$ and a group of N_B people who each expect to have a given amount of money. Mathematically, player i of group A offers a_i, while player j of group B expects at least b_j. The offers and expectations are grouped into

$$A = \sum_{i=1}^{N_A} a_i \quad \text{and} \quad B = \sum_{j=1}^{N_B} b_j. \tag{4.8}$$

A transaction takes place if $A > B$. In that case, player i of group A receives $a_i(N_A - A)/N_A$, and player j of group B receives $b_j B/N_B$. Consider the very likely situation where $A \neq B$. Two types of player can be distinguished: nice ones who want to make A and B converge, and those who try to make them diverge. The nice players of group A have therefore a payoff that rewards small a if $A > B$ and large a if $A < B$; conversely for group-B players. Nice player i of group A rewards decision a_i according to

$$(1 - a_i)\Theta(A - B) + a_i\big(1 - \Theta(A - B)\big) = a_i\big[1 - 2\Theta(A - B)\big] + \Theta(A - B)$$
$$(4.9)$$

Linearizing this payoff gives $-2a_i(A-B)+(A-B)+a_i$. Now, if player i has $S > 1$ strategies $a_{i,1}, \ldots, a_{i,S}$, he will use them according to the difference between their cumulated payoffs. We can therefore drop the last two terms $(A - B) + a_i$; we conclude that strategies can be rewarded according to

$$-a_{i,s}(A - B) \quad s = 1, \ldots, S \tag{4.10}$$

Similarly, player j of group B rewards his strategies with $-b_{i,s}(B - A)$. This is nothing else but a minority mechanism with resource level B. Therefore, B is the resource level of the group A, and A is the resource level of group B: commodity markets are two coupled minority games, whose resource levels vary in time. The fact that each group plays internally a minority game means that the players of a given group compete with each other. For instance, in group A, each agent has an incentive to lower his offer, hoping that his fellows will be generous enough to ensure that $A > B$. But if his fellows offer too little, he has better to offer more. Exact results of the EFPB are readily extended to models with this kind of payoff and have very similar behaviour.

4.3. Speculation

The use of the MG as a model of financial speculation is historically due to the appealing relationship between the predictability H and the fluctuations $\langle A^2 \rangle$, suggesting that wild price fluctuations occur because financial markets are hard to predict. It is therefore tempting to redefine action $+1$ as meaning *buy one share* and -1 as *sell one share*, and to study the properties of A regarded as an excess demand and causing a price change; a common assumption is that of linear price impact function

$$\log p(t + 1) = \log p(t) + A(t). \tag{4.11}$$

$P(A)$ in the standard MG has unremarkable properties in comparison to those of financial markets: (anti-)crashes in the latter are reflected into fat tails $P(A) \propto |A|^{-\gamma}$ for large $|A|$, with $\gamma \simeq 4$ [6, 24, 35], whereas $P(A)$ in the standard MG is essentially Gaussian, possibly with additional Gaussian peaks centered symmetrically around the central peak. What it takes to produce power-law tailed $P(A)$ is to give to the agents the liberty of not playing the game if they perform poorly [17, 33, 46].[10] The resulting model can also be solved exactly [14, 19].

---

[10]This family of models is known under the name of grand canonical MG, in an analogy with models of statistical physics where the number of particles in a system is not constant.

This may be surprising: intuitively, when some agents do not play at each time step, the mean-field nature of the model seems at risk; this is not the case because all the agents *update* their score at each time step: keeping up to date is a way of interacting, although indirectly.

These studies have emphasized the role of information: since the MG is a negative sum game, all the agents eventually withdraw from the game as there is nothing to gain. If the agents have too poor cognitive abilities, for instance if their score memory is finite [19, 33], they measure their own performance in a noisy way; these score fluctuations may trick them sometimes into thinking that they are doing well. Assuming that the agents have infinite score memory, the only possibility to maintain the game alive is to add another type of market participants who are less adaptive and bring some information content to the market that standard agents can exploit: for the sake of simplicity, the new agents only have one strategy and always play it; the rationale behind their behaviour is that they need the market and use it for other purposes than speculation, therefore devoting less attention than speculators to market timing; following Ref. [16,48], we shall call them producers; thinking about them as external fields is a good physical representation of their action. It it clear that they are a priori exploited by the adaptive agents However, the ecology of information (who exploits whom) is subtle. The relationship between the two types of agents is best described as a symbiosis. Indeed, by exploiting the information content left by the producers, the speculators reduce the fluctuations, thereby reducing the losses of the producers. When too many speculations are in the market, they over-exploit the gain opportunities and suffer. Grand canonical games ensure that the average number of active speculators adapts to the initially exploitable arbitrage.

These simple models give a scenario for the emergence of large fluctuations of market prices. They suggest that large fluctuations are made possible by the absence of exploitable information. When the signal-to-noise ratio H/σ^2 is too small, large fluctuations arise [14, 18]. Since $H \to 0$ in standard games, these models link market crashes to a phase transition which corresponds to the frontier of efficient markets. Other agent-based models (e.g. [21]) of financial markets produce fat-tailed price changes when tuned to a phase transition, but MG-based models give a crucial insight about the nature of the phases.

Price return prediction is an interesting part of the research on MG and financial markets. Clearly, fitting an agent-based model to a market is more complex than a normal model. Using an agent-based model to predict future price behaviour first requires the ability of reverse-engineering the model itself. The method proposed by Ref. [34] consists in determining the parameters and reconstructing the strategies of the agents (which is possible only to a limited extend) from a given time series of the model. When one manages to achieve it satisfactorily, one assumes by a leap of faith that the MG is a model of financial markets and then reverse-engineer the time-series of real markets. Remarkably, there periods when the market, as viewed by the model, is truly unpredictable, and periods where large predictable price changes occur over a few time steps [3, 34]; sadly, half of the time, only the amplitude of the change is really known, not the sign. Nevertheless, this idea has been used in order to predict when to discount ketchup [41].

A slightly embarrassing feeling should remain at this stage: are financial markets really minority games? This question has been debated for a while. Two preliminary answers were found: first, Ref. [36] showed that the traders *believe* that the market is a minority game if they are fundamentalists. In other words, if they believe that after a deviation, the price will revert to its 'fundamental' value; trend followers on the other hand believe in majority games. This is a first clue that minority games do not describe markets as a whole. Refs [2, 5] note that if one buys at time t, one should be rewarded if the price increases as time $t + 1$, hence, the payoff should rather be $-a_i(t)A(t + 1)$. However, the most correct way to characterize a market is that of *escape game* [11, 46]: one wishes to anticipate twice the crowd, first by buying before the majority, and then by selling before the majority: the competition lies in both the sign and the timing of the transactions. Still, intuitively, the minority game and its parabola of competition must be present in a market. The final answer comes from writing down explicitly the gain obtained from a transaction [12]. If a trader submits an order to buy a shares at time t, his transactions are executed at price $t + 1$. Now assume that he closes his position by selling a shares at time t'. His gain is given by

$$a\big[p(t' + 1) - p(t + 1)\big] = -aA(t) - (-a)A(t') + a \sum_{\tau=1}^{t'-t-1} A(t + \tau) \quad (4.12)$$

One immediately recognizes a minority game in the first two terms of the right hand side, which correspond to the transactions: one obtains a worst than expected gain if the majority of people are submitting orders of the same kind at time t. The last r.h.s term denotes a *delayed* majority game, that is, a majority game to which one does not take part, since $A(t + \tau)$ does not include any transaction of the trader.[11]

It is clear at this stage that one cannot use the strategy space of the minority game any more. Indeed, the MG, as the majority of current agent-based models, has a strategy and reward space that does not allow explicitly correlated actions, such as buy, hold, sell. The question of the strategy space is therefore of crucial importance if one wishes to implement the realistic payoff of Eq. (4.12). Ref. [11] proposed to keep the notion of states of the world, the heterogeneity lying in the

---

[11]For an extension of this discussion to asynchronous actions, including limit orders, see Ref. [11].

subset of states that each trader can recognize. Remarkably, the resulting model, although having little to do with the MG, shows similar information ecology. In my opinion, this means that the MG is a model of gain opportunity exploitation, not of price returns. In other words, A is merely a deviation from perfect exploitation of a given opportunity, $A < 0$ meaning that the latter is under-exploited, and vice-versa. In this sense, the MG is a real model of information ecology and can be used as such. The states of the world μ must therefore be understood as labels of given opportunities, which are inevitably inter-temporal (cf. Eq. (4.12)).

5. Conclusions

The Minority Game combines two important features: its mechanism makes it the prototype model of competition, and it is exactly solvable. Even more, any system where the agents learn collectively a resource level, be it explicit or implicit, contains a minority mechanism.

References

[1] A. De Martino, I. Perez Castillo and D. Sherrington. On the strategy frequency problem in batch minority games. 2006. preprint physics/0611188.

[2] J. V. Andersen and D. Sornette. The $-game. *Eur. Phys. J. B*, 31:141, 2003. cond-mat/0205423.

[3] J. V. Andersen and D. Sornette. A mechanism for pockets of predictability in complex adaptive systems. *Europhys. Lett.*, 70:697, 2005.

[4] B. W. Arthur. Inductive reasoning and bounded rationality: the El Farol problem. *Am. Econ. Rev.*, 84:406–411, 1994.

[5] J.-P. Bouchaud, I. Giardina, and M. Mézard. On a universal mechanism for long ranged volatility correlations. *Quant. Fin.*, 1:212, 2001. cond-mat/0012156.

[6] J.-P. Bouchaud and M. Potters. *Theory of Financial Risks*. Cambridge University Press, Cambridge, 2000.

[7] J. L. Casti. Seeing the light at El Farol. *Complexity*, 1:7, 1995/1996.

[8] A. Cavagna. Irrelevance of memory in the minority game. *Phys. Rev. E*, 59:R3783–R3786, 1999. cond-mat/9812215.

[9] A. Cavagna et al. A thermal model for adaptive competition in a market. *Phys. Rev. Lett.*, 83:4429–4432, 1999.

[10] D. Challet. Minority mechanisms in models of agents learning collectively a resource level. *Physica A*, 344:24, 2004.

[11] D. Challet. Inter-pattern speculation: beyond minority, majority and $-games. *to appear in J. Econ. Dyn. and Control.*, 2006. physics/0502140.

[12] D. Challet and T. Galla. Price return auto-correlation and predictability in agent-based models of financial markets. *Quant. Fin*, 2005. submitted, cond-mat/0404264.

[13] D. Challet and M. Marsili. Relevance of memory in minority games. *Phys. Rev. E*, 62:1862, 2000. cond-mat/0004196.

[14] D. Challet and M. Marsili. Criticality and finite size effects in a realistic model of stock market. *Phys. Rev. E*, 68:036132, 2003.

[15] D. Challet, M. Marsili, and R. Zecchina. Statistical mechanics of heterogeneous agents: minority games. *Phys. Rev. Lett.*, 84:1824–1827, 2000. cond-mat/9904392.

[16] D. Challet, M. Marsili, and Y.-C. Zhang. Modeling market mechanisms with minority game. *Physica A*, 276:284, 2000. cond-mat/9909265.

[17] D. Challet, M. Marsili, and Y.-C. Zhang. Stylized facts of financial markets in minority games. *Physica A*, 294:514, 2001. cond-mat/0101326.

[18] D. Challet, M. Marsili, and Y.-C. Zhang. *Minority Games*. Oxford University Press, Oxford, 2005.

[19] D. Challet, A. D. Martino, M. Marsili, and I. P. Castillo. Minority games with finite score memory. *J. Stat Mech: Exp and Theory*, 2005. cond-mat/0407595.

[20] D. Challet and Y.-C. Zhang. On the minority game: analytical and numerical studies. *Physica A*, 256:514, 1998. cond-mat/9805084.

[21] R. Cont and J.-P. Bouchaud. Herd behaviour and aggregate fluctuation in financial markets. *Macroecon. Dyn.*, 4:170, 2000.

[22] A. A. C. Coolen. Generating functional analysis of minority games with real market histories. *J. Phys. A: Math. Gen.*, 38:2311–2347, 2005.

[23] A. A. C. Coolen. *The Mathematical Theory of Minority Games*. Oxford University Press, Oxford, 2005.

[24] M. M. Dacorogna, R. Gencay, U. A. Müller, R. B. Olsen, and O. V. Pictet. *An Introduction to High-Frequency Finance*. Academic Press, London, 2001.

[25] D. Fudenberg and J. Tirole. *Game Theory*. MIT Press, 1991.

[26] T. Galla, A. Coolen, and D. Sherrington. Dynamics of a spherical minority game. *J. Phys. A: Math. Gen.*, 36, 2004.

[27] T. Galla and D. Sherrington. Stationary states of a spherical minority game with ergodicity breaking. *J. Stat Mech: Exp and Theory*, 2005.

[28] J. P. Garrahan, E. Moro, and D. Sherrington. Continuous time dynamics of the thermal minority game. *Phys. Rev. E*, 62:R9, 2000. cond-mat/0004277.

[29] W. Guth, R. Schmittberger, and B. Schwarz. An experimental analysis of ultimatum bargaining. *Journal of Economic Behavior and Organization*, 1982.

[30] J. A. F. Heimel and A. A. C. Coolen. Generating functional analysis of the dynamics of the batch minority game with random external information. *Phys. Rev. E*, 63:056121, 2001. cond-mat/0012045.

[31] J. A. F. Heimel and A. D. Martino. Broken ergodicity and memory in the minority game. *J. Phys. A: Math. Gen.*, 34:L539–L545, 2001. cond-mat/0108066.

[32] K. H. Ho, W. C. Man, F. K. Chow, and H. F. Chau. Memory is relevant in the symmetric phase of the minority game. *Phys. Rev. E*, 71:066120, 2005.

[33] P. Jefferies, M. Hart, P. Hui, and N. Johnson. From market games to real-world markets. *Eur. Phys. J. B*, 20:493–502, 2001. cond-mat/0008387.

[34] N. F. Johnson, D. Lamper, P. Jefferies, M. L. Hart, and S. Howison. Application of multi-agent games to the prediction of financial time series. *Physica A*, 299:222–227, 2001. cond-mat/0105303.

[35] R. Mantegna and H. G. Stanley. *Introduction to Econophysics*. Cambridge University Press, 2000.

[36] M. Marsili. Market mechanism and expectations in minority and majority games. *Physica A*, 299:93–103, 2001.

[37] M. Marsili and D. Challet. Continuum time limit and stationary states of the minority game. *Phys. Rev. E*, 64:056138, 2001. cond-mat/0102257.

[38] M. Marsili and D. Challet. Trading behavior and excess volatility in toy markets. *Adv. Complex Systems*, 3(I):3–17, 2001. cond-mat/0011042.

[39] M. Marsili, D. Challet, and R. Zecchina. Exact solution of a modified El Farol's bar problem: Efficiency and the role of market impact. *Physica A*, 280:522, 2000. cond-mat/9908480.

[40] M. Mézard, G. Parisi, and M. A. Virasoro. *Spin glass theory and beyond*. World Scientific, 1987.

[41] P. A. D. M. Robert D. Groot. Minority game of price promotions in fast moving consumer goods markets. *Physica A*, (350):553–547, 2004.

[42] N. Shayeghi and A. Coolen. Generating functional analysis of batch minority games with arbitrary strategy numbers. 2006. preprint cond-mat/0606448.

[43] D. Sherrington and T. Galla. The minority game: effects of strategy correlations and timing of adaptation. *Physica A*, 324:25–29, 2003.

[44] D. Sherrington, E. Moro, and J. P. Garrahan. Statistical physics of induced correlation in a simple market. *Physica A*, 311:527–535, 2002. cond-mat/0010455.

[45] H. Simon. *The Sciences of the Artificial*. MIT Press, 1981.

[46] F. Slanina and Y.-C. Zhang. Dynamical spin-glass-like behavior in an evolutionary game. *Physica A*, 289:290–300, 2001.

[47] Y.-C. Zhang. Modeling market mechanism with evolutionary games. *Europhys. News*, 29:51, 1998.

[48] Y.-C. Zhang. Towards a theory of marginally efficient markets. *Physica A*, 269:30, 1999.

Course 10

METASTABLE STATES IN GLASSY SYSTEMS

Irene Giardina

INFM-CNR, Department of Physics, University of Rome La Sapienza
and
ISC-CNR, Via dei Taurini 19, 00185 Roma

J.-P. Bouchaud, M. Mézard and J. Dalibard, eds.
Les Houches, Session LXXXV, 2006
Complex Systems
© *2007 Published by Elsevier B.V.*

Contents

1. Introduction

Glassy systems are often said to have a 'complex' landscape, which is responsible for the non trivial behaviour they exhibit. Still, the notion of complex landscape, despite being evocative, remains rather vague and requires a more quantitative characterization.

For mean-field models this can be achieved in a rigorous way. In this case one can compute a mean-field free energy, function of the local order parameter (i.e. the set of local magnetizations $\{m_i\}$ in the case of magnetic systems) whose absolute minima correspond to the thermodynamic states of the system. This functional represents the 'landscape', and the complexity is strictly related to its topological features. The example of spin systems is, in this respect, archetypal. For the Ising model, where local interactions are homogeneous, the topology of the free energy landscape is trivial, even at low temperatures: only two minima exist corresponding to the ferromagnetic phases with positive and negative magnetization. On the other hand, if we consider spin glasses, where local interactions are disordered and glassy behaviour is observed, the topology of the free energy landscape is highly non trivial, with many absolute minima (corresponding to multiple ergodic equilibrium phases), many metastable minima (with free energy *density* larger than the ground state one), and a whole hierarchy of saddle points. As I will describe in these notes, the statistical features of this landscape can be directly connected with the thermodynamic and dynamic behaviour of the system.

When considering finite dimensional systems, a free energy functional cannot be defined in simple way. The notion of 'state' itself may become rather subtle (see [2, 3]). From an intuitive point of view, one can still think of a state as a region of the phase space separated by other regions by a free energy barrier which is *large*. In the mean-field limit this barrier would scale as the size N of the system, and the state, even a metastable one, would remain well separated in the thermodynamic limit. In finite dimension however, a simple argument shows that barriers may scale as N only for the ground states. For metastable states on the contrary they are always finite, even if large. If the system is prepared in a metastable state, sooner or later it will jump in a stable state with lower free energy density via nucleation of the stable phase. In other words, metastable states must have finite lifetime and the definition of state has to include a reference to

377

time. For example, it seems reasonable to consider a metastable region as a state when the equilibration time in that region is much smaller than the nucleation time required to drive the system to the stable phase (which is what happens for supercooled liquids above the kinetic spinodal, see [4]). Or, in an off-equilibrium context, when the nucleation time to the stable phase is larger than the experimental time (as in glassy systems below the glass transition temperature [1]).

One possibility is to look at the energy functional, meaning the potential as a function of the local degrees of freedom, which is well defined also for short-range models. In general, the minima of this landscape are not directly related to states. In some cases however such a link can be established and the topology of this landscape may give important information on the system's behaviour. I will discuss an example in the last section of these notes [5, 6].

2. Mean-field Spin Glasses

Spin Glasses are magnetic systems where the mutual interactions between the spins can be either ferromagnetic or anti-ferromagnetic. This is usually modeled by assuming quenched random variable to represent the interactions. As a consequence of the disorder, frustration effects may be very strong, resulting in multiple ergodic phases at low temperature and off-equilibrium aging behaviour when the system is started from a high energy configuration [7–9].

These models are characterized by the presence of a very large number of metastable states. More precisely, the number of states grows exponentially with the size N of the system, and the number density can be written as

$$\mathcal{N}(f) \sim \exp\left[N\Sigma(f)\right] \tag{2.1}$$

where f is the free energy density of the state, and Σ is an entropic contribution usually referred to as the *complexity*. $\Sigma(f)$ may have different shapes (being monotonous or rather having a maximum) and is generically different from zero in a finite interval $[f_0, f_{end}]$ with $\Sigma(f_0) = 0$.

As anticipated in the Introduction, the states of the system can be identified as minima of a mean-field free energy functional, which in this context is called the TAP free energy $F_{TAP}(m)$ and can be computed via the Gibbs free energy of the system with different techniques [8, 10].

More precisely, a state is defined by the stationarity equation

$$\frac{\partial F_{TAP}(m)}{\partial m_i} = 0 \tag{2.2}$$

together with the the condition that the Hessian matrix $A_{ij} = \partial_i \partial_j F_{TAP}\{m_i\}$, evaluated in the solution of (2.2), has not negative modes.

The existence of a finite entropic contribution due to the presence of many states has interesting consequences on the thermodynamics of the system. Let us indeed compute the partition function. If we label with α one of the stable solutions of (2.2), that is one state, then we have

$$Z = \sum_{\alpha}^{\mathcal{N}} Z_{\alpha} = \sum_{\alpha}^{\mathcal{N}} e^{-\beta F_{TAP}(m^{\alpha})} = \int df \, e^{-\beta N(f - T\Sigma(f))} \tag{2.3}$$

In the thermodynamic limit, the integral is dominated by the maximum f^* of the exponent in the free energy range where the states exist (i.e. where $\Sigma(f) \geq 0$). We thus find for the free energy of the system

$$\frac{F}{N} = f^* - T\Sigma(f^*) \qquad\qquad \beta = \left.\frac{\partial \Sigma(f)}{\partial f}\right|_{f^*} \tag{2.4}$$

If $f^* = f_0$ then thermodynamic equilibrium is dominated by the ground states, which are less numerous than the others ($\Sigma(f_0) = 0$). In this case, which is the standard one in non-disordered models, $F = Nf^*$. On the other hand, if f^* is larger than the lower band edge f_0 many metastable states do happen to dominate the thermodynamics. In this case f^* identifies the free energy density of a single equilibrium state, while the global free energy of the system is decreased by a further finite contribution from the complexity.

Also the dynamical behaviour of the system can be heavily influenced by the presence of metastable states and the asymptotic dynamics can remain trapped above the equilibrium energy level [7].

Overall, when considering the role of metastable states, one can distinguish two distinct typologies of behaviour:

1) Metastable states, despite having a finite complexity, are *not* relevant in any respect.

From a static point of view, the partition function is dominated by the lower band edge and $f^* = f_0 = F/N$ indicating that only the ground states are thermodynamically relevant. For what concerns the dynamics, it approaches asymptotically the equilibrium energy level (even in the out of equilibrium regime) and one has $\lim_{t \to \infty} E(t; T) = U_{eq}(T)$.

As I will argue, this behaviour is related to the topological features of the metastable states. Immediately above the lower band edge, the states turn out to be marginally unstable in the thermodynamic limit, while their structure as a whole is extremely *fragile* to external perturbations.

One model that displays this sort of behaviour is the well-known Sherrington-Kirkpatrick model [11], paradigmatic example of a disordered system, where spins interact in pairs through random couplings. Its Hamiltonian is given by

$$\mathcal{H} = -\sum_{i<j} J_{ij} S_i S_j \tag{2.5}$$

with $S_i = \pm 1$ and the J_{ij} are Gaussian variables with zero mean and variance proportional to N.

2) Metastable states are relevant both for the statics and the dynamics.

The partition function is, in some temperature range, dominated by states with finite complexity and one has $f^* > f_0$; $\Sigma(f^*) > 0$. Typically, at high energy the system is paramagnetic and $f^* = f_{para}$; $\Sigma(f^*) = 0$. As the temperature is decreased metastable states with finite complexity become relevant such that $\Sigma(f^*) > 0$ for $T < T_{MC}$ (the meaning of this temperature will be clarified in the next section). Still decreasing the temperature f^* decreases to hit the lower band edge at the static transition temperature T_c where an entropy crisis (Σ becomes zero) occurs.

The dynamics also 'feels' the presence of metastable states, and the asymptotic energy at low temperature never reaches the equilibrium landscape. Rather, one finds $\lim_{t\to\infty} E(t; T) = E_{th}(T) > U_{eq}(T)$ where $E_{th}(T)$ indicates the energy density of the metastable states with largest complexity, the so-called *threshold* states.

Contrary to the previous case, metastable states are topologically stable and their global structure is *robust* to external perturbations.

A model that displays this behaviour is the p-spin spherical model [12], that has been deeply investigated in the last years due to its simplicity (analytical computations can be easily performed for the statics and the dynamics) and for its phenomenological resemblance to fragile glass forming systems. Its Hamiltonian is given by

$$\mathcal{H} = -\sum_{i_1<i_2<\cdots<i_p} J_{i_1\cdots i_p} S_{i_1} \cdots S_{i_p} \tag{2.6}$$

the spins being real variables with $(1/N)\sum_i S_i^2 = 1$, and the random coupling Gaussian variables with variance proportional to $p!/N^{p-1}$.

3. The complexity

The standard way to investigate the individual and global features of metastable states is to compute the number density of TAP solutions, or, rather, the corresponding complexity.

The number density reads

$$\mathcal{N}(f) = \sum_{\alpha=1}^{\mathcal{N}} \delta\big[F_{TAP}(m^{\alpha}) - Nf\big], \tag{3.1}$$

This expression can be written also in integral form as

$$\mathcal{N}(f) = \int \prod_i dm_i \, \delta(m_i - m_i^{\alpha}) \, \delta\big[F_{TAP}(m)\big]$$

$$= \int \prod_i dm_i \, \delta(\partial_i F_{TAP}(m)) \, \big|\det(A)\big| \, \delta\big[F_{TAP}(m) - f\big] \tag{3.2}$$

where the delta functions implement the TAP equations $\partial_i F_{TAP}\{m_i\} = 0$, and the determinant of the Hessian matrix $A_{ij} = \partial_i \partial_j F_{TAP}(m)$ is needed as normalization. Note that with Eq. (3.2) one is actually counting *all* the TAP solutions with free energy density f (meaning minima and saddles). To infer properties of the states one needs to focus on minima and check a posteriori the stability of the investigated solutions.

Since the systems we are dealing with have quenched disorder in the Hamiltonian, one needs at some stage to perform averages over the disorder distribution. The correct way to do that is to average self-averaging quantities, that is quantities which do not fluctuate much from sample to sample [8,9]. In the present case, one should average the complexity, i.e. the logarithm of the number (quenched average). The alternative procedure, averaging the number first and then taking the logarithm (what is called the annealed average) is only approximate but often much simpler. In any case, expression (3.2) represents the starting point for the analytic computation of the number and the complexity.

In (3.2) the modulus of the determinant is quite hard to handle, from a mathematical point of view. For this reason it may be convenient to consider an approximation where the modulus is dropped from (3.2) and ask to what extent this gives a faithful estimate of the number density. This approximation is *a priori* a dangerous one, since without the modulus each TAP solution is weighted with the sign of the Hessian determinant, with the risk of uncontrolled cancellations. Indeed, if we do not impose any constraint on the free energy density and count all the TAP solutions, dropping the modulus in Eq. (3.2) gives a topological constant by virtue of the Morse theorem, instead of the correct result $\mathcal{N} \sim \exp[N\Sigma]$. There are however a few cases where the approximation is under control. For example, if, at a given free energy density, only a given class of TAP solutions (e.g. minima, or saddles of a given order) dominate, then the Hessian has a well defined sign (up to vanishing corrections), the modulus becomes redundant and can be dropped. This is what happens for models of the second typology previously discussed, where minima are dominant at low energies and saddles at larger energies [13]. A second possibility is that the sign of the determinant is somehow connected to sub-leading terms (in N) and can therefore be disregarded when computing the leading term of (3.2). This is a more subtle case and I will comment briefly on that later on [15].

3.1. The BRST supersymmetry

Let us now proceed and further develop Eq. (3.2). An exponential representation both for the delta functions and the determinant can be used:

$$\prod_i \delta(\partial_i F_{TAP}) = \int_{-i\infty}^{+i\infty} \prod_i Dx_i \exp\left[\sum_i x_i \partial_i F_{TAP}(m)\right]$$

$$\delta(F_{TAP} - f) = \int_{-i\infty}^{+i\infty} Du \exp\left[u(F_{TAP}(m) - f)\right]$$

$$\det(A) = \int_{-\infty}^{+\infty} \prod_i D\bar{\psi}_i D\psi_i \exp\left[\sum_{ij} \bar{\psi}_i \psi_j A_{ij}\right], \qquad (3.3)$$

where $\{\bar{\psi}, \psi\}$ are anti-commuting Grassmann variables, and Dx_i, Du etc. stand for $dx_i/\sqrt{2\pi}$, $du/\sqrt{2\pi}$ etc. [14]. In this way we can write

$$\mathcal{N}(f) = \int Du e^{-\beta uf} \int \mathcal{D}m \, \mathcal{D}x \, \mathcal{D}\bar{\psi} \, \mathcal{D}\psi \, e^{\mathcal{L}(m,x,\bar{\psi},\psi)}, \qquad (3.4)$$

with the effective Lagrangian $\mathcal{L}$ given by,

$$\mathcal{L}(m, x, \bar{\psi}, \psi) = \sum_i x_i \partial_i F_{TAP}(m) + \sum_{ij} \bar{\psi}_i \psi_j A_{ij} + u F_{TAP}(m). \qquad (3.5)$$

A key property of the Lagrangian (3.5) is its invariance under a generalization of the Becchi-Rouet-Stora-Tyutin ($BRST$) supersymmetry [16] if ϵ is an infinitesimal Grassmann parameter, it is straightforward to verify that (3.5) is invariant under the following transformation,

$$\delta m_i = \epsilon \psi_i \qquad \delta x_i = -\epsilon u \psi_i \qquad \delta \bar{\psi}_i = -\epsilon x_i \qquad \delta \psi_i = 0. \qquad (3.6)$$

The $BRST$ invariance has interesting consequences on average values. The fact that $\delta\mathcal{L} = 0$ under the $BRST$ supersymmetry implies that the average of any observable of the same variables performed with this Lagrangian must be invariant too. For an observable $\mathcal{O}$, one has $\langle\delta\mathcal{O}\rangle = 0$, where brackets indicate an average with the measure defined in (3.4). This property can be used to generate some useful Ward-Takahashi identities. For example, by setting $\mathcal{O} = m_i \bar{\psi}_j$ and $\mathcal{O} = x_i \bar{\psi}_j$ one gets

$$\langle \bar{\psi}_i \psi_j \rangle = -\langle m_i x_j \rangle \qquad - \beta u \langle \bar{\psi}_i \psi_j \rangle = \langle x_i x_j \rangle. \qquad (3.7)$$

These $BRST$ identities do play a crucial role in the computation of the complexity and of TAP averages. More than that, they not only are mathematical relations between abstract parameters, but also have a deep physical meaning which is crucial to understand the structure of metastable states.

4. Supersymmetry breaking and structure of the states

Starting from from Eq. (3.2) the computation can be performed by using standard techniques to deal with the disorder average [8]. Eventually, thanks to the long range nature of the model, the saddle point method can be used to get an explicit expression for $\Sigma(f)$. The same computation also allows to evaluate average values with the measure defined by the Lagrangian (3.5), that is, going backward to Eq. (3.2), averages over the set of metastable states with free energy density f. Among other things, the identities (3.7) can be explicitly checked. Here comes the surprise. Indeed it turns out that, while for some models these identities are satisfied – as expected, in other ones they are not [19]. In other words, there are cases where the symmetry exhibited by the Lagrangian (3.5) is *spontaneously broken* by the saddle point.

To better understand why the supersymmetry is broken and how this is related to the structure of states one needs to go back to the Ward-Takahashi Identities. Let me focus for example on the first equation in (3.7). From the Lagrangian as in (3.5) one can write $\langle \bar{\psi}_i \psi_j \rangle = \langle A_{ij}^{-1} \rangle$, identifying the second term of the first equation in (3.7) as one component of the inverse Hessian. A direct physical interpretation to the first term of this equation can also be given. Let's add a small magnetic field h_i to the stationarity equations (2.2) (but still weighting the states with the unperturbed free energy). Then, by taking the derivative of the average local magnetization with respect to the field h_j, one gets $(d \langle m_i \rangle / dh_j)|_{h=0} = -\langle m_i x_j \rangle$. Here, again, all averages are performed with the measure appearing in Eq. (3.4), that is they are averages over all the TAP states with free energy density f.

At this point the Ward-Takahashi identity can be re-casted into a relationship between physical quantities:

$$\left. \frac{d \langle m_i \rangle}{dh_j} \right|_{h=0} = \langle A_{ij}^{-1} \rangle, \qquad (4.1)$$

which is nothing but the *average* static fluctuation-dissipation theorem.

Note that for each individual state (i.e. minimum) this is the very natural relationship between the susceptibility and the local curvature. It is immediately obtained by deriving twice with respect to an external magnetic field the TAP equation for an individual solution, and there is no a priori clear reason why it should be violated. The non-trivial feature of Eq. (4.1) is that it represents a relationship between observables *averaged* over the ensemble of metastable states, and not between intra-state observables. This, as I will discuss, is the crucial point to understand why supersymmetry may be broken and the Ward-Takahashi identities not verified.

Interestingly, there is a strict correspondence between the two classes described in section 2 and the breaking or not of the supersymmetry. It may then be useful to describe more in details what happens for the two archetypal models quoted for each of the two classes.

4.1. The case of the SK model

The complexity of the SK model had been computed more than twenty years ago [18] leading to a well-defined $\Sigma(f)$ in a range $[f_0(T), f_{end}(T)]$. Still, it remained as a sort of mystery why, despite the abundance of metastable states above the lower band edge, only states with $f = f_0$ do in fact matter for the statics and the dynamics. More recently [19], it has been pointed out that the SK complexity breaks the BRST supersymmetry at all free energy densities but f_0, thus violating the Ward-Takahashi identities.

More detailed analysis both analytical [20] and numerical [21], have shown that at low temperatures all the stationary points of $F_{TAP}(m)$ are organized into minimum-saddle pairs. The minimum and the saddle are connected along a mode that is the softer the larger the system size N. Moreover, the free energy difference of the paired stationary points decreases with increasing N. In other words, metastable states, despite being topologically stable at any finite N, become *marginal* (more precisely, marginally unstable) in the thermodynamic limit, having at least one zero mode. This means that, at any $f > f_0$, minima and order one saddles are completely mixed and the Morse theorem is obeyed for any free energy density. One may wonder whether the approximation of dropping the modulus would be reasonable in this case. Surprisingly however, the complexity computed in this way does actually lead to the correct result. All the information on the sign of the counted stationary points is indeed encoded in the prefactor which links $\mathcal{N}(f)$ to $\exp[N\Sigma(f)]$ and goes beyond the saddle point contribution [15, 22].

Let me now focus on the global structure of the states. Even if for any large but finite N there are exponentially many stable states, since they have one soft mode, an infinitesimal $O(1/N)$ external field may destabilize some of them, causing the merging of a minimum with the paired saddle and making the states disappear. On the other hand, virtual states, i.e. inflection points of the free energy with a very small second derivative, may be stabilized by the field, giving rise to pairs of new states. In other words, the structure of metastable states is fragile and extremely sensitive to external perturbations.

In such a situation the validity of equation (4.1) may be compromised. Even if the fluctuation-dissipation relation holds inside each given state, the same may not be true when averages over the states are considered, since the number of metastable states can vary dramatically when a small field is applied. If one considers all the states with free energy density f, the average magnetization can be written as $\langle m_i \rangle = (1/\mathcal{N}(f,h)) \sum_\alpha m_i^\alpha$, where $\mathcal{N}(f,h)$ is the number of states with free energy density f, in presence of a magnetic field h. Therefore, at the l.h.s. of equation (4.1) we differentiate with respect to an external field a sum over all metastable states. The problem is that, due to marginality, some elements in this sum may disappear or appear as the field goes to zero. More precisely, we have,

$$\frac{d\langle m_i \rangle}{dh_j}\bigg|_{h=0} = \lim_{h \to 0} \frac{1}{h_j} \left\{ \frac{1}{\mathcal{N}(f,h)} \sum_\alpha m_i^\alpha(h) - \frac{1}{\mathcal{N}(f,0)} \sum_\alpha m_i^\alpha(0) \right\}$$

$$\neq \lim_{h \to 0} \frac{1}{\mathcal{N}(f)} \sum_\alpha \frac{1}{h_j} \{ m_i^\alpha(h) - m_i^\alpha(0) \} = \langle A_{ij}^{-1} \rangle. \tag{4.2}$$

The key point is that the elements in the two summations of (4.2) may be different, because of the action of the field. Therefore, an anomalous contribution arises due to the instability of the whole structure with respect to the field and the fluctuation-dissipation relation (4.1) is *violated*.

4.2. The case of the p-spin model

For this model the supersymmetry and the related identities are always satisfied [23]. One can investigate in much detail the structure of metastable states, also thanks to the special feature that states can be transposed in temperature in the low temperature region (no mixing of states in energy occur).

Metastable states occur within a finite range of free-energies $[f_0(T), f_{th}(T)]$. Here $\Sigma(f_0) = 0$ and $\Sigma(f_{th}) = \Sigma_{max}$. These states correspond to minima of the TAP mean-field free energy, and their geometrical stability can be explicitly checked by looking at the TAP Hessian. Indeed the density of eigenvalues always has positive support, touching the zero only at the threshold level f_{th}, where marginal modes occur. Above f_{th} minima are sub-dominant and saddle points are entropically the relevant TAP solutions [17,25].

Since individual states can be mapped to zero temperature configurations, the same scenario holds for the energy landscape, that is for the Hamiltonian as a function of the configuration point. An intuitive way of understanding this is that states are like valleys in the energy landscape: the zero temperature limit gives back the bottom of the valley, a *bare* value for the energy, while temperature 'dresses' it up with thermal fluctuations, broadening the measure on configurations higher than the bottom and giving larger entropy.

Thinking in terms of energy rather than free energy has the advantage that certain features may be generalized to short-range models, where only the energy landscape can be directly investigated (see section 5.2). Also in terms of energy, the model is characterized by a sharp decoupling of the stationary points, with minima dominating above the threshold level $E_{th} = f_{th}(T = 0)$, and saddles dominating above it. One quantity that well describes this behaviour is the index density $k(E)$ of the typical stationary point at given energy density E, as a function of the energy. More precisely, this quantity is defined as follows. One considers all the stationary points of the Hamiltonian that have energy density equal to E. For each stationary point the density of eigenvalues of the Hessian is computed: the number of negative modes defines the index K in that point (zero index corresponding to minima, non-zero index to the order of the saddle, index equal to N to maxima). Then the average index density is obtained by averaging $k = K/N$ over all the stationary points at that energy. The index function can be easily computed analytically and one gets the behaviour represented in the upper panel of Fig. 1. The whole distribution of the index can be computed analytically for this model [24] showing that the two regions above and below the threshold are statistically different. All this indicates that a well defined *topological* transition takes places in energy at the threshold level.

Interestingly, this topological transition has a strict connection with the dynamical transition exhibited by this model. The p-spin Langevin dynamics can be analytically computed [7], it turns out that two-points functions obey exactly Mode Coupling equations and a purely dynamical transition occurs at the Mode Coupling temperature T_{MC}.

To find the link between the topological transition and the dynamical one, one needs to associate to each energy (the variable driving the topological transition in the energy landscape) a temperature (the parameter driving the dynamical transition). To this aim, one must ask what are the typical stationary points that populate the landscape which is asymptotically visited by the system. If the system was to equilibrate, this landscape would be the equilibrium one, which, at low temperature, is dominated by metastable states. The typical stationary points would then be the minima of the Hamiltonian corresponding to the bottom of the valleys representing such states. The energy we are looking for is then the bare energy of these minima which is obtained by subtracting from the equilibrium energy the vibrational contribution of thermal fluctuations. This argument can be generalized also to the case (larger temperatures) where the equilibrium landscape is populated mostly by saddle points. These saddle points are the closest stationary points to equilibrium configurations and, again, their energy is obtained by subtracting from the equilibrium value a contribution due to thermal fluctuations (which can be analytically evaluated [27]). The mapping then associates T with $E_{bare}(T) = U_{eq}(T) - E_{therm}(T)$.

With this mapping, one finds that the threshold energy E_{th} corresponds to the critical temperature T_{MC}, that is

$$E_{bare}(T_{MC}) = E_{th} \qquad (4.3)$$

Larger temperatures correspond to larger energies, and lower temperature to lower energies. That is, the topological transition occurring in energy corresponds to the dynamical transition occurring in temperature.

The physical interpretation of this fact is the following. If the system is prepared in a high energy configuration, its dynamical evolution initially explores regions of the landscape rich in descending directions (recall that at high energies the typical stationary points are saddles with large index). Asymptotically the system wishes to equilibrate. If the temperature at which the dynamics takes place is larger than T_{MC}, the equilibrium manifold still lies in the region dominated by saddles. Thus, the dynamics always finds paths to reach it and $\lim_{t \to \infty} E(t; T) = U_{eq}(T)$. On the other hand, if $T < T_{MC}$ the equilibrium landscape is in the region dominated by minima. During its evolution in the landscape, while decreasing the energy, the system reaches the threshold level, where minima starts becoming relevant and marginal modes are abundant [26] – *before* approaching the equilibrium manifold, and remains trapped. This leads to $\lim_{t \to \infty} E(t; T) = E_{th}(T) > U_{eq}(T)$.

5. Models in finite dimension

So far, I have discussed in details the relevance of metastable states for mean-field models of spin-glasses. As shown with explicit examples, two possible behaviours may arise:

i) A first one where a mixing of stationary points at any energy level occurs, metastable state have a quasi-soft mode at finite N becoming marginally unstable in the thermodynamic limit and their global structure is extremely fragile to external perturbations. Metastable states are *not* relevant and have no influence on the behaviour of the system.

ii) In the second case metastable states are topologically stable even in the thermodynamic limit, their global structure is robust and the static fluctuation-dissipation theorem holds also on average (the supersymmetry is not broken). They play a crucial role in the static and dynamics of the system. In particular, there is a topological signature of the dynamic transition.

One may wonder whether some of these features survive and can be investigated also in finite dimensional models. In the following I will describe two examples where this is actually the case.

5.1. A simple one-dimensional model

Let me consider the simple one-dimensional Hamiltonian

$$\mathcal{H}(x) = \frac{1}{2}mx^2 + V(x), \tag{5.1}$$

where the position x is a real variable and the mass m is a parameter. $V(x)$ is a Gaussian random potential, with zero average and variance $\overline{V(x_1)V(x_2)} = G(x_1 - x_2)$, with $G(x) = G(-x)$. The physical properties of this model crucially depends on the behaviour of the average displacement $[\Delta(d)]^2 = \overline{[V(x_1) - V(x_2)]^2} = 2G(0) - 2G(d)$, where $d = (x_1 - x_2)$ is the distance. If $\Delta(d)$ goes to a finite value $\Delta(\infty)$ for $d \to \infty$, then the memory is lost after a finite distance and V is called *short range* (SR). On the other hand, if $\Delta(d) \sim d^\gamma$ ($\gamma > 0$), then the displacement grows indefinitely with d and the potential is *long range* (LR). The N-dimensional mean-field version of this model (where x becomes a N dimensional vector with N going to infinity) has been studied long ago [28], and shows that in the SR case the model exhibits the thermodynamic and dynamic behaviour of typology 2) (p-spin like), while in the long-range case belongs rather to the first class (SK-like).

For the one-dimensional model, when the mass term is small, the number of stationary points (minima and maxima) of the Hamiltonian becomes large and can be computed starting from the analogous of Eq. (3.2), with $\mathcal{H}$ in the place of F_{TAP}. Interestingly, in this simple case the modulus can be handled exactly and the number densities of minima and maxima computed without approximations [29]. The result is consistent with the mean-field scenario: in the SR case, the two distributions are partially decoupled and at any given energy only one stationary point dominate; in the LR case on the other hand the two distributions overlap indicating that a complete mixing of minima and maxima occurs at any energy.

5.2. The case of supercooled liquids

When a liquid is cooled fast enough below its melting point, crystallization is avoided and the system enters the supercooled phase. Relaxation time increases rapidly in this temperature regime, and when it becomes comparable with the largest experimentally accessible time the system falls out of equilibrium, remaining stuck in a disordered phase called *structural glass* [1]. For fragile glass-forming systems the dynamics is well described by the Mode Coupling Theory [32] down to a temperature T_{MC} (close to the experimental glass temperature where the systems gets out of equilibrium) where this theory would locate a dynamical transition and where in fact the system exhibits a sharp dynamical crossover.

An old and well accepted idea [31] is that below this temperature the equilibrium landscape is dominated by minima, or valleys, and dynamics starts being heavily influenced by activation processes. Still decreasing the temperature the system may not have enough thermal energy to surmount barriers and remains stuck in a confined region of the landscape which would correspond to the (off-equilibrium) glass state. These valleys, or confining regions, are extremely numerous and where the system ends in depends on the cooling procedure and entropic arguments. This picture suggests that the features of the landscape, especially at low temperature, may have an important role. *Mutatis mutandis* the scenario resembles what observed for the p-spin model, where MCT is exact and where, below T_{MC}, there are many relevant metastable states (valleys). There are of course crucial differences due to the mean-field nature of the p-spin: metastable states have in this case infinite lifetime, while a glass-former is able to escape any confined region if waiting long enough. Still, one can think that for these systems the structure of the landscape is somewhat similar.

With these premises, one can think to characterize in a quantitative way the link between the landscape and the dynamics in the same way it has been done for the p-spin model and as described in section 2. To this aim one must focus on the index density $k(E)$, and look at its behaviour as a function of the potential energy density E of the saddle point. The way to proceed is conceptually the same as for the p-spin, but technically different, since for most of fragile glass-former models analytical computations of this sort are still far away and one has to resort to numerical simulations. Again, one has to determine the index K of the typical stationary points of the potential energy at given energy E. In practice this is done by a sampling of the potential energy function, where stationary points of any order are collected and classified according to their index and energy. The results for a Lennard-Jones binary mixture of particles [5] and for a soft-spheres binary mixture [6] are shown in Fig. 1, where also the behaviour for the p-spin model is reported for comparison. (Note that in this figure the scale of the energy axes is different in the three panels in order to have E_{th} in the same point; besides the value of E_0 is only for visual reference, since a precise evaluation of this point is numerically demanding).

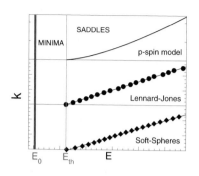

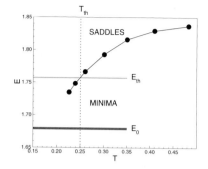

Fig. 1. Index density as a function of the energy density (see text).

Fig. 2. Bare potential energy density as a function of temperature.

As for the p-spin also for these glass formers there exist a sharp decoupling between stationary points and a topological transition occurs in energy between a minima dominated region below E_{th} and a saddle dominated region above E_{th}. The natural question at this point is to understand whether the topological transition represents a signature of the dynamical crossover arising at T_{MC} in the same way it signaled the occurring of the dynamical transition in the p-spin. To ascertain this, one proceeds with mapping each energy to a temperature as in section 2. At low enough temperature, the glass former remains for long timescales confined in one of the numerous 'equilibrium' potential valleys, to eventually jump into another one. The equilibrium energy can then be thought as the bare energy of the bottom of the well plus a vibrational contribution proportional to $K_B T$. Thus, one associates to each temperature T the bare energy $E_{bare} = U_{eq}(T) - 3/2K_B T$. Then, one can determine what is the temperature T_{th} that through this mapping corresponds to the critical value E_{th}. This is shown for a soft-sphere model in Fig. 2 where E_{bare} is plotted as a function of T and compared to the threshold, and the temperature T_{th} is determined.

Molecular dynamics simulations can be performed to determine from the behaviour of the dynamical self-correlation function the temperature T_{MC} where the dynamical crossover and the sharp increase of the correlation time occur. One finds that T_{MC} and T_{th} are consistent one with the other ($T_{th} = 0.24$ and $T_{MC} = 0.242$ for the soft-sphere model and $T_{th} \sim 0.44 \sim T_{MC}$ for the Lennard-Jones mixture). Besides, the index curve can also be used to estimate the average height of the barriers between minima at the threshold level. If we assume that nearby minima are separated by order one saddles then we immediately get $\Delta E = 1/[3k'(E)]|_{E_{th}}$, which gives $5K_B T < \Delta E < 10K_B T$ and indicates that typical barriers between threshold minima are large with respect to thermal fluctuations.

These results show that also for these glass formers the dynamical crossover corresponds to the temperature where the system for the first time approaches the minima dominated landscape. Above T_{MC} the energy landscape visited by the system is rich in descending directions, the main dynamical mechanisms are similar to those arising in the p-spin and MCT well describes the behaviour of the system. Below T_{MC}, the system wishes to equilibrate in a minima dominated region. However it remains dynamically trapped in valleys at the threshold level and, since these valleys are separated by large barriers, it has to resort to thermal activation to escape. T_{MC} therefore represents the first temperature where the system needs to use activation, because negative modes are no longer present and the typical energy barriers are already large. As a consequence a crossover to activated dynamics occurs at this temperature [30].

6. Conclusion

In these notes I have discussed the relevance of metastable states for mean-field spin glass models and some short-range systems.

In the mean-field case many results can be obtained analytically, indicating two quite general scenarios for the structure of the metastable states. Mathematically, these two scenarios correspond to the breaking or not of a supersymmetry arising in the complexity computation. As I have discussed, this supersymmetry has a well defined physical interpretation which is related to the average static fluctuation dissipation theorem and the robustness of the structure of states.

When the supersymmetry is broken, as in the SK model, the structure of states is very fragile to external perturbations. When the perturbation is represented by an external magnetic field, this fragility implies the violation of a Ward identity. Another interesting case arises when the external perturbation is represented by the addition of one spin to the system. This procedure is at the basis of the *cavity method* [8, 33], which has been widely applied in recent years to analyze several kinds of disordered models. If the system is fragile, as for the SK, the cavity method cannot be applied in its standard form, but has to be appropriately modified [34, 35]. Further analysis of this issue for random graphs, diluted systems and short-range models is a program for future research.

When the supersymmetry is not broken, the structure of metastable states is robust and these states are thermodynamically and dynamically relevant. Thanks to numerical simulations, some of the analysis performed on mean-field prototypes can be applied also to some short-range models and I have discussed in these notes the case of glass forming liquids. It appears that for these systems the mean-field scenario of a dynamical transition/crossover at T_{MC} related to a topological transition in the energy still holds.

Acknowledgments

I would like to thank all the people who have worked with me on these subjects, and, in particular, Andrea Cavagna and Giorgio Parisi.

References

[1] C.A. Angell (1995), *Science* **627**, 1924 (1995); F.H. Stillinger (1995), *Science* **267**, 1935; S. Sastry, P.G. Debenedetti and F.H. Stillinger (1998), *Nature* **393**, 554.

[2] C.N. Newman and D.L. Stein (1994), *Phys. Rev. Lett.* **72**, 2286; (1996), *J. Stat. Phys.* **111**, 535.

[3] E. Marinari, G. Parisi, F. Ricci Tersenghi, J. Ruiz-Lorenzo and F. Zuliani, (2000), *J. Stat. Phys.* **98**, 973.

[4] A. Cavagna, I. Giardina and T. Grigera (2003), *Europhys. Lett.* **61**, 74; *J. Chem. Phys.* **118**, 6974.

[5] K. Broderix, K.K. Bhattacharya, A. Cavagna, A. Zippelius and I. Giardina (2000), *Phys. Rev. Lett.* **85**, 5360.

[6] T. Grigera, A. Cavagna, I. Giardina and G. Parisi (2002), *Phys. Rev. Lett.* **88**, 055502.

[7] J.-P. Bouchaud, L.F. Cugliandolo, J. Kurchan and M. Mézard (1997), in *Spin-glasses and random fields*, A.P. Young, Ed. (Singapore, World Scientific).

[8] M. Mézard, G. Parisi and M.A. Virasoro (1987), *Spin Glass Theory and beyond* (Singapore, World Scientific), Vol. 9.

[9] C. De Dominicis and I. Giardina (2006), *Random Fields and Spin Glasses* (Cambridge, Cambridge University Press).

[10] D.J. Thouless, P.W. Anderson and R. Palmer (1977), *Phil. Mag.* **35**, 593; C. De Dominicis (1980), *Phys. Rep.* **607**, 37; T. Plefka (1982), *J. Phys. A* **15**, 1971; A. Georges and J. Yedidia (1991), *J. Phys. A* **24**, 2173; R. Monasson (1995), *Phys. Rev. Lett.* **75**, 2847.

[11] D. Sherrington and S. Kirkpatrick (1975), *Phys. Rev. Lett* **32**, 1792.

[12] A. Crisanti and H.J. Sommers (1992), *Z. Phys. B*, **5**, 805.

[13] A. Cavagna, I. Giardina and G. Parisi (1998), *Phys. Rev. B*, **57**, 11251.

[14] J. Zinn-Justin (1989), *Quantum Field Theory and Critical Phenomena* (Oxford, Clarendon Press).

[15] See also G. Parisi (2006), Preprint cond-mat/0602349, Les Houches Lecture 2005.

[16] C. Becchi, R. Rouet and A. Stora (1975), *Comm. Math. Phys.* **42**, 127; I.V. Tyutin (1975), *Lebedev preprint* FIAN 39.

[17] See T. Castellani and A. Cavagna (2005), *J. Stat. Mech.* P05012 and references therein.

[18] A.J. Bray and M.A. Moore (1979), *J. Phys. C* **12**, L441; A.J. Bray and M.A. Moore, (1980), *J. Phys. C* **13**, L469.

[19] A. Cavagna, I. Giardina, G. Parisi and M. Mezard (2003), *J. Phys. A* **36**, 1175; A. Annibale, A. Cavagna, I. Giardina and G. Parisi (2003), *Phys. Rev. E* **68**, 061103.

[20] T. Aspelmeier, A.J. Bray and M.A. Moore (2004), *Phys. Rev. Lett.* **92**, 087203.

[21] A. Cavagna, I. Giardina and G. Parisi (2004), *Phys. Rev. Lett.* **92**, 120603.

[22] J. Kurchan (1991), *J. Phys. A* **24**, 4969; G. Parisi and T. Rizzo (2004), *J. Phys. A* **37**, 7979.

[23] A. Cavagna, J.P. Garrahan and I. Giardina (1998), *J. Phys. A* **32**, 711.

[24] A. Cavagna, J.P. Garrahan and I. Giardina (2000), *Phys. Rev. B* **61**, 3960.

[25] A. Crisanti, L. Leuzzi and T. Rizzo (2004), *Eur. Phys. J. B* **36**,129.

[26] J. Kurchan and L. Laloux (1996), *J. Phys. A* **29**, 1929.

[27] Andrea Cavagna, Irene Giardina, Giorgio Parisi (2001), *J. Phys. A: Math. Gen.* **34**, 5317.

[28] M. Mézard and G. Parisi (1992), *J. Phys. I France* **2**, 2231; L.F. Cugliandolo and P. Le Doussal (1996), *Phys. Rev. E* **53**, 1525.

[29] A. Cavagna, J.P. Garrahan and I. Giardina (1999), *Phys. Rev. E* **59**, 2808.

[30] A. Cavagna (2001), *Europhys. Lett.* **53**, 490.

[31] M. Goldstein (1969), *J. Chem. Phys.* **51**, 3728.

[32] U. Bengtzelius, W. Götze and A. Sjölander (1984), *J. Phys. C* **17**, 5915.

[33] M. Mézard and G. Parisi (2001), *Eur. Phys. J. B* **20**, 217; *J. Stat. Phys.* **111**, 1.

[34] A. Cavagna, I. Giardina and G. Parisi, (2005), *Phys. Rev. B* **71**, 024422.

[35] T. Rizzo (2005), *J. Phys. A* **38**, 3287; G. Parisi and T. Rizzo, Preprint cond-mat/0411732.

Course 11

EVOLUTIONARY DYNAMICS

Daniel S. Fisher

Lyman Laboratory, Harvard University
Cambridge, MA 02138, USA

J.-P. Bouchaud, M. Mézard and J. Dalibard, eds.
Les Houches, Session LXXXV, 2006
Complex Systems
© *2007 Published by Elsevier B.V.*

Contents

1. Introduction and Questions

The basic laws of evolution have been known for more than a century: heritable variation (Mendel), selection (Darwin's survival of the fittest) and random mutations and sexual recombination to produce the variation. The *genome* of an organism stores the primary information, but selection acts on the *phenome*: the collection of its properties, behavior, etc. With these laws, there is no basic puzzle: given *enough* time *anything* can evolve. Even a cell could arise spontaneously from an extremely rare fluctuation. But somewhere between such absurdly improbable events and small, fast, evolutionary changes in microbes that can be directly observed, are evolutionary processes that can occur on a broad spectrum of time scales: from days in the laboratory to billions of years.

The fossil record and the diversity of existing species illustrates the type of phenomes that can evolve on million and billion year time scales, and recent sequencing data provides the associated genomes. The understanding of *phylogeny* — how organisms are related to and descended from others — is impressive. Yet the understanding of the *dynamics* of evolution, and even what sets the time scales is very poor. Indeed, one could make a good case that the most fundamental puzzles about evolution are the *quantitative* ones, most basically: How is evolution of complex functions, body plans, etc. so fast? Of course, this depends on what one means by "fast": Compared to what expectations?

These lectures start with general questions to motivate and set the stage. The focus then shifts to analysis of some of the simplest aspects of evolutionary dynamics before coming back to broader issues at the end.

1.1. Difficulties

A classic problem, going back to Darwin, is the evolution of an eye. The general view among biologists emphasizes the long times involved. This is expressed quantitatively by Richard Dawkins, one of the great expositors of evolution, who says that an eye could not evolve in a thousand generations, maybe not in a million generations, but clearly could evolve in a billion generations. But where do these numbers come from? The time scales are known from the fossil record, but understanding the ages of fossils comes solely from geology, radiochemistry and physics. Thus Dawkins' and other such statements rely on knowing the answer: there is no understanding from biology or evolutionary theory of the time scales. Recently, there has become available a crude measure of the time scales from biology: information from rates of neutral mutations — ones that do not change proteins – provides estimates of time scales for evolutionary history consistent to within an order of magnitude or better with dating of fossils. But such neutral changes, *ipso facto*, are not evolutionary, thus again they only provide a clock.

A way I like to phrase the primary question to evolutionary biologists is to ask what their reaction would be if they learned that the physical scientists had messed up and really life was 10^{100} years old instead of about 10^{10} years, or, for that matter, $10^{10^{10}}$ years. Would they expect that there would have been much greater diversification? Or evolution of completely new abilities of cells or whole organisms? Some have honestly said that they would not know how to react: the lack of understanding of such quantitative issues is poor enough that they would not have even rough expectations. Indeed, it is not even clear that the total time available — although this is what tends to get emphasized — is the most important quantity. Since the number of evolutionary "experiments" is roughly proportional to the total number of organisms that have ever lived, perhaps this is a better parameter. So if life on earth was a thousand times younger, but the earth a thousand times larger would as complex and diverse organisms have evolved? Or, conversely, if it were a thousand times older but a thousand times smaller (with a similar diversity of environments)? To avoid the issue of things that may have only happened once (for which talking about probabilities is seriously problematic), one can best ask these questions about the time since the first cells. Or, at a later stage, since the origins of multicellular life.

Quantitative questions about evolution on geological time scales are surely very hard to answer. But one can ask similar questions on much shorter time scales and for more modest evolutionary changes. Crudely, what combinations of parameters — and other features — determine what types of evolutionary processes can occur? This, of course, depends both on the biology of the organisms as they exist now, and on the evolutionary history which gave rise to them.

Theodosius Dobzhansky's famous dictum is that "nothing in biology makes sense except in the light of evolution". In contrast to physics, in biology a simpler explanation is no more likely to be right — unless it is simpler in evolutionary rather than functional terms. Thus evolution is the only Ockham's razor in biology — but it is hardly ever used quantitatively. The difficulty of the field and its state of development account for this. Most evolutionary theory — including quantitative modeling — deals with phenomes. Yet these are controlled by genomes which are completely different beasts. And the mapping between genomes and phenomes is extremely complicated containing all the richness of biology and ecology. Population genetics focuses on genomic changes, but mostly either specific small changes or statistical analyses of widespread nearly-neutral variation with information on the phenotypic effects lost.

It is often said that genomic sequence data is "like reading the lab notebooks of nature". But this is an extremely misleading analogy. By the time species (or even clearly unidentifiable strains) have diverged, the number of genetic differences is so large that one cannot extract the significant changes, nor whether these arose as a long series of small phenotypic changes or by a few drastic changes via "hopeful monsters". Thus a much better analogy is that genomes are like indexes of successful textbooks — indiscriminate indexes that randomly mix useful and useless entries — with hints of the original evolutionary "ideas" very hard to extract.

The focus on phenomic evolution can give rise to major misconceptions. Often the genetic bases of phenotypic changes are only considered implicitly: as giving rise to the variability of phenotypic traits on which selection can act. A striking example is a recent estimate of the time to evolve a vertebrate eye. [1,2] This "pessimistic-at-every-step" estimate of a few hundred thousand generations sounds encouraging, but it has a fatal flaw: nowhere does the population size or the mutation rate, or indeed, the genome at all, enter the analysis. All that is considered is selection on *assumed* phenotypic variation, and this is assumed to be sufficient for changes in quantitative traits equivalent to hundreds of standard deviations. Yet this calculation is cited by Richard Dawkins as "stilling Darwin's shudder"! [3] I, for one, find this highly disturbing, especially as it comes together with the dismissal of questions about the lack of quantitative understanding of evolution as an "argument from personal incredulity".

1.2. Prospects

Thanks to the enormous advances in molecular and cell biology, the ability to observe and manipulate — genetically, chemically, and physically — organisms in the laboratory, and the explosion of DNA sequencing technology, we are presented for the first time with the opportunity to greatly expand our understanding of the dynamics of evolution. The goal is to take it from a largely-historical field — which some argue it is intrinsically [4] — to a more fully developed field of science.

1.2.1. Experiments

In order to make progress, a broad spectrum of laboratory experiments are crucial: this necessitates focusing on microbes. Bacteria — more accurately eubacteria and archaea, two very different groups that are often lumped together —

have limited morphological diversity but tremendous catabolic, metabolic, and sensory diversity and in almost any environment some can live. Concomitantly, the genetic diversity of bacteria is enormous (by some measures, the genetic diversity just among *E. coli* is greater than that among all vertebrates). Because many species of bacteria can be grown in the lab and population sizes are large, evolution of a variety of functions and ecologies of interacting species can be studied in the laboratory.

Experimental evolution has, thus far, not been a large field. But there have been a variety of experiments over the years and considerably more recently. Evolutionary experiments on multicellular organisms are primarily selective breeding: sexual recombination of genes in the existing gene pool, sometimes together with a limited number of mutational changes, can be selected on to produce remarkable variation. A the opposite extreme are viruses which have high mutation rates, can evolve rapidly, and have small enough genomes that re-sequencing of multiple strains can be done efficiently. Laboratory evolution of phages, viruses that infect bacteria, is a growing field. But bacteria have the greatest potential for laboratory evolution: they have short generation times and high population densities, tremendous natural diversity, and they can be manipulated genetically in many ways — including adding genes and selectively mutating parts of their genomes. And sequencing costs are now becoming low enough to re-sequence the whole genomes of evolved strains. Combinations of evolving and engineering bacteria as well as individual proteins are being used to develop useful bacterial functions, such as to clean up environmental waste or manufacture particular chemicals.

Laboratory evolution of bacteria with the goal of understanding evolutionary dynamics is beginning to burgeon — if perhaps less so thus far than one might hope. Such experiments go back to Leo Szilard's chemostats in the early 1950s. [5] And the Delbruck-Luria experiments to directly observe the effects of new mutations that occur in the lab probe the most basic evolutionary process. [6] In recent years, Richard Lenski and his collaborators have taken the lead with a spectrum of experiments, mostly with *E. coli* — molecular and cellular biologists' favorite bacterium. Their primary experiment, evolving *E. coli* to grow better in low glucose, has gone on for almost twenty years and 40,000 generations. [7] A wide range of interesting results have come from this one experiment. Some of these could not have even been found without the leaps in biological methods and knowledge that took place while the experiment was being carried out. Unfortunately, there is not the space here for even a cursory review of these or other experiments.

For interpreting and guiding laboratory experiments, theory has already played an important role. But much more is needed, both for this, and more generally to develop a broader understanding of evolution.

1.2.2. Types of theory

There are three general types of theory that are needed to understand evolutionary dynamics. First is phenomenological theory which starts from a mapping — or statistical aspects of it — between some set of genetic changes and the corresponding phenotypic changes: specifically, the effects of these on fitness in some defined set of contexts, including interactions between organisms. For such approaches, the biology is assumed given and the focus is the evolutionary dynamics that this drives. A second type of theory incorporates — and strives to inform — some understanding of aspects of the molecular and cellular biology: for example, evolution of signaling pathways or metabolic or regulatory networks. A third type of theory is abstract modeling: formulation and analysis of simple models that incorporate a few essential features with the goal of developing concrete understanding of these, and — especially crucial for evolution — how to extrapolate over broad ranges of parameters. Such models need have no connection to biology: understanding evolutionary processes in far simpler contexts — e.g. "genetic" algorithms in computer science — should enable a focus on aspects that are well beyond our ability to even model in biology. The hope, of course, is applicability — at least of the gains in understanding — beyond the specific models. Analogies of all these types of theory have played crucial roles in physics, especially condensed matter physics. In evolutionary dynamics, phenomenology has dominated, molecular-interaction based theory is just now developing, and instructive abstract modeling is almost nonexistent.

Simulations can also play a role in understanding evolutionary dynamics, especially for exploring different scenarios and processes. But there are fundamental and ubiquitous difficulties with simulations of many interacting components (here individuals, genes, etc). These are particularly problematic for evolution because of its crucial dependence on rare events and the very broad spectrum of time scales involved. In the absence of a good theoretical framework, it is impossible to extrapolate reliably from the ranges of parameters that can be studied in simulations to other much larger or smaller parameters — even if no new qualitative features arise in the more realistic regimes. Furthermore, as soon as there are more than a few features and parameters in a model, it is hard to infer which are the essential aspects on which some observed behavior depends. Thus most simulations are analogous to macroscopic evolutionary theory and yield little useful quantitative information. It is too easy to find evolution in simulations — but too hard to learn much beyond the specific model.

The focus of these lecture notes is almost exclusively phenomenological theory — as we shall see, even this rapidly becomes difficult. But we conclude with a few comments about broader needs: for abstract modeling at the one end and for incorporation of biological architecture and organization — molecular and cellular — at the other.

1.3. Numbers

Before trying to develop any quantitative theory, we start with some numbers. At this point, it is not at all clear which numbers are important: understanding this is one of the long term goals.

1.3.1. Genomes, genetic changes, and genetic differences
First, some typical sizes of genomes: the number of genes, and the size of the genome in base pairs: bp. Note that these vary substantially within groups of organisms, and in some species can be much larger than the sizes given here.

Small viruses: 10^4 bp. 10's of genes

Bacteria: 10^{6-7} bp. 500 to thousands of genes

Budding yeast: 2×10^7 bp. 6,000 genes

Humans: 2×10^9 bp. about 25,000 genes

Viruses cannot reproduce on their own — and thus are not really alive: they are basically parasitic bits of genetic material. Bacteria and archaea are *prokaryotes* which do not have nuclei. They normally reproduce asexually but can exchange DNA. [8] Budding yeast (wine-making yeast), one of the best studied laboratory organisms, is a single celled *eukaryote*: it has a nucleus and other organelles similar to all animals. Yeast can reproduce either sexually or asexually. In humans, a good fraction of the genes have a clear homolog — a common ancestor — in the yeast genome. It is remarkable that with only a factor of four more genes than yeast, all the complexity of higher animals can exist. Some of the non-protein-coding parts of genomes are involved in gene regulation which is successively more complicated going from bacteria to yeast to multicellular organisms. Yet unlike in the single celled organisms, most of the much larger genomes of vertebrates — and the even larger ones of plants — has no known function. Thus what the essential size of the information in these genomes is is unclear — quite possibly an order of magnitude smaller than their total size.

Mutation rates are remarkably small. The simplest mutations are single base mutations: changing, for example, from an *A* to a *G*. The rates of a subset of these are found directly from observations, although mutation rates can vary substantially throughout a single genome because of the local context and other factors. There are many other types of mutational changes: insertions and deletions, duplications (including of whole genes and even whole genomes), transposable elements that move around the genome, etc. Far less is known about the rates of these, but cumulatively the number that occur is in a similar range to the total number of point mutations that occur.

Point mutation rates in units of per base pair per generation and per genome per generation give a sense of the numbers. For viruses, mutation rates are very high, as much as 10^{-4} per bp, and 10^0 per genome. Bacteria replicate their DNA with remarkable accuracy with point mutation rates reported as low as $10^{-9\frac{1}{2}}$ per bp, although an order of magnitude higher may be more typical. These correspond to rates of any error at all in the whole genome of 10^{-2} or less per cell division! Humans reproduce DNA less accurately than bacteria about 10^{-8} mutations per bp corresponding to 10^1 mutations per genome — although the comparison is somewhat unfair as it takes many cell divisions for a human egg to produce another human egg.

The magnitude of *genetic differences* between individuals within a species and between species are also instructive. Human genomes differ from each other by about one part in a thousand, chimpanzee and human by about a part in a hundred, and human and mouse by about fifteen percent.

Bacterial genomes vary enormously. And in many bacterial phyla — the highest level classification — no organisms are known! Their existence is inferred from ribosomal RNA (rRNA) sequences which differ considerably from those of previously known phyla. [Ribosomes, the most basic machine crucial to all life, convert DNA sequences, via messenger RNA, to proteins. Their functional core is itself RNA that is coded for by genomic DNA.] As bacteria normally reproduce asexually, species are not well defined. But in recent years similarities in rRNA have been used to loosely define bacterial species: e.g., if these differ by less than about 3%. But even with a tighter definition of 1% — comparable to human-mouse differences in rRNA sequences — a single species of bacteria can have widely varying sizes and contents of their genomes, with a core of half or so genes in common, and the others completely different.

Beyond mutational changes within an individual organism's DNA, genomes can change by recombination of DNA and other mechanisms of DNA transfer between organisms. Most species of eukaryotes reproduce sexually at least some of the time: in some cases, always, in others — such as yeast — only occasionally with many generations of asexual reproduction in between. Bacteria, while normally asexual, have various mechanisms for acquiring DNA from other bacteria, both from members of the same species, and from unrelated species. [8] Little is known about the rates of these processes except in particular circumstances.

1.3.2. Populations and generations

Mutation and recombination provide the genomic variability and thus the kind of experiments nature can perform. But the number of such experiments is determined by population sizes and the numbers of generations available.

In a human body, there are of order 10^{14} cells. But a human is host to an order of magnitude larger number of bacteria. World wide, the number of cells is even more dominated by bacteria.

Total number of bacteria: Good estimates are difficult as many of the environments in which bacteria live, especially deep into the earth, are hard to sample. A recent upper-range estimate of the total number in all environments is 10^{31}. [9]

Number of bacterial generations: The time between cell divisions in bacteria varies widely. The conventional figure for *E. coli* is twenty minutes: 10^3 seconds. But this is in optimal conditions in the lab. In human guts, the turnover time, and hence average division time, is a few days, so 10^{5-6} seconds is more realistic. In other environments, divisions may be far less frequent, even many years: $> 10^8$ seconds. If we take an optimistic value of 10^5 seconds for a mean generation time, then in the few billion years since the first bacteria, the average number of bacterial generations is 10^{12} — roughly the evolutionary time in dimensionless units.

Total number of bacterial cell divisions: From the above estimates, the total number of cell divisions since the first bacteria is of order 10^{43} — although this may well be an overestimate by a few orders of magnitude. For those who like natural logs, this is about e^{100}, an easy number to remember.

Vertebrates: I do not know of estimates of the total number of vertebrates, but 10^{15} is likely an overestimate. Even with ten generations per year, this would give less than 10^{25} total vertebrate births ever. Since chimpanzee and man diverged, there have been perhaps of order 10^{12} individuals. Thus any given base pair has mutated only about a total of 10^4 times in all these in individuals together. In one lineage, only about 1% of the base pairs have mutated — 10^6 generations at a rate of 10^{-8} per generation — the observed differences between humans and chimps.

1.3.3. Explorations of genome space

The total size of genome space is enormous: even with a few megabases of DNA for bacteria: of order 10^{10^6} possible sequences. A drastic overestimate of the number of sequences ever explored by nature (since early cells) is from the total amount of DNA ever produced $10^{43} \times 10^7 \sim 10^{50}$. Assuming this was completely random, it would still provide less than the number of possible sequences of 90 nucleotides (four types). Thus only sequences of at most 30 amino acids — not much larger than a single functional domain of a protein and too small to be considered a protein on its own — could have been fully explored. And the actual extent of the exploration of sequence space is surely far less.

A better way to think of these numbers is in terms of the size of steps that can be taken in genome space. Because mutation rates are low, complex mutations, in which K changes happen in one cell division, are very rare. But beyond single mutations that increase fitness or are at worst neutral in the present environment, efficient exploration of genome space would seem to require genetic changes that involve downhill steps. If K changes are needed for the new genome to be fitter, a crude estimate of the rate of this process is K mutations in the same generation: i.e., mutation rate to the Kth power. Even with point mutation rates of 10^{-6} per bp — which bacteria can often not survive — the maximum K for a multiple-point mutation that could ever have occurred in any bacterium is about $K = 7$. In a more concrete context: for all *E. coli* in all humans ever, at rates of 10^{-9} per bp, all possible three-point mutations could have taken place, but almost none of the possible four-point mutations. Of course, these may be large underestimates because the multi-point mutations need not happen in one cell division if the intermediaries are not lethal: we discuss this point later.

How to think quantitatively about the effects of sexual recombination is less clear. If this acts primarily to move around already evolved genes, then one needs to consider recombinations at this level and consider how efficiently gene-combination space is explored. Or perhaps protein-domain space is better to consider.

1.3.4. What numbers matter?
Whether one thinks of 10^{12} generations, or even e^{100} cell divisions ever, as enormous enough numbers to obviate the need for quantitative thinking about evolutionary dynamics, depends, perhaps, on one's background. But, as we shall see, even in the simplest idealized situations, it is not known which combinations of parameters are most important for determining the evolutionary potential. My own belief is that the current lack of understanding of evolutionary dynamics is high enough that, except for knowing answers from nature, one cannot have concrete expectations. And from nature we only know about long time scales with natural mutational processes and population sizes. To understand evolution on shorter time scales in the lab, especially if genetic changes and selection can be made far more efficient than in nature, surely requires far better quantitative understanding.

2. Analysis of phenomenological models

The primary focus of these lectures is phenomenological theory with the mapping of possible genetic changes to fitness assumed. How this is determined by the biology, we only discuss briefly at the end. Due to both time limitations and the difficulty of the problems, we only consider some of the simplest situations. And we focus on asexual reproduction which is much easier to analyze. The introductory aspects are well known [10, 11], although I hope the way they are discussed here will provide additional insights: these are needed for even the slightly more complicated situations that are discussed later.

2.1. General formulation

We are interested in the dynamics of interacting populations of *asexual* organisms which reproduce or die and can mutate to other genotypes. Defining the population with genome α to be n_α, we consider the simplest situation in which there is no spatial structure. For the aspects we are interested in the details of the cell division and death processes do not matter much. A simple model is to consider these to be continuous time processes with birth rate, B_α, death rate D_α, and mutation rate from genome β to genome α, $M_{\alpha\beta}$. Because of the stochasticity, we need to study the joint probability distribution of all the $\{n_\alpha\}$. This changes with time:

$$\mathrm{Prob}\big[n_\alpha(t+dt)_\alpha(t)+1\big] = dt\left[B_\alpha n_\alpha - n_\alpha \sum_\beta M_{\beta\alpha} + \sum_\beta M_{\alpha\beta} n_\beta\right] \quad (2.1)$$

$$\mathrm{Prob}\big[n_\alpha(t+dt)\big]_\alpha(t)-1] = dt\, D_\alpha n_\alpha. \quad (2.2)$$

The ecology enters in the dependences of the birth and death rates on the environment and the other organisms:

$$B_\alpha = B_\alpha\big(n_\alpha, \{n_\beta\}, t\big) \quad (2.3)$$

and similarly D_α, with the explicit time dependence from changes in the environment. More realistically the populations and environment also depend on spatial location and the mobility of the organisms is then also be important. But even without this, the system is complicated enough. The dynamical evolution equations are very general but, like the many body Schrodinger equation in physics, almost totally useless!

The simplest ecology is when all individuals of all species are competing for the same resources: i.e., with competition only with the total population

$$N(t) \equiv \sum_\alpha n_\alpha(t). \quad (2.4)$$

Except for initial transients, in a constant environment this competition can be taken into account by ensuring that for each birth there is a death and *vice versa* to keep N constant at the carrying capacity of the environment. This is a simple enough situation to analyze various aspects of, and will be the focus of these lectures.

2.2. Deterministic approximation

If, as is usually the case with microbes, the populations are large, it is tempting to approximate the dynamics as *deterministic*. In this limit the populations can be treated as continuous variables with

$$\frac{dn_\alpha}{dt} \approx \Phi_\alpha n_\alpha + \sum_\beta \left[M_{\alpha\beta} n_\beta - M_{\beta\alpha} n_\alpha \right] \tag{2.5}$$

with $\Phi_\alpha = B_\alpha - D_\alpha$ the growth (or decay) rate of population α now reflecting both birth and death processes.

With the simplest competition, N can be kept constant by taking

$$\Phi_\alpha \left(n_\alpha, \{ n_\beta \} \right) = \phi_\alpha - \frac{1}{N} \sum_\beta \phi_\beta n_\beta = \phi_\alpha - \bar{\phi}(t) \tag{2.6}$$

with ϕ_α the (constant) "fitness" of organisms α in this environment — how fast a population n_α, would grow in the absence of any competition. The organisms compete only with the mean fitness, $\bar{\phi}(t)$, of the population. In this simplest ecology, $\{\phi_\alpha\}$, together with connections between genomes given by the non-zero elements of the mutation matrix, can be thought of as a "fitness landscape".

The rate of change in the mean fitness of the population is simple if the mutation rates are small enough that selection dominates. From Eq. (2.5) this is found to depend only on the variance of the fitness within the population:

$$\frac{d\bar{\phi}}{dt} \approx \text{var}[\phi], \tag{2.7}$$

a general result that is valid when the effects of mutations can be neglected: it is known as the "fundamental theorem of natural selection". But a crucial question is then: What determines the variance? Its dynamics will be controlled by the third cumulant, whose dynamics is controlled by the fourth cumulant, etc. And if mutations really can be neglected, the fittest individuals will take over the population and the evolution soon stop. Nevertheless, on short time scales selection on existing variance in a population will increase the fitness at a rate proportional to the variance. This is often used to estimate evolution rates. But, disturbingly, it is often *assumed* that such variance can be maintained and continue to be selected on even though for this to happen *requires* mutations. An extreme example of the dangers of such an assumption was discussed in the introduction.

We now consider more generally the deterministic approximation to the dynamics with mutations included. With the simplest competition, $\bar{\phi}(t)$ plays the role of a Lagrange multiplier and the dynamics, Eq. (2.5) is effectively linear.

Thus at long times, the behavior will be determined by the largest eigenvalue of the fitness-plus-mutation matrix with a steady state represented by the largest eigenvector eventually being reached.

In situations in which there is an optimum genome — the highest peak in the fitness landscape — the equilibrium population will be distributed among this optimum, say $\alpha = 0$, and genomes a few mutations away from it which are nearly as fit. For example, the relative population size of a deleterious mutant, β, with fitness lower by $\delta_\beta = \phi_0 - \phi_\beta$ which is produced from the optimum genome at a rate $M_{\beta 0}$ will have an equilibrium population $n_\beta \approx n_0 M_{\beta 0}/\delta_\beta$. [Here and later, we ignore "back" mutations, here $\beta \to 0$. This can be justified in many, although by no means all, situations.] An equilibrium distribution around a fitness maximum is known as an Eigen quasispecies — Eigen rather than eigen after Manfred Eigen — and is a useful concept in situations with high mutation rates, especially viruses. But if — as is always the case in nature — fitter genomes exist but are far away from the genomes of the existing populations — i.e. requiring many mutations to reach — then the deterministic approximation will (almost always) give complete nonsense!

In practice, or course, any fitness maximum is only a local maximum: there will always be fitter genomes further away. To illustrate the effects of this, it is useful to analyze a simple example. Consider an effectively one dimensional landscape in which the genomes are labeled by an integer x with a peak at $x = 0$ with $\phi_0 = 0$, a valley of width W of depth δ, i.e. $\phi_x = -\delta$ for $x = 1, 2 \ldots W$, separating it from a higher peak at $x = K \equiv W + 1$ with $\phi_K = s > 0$. Mutations occur at rate m that take x to $x + 1$ for $x \leq W$. Again we ignore the effects of back mutations that decrease x: here this is justified if $m \ll \delta$, as we assume. Population losses from the "forward" mutations are equivalent to shifting all fitnesses by m, which we can hence also ignore. If the population of size N initially all has genome $x = 0$, the dynamics is straightforward to analyze in the deterministic approximation. One finds

$$n_x = N \left(\frac{m}{\delta}\right)^x \left[1 - e^{-\delta t}\left(\sum_{y=0}^{x-1} \frac{(\delta t)^y}{y!}\right)\right] \tag{2.8}$$

for $x \leq W$. There are two regimes: for $t \ll x/\delta$, $n_\delta \approx N e^{-\delta t}(mt)^x/x!$ while for $t \gg x/\delta$, the steady state distribution in the valley is achieved: $n_x \approx N(m/\delta)^x$. The steady state thus "propagates" at speed $1/\delta$. Analysis of the population at the peak, n_K, shows that already at times, $t > \tau_{\text{nuc}} \approx W/(\delta + s)$, n_K grows exponentially at rate s. The time τ_{nuc} is the dominant time at which the mutations to the fittest genome appear: the apparent nucleation time (determined by a balance between the exponentially small rate of mutations to the fittest population and its exponential growth). But something is worrying: τ_{nuc} does not depend on the mutation rate! A "sweep" time $\tau_{sw} \approx \frac{W}{s} \ln[(\delta + s)/m]$ after τ_{nuc}, the fittest population will take over and become the dominant population. This does depend weakly on m but only because the fitter population at $t = \tau_{\text{nuc}}$ is of order $N[m/(\delta + s)]^K$ because of the K mutations needed to reach it from the original genome: after it has been reached, the fitter population grows exponentially.

2.3. *Failures of deterministic approximations*

Are the above results reasonable? That depends on the question. For times shorter than $\tau_{\text{nuc}} + \tau_{\text{sw}}$, the population is still dominated by the original unmutated population. Thus $\bar{\phi}(t)$ has changed very little and ignoring the competition entirely should be legitimate. In this case, averaging the stochastic dynamical equations will result in exactly the deterministic approximation which would then correctly yield $\langle n_K(t) \rangle$ and the time at which this average becomes of order N would be correctly given by the above analysis. But is this the right question? We must examine how the average arises. From the above analysis we see that at the time τ_{nuc} when the last mutations occur that dominate the later behavior of the fittest population, this average population $\langle n_K \rangle$ is a very small fraction of the total. Indeed, unless $N > [\delta/m]^K$ — which is huge for small mutation rates and broad valleys — there is, on average, much less than one fit individual at the time τ_{nuc}. But this must mean that something has gone wrong with the analysis — or that the average is a very misleading quantity. The latter is indeed the case unless the population sizes are really enormous, but how enormous will depend on the specific context. In general, as we shall see, taking averages is very misleading for evolutionary dynamics.

There is a very important lesson from the failures of the deterministic approximation. In physics, when the number of constituents (atoms, etc.) is large, thermodynamics and other approaches that focus on average quantities with small fluctuations around these are good. The basic starting points for theoretical treatments are often called mean field theories, and fluctuations can usually be added systematically to these. More generally, the enormous power of the renormalization group framework relies on —- and leads to understanding of — the simplifications that occur with large numbers and long time scales. In a broad range of contexts, this justifies the ideas of universality: dependence of many features of large systems on only a few aspects of the microscopic structure and interactions. But evolutionary dynamics has a crucial component that is very different. Mutations that arise initially in *one* individual can take over the whole population. And individuals that migrate to a new environment can give rise to new populations. Thus rare events are essential. Much of the difficulty in understanding evolutionary dynamics — even when the genome-to-phenome mapping is known — comes from the interplay between such rare individual events and the approximately deterministic dynamics of populations once they become large. Conventional "mean-field" like approaches are thus doomed to failure.

To go beyond the dangerous deterministic approximation, and to gain some intuition we turn to the simplest evolutionary system: a fitter population being fed by stochastic mutations at a fixed rate Nm from an original population (this is equivalent to $W = 0$ in the above). But first we have to understand individual populations with stochastic dynamics. The remainder of this section is all standard material but we emphasize heuristic arguments rather than exact results to gain intuition which is needed for more complicated situations for which exact analysis is too difficult.

2.4. Single population

A single population without mutations is the simplest context in which to understand the stochastic dynamics of the birth and death processes. For now we ignore any limits on the population size. [The specific continuous-time model we study is somewhat different than the conventional ones with discrete generations. [10, 11] Nevertheless the key aspects of the behavior are the same; we comment later on the differences.]

Define $p_n(t)$ as the probability that there are n individuals at time t. Then

$$\frac{dp_n}{dt} = B(n-1)p_{n-1} + D(n+1)p_{n+1} - (B+D)np_n \tag{2.9}$$

so that $B - D$ is the growth rate of the mean population:

$$\frac{d\langle n \rangle}{dt} = (B - D)\langle n \rangle. \tag{2.10}$$

The variance grows proportional to $\langle n \rangle$

$$\frac{\text{var}[n]}{dt} = (B + D)\langle n \rangle \tag{2.11}$$

so that the effective diffusion constant of the population distribution around its mean is $\frac{B+D}{2}\langle n \rangle$. This should be expected given the stochastic birth and death rates proportional to n so that variations around the mean after one generation are of order $\sqrt{n}$.

Before analyzing the dynamics of the distribution p_n in detail, it is instructive to try making some heuristic arguments to see if we can guess the behavior. For more complicated situations, we will have to rely on such arguments, so we had better get some practice! It is convenient to work in units of generations defined, for convenience, by setting the death rate $D = 1$. The net growth (or decay) rate of the population per generation is defined as

$$r \equiv \frac{B - D}{D}. \tag{2.12}$$

Consider a large initial population of size $n_0 \gg 1$. There are then two characteristic time scales. The first is the obvious one: $1/|r|$ for growth or decay of the mean. The second is associated with fluctuations, referred to in the context of evolutionary dynamics as *drift*. Drift is simplest in the *neutral* case: i.e. the mean growth rate, r, is zero. As long as the deviations from n_0 are small compared to n_0, we can approximate the fluctuations around the mean by a diffusion constant n_0 so that

$$n(t) \approx n_0 \pm \mathcal{O}\left(\sqrt{n_0 t}\right) \tag{2.13}$$

valid until $t \sim n_0$ by which point the variations will have either decreased or increased in magnitude due to the accumulated changes in n. Thus the characteristic time scale for the fluctuations to substantially change the population is of order n_0 generations. If the population happens to decrease by, say, a factor of two over this time interval, then the time scale for another factor of two decrease, if it occurs, will be of order $n_0/2$, the next factor of two decrease in $n_0/4$ generations, etc. From this we can guess that the characteristic time in which the population can fluctuate away entirely — i.e. become extinct — is of order n_0. Although this argument is sloppy, we will see that it is basically correct: a population, n_0, with no mean growth or decay is likely to die out in of order n_0 generations with substantial probability. If it has not died out, then it is likely to have increased: this must be the case as the mean population is unchanged. The above argument suggests that, if it has not died out, the population will typically have increased in n_0 generations by of order a factor of two.

For a population with a small mean growth rate, $r \ll 1$, the time scale for fluctuations to drive the population to zero will be shorter than that for it to grow on average if $n_0 \ll 1/r$. Thus we expect that there is a characteristic population size of order $1/|r|$: for $n \gg 1/|r|$ the population is very likely to grow (or decay if $r < 0$), while for $n \ll 1/|r|$ its behavior is dominated by fluctuations and it might go extinct.

What if the initial population is $n_0 = 1$? Again, first consider the neutral case. In spite of this neutrality, after a time of order unity there is a substantial chance that the population will have died out. If it has not, it is likely to be larger, say of order 2. A time of order 2 later, there is again a good chance that is will have died out, if it has not done so, then it will be of order 4, etc. If it survives up to time t, then we can guess that $n(t)$ will be of order t, but with a broad distribution around this, presumably with width of order t. What we cannot get from this argument is the probability that the population will survive to a time t. But this argument does suggest that the conditional probability, q, of surviving for a further time t, given that it has not until time t, is roughly independent of t. If the conditional survival probability from t to $2t$ is roughly independent of that from, say, $16t$ to $32t$, then the probability of survival up to a long time t is roughly the product of $\log_2 t$ factors of q. This suggests that the probability of survival is of order $e^{-\kappa \ln t} = 1/t^\kappa$ with some exponent $\kappa \sim \ln(1/q)$. Such behavior is reminiscent of a conventional one-dimensional diffusion process with an absorbing boundary at zero. Indeed, the behavior could have been guessed by considering the variable $y \equiv 2\sqrt{n}$ which fluctuates with an effective diffusion coefficient of unity [since $n/(\sqrt{n})^2 = \mathcal{O}(1)$]. The survival probability for a time t of a such a random walk in y is of order $1/\sqrt{t}$ suggesting $\kappa_{RW} = \frac{1}{2}$.. This argument is tempting but wrong!

But we can get the right answer for the survival probability using some additional information. We know that the average $\langle n(t) \rangle = 1$ for all t. From the above discussion we expect that $n(t)$ is either of order t, or exactly zero. This is only consistent if the probability of it being non-zero is of order $1/t$: i.e. $\kappa = 1$. Note that we have here combined heuristic arguments with an exact result (which by itself was somewhat misleading) to obtain a concrete prediction.

Let us now analyze the behavior more carefully. The evolution of the probabilities can be analyzed exactly from Eq. (2.9) by using the generating function

$$Q(z) \equiv \sum_n z^n p_n, \tag{2.14}$$

but the results are somewhat messy. For the simplest case, starting with a population $n_0 = 1$, the solution can be guessed:

$$p_0 = 1 - h(t) \tag{2.15}$$

and

$$p_n = h(t)[(1 - a(t))[a(t)]^{n-1} \quad \text{for } n \neq 0 \tag{2.16}$$

with

$$a = \frac{(1+r)\left(1 - e^{-rt}\right)}{1 + r - e^{-rt}} \tag{2.17}$$

and the survival probability

$$h = \frac{r}{1 + r - e^{-rt}}. \tag{2.18}$$

If r is negative, the survival probability decays to zero exponentially as expected, but with a small coefficient that is not obvious: $h \approx (-r)e^{-(-r)t}$ for $r < 0$. For $r > 0$, the population has a non-zero chance of surviving forever: for $t \to \infty$, $h \to r/(1+r)$. *If* it survives, the population typically grows exponentially: the *conditional mean* given non-extinction is

$$\langle n|n \neq 0 \rangle = \frac{1}{1-a} = \frac{(1+r)e^{rt} - 1}{r} : \tag{2.19}$$

note the prefactor which is large, $\approx 1/r$, for small growth rate. We call the probability that the lineage descended from one individual survives to grow exponentially, the *establishment probability*, ϵ. For this simple dynamics, $\epsilon = r/(1+r)$. The result Eq. (2.19) should be contrasted with the mean population $\langle n \rangle = e^{rt} = h\langle n|n \neq 0 \rangle$ in which there are two — misleadingly — canceling factors of r: arguing on the basis of the overall mean population is thus very dangerous! It is the (much larger) conditional mean that reflects the typical size of the population given that it survives. But even if it does survive, the distribution of n is still broad: it is exponentially distributed with, e.g. standard deviation around the mean that is of order the mean itself. But it does have a "typical" scale which is well characterized by the conditional mean.

For the neutral case, the survival probability is

$$h = \frac{1}{1+t} \tag{2.20}$$

so that as we guessed, the exponent $\kappa = 1$ determines its decay The conditional mean population if it survives is

$$\langle n|n \neq 0 \rangle = 1 + t \tag{2.21}$$

again confirming the heuristic argument given above. We shall see shortly why the random walk analogy, which suggested $\kappa_{RW} = \frac{1}{2}$. fails.

We can use the results for the neutral case to better understand the behavior in the growing case with small r. For a time, τ_{est}, of order $1/r$, the fluctuations dominate the dynamics. Through this time, the survival probability is only of order $r \sim 1/\tau_{est}$. But if the population survives this long, it is likely to be of size of order $\tau_{est} \sim 1/r$ (as from Eq. 2.19). Then the deterministic growth takes over and and the population starts to grow exponentially. Thus $\langle n|n \neq 0 \rangle \sim \frac{1}{r}e^{rt}$ at later times. This process is the establishment of the population: it has survived long enough — for $t \sim \tau_{est}$ — to become large enough, $n \sim 1/r$, to grow exponentially. Eventually, of course, such exponential growth must slow down and the population saturate. But we have been ignoring, so far, any limits on the population size: we will discuss these shortly.

Once we have the results for $n_0 = 1$, we can find the behavior for general n_0 by observing that each of the n_0 initial lineages behaves independently and $n(t)$ is the sum of the sizes of each of these lineages. Thus $\text{Prob}[n(t)|n_0]$ is the convolution of $\text{Prob}[n(t)|n(0) = 1]$ with itself n_0 times. In the neutral case for large n_0, we can check the heuristic argument that after a time of order n_0 there is a substantial chance that the population has died out. The probability that a particular one of the lineages survives is of order $1/t$. Thus the probability *none* survive is $\approx (1 - 1/t)^{n_0} \approx e^{-n_0/t}$ which indeed starts to grow substantially when $t \sim n_0$ — as expected. For $r > 0$, the survival probability for infinite time is similarly found to be large for $n_0 \gg 1/r$: this supports the basic picture that once a population reaches $\sim 1/r$ it is likely to become established and grow exponentially from then on.

At this point, we should pause and ask which of our results are general and which are specific to the detailed model of the dynamics. We have analyzed the case in which reproduction and death are continuous time processes that occur at some rates, B and D, respectively. In many situations, a more realistic model is discrete generations with a distribution of number of offspring in the next generation, and death of the parent. The neutral case corresponds to a mean number of offspring being unity. Slow growth corresponds to mean number of offspring of $1 + r$, resulting in a growth in the mean population of r per generation. But the fluctuations with this dynamics is somewhat different: in particular, it will depend on the variance in the number of offspring when the population is large, and, when it is small, can depend on the whole distribution. Nevertheless, the overall behavior is very similar: the diffusion of the population is proportional to n for large n, in the neutral case the survival probability to a long time t is proportional to $1/t$, and the probability of establishment of the population starting with one individual is, for small r, proportional to r. But the coefficients of these depend on the details of the dynamics. In particular, the establishment probability for small r is in general cr with the constant c depending on the reproduction and death processes.

2.5. Continuous n diffusion approximation

If the growth or decay rates are small and one is primarily interested in the dynamics of populations over many generations, the behavior simplifies. Most populations will be either relatively large or zero: as we have seen, if n is small, it is likely to become either zero or substantially larger in a relatively short time. It is thus natural to try and approximate the population as a continuous variable — but with zero playing a special role. Surprisingly, this is a good approximation. We can guess the appropriate stochastic Langevin equation, but have to be careful what it means:

$$\frac{dn}{dt} = rn + \sqrt{n}\eta(t) \tag{2.22}$$

with η gaussian white noise with covariance $\langle\eta(t)\eta(t')\rangle = 2\delta(t-t')$. [In general, the strength of the noise will depend on details of the birth and death processes, we choose the value that corresponds to the continuous time dynamics analyzed above.] The correct interpretation of Eq. (2.22) is the Ito one with $n(t+dt)(t) + dt[rn(t) + \sqrt{n(t)}\eta(t)]$. This means that the probability density $p(n, t)$ satisfies

$$\frac{\partial p}{\partial t} = -\frac{\partial(rnp)}{\partial n} + \frac{\partial^2(np)}{\partial n^2} \tag{2.23}$$

with the n inside *both* derivatives in the diffusive-like term. As can be checked, this is necessary for the mean $\langle n\rangle$ to grow proportional to $\langle n\rangle$.

To see how the naive argument for the survival probability from diffusion of the variable $y = 2\sqrt{n}$ goes wrong, we need to contrast the Ito convention with the less-physical Stratonovich one often used by physicists. In the Stratonovich convention, the noise effectively acts at time $t + dt/2$ so that in our case its coefficient is $\sqrt{[n(t+dt)+n(t)]/2}$. The diffusion coefficient proportional to n then appears in the form: $\frac{\partial}{\partial n}\left[n\frac{\partial p}{\partial n}\right]$, But this would yield $d\langle n\rangle/dt = r\langle n\rangle + 1$ which is not correct. In the Stratonovich convention, variables can be changed straightforwardly and a diffusion equation for n with diffusion coefficient n is equivalent to that for y with diffusion coefficient unity. But with the Ito convention, one must be careful changing variables: if $q(y, t)$ is the probability density of y, then $\frac{\partial q}{\partial t} = -\frac{\partial}{\partial y}\{[\frac{(ry)}{2} - \frac{1}{y}]q\} + \frac{\partial^2 q}{\partial y^2}$ which corresponds to an extra term that drives y to zero. If this term had not been there, the survival probability would have decayed as $1/\sqrt{t}$. But in its presence, the pushing of y towards zero has a comparable effect to the stochastic parts for any y: it changes the survival probability to $1/t$.

A simple solution for $p(n, t)$ can be found which is analogous to that we found above for the discrete-n distribution. For the neutral case this is

$$p = \frac{1}{t^2}e^{-n/t} + \left(1 - \frac{1}{t}\right)\delta(n). \tag{2.24}$$

If we change t to $t + 1$, this roughly corresponds to starting with n around 1, although we cannot take this solution — or the continuous n approximation — seriously for $n = \mathcal{O}(1)$. We leave it as an exercise to find the corresponding solution for $r \neq 0$ of the form

$$p = h\gamma e^{-\gamma n} + (1 - h)\delta(n) \tag{2.25}$$

with $dy/dt = -ry - y^2$ and the survival probability $h = \exp(-\int_0^t dt' y)$ and to show that it yields the same behavior as the proper discrete analysis in the appropriate regimes of r, n, and t. This, and the general solution, can be found by Laplace transforming Eq. (2.23) in n to λ: the resulting first order partial differential equations in t and λ can be solved by method of characteristics. The Laplace transform is analogous to the generating function, $Q(z)$, for discrete n: using the latter results in a similar PDE.

2.6. Problems with averages

It is, perhaps, somewhat surprising that the continuous n approximation works so well. As discussed above, the problem with the deterministic approximation appears, naively, to do with the role of fractional individuals which are also present in the continuous approximation. But the latter does include fluctuations whose form, being proportional to $\sqrt{n}$, is related to the crucial role of zero populations. In the continuous approximation, in contrast to the deterministic approximation, strictly-zero populations exist and non-zero populations smaller than unity are rare. This is the primary reason that average populations are often a very poor characterization of the distribution. As we have seen in the simple situation of a single population with a slow mean growth rate, the average is completely dominated by rare instances in which the population is anomalously large — the population is usually zero — and in those rare cases it is much larger than would be guessed from the average. Thus in our earlier example of evolution in a fitness landscape with a broad valley separating two peaks, we can guess that, typically, after time $\tau_{\text{nuc}} + \tau_{\text{sw}}$ when the average of the fittest population becomes of order N, in actuality it will almost always be zero. This means that in the very rare cases in which it is non-zero it must be much larger than its average. When this occurs, the total population will have already become much large than N and we are no longer justified in ignoring the fixed total N constraint: the role of the mean fitness of the population, $\bar{\phi}(t)$, becomes very important. But since this dependence makes the birth and death rates depend on the $\{n_\beta\}$, averaging of the dynamical evolution equations can no longer be done straightforwardly: indeed, it becomes very tricky, involving all higher moments — and not very useful.

2.7. Mutant population and selective sweeps

Thus far, we have only analyzed simple population dynamics without selection or mutations or even limited total resources: we have allowed n to become arbitrarily large. If there are limited resources so that the birth and death rates depend on the population size, then the dynamics is already much harder to analyze exactly. A more interesting situation is two populations that are competing for total resources. If the competition results in the total population size being fixed, this is equivalent to a particular form of population-size dependence of the birth and death rates.

Consider a population of *fixed* total size N, that consists of one population of size n — which for future purposes we will call the mutant population — with birth rate b and another — conventionally called the "wild-type" population — of size $N - n$ with birth rate B. To keep N fixed, for each birth one randomly chosen individual dies. With probability n/N this will be a mutant: thus n will increase only if the birth is a mutant and the death a wild-type, while it will decrease only if the birth is a wild-type (which occurs at overall rate $B(N - n)$) and the death a mutant. Thus the effective birth rate for the mutants is $b(1 - n/N)$ and the effective death rate is $B(1 - n/N)$. Even in the continuous approximation, the dynamics now becomes much harder to analyze exactly (special functions, etc. are required). But using a combination of the intuition gained for the simplest case, and a few results that are easy to derive, we can understand as much as we want.

First, consider the neutral case in which both populations have the same birth rate $b = B = 1$ (and hence also the same death rate). In this case, the individual lineages are all equivalent, we hence need consider only one of them. There are two possibilities: either this particular lineage will die out, or it will fluctuate large enough that all the others will die out: i.e. it will reach N after some time: *fixation* of this lineage. For $n \ll N$, the dynamics are similar to the case we have already studied: the probability of surviving until a time t is about $1/t$, and the typical population if it does survive is of order t. This should be valid until $n \sim (N/2)$ which will occur with probability of order $1/N$ and in time $t = \mathcal{O}(N)$. Once $n = N/2$ it has, by symmetry, a fifty percent chance of fixing. Thus we expect the probability of fixation is of order $1/N$ and, if it does fix, this will take a time of order N.

A simple argument shows that the fixation probability is exactly $1/N$: in the neutral case, each of the N original individuals has an equal chance of fixing; since only one can fix, the probability of a particular one doing so is $1/N$. This immediately implies that if the initial mutant population is n_0, the probability it fixes is n_0/N. Unless $N - n_0$ is much less than N — i.e. that the wild-type population is a small minority — the average time to fixation will be of order N. If $N - n_0$ is small, then the typical time to fixation will be of order $N - n_0$. But the average fixation time in this regime is more subtle. It is equivalent to the average time to extinction of an initially small mutant population, given that it goes extinct, which it is very likely to do. Since the probability of going extinct in a time interval dt is $dt\, n_0/t^2$ for $t \gg 1/n_0$, the mean time to extinction would be infinite but for the upper bound on n. Cutting off the integral over t at $t = \mathcal{O}(N)$ (corresponding to $n \sim N$) yields a conditional mean extinction time of $n_0 \ln N$. Thus the conditional mean fixation time for $N - n_0 \ll N$ is $(N - n_0) \ln N$

by the same argument. Note that results for mean fixation times can be found exactly from a recursion relation that connects different values of n_0, without fully analyzing the dynamics. But, as is becoming a pattern, the average fixation time is often not a very useful characterization of the behavior. Thus one learns rather less from these exact results than from the heuristic arguments!

We now consider a mutant population that has a small *selective advantage*, s, over the wild-type: i.e. s is its differential growth rate per generation of the wild-type, $s = (b - B)/B$. If $s \ll 1/N$, the mean growth will be swamped by the fluctuations and the behavior is essentially that of the neutral case. But for $s \gg 1/N$, the selection is strong enough that a sufficiently large mutant population will rise to fixation: this is called a *selective sweep*. But with a single individual initially, the mutant population is likely to die out. Yet with probability s It will reach a size of order $1/s$ and become established. If it does become established, say in a time τ_{est}, then it will grow roughly deterministically as

$$n \approx \frac{1}{s} e^{s(t - \tau_{est})} \tag{2.26}$$

until it becomes large enough to become a majority of the population: this will take a time, $\tau_{sw} \approx \frac{1}{s} \ln Ns$, the *sweep time* (although it will take twice as long for the full sweep to when the wild-type population disappears completely). After the mutant population has reached $N/2$, the wild-type population will decrease exponentially with $N - n \sim e^{-st}$ until it reaches of order $1/s$ and fluctuations take over to drive it to extinction a time of order $1/s$ later. In the intermediate regime the dynamics is close to deterministic with the "logistic" form $n \approx v e^{st}/[1 + v e^{st}/N]$ with, in the case of a single individual at time zero, the appropriate $v \approx e^{-s\tau_{est}}/s$. It is convenient to *define* τ_{est} in terms of the population in the deterministic regime pretending that it had been deterministic back to a time τ_{est} at which it was $1/s$. The fluctuations at early times, plus the relatively weak fluctuations for times after τ_{est}, can then all be incorporated into τ_{est} which is thus a stochastic quantity with a distribution of values of order $1/s$. The advantages of speaking in term of a stochastic τ_{est} is that this correctly matches together the fluctuation regime for small n with the deterministic regime in which the fluctuations are negligible.

The analysis we have sketched here, in particular the use of τ_{est}, is an example of a *matched asymptotic expansion* in which the existence of a small parameter, here $1/Ns$, enables different regimes to be handled separately and matched together. Here the regimes are $n \ll N$ for which the non-linearities from the dependence of the growth rate on n can be ignored, and $n \gg 1/s$ for which the fluctuations can be ignored. Understanding the scales involved — e.g. here the time scale $1/s$ in the small n regime — and being able to separate the regimes is essential for understanding more complicated situations such as those we consider later. If the regimes and scales are understood, then for many purposes, cruder methods of matching are sufficient. In this context one could pretend that there is a strict separation of regimes: $n < 1/s$ with neutral drift only, $1/s < n < N/2$ with deterministic exponential growth of n, and $N/2 < n < N$ with deterministic exponential decay of $N - n$. The only things one would really miss are numerical factors of order unity. But since in biology all equations are wrong (in contrast to what is often pretended in physics!), such errors are likely to be less significant than those we have made in writing down the model (e.g. the particular form of the stochastic birth and death processes).

Before proceeding, we briefly consider the case of a deleterious mutant with $s = -\delta$ negative. In this case, the dynamics is stochastic until a time of order $1/\delta$ and the chances of surviving that long are of order δ. After that, even the lucky survivors, which will have reached population sizes of order $1/\delta$, will tend to die out, with the probability of survival decaying exponentially at later times. We shall need these results when we consider deleterious intermediaries.

2.8. Mutation and selection

We now turn to the combination of mutation and selection. The simplest situation is to start with a single population of size N which mutates at rate m per generation to a fitter mutant population, n, with selective advantage s which competes with the original population so as to keep the total population size fixed at N. This is just the situation analyzed above with the addition of the mutations. We focus on the case of s small but N large enough that $Ns \gg 1$.

There are several important time scales. The first is the growth time: $1/s$. The second is the time to drift from a single individual to $n \sim 1/s$: this is also of order $1/s$. The third is the sweep time: $\tau_{sw} = \frac{1}{s} \ln Ns$ which is substantially longer. The fourth is the typical time between mutations: $1/(Nm)$ which for now we will assume is long. But there is a fifth, less obvious, time scale which is more important: the time for the mutant population to become established. The basic process is simple. Each new mutant has a probability of order s of surviving drift. Eventually one of these will definitely become established. As the establishment probability is $\epsilon \approx s$, of order $1/s$ new mutants are needed for one to establish and eventually one will. The stochastic establishment time is thus $\tau_{est} \sim 1/(Nms)$ which is much longer than $1/s$ if $Nm \ll 1$. Therefore when the mutations are limiting, $\tau_{sw} \ll \tau_{est}$ and the stochastic establishment dominates the total time until the mutant population fixes.

If we tried to ignore the constant total population constraint — which should be alright at short times, — then the average mutant population would be

$$\langle n \rangle \approx \frac{Nm}{s}[e^{st} - 1]. \tag{2.27}$$

If Nm is small, $\langle n \rangle$ becomes of order $1/s$ at a time $\frac{1}{s} \ln Nm$ — but this is *not* one of the characteristic time scales. How does this happen? Consider the distribution of establishment times. As these are typically of order $1/Nms$, define a stochastic variable $\beta \equiv Nms\tau_{est}$ which is typically of order unity. After establishment, the mutant population is

$$n \approx \frac{1}{s}e^{st-s\tau_{est}} = \frac{1}{s}e^{st}e^{-\beta/Nm}. \qquad (2.28)$$

For atypically small β, the probability density $\rho_{est}(\beta) \approx 1$ since the establishment is effectively a Poisson process with rate Nms. Thus $\langle e^{-\beta/Nm} \rangle \approx Nm$ being dominated by anomalously small $\beta = \mathcal{O}(Nm) \ll 1$. For $Nm \ll 1$, we see that the average reflects very atypical instances..

If the overall mutation production rate, Nm, is *large*, then $\langle n \rangle$ becomes $1/s$ when $t \approx 1/(Nms)$ still the typical establishment time. In this regime, $\tau_{est} \ll \tau_{sw}$ so that many mutant lineages become established before any sweeps to fixation. The dynamics is thus close to deterministic with the time to half-fixation of the mutant population $\frac{1}{s} \ln(s/m)$ as suggested by the deterministic approximation.

All the results we have discussed thus far are well known, although, as we have seen, the readily calculable quantities can be very misleading. Armed with some understanding and experience with heuristic arguments, we now turn to more interesting situations.

3. Acquisition of multiple beneficial mutations

In most environments there are likely to be many potentially beneficial mutations available and by acquiring multiple such mutations, the fitness can continue to increase. How fast does this happen? Surprisingly, even in the simplest possible model, this is not an easy question to answer and, in spite of much literature on the subject, the correct behavior has only been derived very recently — and then by statistical physicists. In order to make progress, the heuristic understandings of the simple situations we have already discussed are invaluable.

The simplest model of multiple beneficial mutations is a *staircase model*: a fitness "landscape" that consists of a long regular staircase with each step representing a single beneficial mutation that increases the fitness by the *same* small amount, s. The effects of the mutations are considered *additive*, so that acquiring x of them increases the fitness, ϕ, by sx. The competition is for total resources which keeps the total population fixed at N. The mean growth rate of the sub-population with x mutations is $s[x - \bar{x}(t)]$ with $\bar{\phi} = s\bar{x}(t)$ the average fitness of the population at that time — not the average over all histories, but that of the particular populations at time t: $\bar{x}(t)$ is thus a stochastic variable whose dynamics we are particularly interested in. Being physicists, we can consider the staircase to be infinitely long, with the beneficial mutations occurring at rate m and never being depleted. There are then just three parameters, N, m, and s.

We are interested in the mean *speed of evolution*

$$v \equiv \frac{d}{dt}\langle\bar{\phi}\rangle \tag{3.1}$$

— assuming this exists — and more generally in the dynamics of the fitness distribution within the population. How do these depend on the parameters? From the discussion of a single mutant population arising and fixing, and of the continuous n approximation discussed earlier, we can guess that with s small the parameters will enter in combinations such as Ns and Nm.

We will focus on the regime in which the there is *strong selection*

$$s \gg \frac{1}{N} \tag{3.2}$$

and the mutation rate is small relative to the selection:

$$m \ll s. \tag{3.3}$$

This is applicable in almost all contexts for single-celled organisms and for all but small populations of multicellular organisms: in very small populations, ($Ns < 1$), drift can dominate over selection for weakly beneficial mutations. The analysis we outline here was done in collaboration with Michael Desai [12]. A different regime, $m \gg s$, can obtain for viruses and for almost neutral mutations more generally; the staircase-model in this regime has been studied by Rouzine et al [13].

3.1. Deterministic approximation?

For very large population sizes, we can hope to use the deterministic approximation. This is straightforward to analyze. Starting from a single population of size N at $x = 0$, the subpopulation with x mutations, n_x, is found to be

$$n_x(t) = N \frac{[b(t)]^x e^{-b(t)}}{x!} \tag{3.4}$$

with mean number of mutations,

$$\bar{x}(t) = b(t) = \frac{m}{s}[e^{st} - 1] \tag{3.5}$$

so that the evolution speed is

$$v_{\text{det}} = mse^{st}. \tag{3.6}$$

The evolution is thus exponentially accelerating in the deterministic approximation. Concomitantly, the distribution is getting broader and broader with standard
deviation of the fitness $s\sqrt{b(t)}$. But we should be highly suspicious: from our
earlier discussion of evolution from one fitness peak to a higher one via an in-
tervening valley, we can guess that the dominant mutations that give rise to the
population at a large x will arise from an exponentially — or smaller — population at $x - 1$. Thus near the "front" of the fitness distribution the discreteness of
the individuals is crucial: again, $\langle n_x \rangle$ is very misleading when it is less than unity.
No matter how large N is, we will eventually — actually very quickly given the
exponentially growing speed — run into this problem.

Thus we are faced with a situation in which there is *no* well-defined speed
in the limit of large N: everything must be controlled by fluctuations (except
perhaps at early times, although even then the deterministic approximation is
dangerous).

3.2. Successional sweeps: modest population sizes

To proceed, we first consider the simplest regime in which the population is not
very large. We can then use results we have already obtained for mutations and
selective sweeps. If the total mutation production rate, Nm, is small, the dynamics is mutation limited. From an initially monoclonal population with $x = 0$, (i.e.
$n_{x=0}(t = 0) = N$), mutations will occur to $x = 1$. After a stochastic establishment time, τ_{est}, of order $1/Nms$, one of the mutants will become established. It
will then sweep to dominate the population in a time $\tau_{\text{sw}} \approx \ln(Ns)/s$. If

$$Nm \ll \frac{1}{\ln Ns}, \tag{3.7}$$

$\tau_{\text{sw}} \ll \tau_{\text{est}}$ so that the establishment process will dominate the time for the population to increase its mean fitness by s. Such a sweep is fast enough that it is
unlikely there will be further mutations established, either from the original population or from the new population with $x = 1$, until the sweep is essentially
complete and the mean fitness becomes $\bar{\phi} \approx s$. The process will then begin
again with an establishment and sweep of an $x = 2$ mutant after which $\bar{\phi} \approx 2s$,
etc.. In each round, the mutant offspring will succeed their parents: we thus call
this process *successional mutations*. In this regime, the distribution of the sub-
populations will usually be concentrated almost entirely at a single value of x.

But occasionally, in mid-sweep, it will be bimodal concentrated on two successive values of x. The average speed of evolution in the successional mutations regime is given by

$$v \approx \frac{s}{\langle \tau_{est} \rangle} \approx Nms^2 \tag{3.8}$$

since the establishment is a approximately a Poisson process with rate Nms: i.e. the probability of an establishment in the interval $(t, t + dt)$ is simply $dt\, Nmse^{-Nmst}$. [Note that more generally, as discussed earlier, the establishment probability of a single mutant will be cs rather than s with c depending on the birth and death processes. The result for the speed will thus in general be modified by a multiplicative factor of c.]

3.3. Multiple mutations in large populations

If the population is large enough that there are many new beneficial mutations each generation, $Nm \gg 1$, then the behavior is very different. For such large populations the first establishment time is much less than the sweep time for a single mutant. This means that after establishment of the first mutant, but before it can sweep, there will be further establishments of other mutations from the original population. And, more importantly, there will be new mutations from the already fitter $x = 1$ population. These double mutants will be fitter than the single mutants and can out-compete them. But before they fix, they can themselves give rise to even-fitter triple mutants, etc. Eventually, one of the mutant populations will takeover and become the majority population. But by then there will already be mutants with several more mutations that are destined to fix.

We make the *Ansatz* that there is a roughly steady state distribution of the populations $\{n_x\}$ around some mean value $\bar{x}(t)$ which advances step by step at a mean speed v, with $\bar{x}(t)$ most of the time an integer, and $n_{\bar{x}(t)} \approx N$ dominating the population at time t. At any time, there will be a fittest mutant in the population, at some $x = \bar{x}(t) + q(t)$: we define q as the *lead* of these fittest mutants. They are fitter than the average members of the population by qs and thus their population, once they have become established, will grow as e^{qst} until the mean fitness of the population increases. We assume that the fitness advantage of each mutation is small enough that $qs \ll 1$.

With small mutation rate, the lead population will become established and be growing exponentially before the next fitter mutants establish. We take time zero as the time at which the next-most fit population became established, and we label the populations by $x - \bar{x}$ rather than x. For simplicity, consider the situation at which the mean fitness increases by s at the same time zero. Then we have for some time interval,

$$n_{q-1} \approx \frac{e^{(q-1)st}}{qs} \tag{3.9}$$

(qs rather than $(q-1)s$ in the denominator because this population became established while it had lead qs). The rate of mutations into n_q is $mn_{q-1}(t)$. As each new mutant has a chance qs of becoming established, we expect that one of them will become established when

$$\int_0^\tau mn_{q-1}(t)dt \sim \frac{1}{qs}. \tag{3.10}$$

Assuming this takes a time long compared to $1/qs$, the integral is $e^{(q-1)st}/[q(q-1)s^2]$ which means that the time, τ_q, for establishment of the new lead population is

$$\tau_q \approx \frac{1}{(q-1)s} \ln(s/m). \tag{3.11}$$

Indeed, because of the exponential increase in the rate of mutations, if no mutant has become established by τ_q one is very likely to be in another, smaller, time of $1/(q-1)s$ or so later. Thus the variations in τ_q are small compared to τ_q itself by a factor of $1/\ln(s/m)$: we assume that s/m is very large so that $\ln(s/m)$ is itself a relatively large parameter.

A more detailed analysis shows that many additional similar mutant populations will be established soon after the first. Although these start growing later and are each typically substantially smaller than the first-established, collectively they decrease the effective establishment time by about $\ln(q-1)/(q-1)s$ canceling a factor of $q-1$ that would have appeared inside the logarithm had we used the first establishment alone. In practice, such corrections of order unity are comparable to the errors we are making from the approximations, in particular by assuming $\ln(s/m)$ is large. In particular, changing the establishment probability from qs to cqs to reflect different birth and death processes would result in similar small corrections.

We have derived the time for the front of the distribution to advance one step. For consistency, this must also be the time in which the mean fitness advances by s. Thus

$$v \approx \frac{s}{\langle \tau_q \rangle} \tag{3.12}$$

in terms of the mean τ_q (although, as noted, τ_q does not vary much: it is typically close to its mean).

But we now need to find q. After the lead population has become established, it will grow essentially deterministically with mutations into it from less fit populations no longer playing a substantial role. Thus the lead population proceeds from establishment by mutation to growth by selection. Under the conditions assumed above that the mean $\bar{x}$ advances by one around the same time as the front advances — i.e. the new lead becomes established — the (soon-to-be) second-fittest population grows at rate $(q - 1)s$ for a time $\tau_q = s/v$, after which $\bar{x}$ advances and it grows more slowly at rate $(q - 2)s$ for a further time interval s/v. Its fitness advantage over the mean decreases step by step until it becomes the dominant population. This takes a total time

$$\tau_{sw} \approx (q - 1)\tau_q \approx \frac{\ln(s/m)}{s},\tag{3.13}$$

the steady-state sweep time for new mutations. During this time, the formerly-lead population has grown to a size that, for consistency, must be about N so that:

$$\frac{1}{qs}\exp\left[\frac{q(q - 1)s^2}{2v}\right] \sim N\tag{3.14}$$

yielding, after plugging in for τ_q, the lead

$$q \approx \frac{2\ln(Ns)}{\ln(s/m)}\tag{3.15}$$

(ignoring the factor of q inside $\ln Ns$ which is in any case comparable to other factors we are ignoring).

The speed of evolution is obtained from the consistency condition that the lead population sweeps to become the dominant population of size N, yielding

$$v \approx s^2\frac{2\ln(Ns) - \ln(s/m)}{\ln^2(s/m)}.\tag{3.16}$$

Several aspects of these results are important to note. Most dramatically, the dependence of v on N has gone from linear in the successional mutations regime to logarithmic at higher populations. Almost all the mutations that occur in the large populations are wasted: only those occurring near the front of the distribution — on the already fittest multiple mutants — are important, the others are destined to die out after being out-competed by the fitter ones. The primary role of the bulk of the population is to lower the mean fitness. Away from the front, mutations have little effects on the dynamics: selection completely dominates. Thus the steady state distribution and its speed are determined by the balance between mutations at the front and selection in the bulk of the distribution. This is why the overall production rate of mutations, Nm, does not enter: m enters rather in the combination s/m and the behavior is only logarithmically dependent on the mutation rate.

Because selection dominates most of the distribution, the evolution speed is very well approximated by the the variance of the fitness, the general result mentioned earlier:

$$v = \frac{d\bar{\phi}}{dt} \approx \text{var}[\phi]. \tag{3.17}$$

But as we have seen, what really matters is the front of the distribution which is many standard deviations away from the mean: the variance is not a very useful characterization of the distribution. It is the balance between the mutational dynamics at the front — a tiny fraction of the population — and selection in the bulk that determines both the steady state distribution (including the variance) and the speed. On the simple fitness staircase we are considering, the distribution of fitness is close to gaussian many standard deviations away from the mean, indeed, until the sub-populations are of order $1/qs$. [Note that this is rather unusual: distributions tend to be gaussian (if at all) only near their mean — the *central* in the "central limit theorem" — with tails of different forms.]

The evolution of the population distribution is, of course, not really steady. But the nature of the dynamics at the front implies that it does not fluctuate much. A proper analysis of the fluctuations is rather complicated, but can be done along similar lines to the above heuristic analysis. [12]

We briefly mention several minor caveats about the above results. Strictly speaking, Eq. (3.16) is only valid for particular values of, say, $\ln(Ns)$ for which the lead advances at the same time as the mean advances. In general, the dependence on q, and hence on the two logarithmic factors is more complicated. In addition, for very large N, v becomes comparable to s^2 and larger. In this regime, significant new establishments continue in the second-fittest population while establishments of the new fittest population are occurring. The mean fitness in this regime advances more smoothly, and there are several sub-populations around $\bar{x}$ that contribute to the total as the standard deviation of the distribution is larger than s. But Eq. (3.16) is a very good approximation for the whole regime with $Nm \gg 1$.

Crossover between regimes The multiple-mutations analysis is valid when $Nm \gg 1$. The border of its validity, $N \approx 1/m$, corresponds to $q = 2$: this is when the fittest, sweeping, population produces new mutants soon before it become the majority population. For not-much larger N, there thus appears a small population two steps above the mean. For smaller N, the condition for validity of the successional mutations regime is that $Nm \ll \ln(s/m)$. Between these regimes there is a crossover that is straightforward to work out. Especially as this occurs only in a narrow range of N, we ignore this here and refer the reader to reference [12].

3.4. Beyond the staircase model

The staircase model we have been discussing is very unrealistic. It has several key simplifications: first, that all beneficial mutations have the same fitness advantage; second, that there are no deleterious mutations; third, that the effects of the beneficial mutations are additive; and fourth, that there is an infinite supply of beneficial mutations so that they are not depleted. But one of the advantages of starting with such a simple model is that additional effects can be added and understood one by one. We briefly discuss relaxing each of the assumptions and some of the additional features that can then occur.

3.4.1. Distribution of beneficial mutations

One could argue that when all mutations confer the same selective advantage and their effects are additive, a rough result for the speed could be guessed without any calculations: with important mutations acting only in the small front of the distribution where the populations are small, logarithmic dependences on population size and mutation rates could have been anticipated. As the basic scale of the speed is s^2, this suggests v will be equal to s^2 times logarithmic factors. But when there is a distribution of fitness increments, one can no longer make such an argument: what s would one use for the basic scale of the speed?

In reality, different beneficial mutations will give rise to different increases in the fitness. If there are many possible such mutations, each individually with a very low rate, these can be modeled by a distribution of mutation rates: $\mu(s)ds$ for mutations with fitness increments in the range $(s, s + ds)$. As mutations of large effect are likely to be fewer in number — of if more complicated mutational processes (see later) are involved, would occur at much lower rates — we expect $\mu(s)$ to fall-off with increasing s. Which range of s is most important for the evolution?

If the population size is sufficiently small that a mutation arises, becomes established, and fixes before others can become established, the evolution occurring via the one-by-one establishment of a succession of mutants with a distribution of strengths. Given an establishment probability of s and a fitness increment of s, the speed of evolution is

$$v \approx N \int_0^\infty s^2 \mu(s)ds \tag{3.18}$$

so that in this successional mutations regime it is the mean-square s that controls the behavior.

But when the population size is larger — roughly when the total beneficial mutation production rate, $N \int_0^\infty \mu(s)ds$, is large — then new mutants can establish before earlier ones fix. This can have large effects. For example, if a mutation, A, with s_A becomes established, its population will grow exponentially. But if, before it takes over the population, another more beneficial mutation, B, occurs with $s_B > s_A$, B can out-compete A even though it arose later. If this occurs, then the mutation A is wasted. This process is known as *clonal interference* between the different mutant lineages. [14] As mutant populations with larger s grow exponentially faster than those with smaller s, this interference suggests that the evolution will be dominated by mutations with anomalously large s. But if s is too large, the mutations will be so rare that other smaller ones will arise and fix first. Thus there should be some dominant range of s not too large and not too small. Various authors have considered this effect and tried to estimate the dominant s and the resulting behavior.

But there is a crucial complication: while mutation A may be out-competed by a stronger mutation B, the population with A could itself produce a mutation C: if $s_A + s_C > s_B$, the double-mutant AC can out-compete B. Indeed, when clonal interference occurs, such double mutants will also: if the original population can produce many new further mutants before earlier ones fix, a mutant population can produce some double mutants before it fixes. Thus whenever clonal interference is important, *multiple mutations* are likely to be important as well.

Analyzing the interplay between multiple-mutation and clonal interference is well beyond the scope of these lectures and is still only partially understood. But it can be shown that a simple approximation works rather well. For each s, acting alone, the speed, v_s, can be estimated from the constant-s model using an effective mutation rate

$$m_s \sim s\mu(s). \tag{3.19}$$

Then v_s is maximized to find the most effective s: the strength $\tilde{s}$ of these *predominant mutants* gives

$$v \approx \max_s v_s = v_{\tilde{s}}. \tag{3.20}$$

This predominant mutants approximation turns out to be surprisingly good. As long as the distribution $\mu(s)$ falls off sufficiently rapidly (faster than a simple exponential), the predominant s is roughly independent of N for large N. Although the predominant mutants approximation suggests that a broad range around $\tilde{s}$ is likely to contribute, it turns out that this is not the case: the important range of s around $\tilde{s}$ is narrow compared to $\tilde{s}$. [12] But $\tilde{s}$ does depend weakly on the overall mutation rate. The simplest to consider is increasing the rates of all types of mutations uniformly: i.e., multiplying $\mu(s)$ by a factor of g. This results in a decrease in $\tilde{s}$ and a somewhat weaker dependence of v on g than in the simple staircase model. But this is hard to distinguish from various other effects (noted below). Thus detailed predictions of the effects of increasing mutation rates are not robust. Yet the weak logarithmic dependence on the mutation rate in the multiple mutations regime is robust, and is in striking contrast to the linear dependence in the simple successional mutations regime.

3.4.2. Deleterious mutations and optimal mutation rate

Most mutations are not beneficial: far more are deleterious. Thus the distribution of mutation rates, $\mu(s)$, should have most of its weight at negative s. In the absence of beneficial mutations there will be an equilibrium distribution of deleterious mutations present in the population: as discussed earlier, the deterministic approximation is usually good in this case. When beneficial mutations are present and there is continual evolution, the deleterious mutations still play a role, in particular altering the shape of the fitness distribution and slowing down the evolution somewhat. These effects are small unless the deleterious mutation rate is rather large. But if it is large, the mean fitness can actually decline. This phenomena, known as Muller's ratchet, occurs if the fittest genome in the population disappears because of deleterious mutations, then the next fittest, and so on. In this situation, even without beneficial mutations, the deterministic approximation fails and fluctuations — most crucially of the fittest remaining population — dominate.

We thus see that there are two competing effects of increasing the overall mutation rate: more beneficial mutations increases the speed of evolution but more deleterious mutations decreases the speed. This raises an important general question: what is the optimal overall mutation rate? It is not at all clear how to frame this question in any general situation. But in the specific context of continual evolution with a distribution of beneficial and deleterious mutations whose effects are additive, it can be addressed. Specifically: if the overall mutation rate is increased by a factor of g, by $\mu(s) \rightarrow g\mu(s)$, the speed of evolution changes. For small g, v increases with g — first linearly, then logarithmically. But for large enough g, the deleterious mutations start to dominate and v decreases. [13] This implies that there is an optimum g which depends on the population size and the distribution $\mu(s)$.

3.4.3. Interactions between mutations

The effects of different mutations are generally not additive. Specifically: the selective advantages (or losses) of a mutation A, a mutation B and the double mutant AB are not simply related:

$$s_{AB} \neq s_A + s_B. \tag{3.21}$$

Such *interactions* between the effects of mutations is known as *epistasis*. It surely plays crucial roles for long term evolution — including speciation via of separated sexual populations. But more simply, in asexual populations, interactions between mutations would appear to invalidate the scenario for the acquisition of multiple beneficial mutations that we have been analyzing. One effect of interactions is conditionally beneficial (or deleterious) mutations for which a first beneficial mutation, A, changes whether or not a second mutation B is beneficial, or how beneficial it is. But as long as $s_{AB} > s_A$, the second mutation can add to the first, whether or not it would have been beneficial on its own. Similarly, mutation A could eliminate the potential of an otherwise-beneficial mutation B. An important example of this is mutations that are in some sense in the same class: if any of a number of different mutations results in the same phenotypic changes with a second mutation in the same class giving no further effect, then this class of mutations can be considered as one type of mutation with a rate that is the total rate of all the mutations in the class.

How do these various forms of epistasis affect the dynamics of asexual evolution via acquisition of multiple beneficial mutations ? What is needed for the scenario we have analyzed to obtain is *not* that the effects of mutations are actually additive. The crucial feature is that there are a large number of beneficial mutations always available with the *distribution* of their selective advantages, $\mu(s)$, roughly independent of earlier mutations — even though which mutations are available depends on the past history. If this is the case, then the scenario we have analyzed is a good approximation to the dynamics and our quantitative results should be applicable.

But there are other effects of interactions between mutations that do not play a role in the uphill climbs we have discussed but could nevertheless be important: *deleterious intermediaries*. For example, if s_A and s_B are both negative, but s_{AB} is positive: two mutations are then needed to produce the beneficial combination with the first step downhill in fitness. As this requires two mutations, for any particular such *two-hit* process, the rate will be very small. But a crucial question then arises: how many such potentially beneficial two-hit processes are there relative to the number of beneficial single mutations? And how small are the rates? We return to these questions later.

3.4.4. *Depletion of beneficial mutations*
In a constant environment with only the simplest genome-independent competition which does not change as the organisms evolve, one would expect there to be

locally optimal genomes whose fitness cannot be increased by single mutations — i.e., fitness peaks. If such a peak is reached, there will be no more beneficial mutations available — except more complicated processes with deleterious intermediaries such as the two-hit process discussed above. Before a peak is reached, the supply of beneficial mutations is likely to decrease and the rate of increase of the fitness slow down. If the effects of beneficial mutations are additive, they will simply be depleted, although how long this takes depends on whether there are a modest number of available beneficial mutations — or classes of such mutations — with relatively high rates, or many more available mutations but each with much lower rates. With interactions between mutations the situation is more complicated: if on average each beneficial mutation acquired enables one other to become available, the evolution can continue — unless an unlucky route that ends in a local fitness maximum is taken. And if two-hit processes with deleterious intermediaries can occur, the chances of becoming stuck at a local fitness maximum is far lower. Understanding the possible behaviors even with a constant environment requires far more knowledge of local fitness landscapes. And these depend on many aspects of the biological architecture as well as particulars of the past history and the type of selective pressures in the current environment.

3.5. Experiments on the speed of asexual evolution

To test the basic results of the theory outlined above for acquisition of multiple beneficial mutations, Michael Desai undertook experiments on asexual evolution of budding yeast in Andrew Murrray's lab. [15] The goal was to investigate the dependence of the evolutionary dynamics on the mutation rate and population size. The environmental conditions used were low glucose in which the yeast divided about 70% as fast as in high glucose: these conditions should cause sufficiently broad stresses that many potentially beneficial mutations are available. To eliminate all but the simplest competition, the yeast were kept in exponential growth phase at low enough densities that interactions between them were not important. The selection was simple: at regular intervals all but a small fraction of the cells were discarded, so that lineages that divide faster yield a larger fraction after the next dilution: this is roughly equivalent to keeping a fixed effective population size, N (approximately the geometrical mean of the time-dependent population size). Three different population sizes, with N spanning a factor of 2500, were used and two strains with different mutation rates, differing by about a factor of ten: the higher rate was a "mutator" strain which had one of its mutation repair mechanisms knocked out. After each of the populations evolved for some time, it was mixed in with a marked unevolved strain and the difference between their fitnesses measured by direct competition. At the end of 500 generations, some of the evolved populations were sampled and the fitnesses of

96 individuals from each measured to obtain the fitnesss distribution within the population. We will not reproduce the results here, but summarize some of the salient features. [15]

Even in the largest population of mutators, there is no sign of depletion of the supply of beneficial mutations: the rate of increase, v, of the mean fitness stays roughly constant over the 500 generations. The dependence of the evolution speed on population size is far weaker than linear — as would have been the case in the successional mutations regime — and consistent with logarithmic in N. The dependence of v on the mutation rate is also much weaker than linear. These suggest that the multiple mutations scenario does indeed apply in this experimental context. An important check on this is provided by the fitness distributions: in contrast to the usually-monoclonal, sometimes-bimodal behavior expected in the successional mutations regime, in the large populations the distributions have a single peak with a substantial width consistent with expectations.

All of the data can be well-fitted by the simple staircase model with a single value of s — about 2% per mutation — and two values of beneficial mutation rate m differing by the expected factor of ten. With these parameters, the expected lead of the largest mutator populations is about $q = 4$ so that quadruple mutants above the mean sweep together — far faster than individual 2% mutants could on their own. Deviations of the measured speeds from predictions are within the ranges expected from fluctuations. There is some excess width to the fitness distributions of the mutator populations which can reasonably be attributed to deleterious mutations ignored in the simple model. Other scenarios for the evolution, in particular by a series of small successional mutations or by one or two large ones, are ruled out (except perhaps by appealing to fortuitous circumstances).

That the data appear to be described so well by the highly overly simplified staircase model is *a priori* surprising. But in light of the discussion above about distributions of beneficial mutations, it is reasonable to expect that there is a characteristic strength, $\tilde{s}$, of beneficial mutations that dominates the dynamics. If the probability of mutations with $s > \tilde{s}$ falls off rapidly, then the continual evolution will happen via multiple mutations of the predominant size $\tilde{s}$ with $\tilde{s}$ depending only weakly on $\ln N$. Some dependence of $\tilde{s}$ on the overall mutation rate is expected, but with a factor of ten difference of mutation rates between the normal and mutator strain this probably has little effect beyond that predicted by the simple model with fixed s. In any case, as noted above, increasing the overall mutation rate also changes other aspects of the dynamics: the distribution $\mu(s)$ will not be increased uniformly (different types of mutations are affected differently by knocking out a particular mutation repair system); the relative likelihood of two-hit mutations increases; the depletion of some of the beneficial mutations speeds up; and deleterious mutations play a larger role.

The quantitative agreement of the experiments on yeast evolution is satisfying — but we must remember that the experiments were explicitly designed to test the simplest situation beyond acquisition of a single beneficial mutation. And one might argue that the results are biologically boring: Why should one care about many small changes in a mildly stressful non-interacting environment not that different from environments the organisms have experienced before? Such an attitude may be reasonable as far as understanding *current* biological function. But for understanding evolution, it is totally unreasonable: How can one even begin to understand the dynamics of interesting evolutionary processes without understanding the simplest? And if nothing really makes sense except in light of evolution ...

We now turn to other — and surely more interesting — aspects of evolutionary dynamics. But even to frame good questions, the insights from the simplest evolutionary processes is essential.

4. Recombination and sex

Sex, in the general sense of combining some of the DNA from two organisms, surely plays a crucial role in evolution. Almost all successful groups of multi-cellular organisms reproduce sexually — some always, others only occasionally. The long-term benefits of sex must thus outweigh the shorter term costs, including maintaining the complicated mechanisms for sexual reproduction and the "wasting" of reproductive effort by females when they produce males rather than reproducing parthenogenetically: sex requires producing more offspring to pass on their genetic material. [16]

Bacteria, which reproduce asexually by division, nevertheless have various mechanisms for picking up DNA from other bacteria (and viruses). A well known benefit — to the bacteria! — is acquisition of antibiotic resistance by acquiring a functional group of genes — an "operon" — from another bacterium that has evolved mechanisms for dealing with similar chemicals. [8]

Many potential benefits of sex have been discussed in the literature for both single-cell and multi-cellular organisms. [16] But which are the dominant benefits in which circumstances is controversial and little understood.

A concrete benefit of sex can occur in large populations with many potentially beneficial mutations available: the situation we have been analyzing. As we have seen, with purely asexual reproduction most mutations are wasted in large populations: only those in the already-fittest individuals tend to matter and the rate of evolution increases only logarithmically with the population size. With sex, the evolution rate could be much faster. An extreme model is instructive: in each generation assume all the genes (or even parts of genes) recombine randomly so that an individual genome includes one of each gene but with the specific allele chosen independently from the pool of variants of that gene. If the effects of mutations are roughly additive, then — at least naively — each gene will evolve separately. The rate of change of the fitness would then be the sum of the rate of changes of the fitnesses of each gene and the overall speed of evolution would continue to be mutation limited. Thus v would grow linearly with the population size even when a large numbers of different mutations are simultaneously present in the population. This suggests an enormous advantage of sexual reproduction in large populations. But what happens in more realistic models of recombination?

Various types of recombination can be studied in the simple staircase model: a large supply of beneficial mutations each with the same fitness advantage and their combined effects additive. The diversity within the population now plays a crucial role: it is not sufficient to know how many individuals there are with a given number of mutations. As they can recombine to be in the same individual, one needs to know the distributions of the *different* specific mutations among the subpopulations. There have been various efforts to analyze some forms of recombination in this model. [17–19] But it is still far from understood. For example: with a fixed recombination rate, for very large populations is the evolution speed proportional to N, to a power of $\ln N$, or something in between? Indeed, is the linear dependence on N correct even in the extreme model?

The applicability of the additive approximation for beneficial mutations is much more questionable when there is recombination. In the asexual case, the crucial feature is that the *distribution* of potentially beneficial mutations does not depend much on the past history: i.e., on which mutations have already been acquired. Thus while interactions between mutations are important, they are primarily so in a certain statistical sense: how the interactions affect the evolution of the distribution of available beneficial mutations. But as soon as there is recombination, the interaction between the specific mutations that have accumulated in two different individuals is crucial. In general, there are likely to be incompatibilities. For example, if in lineage 1 there is a mutation B_1 that was conditionally beneficial on an earlier beneficial mutation A_1 (i.e. $s_{B_1} < 0$, $s_{A_1} > 0$, and $s_{A_1 B_1} > s_{A_1}$ and in lineage 2 mutation D_2 is similarly conditionally beneficial on C_2, then it is more likely than not that, e.g., the recombinations $A_1 D_2$ and $B_1 C_2$ are less beneficial or deleterious. Thus, in this case, sex breaks up beneficial combinations — one of its negative effects. Indeed, it is just such an effect that can be a source of speciation in separated populations: these can accumulate different mutations which are incompatible so that the populations can no longer productively mate. But for long term evolution, what matters most (as we have seen for large populations of asexuals) are the anomalously fit individuals. Thus the rare matings that produce individuals far fitter than average — for example by combining different beneficial mutations even when their effects are simply additive — can be the most important.

In organisms such as yeast which can reproduce either sexually or asexually, sex can provide a valuable probe of asexual evolution, particularly the distributions of beneficial (and approximately neutral) mutations and interactions between these.

To even begin to understand the effects of sex requires far more knowledge of how multiple genetic changes together determine the phenome of organisms: again, this depends crucially on the biological architecture and past evolutionary history. We comment briefly on such issues — whose addressing requires going well beyond phenomenological theory — at the end of these lectures.

5. Deleterious intermediaries and combinatoric possibilities

So far, we have considered the evolutionary effects only of mutations that increase the fitness in the current environment. This corresponds to the conventional picture of asexual evolution by a series of uphill steps. But at least in large microbial populations, two-hit mutations that involve an intermediate downhill step can occur on reasonable time scales. Even if the intermediary is lethal so that the two mutations must happen the same generation, this can occur. For example, if the roughly 10^{15} bacteria in a human body divide every few days, even with point mutations rates as low as 10^{-9} per cell division, in the lifetime of a single human host a large fraction of the possible *simultaneous* two-point mutations are likely to have occurred in the most common species of the human's bacterial ecology. But this is a drastic underestimate of the rates of double mutations.

Consider a beneficial double mutation which increases the fitness by s, but with the intermediary deleterious with loss of fitness δ. We will refer to these as two-hit beneficial mutations even when the two mutations occur in different generations. If the mutation rate for the second mutation is μ, then we need to estimate the probability that a first mutation to the deleterious intermediary gives rise to a second mutation that establishes. Since the fate of each first mutation is independent, this depends only on the probability, ϵ, of a single intermediary mutant individual giving rise to an established favorable double mutant: ϵ is thus the *establishment probability* for the double mutant which plays the role of the establishment probability, $\epsilon = s$, for a single beneficial mutation.

First consider the neutral-intermediary case, $\delta = 0$. The lineage from a typical individual will die out in a few generations, so the probability that a mutation occurs and establishes from one of this lineage is of order μs. But with probability of order $1/\tau$, the lineage will survive for more than τ generations, and if it does so, its population size, $n(\tau)$, will become of order τ. The probability that such a lineage gives rise to an established mutant is $\mu s \int_0^\tau n(t)dt \sim \mu s \tau^2$ until this becomes of order one: lineages that survive longer than $\tau \sim 1/\sqrt{\mu s}$ are very likely to do so. Thus the probability that a single first mutant gives rise to an established beneficial double mutant is

$$\epsilon \sim \sqrt{\mu s} \tag{5.1}$$

This establishment probability is dominated by the rare lucky intermediary that lasts for an anomalously long time.

If the intermediary is weakly deleterious, then (as discussed earlier) it lineage is effectively neutral for times up to of order $1/\delta$. Thus if $\delta < \sqrt{\mu s}$ the neutral result applies. In one specific context, this condition on δ thus provides a concrete answer to the question: How neutral does a mutation need to be to be "neutral"?

The non-neutral regime obtains when

$$\delta > \sqrt{\mu s}. \tag{5.2}$$

The longest a deleterious lineage is likely to last is of order $1/\delta$, in which case it will reach a population size of order $1/\delta$ and the probability that such a lineage gives rise to an established second mutant is $\sim \mu s/\delta^2 \ll 1$. Since this happens with probability of order δ, we conclude that

$$\epsilon \sim \frac{\mu s}{\delta}. \tag{5.3}$$

[In this regime, the process is loosely analogous to quantum mechanical tunneling through an intermediate state with energy higher by δ.] Note that for $\delta \sim \sqrt{\mu s}$ this becomes the neutral result as it should.

The establishment rate for a two-hit mutation in a population of size N with first mutation rate μ_A, second mutation rate μ_B, a least-deleterious intermediary mutant, A, with $s_A = -\delta_A$, and selective advantage of the double mutant, s_{AB}, is thus

$$N \mu_A \epsilon_{AB} \sim N \mu_A \min \left[\frac{\mu_B s_{AB}}{\delta_A}, \sqrt{\mu_B s_{AB}} \right]. \tag{5.4}$$

This result should obtain as long as the population size is large enough that the maximum intermediary population size needed for the above argument is $\ll N$: this is not a stringent condition, at worst requiring $N \gg 1/\sqrt{\mu_B s_{AB}}$.

For three-hit mutations, ABC, with two deleterious intermediaries, A and AB, the above argument can be iterated using the establishment probability ϵ_{BC} for the a BC double-hit mutation from an A individual (in place of s_B in the above), to obtain the establishment probability ϵ_{ABC}. For the case with both intermediaries almost neutral — now requiring A to be extremely close to neutral and the population sufficiently large — the overall establishment rate of the triple mutant is

$$N\mu_A \epsilon_{ABC} \sim N\mu_A \sqrt{\mu_B} [\mu_C s_{ABC}]^{\frac{1}{4}} \tag{5.5}$$

while if the intermediaries are more deleterious:

$$N\mu_A \epsilon_{ABC} \sim N\frac{\mu_A \mu_B \mu_C s_{ABC}}{\delta_A \delta_{AB}}. \tag{5.6}$$

The above results for two-hit mutations were recently obtained from exact calculations [20], but without the heuristic arguments that aid their understanding and are needed for more complicated situations.

For a K-hit beneficial mutant with deleterious intermediaries of typical strength δ that is not tiny, and mutation rates of order μ, the establishment rate is similarly

$$\frac{1}{\langle \tau_{\text{est}} \rangle} \sim Ns\mu \left[\frac{\mu}{\delta}\right]^{K-1}. \tag{5.7}$$

Note that this is just μs times the average steady-state number in the population, n_{K-1}, of the multiple mutant that gives rise to the beneficial mutant.

The problem with the deterministic approximation for this process (discussed earlier) is now apparent. Because of the exponential growth as $n_K(t) \approx e^{s(t-\tau_{\text{est}})}/s$ of the fitter final mutant once it is established at time τ_{est}, averaging $n_K(t)$ over the distribution of τ_{est} is dominated by extremely rare anomalously fast establishments for which $\tau_{\text{est}} \ll \langle \tau_{\text{est}} \rangle$. At such an establishment time which dominates the average, the actual n_K is almost always zero, but in the extremely rare cases when it is non-zero, this quickly gives rise to a population that dominates the average and rises to be much larger than N while the average is still small and the chance of establishment still tiny.

We have found that he rate for any particular multi-hit mutation is very low, even with weakly deleterious intermediaries. But this brings us to a crucial question: how many potentially beneficial K-hit mutations are there? For point mutations, in a genome of length G there are of order G^K K-tuples. But surely most of these are unlikely to be beneficial. Nevertheless, it is reasonable to expect that the number of potentially beneficial K-hit processes grows exponentially in K, say as Q^K with Q a large number. This means that that the *total* rate for K-hit processes is proportional to

$$Ns\left(\frac{Qm}{\delta}\right)^K \tag{5.8}$$

a product of a large and a small number to a power. Whether the product Qm/δ is a large or a small number is one measure of whether the mutation rate should be considered "large" or "small" in this environment. Loosely speaking, this determines whether the exploration of genome space in an evolving population is local or far-reaching: surely a crucial question.

Even for single-point beneficial mutations in microbes, how many there are in a typical environment with broad stresses is not known — although the experiments on yeast discussed above and other experiments on *E. coli* give some indications. About multi-hit possibilities, nothing is known. Phenomenological analysis is useful for raising such questions and considering their potential consequences. But, once again, one cannot even hope to answer them without far more understanding of the biology and of evolutionary histories.

6. Beyond the simplest questions

In these lectures we have focused exclusively on understanding evolutionary dynamics when the mapping between the genome and phenome is given. And even that only in constant environments in which the fitness is a single quantity which depends on specific aspects of the phenome and is hence a function of only the organism's own genome. We have only considered the simplest interactions between organisms: competition for total resources which implicitly also assumes no spatial structure of the populations. We have seen that even with these gross simplifications the evolutionary dynamics can be subtle. And, as soon as there is sexual recombination, very little is understood.

We end by discussing briefly three general directions in which far more theory is needed. For all but the last, laboratory experiments and close interactions between theory and experiments are essential.

6.1. Space, time and ecology

Most theoretical studies of population dynamics focus on phenotypes with genetic variability assumed. This is already an enormously rich subject even if the physical environment is constant. But once mutational processes are included, very little is understood.

Temporal variations of environments are surely crucial for long term evolution: the many different "tasks" organisms face — as individuals during their lifetimes and as populations on longer time scales — mean that their "fitness" is a poor concept. One has to, at a minimum, consider multiple aspects of fitness: most simply, fitnesses in different contexts. As far as one type of organism is concerned, time-dependence of the physical environment and of the biological environment are similar (although the stresses they cause may be very different). And once genetic changes are considered, little is known: even with only two distinct environments, the interplay between the time scales for the environmental variations and the genetic changes on evolutionary dynamics is only beginning to be explored. Of course, organisms feed back and change their own environments. Once there are several types of organisms, this gives rise to ecologies.

Simple ecological interactions between organisms which depend on the species or strains involved can give rise to stable coexistence, to oscillations, and to chaos (studies of these phenomena by Robert May and others played an important role in the developments of understanding of chaotic dynamics). The interplay between these effects and genetic changes — i.e. the evolution of simple ecologies — has been little explored and is a ripe for both experiments — some underway — and theory.

Another essential complication is spatial variation and mobility of populations. Even in the simplest models with organisms that are phenotypically identical but differ by some neutral mutations (most commonly, mutations in protein coding regions that do not change the amino acids because of the redundancies in the three-nucleotide to amino-acid genetic code), the spatial dynamics of populations is interesting and subtle. In recent years, such neutral genetic differences have been used to track human migrations. But even some of the simplest questions in the simplest models with stochastic spatial motion are not yet answered. Again, once genetic changes occur, additional complications arise. And of course, phenotypic variability brings in the full richness of evolution.

Even within the simplest model of acquisition of beneficial mutations that we have discussed, the interplay between sexual recombination and spatial variation is essential to understand. When mating is within separate fractions of the population on short time scales but there is mixing of the populations on longer time scales, the evolution can be very different than in fully mixed populations. And, of course with interactions between mutations. . .

6.2. *Biological architecture*

To even begin to address questions about how interactions between genetic changes combine to give phenotypic changes, one needs to understand many aspects of biological organization of cells and organisms.

There are two extreme caricatures of the architecture and functioning of a cell. One is that each protein (and its regulation) has a well defined function or functions and that these are grouped into modules which are themselves linked together to perform higher level functions, etc. With this as a paradigm, molecular and cell biology (aided recently by genomics) have established many connections between genes (and gene regulation) and phenotypic traits. But these successes have given rise to a bias towards thinking of evolutionary processes in too restrictive terms.

The opposite extreme is a holistic network of multiple interactions between proteins and other components. If this were the structure, changes of almost any part would affect much of the rest — and therefore multiple aspects of cell behavior. This scenario is closer to the macroscopic paradigm of evolution via selection on quantitative traits that are affected by many genes. In this paradigm, most genetic changes have side-effects — pleiotropy. And microscopically many genes have multiple functions at present, different functions in the past, and, potentially, new functions in the future.

These caricatures represent very different views of genotype to phenotype mappings. How does the extent to which each of these is true affect evolutionary potential — the *evolvability*? And how does it affect evolutionary dynamics? Conversely, how does evolutionary history affect the extent to which these caricatures represent reality? Even rough answers to these are needed to address some of the questions we have raised earlier about interactions between mutations — most crucially impacts of changes on each other and thus the combinatoric possibilities.

An advantage of modular architecture is that changes in one module are less likely to result in deleterious effects on others and thus incompatibilities between different beneficial mutations. But the network picture suggests enormously more combinatorial possibilities of changes, even if few of these are beneficial. Laboratory evolution experiments together with genetic methods and re-sequencing to track down changes are just beginning to start addressing these questions: for now in a small number of specific contexts, but in the future potentially in a wide enough spectrum of evolutionary contexts that general lessons can be learned. Better theoretical understanding of the evolutionary dynamics is needed both to design and to interpret such experiments. And such experiments will enable more useful general modeling and analysis.

6.3. Abstract models

To develop understanding — qualitative and quantitative — of broad issues in evolutionary dynamics, abstract modeling is also needed. Scenarios can be studied in toy models that are crude caricatures of a few potentially important features. And general issues of dependence of key quantitative parameters — e.g. population and genome sizes — can be analyzed.

As discussed in the introduction, to many — from Darwin on — the biggest puzzle in evolution is the evolution of complex functions. But to even begin to think about how this occurs so — seemingly to many — fast, one needs some quantification of degree of complexity and of how fast is "fast".

Independent of their motivations for doing so, some advocates of "intelligent design" as an "alternative" to evolution have tried to introduce notions and questions that should be taken far more seriously than they have been. In particular, some have focused on the apparent "irreducible complexity" of certain biological functions — loosely, how many components and interactions these need to function at all. Whatever the legitimate criticisms of the examples they have chosen (and the lack of understanding of biology these might represent) the general issue cannot be waved away. This is not an issue of marginal improvements of existing functions, but of the evolution of "new" — whatever that means — functions. It is surely true that even many of the simple functions that cells perform could not be evolved by a series of purely beneficial mutations or other genetic changes *in their current biological context*. Perhaps they could occur by a route that also changes many other functions of the cell. But to invoke such an explanation relies on murky assumptions about the biological architecture and dependence of fitness on multiple functions. And this is a major part of what one is trying to understand. In the real world, the evolution of complex functions surely relies crucially on past evolution of other functions. This issue is thus at the heart of evolvability.

The state of understanding of the difficulty of evolving any even moderately complex processes is so poor that almost any progress on abstract models that include some of the essential aspects of evolution would be valuable.

A long term goal is to formulate and address questions about how the difficulty of evolving functions depends on their complexity, and how this depends on the basic biological architecture — crudely, modular or holistic network — on recombination, and on quantitative parameters. For a class of functions that have increasing complexity loosely parametrized by some H, how does the difficulty of evolving these grow with H? To make sense of this question, one first needs well defined classes of biological-like functions, such as pattern recognition, for which knowledge from computer science should be invaluable. For these, one then wants a definition of H — such as the minimum number of components needed, although this may not be a good measure. And one has to consider definite classes of architectures and types of mutational or recombinational processes. An obvious measure of difficulty, D, is the total number of "cell divisions", NT, needed. However, as discussed in the introduction, it is not even clear that this is the most relevant combination of population parameters: indeed, the analysis of the staircase model and ensuing discussion suggests it is not. If the difficulty, D, was proportional to a power of H, highly complex functions could evolve readily. But the naive expectation is that D grows exponentially with H. This would be expected if all H components were needed for increased fitness and the probability of these arising were the product of H small factors (as in the estimate of rates of K-hit processes). If this were the case, the largest evolvable H would be modest even with enormous NT. But high evolvability might mean that how D depends on H is intermediate between these behaviors.

This could only arise from the nature of the architecture and concomitant maps between genome and phenome that are the *least unlikely* to have evolved in an *ensemble* of environmental histories.

The crucial issue here, as already arose in the simple processes that we considered in these lectures, is extrapolation. For example: for many purposes, once the difficulty, D, of a problem grows faster than a power of its size — here H — whether D grows as e^{CH} or as, say $e^{cH^{\frac{1}{3}}}$ does not much matter. But with $NT \sim e^{100}$, it matters a great deal. As emphasized earlier, this means that simulations are of little use in the absence of a theoretical framework: extrapolation over a wide range of parameters is not possible.

Progress in developing abstract models and in framing quantitative questions would be major steps forward conceptually and certainly advance our quantitative understanding.

7. The state of the field

If nothing else, I hope these lectures have made the case that there is a huge amount to be done to even begin to understand evolutionary dynamics. I would thus like to end with a comment on the state of the field. As this author, this school, and much of the audience are physicists, it will take the form of an analogy with the field of physics. It is loose, but I think instructive.

But first a quote from Richard Lenski from a review of the late great evolutionary biologist Ernst Mayr's last book: [21]

"Mayr argues that the precise mathematics that underlie physics are not applicable to biology, in which determinism, typological thinking, and reductionism have limited utility. . . . [H]e builds on this point by splitting biology into two distinct domains, functional and historical. While functional biology may fit within a framework similar to that of physics, Mayr argues that the historical domain of biology–in essence, evolution–requires a different framework. My [Lenski's] own view is that evolutionary history, reflects dynamical processes (e.g., mutation and natural selection) that can be described mathematically and tested experimentally (as indeed they often are), although evolving biological systems are more complicated than what physicists study. [Arguing for a] distinction between the functional and historical domains of biological understanding may reflect [a] limited interest in evolutionary dynamics per se."

If the basic laws of evolution are analogous to the laws of quantum mechanics, then the simplest evolutionary process (well described by population genetics) is like the hydrogen atom. The most complicated evolutionary processes directly observed are like simple molecules. And the statistical dynamics of multiple neutral mutations is analogous to ideal gasses. At the opposite extreme is most of evolutionary theory. This is much like geology: the constituents are known, many patterns are observed — some qualifying as laws — with varying degrees of understanding, and historical scenarios are well developed and can be predictive.

But there are many levels between simple chemistry and geology. Most (although by no means all) of these are understood, each level in terms of lower levels: largely from condensed matter physics and geophysics. And these enable extrapolation over a huge range of length and time scales.

In contrast, for evolutionary dynamics understanding of most of the intermediate levels — or even what these are — is very limited. And the lack of quantitative understanding masks, I believe, severe limitations of the qualitative understanding. Genomic data is, perhaps, starting to provide the more complicated chemistry. And the simple models of acquisition of multiple beneficial mutations are perhaps like ideal periodic solids or one-dimensional Ising models. But surely a major effort combining experiments, sequencing data, observations, and theory is needed.

One can hope that in the near future far more interest will be sparked in evolutionary dynamics, *per se*.

Evolution is — in contrast to popular perceptions in the United States — a fact. The evidence, reinforced by understanding of the basic laws, is overwhelming. But it will take far better understanding on multiple levels for evolution to become a fully fledged theory.

8. Acknowledgments

My understanding of evolution — albeit very limited — owes much to colleagues too numerous to mention. Thanks are due to all these. Almost all the original work summarized in these notes was carried out in collaboration with Michael Desai to whom I am most grateful. He and Olivia White graciously provided many useful comments on these notes. This work was supported in part by the National Science Foundation, Merck Corporation, and the National Institutes of Health.

References

[1] There is an extensive literature on most of the topics discussed here. Below are only a few general and historical references, together with some particularly pertinent recent papers.

[2] D.E. Nilsson and S. Pelger, Proc. Roy. Soc. B 256, 53 (1994).

[3] R. Dawkins, Nature 368, 690 (1994).

[4]   E. Mayr, *What Makes Biology Unique? Considerations on the Autonomy of a Scientific Discipline* (Cambridge, New York 2004).

[5]   A. Novick and L. Szilard, Proc. Natl. Acad. Sci. USA 36, 708 (1950).

[6]   S.E. Luria and M. Delbruck, Genetics 28, 491 (1943).

[7]   S.F. Elena and R.E. Lenski, Nature. Revs. Genetics 4, 457 (2003); T.F. Cooperl, D.E. Rozen and R.E. Lenski, Proc. Natl. Acad. Sci. USA 100, 1072 (2003).

[8]   W.F. Bodmer, in *Prokaryotic and Eukaryotic Cells*, edited by H.P. Charles and J.G. Knight, (Cambridge University, Cambridge 1970); M.G. Lorenz and W. Wackernagel, Microbio. Review 58, 563 (1994).

[9]   W.B. Whitman, D.C. Coleman and W.J. Wiebe, Proc. Natl. Acad. Sci. USA 95, 6578 (1998).

[10] J.H. Gillespie, *Population Genetics: A Concise Guide*. (Johns Hopkins, Baltimore 1998).

[11] W.J. Ewens *Mathematical Population Genetics: I. Theoretical Introduction* (Springer, New York 2004).

[12] M.M. Desai and D.S. Fisher, submitted to Genetics.

[13] I.M. Rouzine, J. Wakeley, and J.M. Coffin, Proc. Natl. Acad. Sci. USA 100, 587 (2003).

[14] P. Gerrish and R.E. Lenski, Genetica 102/103, 127 (1998).

[15] M.M. Desai, D.S. Fisher, and A.W. Murray, submitted to Curr. Bio.

[16] J. Maynard-Smith, J. Theor. Biol. 30, 319 (1971).

[17] Y. Kim, and H.A. Orr, Genetics 171, 1377 (2005).

[18] N.H. Barton and S.P. Otto, Genetics 169, 2353 (2005) and references therein.

[19] I.M. Rouzine and J.M. Coffin, Genetics 170, 7 (2005).

[20] Y. Iwasa, F. Michor, and M.A. Nowak, Genetics 166, 1571 (2004).

[21] R.E. Lenski, Bioscience 55, 3568 (2005).

Course 12

STATISTICAL MODELLING AND ANALYSIS OF
BIOLOGICAL NETWORKS

Johannes Berg

*Institut für Theoretische Physik, Universität zu Köln,
Zülpicherstrasse 77, 50937 Köln, Germany*

J.-P. Bouchaud, M. Mézard and J. Dalibard, eds.
Les Houches, Session LXXXV, 2006
Complex Systems
© *2007 Published by Elsevier B.V.*

Contents

449

Networks of bio-molecular interactions have recently attracted attention across several disciplines, ranging from molecular biology to statistical physics. One of the reasons for the focus on interactions between genes or proteins is the surprising result that the complexity of an organism is only weakly linked with its number of genes. However loosely defined the concept of "complexity of an organism" is, it is certainly surprising that humans and the round worm *C. elegans* have roughly the same number of genes [1, 2].

This outcome of recent genome sequencing projects has put the spotlight on the interactions between genes or between proteins. With a genome consisting of $N \sim 10^4$ genes, N^2 pairwise interactions between genes are possible; the phase space of interaction networks scales exponentially with the number of potential interactions. Regulatory interactions stem from short binding sites for transcription factor molecules in the regulatory region of genes, protein interactions are mediated by matching protein surfaces. Binding sites can change quickly, leading to new interactions [3]. Such evolutionary changes in bio-molecular networks may be the mainspring of evolutionary flexibility and the emergence of new species [4].

The second reason for the interest in bio-molecular networks is amount of experimental data which has become available in recent years. A wide range of novel techniques has been developed to determine molecular interactions both *in vivo* and *in vitro*. The techniques include microarrays to probe the binding of proteins to short strands of DNA [5]. This is used to determine regulatory interactions between transcription factors and regulatory DNA. Protein interactions are measured using yeast two-hybrid screening [6], pull-down assays, where one protein type is immobilized on a gel, and 'pulls down' binding partners from a solution, or mass spectrometry [7].

Several approaches feed current theoretical activity to analyze this data. One is to construct statistical models of networks and their dynamics, and to compare observables of these models with features of the empirical data [8–10]. A frequently used observable is the distribution of connectivities [11]. Implicit is the assumption that the statistics of the network is homogeneous, i.e. different parts of the network obey the same statistics. A second approach is to search for inhomogeneities in the network statistics, which frequently can be linked with functional aspects of the network [12, 13]. In these lectures a synthesis of these approaches will be discussed, namely using statistical models of networks to quantify local deviations of the network statistics. The hinge connecting the two approaches is Bayesian statistics.

1. A primer in Bayesian analysis

Bayes' theorem quantifies how probable a hypothesis is given certain evidence. Many tools for data analysis are founded on Bayes' theorem, in particular methods for inferring model parameters from data. At the heart of these methods are statistical models which connect the hypothesis with the observed data. Hence Bayesian methods applied to systems with many degrees of freedom are intimately linked with statistical mechanics.

We illustrate Bayesian statistics with a very simple example. Consider a process in which a set $\{x_i\}$ of independent random numbers is generated; $x_i \in \{0, 1\}, i = 1 \ldots N$. $x_i = +1$ is chosen with probability λ, and $x_i = 0$ with probability $1 - \lambda$. Subsequently, a second set of independent random numbers $\{E_i\}$ is generated. If $x_i = 0$, the variable E_i is chosen from the ensemble $P(E)$, if $x_i = 1$, E_i is chosen from the ensemble $Q(E)$. An example of such probability distributions, as well as an instance of random numbers $\{E_i\}$ is shown in Figure 1 a) and b).

This setup can be thought of as mimicking the binding energies of transcription factor (TF) molecules to DNA. Sites on the DNA can either be functional, so a transcription factor binding to this site will affect the expression of a gene. The binding energies of a TF to such a site are under evolutionary selection to be high. Or, on the other hand, sites can be non-functional; binding of a transcription factor to such a site will have no effect. There is no substantial selection pressure on the energy of binding to such a site. As a result, the statistics of the binding energies is different for functional and for non-functional sites.

Now suppose the parameter λ controlling the relative frequency of the two ensembles, and the information x_i of which ensemble the variable E_i was chosen from were both hidden from us. The task is then, as far as possible, to infer the these variables from the data. In the biological example above, the task is to identify an unknown number of functional sites from a 'background' of non-functional sites, given the binding energies. For an evolutionary analysis of empirical binding energy data see for instance [14].

Probability and likelihood. Given the parameter λ controlling the relative frequency of the two ensembles and the functional form of these distributions, the probability distribution of E is

$$Pr(E|\lambda) = \lambda Q(E) + (1 - \lambda)P(E). \tag{1.1}$$

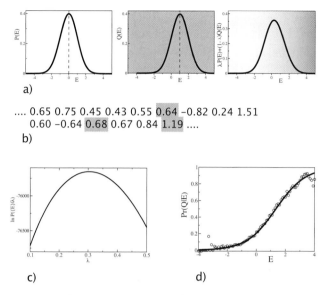

a)

.... 0.65 0.75 0.45 0.43 0.55 0.64 −0.82 0.24 1.51
0.60 −0.64 0.68 0.67 0.84 1.19 ....

b)

c)   d)

Fig. 1. **A simple application of Bayesian statistics.** a) Random numbers are drawn alternatively from distribution $P(E)$ (left) or $Q(E)$ (center). With probability $1 - \lambda$ the former distribution is chosen, with probability λ the latter. The corresponding mixed ensemble $(1 - \lambda)P(E) + \lambda Q(E)$ is shown on the right for $\lambda = 0.3$. In this example, $P(E)$ and $Q(E)$ are both Gaussian ensembles, with variance 1 and means 0 and 1, respectively. b) An instance of random numbers drawn from $P(E)$ (no shading) or $Q(E)$ (shaded in grey). c) The probability of a given instance $\{E\}$ as a function of λ can be computed from $P(E)$ and $Q(E)$, see text. This likelihood of λ is strongly peaked near $\lambda = 0.3$. d) The probability that a given number E was drawn from the Q-ensemble is evaluated numerically (circles). The full line shows the result from Bayes' theorem, see text.

Since the variable x is unknown, both states of x have to be considered with their relative probabilities λ and $(1-\lambda)$. The elements of the set $\{E_i\}$ are independently distributed, so the probability density for $\{E_i\}$ given λ is

$$Pr\big(\{E_i\}|\lambda\big) = \prod_{i=1}^{N}\big[(1 - \lambda)P(E_i) + \lambda Q(E_i)\big]. \tag{1.2}$$

In physics one typically reads this expression as a probability distribution over $\{E_i\}$; $Pr(\{E_i\}|\lambda) \equiv P_\lambda(\{E_i\})$. In this case, λ is considered a constant, such as coupling constant or a mass, and the set of $\{E_i\}$ describes the variables of a theory, such as fields. In the context of statistical inference, the reading is subtly different; the set of $\{E_i\}$ describes the observed data and is fixed, and λ

is the unknown free parameter of our model. As a function of λ at fixed $\{E_i\}$, $Pr(\{E_i\}|\lambda) \equiv P_{\{E_i\}}(\lambda)$ is called the likelihood.

Maximum likelihood estimate. We now ask which value of the parameter λ maximizes the likelihood given a set of empirical observations $\{E_i\}$. In Figure 1 c) we plot the (logarithm) of the likelihood (1.2) against λ. The instance of variables $\{E_i\}$ is taken from a set of 50000 random numbers generated with $\lambda = 0.3$ (however here already a few tens of data points are sufficient for a reasonable estimate of a single parameter). The likelihood is found to be strongly peaked near $\lambda = 0.3$.

The same approach can be used to estimate parameters describing the two alternative distributions $P(E)$ and $Q(E)$ in our example. Of course, the number of data points required to yield a reliable parameter estimate increases with the number of free parameters.

Bayes' theorem and log-likelihood scoring. We now attempt to infer the set of hidden variables $\{x_i\}$, i.e. we ask about the probability that a given E was generated from the ensemble $Q(E)$. This event is denoted as hypothesis Q, the alternative hypothesis of E being generated by $P(E)$ is denoted as P. In other words, we aim to compute $Pr(Q|E)$. The related quantity, $Pr(E|Q) \equiv Q(E)$ is already known. To derive $Pr(Q|E)$ from $Pr(E|Q)$ we use the elementary relation

$$Pr(Q|E)Pr(E) = Pr(Q, E) = Pr(E|Q)Pr(Q) \tag{1.3}$$

yielding

$$Pr(Q|E) = \frac{Pr(E|Q)Pr(Q)}{Pr(E)}. \tag{1.4}$$

Although easily derived and stated, Bayes' theorem (1.4) is both deep and powerful. It links our degree of belief in a hypothesis (left hand side) to the probability of the data occurring under some model (right hand side), thus formalizing mathematically the process at the heart of any scientific inquiry.

We evaluate (1.4) using $Pr(E|Q) = Q(E)$ and (1.1) to obtain

$$Pr(Q|E) = \frac{Q(E)\lambda}{Q(E)\lambda + P(E)(1 - \lambda)}$$
$$= \frac{e^{S(E)}}{e^{S(E)} + 1} \tag{1.5}$$

with

$$S(E) = \log\left(\frac{Q(E)}{P(E)}\right) + \log\left(\frac{\lambda}{1 - \lambda}\right). \tag{1.6}$$

Thus $Pr(Q|E)$ is a monotonously increasing, sigmoidal function of $S(E)$, saturating at 0 for $S(E) \to -\infty$ and tending to 1 for $S(E) \to \infty$. It tells us how much information is gained from knowing E on the alternative hypothesis Q and P. Whenever we use Bayes' theorem to decide between two alternative hypothesis, the so-called *posterior probability* $Pr(Q|E)$ can be written in terms of the so-called *log-likelihood score* $S(E)$.

The log-likelihood score (1.6) has a simple intuitive meaning. It compares the likelihoods of E occurring alternatively in the ensemble $Q(E)$ and in the ensemble $P(E)$, as well as the so-called *prior probabilities* λ and $1 - \lambda$ of hypotheses Q and P respectively. The first term of (1.6) is the only term depending on E. It contains a fraction which is larger than one if E is generated with greater likelihood in the ensemble $Q(E)$ than in the alternative ensemble $P(E)$, and smaller than one in the opposite case. Taking the logarithm of this fraction results in a quantity which is respectively positive or negative. Larger values of E thus lead to an increasing belief in hypothesis Q, with the gain in certainty given by (1.5).

In Figure 1 d) we compare the result for $Pr(Q|E)$ from Bayes' theorem with the direct numerical computation of the relative frequency with which a given value of E was generated by the ensemble $Q(E)$. (This can be done easily; *in silico* the "hidden variables" are not truly hidden. In real life, particularly *in vivo*, such a validation is both more challenging and more rewarding.)

A unified approach. In the previous paragraph, we have inferred information individually from single variables E. The same methodology can be applied to the entire set $\{E_i\}$. This yields a machinery which shows the close relationship between inferring the parameter λ, which controls the relative frequency of ensembles Q and P, and the values of x_i, determining which ensemble a given E_i was chosen from.

We consider the two hypotheses that set $\{E_i\}$ is either generated from a mixed ensemble of the form (1.1), or that it comes from an ensemble where all values are drawn from same distribution $P(E)$ (corresponding to $\lambda = 0$). Both the parameter λ and the values of x_i are hidden, so

$$Pr(\{E_i\}|\text{mixed}) = \int d\lambda \, P(\lambda) \, Pr(\{E_i\}|\lambda) \qquad (1.7)$$

where the prior $P(\lambda)$ encodes any prior information we might have on the value of λ. Using Bayes theorem (1.4) to find the probability that a given realization $\{E_i\}$ stems from this mixed ensemble we find

$$Pr(\text{mixed}|\{E_i\}) = \frac{Pr(\{E_i\}|\text{mixed}) \, Pr(\text{mixed})}{Pr(\{E_i\})}$$

$$= \frac{e^{S}(\{E_i\})}{e^{S}(\{E_i\}) + 1} \qquad (1.8)$$

with

$$S(\{E_i\}) = \log\left(\frac{Pr(\{E_i\}|\text{mixed})}{Pr(\{E_i\}|\text{P})}\right)$$
$$+ \log\left(\frac{Pr(\text{mixed})}{1 - Pr(\text{mixed})}\right). \tag{1.9}$$

$Pr(\{E_i\}|\text{P}) = \prod_i P(E_i)$ is the probability of generating the data under the pure ensemble $P(E)$. We focus on the first term of (1.9), which contains the dependence on the empirical data $\{E_i\}$

$$S'(\{E_i\}) = \log\left(\frac{Pr(\{E_i\}|\text{mixed})}{Pr(\{E_i\}|\text{P})}\right)$$
$$= \log\left(\frac{\int d\lambda \, P(\lambda) Pr(\{E_i\}|\lambda)}{Pr(\{E_i\}|\text{P})}\right). \tag{1.10}$$

Suppose that we have no prior information on the parameter λ, so that the prior on λ is flat, $P(\lambda) \sim$ const. If the likelihood $Pr(\{E_i\}|\lambda)$ is a sharply peaked function of λ (see figure 1c) the integral over λ in (1.10) can be replaced by the maximum of the integrand yielding

$$S'(\{E_i\}) \approx S'(\lambda^\star, \{E_i\}) = \log\left(\frac{Pr(\{E_i\}|\lambda^\star)}{Pr(\{E_i\}|\text{P})}\right). \tag{1.11}$$

where $\lambda^\star$ is given by

$$\lambda^\star = \text{argmax}_\lambda \, Pr(\{E_i\}|\lambda) = \text{argmax}_\lambda \, S'(\lambda, \{E_i\}). \tag{1.12}$$

The last equality follows from the monotonicity of the logarithm. Hence, without prior information on λ, the Bayesian approach gives back exactly the maximum likelihood result derived above. Alternatively, we could have derived the score $S'(\lambda, \{E_i\})$ by calculating $Pr(\lambda|\{E_i\})$. If the likelihood $Pr(\{E_i\}|\lambda)$ is not a sharply peaked function of λ some uncertainty on the value of λ remains, this uncertainty is specified by the width of the distribution $Pr(\lambda|\{E_i\})$.[1]

Analogously, we can extract the optimal estimate for $\{x_i\}$ from the score (1.10) by approximating

$$\int d\lambda \, P(\lambda) Pr(\{E_i\}|\lambda) = \int d\lambda \, P(\lambda) \tag{1.13}$$

---

[1] A subtlety arises when parameterizing λ differently, e.g. through $\lambda' = f(\lambda)$. Since $Pr(\lambda|\{E_i\})$ is a probability density, $Pr(\lambda'|\{E_i\}) = Pr(\lambda|\{E_i\})/f'(\lambda)$, so the maximum of $Pr(\lambda'|\{E_i\})$ does no occur at $f(\lambda')$. However, if the likelihood is sharply peaked, this change in the maximum is small. Moreover, using a flat prior in λ' is different from a flat prior in λ; using a prior flat in λ' gives back the maximum likelihood result, which is invariant under variable transformations.

$$\prod_{i=1}^{N}\left(\sum_{x_i=0,1} x_i \lambda Q(E_i) + (1-x_i)(1-\lambda)P(E_i)\right)$$

$$\approx \int d\lambda \, P(\lambda) \prod_{i=1}^{N}\left(x_i^{\star}\lambda Q(E_i) + (1-x_i^{\star})(1-\lambda)P(E_i)\right),$$

where the optimal estimate of $\{x_i\}$ is given by

$$\{x_i^{\star}(\lambda)\} = \mathrm{argmax}_{\{x_i\}\in\{0,1\}^N} \prod_{i=1}^{N}\left(x_i \lambda Q(E_i) + (1-x_i)(1-\lambda)P(E_i)\right)$$

$$= \mathrm{argmax}_{x_i} S'\left(\{x_i\}, \lambda, \{E_i\}\right) \tag{1.14}$$

with

$$S'\left(\{x_i\}, \lambda, \{E_i\}\right)$$
$$= \sum_i \log\left(\frac{x_i \lambda Q(E_i) + (1-x_i)(1-\lambda)P(E_i)}{P(E_i)}\right). \tag{1.15}$$

The optimization over x_i gives $x_i^{\star} = 1$ if $\lambda Q(E_i) > (1-\lambda)P(E_i)$, or equivalently if the score $S(E_i)$ of (1.6) is positive, and zero otherwise. In the final step, the integral over λ with if performed, yielding $Pr(\{x_i\}, \lambda|\{E_i\})$. $\{x_i^{\star}\} = \int d\lambda \, P(\lambda)\{x_i^{\star}(\lambda)\}$ An alternative derivation of the score (1.15) would have been to compute the posterior probability.

In our simple example the sum over all $\{x_i\}$ in (1.13) is trivial (because the x_i are chosen to be statistically independent). In many applications this is not the case. One way to proceed then is to simultaneously maximize the score (1.15) over λ and $\{x_i\}$. This procedure is known as the Viterbi approximation.

In the preceding paragraphs we have used a very simple example to illustrate basic aspects of Bayesian reasoning. The key step was to use statistical models describing different hypothesis as input into Bayes' theorem (1.4). This approach yields scoring functions which quantify our belief in a certain hypothesis given empirical data. These elements of Bayesian statistics constitute an extremely versatile methodology and appear in different guises across a wide range of applications. A gentle and brief introduction can be found in [15]. For a link with information theory, see [16]. A thorough standard text is [17]. In the following we apply the tools developed to the analysis of networks of molecular interactions.

2. Bayesian analysis of biological networks

Different parts of a network of bio-molecular interactions fulfill different functions, and thus follow a different evolutionary dynamics. The *global statistics* of a network, e.g., its connectivity distribution, provides a background statistics, *local deviations* from this background may represent functional units. In the computational analysis of biological networks, we thus typically have to discriminate between different statistical models governing different parts of the network. A Bayesian approach to exploit these differences requires

• A model describing the global statistics, or background statistics, of the network.

• A model describing the deviation from the background statistics we aim to detect. For instance, network clusters are described by a model with an enhanced number of interactions relative to the background.

• A scoring function discriminating between the two alternatives.

• An algorithmic procedure to identify the parts of the network with high values of the score.

In the following we will discuss models to describe the background statistics of networks, as well as models of clusters, network motifs, and cross-species correlations. We use these models to derive scoring functions and apply them to detect clusters in protein interaction networks, motifs in regulatory networks, and cross-species correlations in co-expression networks.

Definitions. A network is specified by its adjacency matrix $\mathbf{a} = (a_{ii'})$. For binary networks $a_{ii'} = 1$ if there is a link between nodes i and i', and $a_{ii'} = 0$ if there is no link. Networks with undirected links are represented by a symmetric adjacency matrix. The *in* and *out connectivities* of a node, $k_i^+ = \sum_{i'} a_{i'i}$ and $k_i^- = \sum_{i'} a_{ii'}$, are defined as the number of in- and outgoing links, respectively. The total number of directed links is given by $K = \sum_{i,i'} a_{ii'}$.

To focus on a specific part of the network we define an ordered subset $\mathcal{A}$ of n nodes $\{r_1, \ldots r_n\}$ (see Fig. 2a). The subset $\mathcal{A}$ induces a *pattern* $\hat{\mathbf{a}}(\mathcal{A})$ on the network, represented by the restricted adjacency matrix containing only links internal to node subset $\mathcal{A}$. $\hat{\mathbf{a}}$ is thus an $n \times n$ matrix with entries $\hat{a}_{ij} = a_{r_i r_j}$ $(i, j = 1, \ldots, n)$. Together, subset of nodes $\mathcal{A}$ and its pattern $\hat{\mathbf{a}}(\mathcal{A})$ form a *subgraph*.

A null-model for biological networks. The simplest ensemble of networks is generated by connecting all pairs of nodes independently with the same probability w. This ensemble is the well-known Erdős–Rényi-model of random graphs and yields a Poissonian distribution of connectivities. However, in biological networks the connectivity distribution often differs markedly from that of the Erdős–Rényi-model. If biological function is not tightly linked to connectivity at the level of individual nodes, we should include the connectivity distribution in our null model. This ensemble is defined as the unweighted sum over all

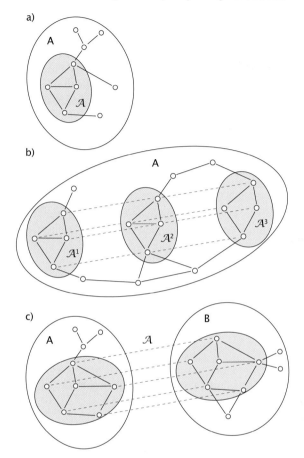

Fig. 2. Deviation from a uniform global statistics in biological networks. a) A network cluster is distinguished by an enhanced number of intra-cluster interactions. b) A network motif is a set of subgraphs with correlated interactions. In a limiting case, all subgraphs have the same topology. c) Cross-species correlations characterize evolutionarily conserved parts of networks.

networks matching the connectivity distribution of the empirical data. In this ensemble, the probability $w_{ii'}$ of finding a link between a pair of nodes i, i' depends on the connectivities of the nodes. Assuming links between different node pairs to be uncorrelated, a given subset of nodes $\mathcal{A}$ has a pattern $\hat{\mathbf{a}}$ with probability

$$P_0(\hat{\mathbf{a}}) = \prod_{i,i' \in \mathcal{A}}^{n} (1 - w_{ii'})^{1-a_{ii'}} w_{ii'}^{a_{ii'}}. \tag{2.1}$$

For $n = N$, when $\mathcal{A}$ includes the entire network, the probability of finding a directed link between nodes i and i' is approximately $w_{ii'} = k_{r_i}^- k_{r_{i'}}^+ / K$, that of an undirected link $w_{ii'} = k_{r_i} k_{r_{i'}} / K$ [18]. If we furthermore impose the constraint that the null model describe the statistics of a *connected* dataset, the probabilities in (2.1) are increased by a factor that can be determined from the data such that the average number of links in ensemble (2.1) equal that of the average number of links in connected subgraphs.

Network clusters. The simplest type of deviation from the null-model is strong inhomogeneities in the network statistics resulting in parts of the network where nodes have more interactions among each other than in the rest of the network, see figure 2a). Such network clusters or network communities also play a role in computer networks and social networks; their detection is a challenging problem [19, 20]. Examples in biological networks are aggregates of several proteins held together by mutual interactions, which show up as highly connected clusters in protein interaction networks and sets co-regulated genes, leading to clusters in co-expression networks.

Clusters are subgraphs with a significantly enhanced number of links compared to the rest of the network; the feature distinguishing network clusters is thus the number of internal links

$$L(\hat{\mathbf{a}}) = \sum_{i,i' \in \mathcal{A}}^{n} \hat{a}_{ii'}. \tag{2.2}$$

An ensemble of random graphs describing a (sub)graph with an enhanced number of internal links is

$$Q_\sigma(\hat{\mathbf{a}}) = Z_\sigma^{-1} \exp[\sigma L(\hat{\mathbf{a}})] P_0(\hat{\mathbf{a}}) \tag{2.3}$$

of the same form as (2.1), but with a bias towards a high number of internal links. The average number of internal links is determined by the value of the link reward σ. We have introduced the normalization factor $Z_\sigma = \prod_{ii'}^{n} \sum_{\hat{a}_{ii'}=0,1} \times \exp[\sigma L(\hat{\mathbf{a}})] P_0(\hat{\mathbf{a}})$, which ensures that $Q_\sigma(\hat{\mathbf{a}})$ summed over all patterns $\hat{\mathbf{a}}$ gives unity. The construction of the ensemble (2.3) is familiar from statistical mechanics, where the unweighted average over all micro-states is biased towards states with a certain energy by a Boltzmann-factor $\exp\{-\beta E\}$. The normalization factor Z_σ corresponds to the partition function.

Based on the statistical model (2.1) describing the network background, and the model (2.3) describing network clusters, we can construct a log-likelihood score to identify network clusters

$$S(\mathcal{A}, \sigma) = \log\left(\frac{Q_\sigma(\hat{\mathbf{a}})}{P_0(\hat{\mathbf{a}})}\right) = \sigma L(\hat{\mathbf{a}}(\mathcal{A})) - \log Z_\sigma. \tag{2.4}$$

A positive score results if it is more likely for the pattern $\hat{\mathbf{a}}(\mathcal{A})$ to arise in the model describing clusters than in the alternative null model. High score values indicate strong deviations from the network background. Patterns with a high score (2.4) are *bona fide* clusters. The first term of the score weighs the total number of links. As expected, a pattern with many internal links yields a high score. The second term acts as a threshold and assigns a negative score to a pattern with a too small number of internal links. This term takes into account the connectivities of the nodes: highly connected nodes have more internal links already in the null model. Node subsets with highly connected nodes tend to give lower scores.

Given the scoring parameter σ, the maximum-score node subset $\mathcal{A}^{\star}(\sigma)$ is defined by

$$\mathcal{A}^{\star}(\sigma) = \operatorname{argmax}_{\mathcal{A}} S(\mathcal{A}, \sigma). \tag{2.5}$$

At this point, the scoring parameter σ is a free parameter, whose value needs to be inferred from the data. The principle of maximum likelihood yields $\sigma^{\star} = \operatorname{argmax}_{\sigma} Q_{\sigma}(\hat{\mathbf{a}}(\mathcal{A})) = \operatorname{argmax}_{\sigma} S(\mathcal{A}, \sigma)$, giving the value of σ which optimally fits the model (2.3) to the pattern $\mathcal{A}$. The maximum-score node subset at the optimal scoring parameter is then determined by the joint maximum of the score over $\mathcal{A}$ and σ

$$S(\mathcal{A}^{\star}, \sigma^{\star}) = \max_{\sigma} S(\mathcal{A}^{\star}(\sigma), \sigma) = \max_{\mathcal{A}, \sigma} S(\mathcal{A}, \sigma). \tag{2.6}$$

Network motifs. Small subgraphs of bio-molecular networks may carry out specific functions. If this function is required in different contexts, and thus in different parts of the network, the corresponding topology may appear repeatedly in different parts of the network. These repeated topologies are called *network motifs* [13,21]. Detecting network motifs is the first step to unravel their potential function.

Network motifs are distinguished from the null model by enhanced correlations between the links of individual subgraphs, see Fig. 2b). To quantify these correlations, we need to specify the parts of the network with correlated patterns. We define a *graph alignment* $\mathcal{A}$ by a set of several node subsets $\mathcal{A}^{\alpha}$ ($\alpha = 1, \ldots, p$), each containing the same number of n nodes, and a specific order of the nodes $\{r_1^{\alpha}, \ldots, r_n^{\alpha}\}$ in each node subset. An alignment associates each node in a node subset with exactly one node in each of the other node subsets. The alignment can be visualized by n "strings", each connecting p nodes as shown in Fig. 2(b).

An alignment specifies a pattern $\hat{\mathbf{a}}^{\alpha} \equiv \hat{\mathbf{a}}(\mathcal{A}^{\alpha}, \mathcal{A})$ in each node subset. For any two aligned subsets of nodes, $\mathcal{A}^{\alpha}$ and $\mathcal{A}^{\beta}$, we can define the *pairwise mismatch* of their patterns

$$M(\hat{\mathbf{a}}^\alpha, \hat{\mathbf{a}}^\beta) = \sum_{i,i'=1}^{n} \left[\hat{a}_{ii'}^\alpha \left(1 - \hat{a}_{ii'}^\beta\right) + \left(1 - \hat{a}_{ii'}^\alpha\right) \hat{a}_{ii'}^\beta \right]. \tag{2.7}$$

The mismatch is a Hamming distance for aligned patterns. The average mismatch over all pairs of aligned patterns is termed the *fuzziness* of the alignment.

An ensemble describing p node subsets with correlated patterns $\hat{\mathbf{a}}^1, \ldots, \hat{\mathbf{a}}^p$ with an enhanced number of links is given by

$$Q_{\mu,\sigma}(\hat{\mathbf{a}}^1, \ldots, \hat{\mathbf{a}}^p) = Z_{\mu,\sigma}^{-1} \prod_{\alpha=1}^{p} P_0(\hat{\mathbf{a}}^\alpha)$$

$$\times \exp\left[-\frac{\mu}{2p} \sum_{\alpha,\beta=1}^{p} M(\hat{\mathbf{a}}^\alpha, \hat{\mathbf{a}}^\beta) + \sigma \sum_{\alpha=1}^{p} L(\hat{\mathbf{a}}^\alpha) \right]. \tag{2.8}$$

The parameter σ quantifies the potentially enhanced number of internal links in network motifs [13, 21], providing the possibility of feedback or other faculties not available to tree-like patterns. The parameter $\mu \geq 0$ biases the ensemble (2.8) towards patterns with small mutual mismatches.

Given the null model (2.1) and the model (2.8) with correlated patterns, we obtain a log-likelihood score for network motifs

$$S(\mathcal{A}, \mu, \sigma)$$

$$= \log\left(\frac{Q_{\mu,\sigma}(\hat{\mathbf{a}}^1, \ldots, \hat{\mathbf{a}}^p, \mathcal{A})}{P_0(\hat{\mathbf{a}}^1, \ldots, \hat{\mathbf{a}}^p, \mathcal{A})} \right)$$

$$= -\frac{\mu}{2p} \sum_{\alpha,\beta=1}^{p} M(\hat{\mathbf{a}}^\alpha, \hat{\mathbf{a}}^\beta) + \sigma \sum_{\alpha=1}^{p} L(\hat{\mathbf{a}}^\alpha)$$

$$- \log Z_{\mu,\sigma}. \tag{2.9}$$

High-scoring alignments $\mathcal{A}$ indicate *bona fide* network motifs. The first and second terms reward alignments with a small mutual mismatch and a high number of internal links, respectively. The term $\log Z_{\sigma,\mu}$ acts as a threshold assigning a negative score to alignments with too high fuzziness or too small a number of internal links.

Again, both the alignment $\mathcal{A}$, as well as the scoring parameters μ and σ are undetermined. For given scoring parameters, the maximum-score alignment

$$\mathcal{A}^\star(\mu, \sigma) = \text{argmax}_{\mathcal{A}} S(\mathcal{A}, \mu, \sigma) \tag{2.10}$$

occurs at some finite value of the number of subgraphs $p^\star(\mu, \sigma)$.

The scoring parameters μ and σ can again be determined by maximum likelihood, which corresponds to maximizing the score $S(\mathcal{A}^\star(\mu, \sigma), \mu, \sigma)$ with respect to the scoring parameters. This yields parameter values which fit the model (2.8) to the maximum-score network motifs $\mathcal{A}^\star$.

Cross-species analysis of bio-molecular networks. The common origin of the known forms of life is reflected at all levels of biological description, from biological sequences to morphology. Also networks of bio-molecular interactions share similarities across different species, as illustrated in Figure 2c). The similarities can be distributed inhomogeneously; part of the network may remain relatively unchanged, forming a conserved core, such as a common metabolic pathway. Other network parts may be characterized by a rapid turnover of both nodes and interactions. We aim to detect the conserved parts of networks in different species, taking into account information both from interactions between network nodes and information on the nodes themselves (such as the nucleotide sequence of a gene, the structure of protein, or the biochemical role of a metabolite).

A log-likelihood score assessing the link statistics of subgraphs in network A and in network B and an alignment $\mathcal{A}$ between them follows directly from (2.9)

$$S^\ell(\mathcal{A}, \mu, \sigma_A, \sigma_B) = -\mu M(\hat{\mathbf{a}}, \hat{\mathbf{b}}) \\ + \sigma\big(L(\hat{\mathbf{a}}) + L(\hat{\mathbf{b}})\big) - \log Z_{\mu,\sigma}. \tag{2.11}$$

This score is termed the *link score* as it refers to the statistics of links. It assumes the dynamics of different links is statistically uncorrelated. Functional contraints may lead to correlations, see for instance [22].

To assess the similarity of nodes we consider a measure θ_{ij} describing the similarity of node i in network A and node j in network B. This node similarity measure may be a percentage sequence identity, a distance measure of protein structure, or a measure of biochemical similarity. To construct a log-likelihood score of node similarity we contrast a null model, describing a random ensemble of node similarities, with a model describing a statistics where node similarity is correlated with the alignment. In the null model node similarities θ_{ij} of different node pairs i, j are identically and independently distributed. Their distribution is denoted by $p_0^n(\theta_{ij})$. The similarities between aligned pairs of nodes follow a different statistics (typically generating higher values of θ), denoted by $q_1^n(\theta)$. The distribution of pairwise similarity coefficients between one aligned node and nodes other than its alignment partner is denoted by $q_2^n(\theta)$.

Assuming that for a given alignment the statistics of links and nodes similarities are uncorrelated we obtain the log-likelihood score

$$S(\mathcal{A}) = S^\ell(\mathcal{A}) + S^n(\mathcal{A}), \tag{2.12}$$

with the information from node similarity contributing a *node score*

$$S^n(\mathcal{A}) = \sum_{i \in \mathcal{A}} s_1^n(\theta_{ii}) + \sum_{\substack{i \in \mathcal{A}, j \neq i \\ j \in \mathcal{B}, i \notin \mathcal{A}}} s_2^n(\theta_{ij}) \qquad (2.13)$$

and $s_1^n(\theta) \equiv \log(q_1^n(\theta)/p_0^n(\theta))$ and $s_2^n(\theta) \equiv \log(q_2^n(\theta)/p_0^n(\theta))$.

Again the scoring parameters entering (2.12) are determined from the maximum-likelihood principle by maximizing the score with respect to the scoring parameters. The maximum-score alignment projects out the network parts which are strongly conserved between the two species. Since two independent sets of information, sequence similarity and network similarity, enter the score, two nodes can be aligned because of (i) large node similarity and hence a high node score, (ii) large topological similarity and hence a high link score, and (iii) both. While depending on the quality of the network data cases (iii) and (i) predominate, unrelated genes can take on similar functional roles in different organisms and can then appear in the same position in bio-molecular networks. This scenario is known as non-orthologous gene displacement [23]. An example will be discussed in section 3.

The statistical and model-based approach to network analysis developed here is not limited to clusters, network motifs, and cross-species correlations. It can be applied, for instance to the connectivity distribution itself. See e.g. [24,25], where the 'scale-free' label for many empirical bio-molecular networks is questioned.

3. Applications

In this section some applications of the scoring functions developed above are discussed. We look at clusters in the protein interaction networks of yeast, motifs in the regulatory network of *E. coli*, and compare co-expression networks between *H. sapiens* and *M. musculus*. Sources of data are the high-throughput dataset of Uetz et al. [6] for protein interactions in yeast, regulatory interactions in *E. coli* are taken from [13], and the expression data of Su et al. [26] was used to construct co-expression networks of ~ 2000 housekeeping genes. Human-mouse orthologs were taken from the Ensembl database [27]. Details on the algorithms to maximize the scores are given in the original publications [28–30].

Clusters in protein interaction networks. We use the scoring function (2.4) to identify clusters in the protein interaction network of yeast. First we explore the dependence of the high-scoring regions on the scoring parameter σ. The subgraphs with the highest score for a given value of σ are shown in Fig. 3a). At low values of σ, subgraphs with many nodes, but few internal interactions per node yield the highest score. At high values of σ, subgraphs with many internal

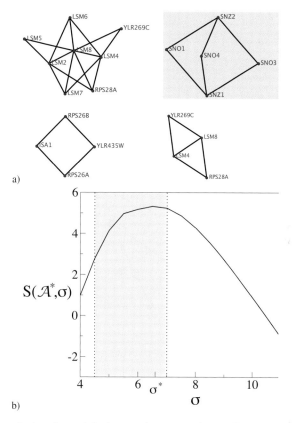

Fig. 3. **Scoring clusters in protein interaction networks.** a) The maximum-score subgraphs for $\sigma < 4.25$, $4.25 < \sigma < 7$, $7 < \sigma < 11$, $\sigma > 11$ (left to right). The subgraph resulting from the optimal scoring parameter $\sigma^\star = 6.6$ is highlighted in grey. The maximum-score subgraphs for $7 < \sigma < 11$ and for $\sigma > 11$ are distinguished by the connectivities of their nodes with the latter having a higher average connectivity. This accounts for the former having a higher score for $7 < \sigma < 11$ despite the smaller number of internal links. b) The score S of the maximum-score node subset $\mathcal{A}^\star(\sigma)$ is shown as a function of the scoring parameter σ. The dotted lines indicate the values of σ where the maximum-score node subset changes. The maximum of the score with respect to σ indicates the optimal scoring parameter. The grey region $4.25 < \sigma < 7$ indicates the values where $\mathcal{A}^\star(\sigma) = \mathcal{A}^\star(\sigma^\star)$.

interactions yield the highest score, however these subgraphs tend to be small. This interplay between subgraph size and internal connectivity leads to a joint score maximum over $\mathcal{A}$ and σ at the optimal scoring parameter $\sigma^\star = 6.6$, see Fig. 3b).

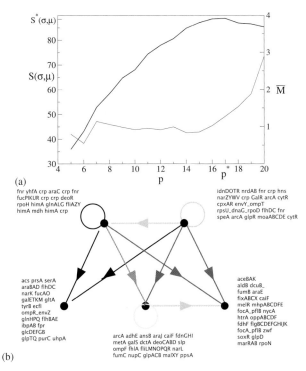

(a)

(b)

fnr yhfA crp araC crp fnr
fucPIKUR crp crp deoR
rpoH himA glnALG fliAZY
himA mdh himA crp

idnDOTR nrdAB fnr crp hns
narZYWV crp GalR arcA cytR
cpxAR envY_ompT
rpsU_dnaG_rpoD flhDC fnr
speA arcA glpR moaABCDE cytR

acs prsA serA
araBAD flhDC
narK fucAO
galETKM gltA
tyrB ecfl
ompR_envZ
glnHPQ flhBAE
ibpAB fpr
glcDEFGB
glpTQ purC uhpA

aceBAK
aldB dcuB_
fumB araE
fixABCX caiF
melR mhpABCDFE
focA_pflB nycA
htrA oppABCDF
fdhF flgBCDEFGHIJK
focA_pflB zwf
soxR glpD
marRAB rpoN

arcA adhE ansB araJ caiF fdnGHI
metA galS dctA deoCABD slp
ompF fhlA fliLMNOPQR narL
fumC nupC glpACB malXY ppsA

Fig. 4. **Motifs in the regulatory network of *E. coli*.** (a) Score optimization at fixed scoring
parameters $\sigma = 3.8$ and $\mu = 4.0$ for subgraphs of size $n = 5$. The total score S (thick
line) and the fuzziness $\overline{M}$ (thin line) are shown for the highest-scoring alignment of p
subgraphs, plotted as a function of p. (b) The consensus motif of the optimal alignment,
and the identities of the genes involved. The alignment consists of 18 subgraphs sharing
at most one node. The 5 grey values correspond to the consensus motif $\overline{a}$ in the range
0.1–0.2, 0.2–0.4, 0.4–0.6, 0.6–0.8, 0.8–0.9.

The maximum-score cluster $\mathcal{A}^{\star} \equiv \mathcal{A}^{\star}(\sigma^{\star})$ consists of the proteins SNZ1,
SNZ2,SNO1,SNO3, and SNO4, highlighted in grey in Fig. 3b). The proteins
in this cluster have a common function; they are involved in the metabolism of
pyridoxine and in the synthesis of thiamin [27, 31]. Furthermore, SNZ1 and
SNO1 have been found to be co-regulated and their mRNA levels increase in
response to starvation for aminoacids adenine, uracil, and tryptophan [32].

Network motifs in the *E. coli* regulatory network. We now apply the scoring function (2.9) to the identification of network motifs in the gene regulatory
network of *E. coli*.

We first investigate the properties of the maximal score alignment at fixed
scoring parameters. Fig. 4(a) shows the score S and the fuzziness $\overline{M}$ (average

mismatch of the aligned subgraphs) for the highest-scoring alignment with a prescribed number p of subgraphs, plotted against p. The fuzziness increases with p, and the score reaches its maximum $S^\star(\sigma, \mu)$ at some value $p^\star(\sigma, \mu)$. For $p < p^\star(\sigma, \mu)$ the score is lower, since the alignment contains fewer subgraphs and for $p > p^\star(\sigma, \mu)$ it is lower since the subgraphs have higher mutual mismatches.

The optimal scoring parameters μ and σ are again inferred by maximum likelihood. The resulting optimal alignment $\mathcal{A}^\star \equiv \mathcal{A}^\star(\mu^\star, \sigma^\star)$ is shown in Fig. 4(b) using the so-called *consensus motif*

$$\bar{\mathbf{a}} = \frac{1}{p} \sum_{\alpha=1}^{p} \hat{\mathbf{a}}^\alpha(\mathcal{A}^\star). \tag{3.1}$$

The consensus motif is a *probabilistic pattern*; the entry $\bar{a}$ denotes the probability that a given binary link is present in the aligned subgraphs. The motif shown in Fig. 4(b) consists of $2 + 3$ nodes forming an input and an output layer, with links largely going from the input to the output layer. Most genes in the input layer code for transcription factors or are involved in signaling pathways. The output layer mainly consists of genes coding for enzymes.

Comparing co-expression networks of *H.* sapiens and *M. musculus*. We compare co-expression networks of *H. sapiens* and *M. musculus*. In co-expression networks, the weighted link $a_{ii'} \in [-1, 1]$ between a pair of genes i, j is given by the correlation coefficient of their gene expression profiles measured on a microarray chip. Genes which tend to be expressed under similar conditions thus have positive links. The score (2.11) can easily be generalized to weighted interactions, see [29].

We focus on strongly conserved parts of the two networks. Figure 5 shows a cluster of co-expressed genes which is highly conserved between human and mouse (link conservation is shown in blue, changes between the links in red).

With one exception, the aligned gene pairs in this cluster have significant sequence similarity and are thought to be orthologs, stemming from a common ancestral gene. The exception is the aligned gene pair human-HMGN1/mouse-Parp2. These genes are aligned due to their matching links, quantified by a high contribution to the link score (2.11) of $S^\ell = 25.1$. The "false" alignment human-HMGN1/mouse-HMGN1 respects sequence similarity but produces a link mismatch ($S^\ell = -12.4$); see Fig. 5(b). Human-HMGN1 is known to be involved in chromatin modulation and acts as a transcription factor. The network alignment predicts a similar role of Parp2 in mouse, which is distinct from its known function in the poly(ADP-ribosyl)ation of nuclear proteins. The prediction is compatible with experiments on the effect of Parp-inhibition, which suggest that Parp genes in mouse play a role in the chromatin modification during development [33].

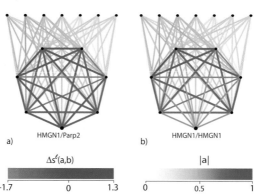

a) HMGN1/Parp2   b) HMGN1/HMGN1

$\Delta s^{\ell}(a,b)$ $|a|$

-1.7   0   1.3   0   0.5   1

Fig. 5. **Cross-species network alignment shows conservation of gene clusters.** (a) 7 genes from a cluster of co-expressed genes (circle) together with 7 random genes outside the cluster (straight line). Each node represents a pair of aligned genes in human and mouse. The intensity of a link encodes the correlation value a in human. The color indicates the evolutionary conservation of a link, with blue hues indicating strong conservation. The conservation is quantified by the excess link score contribution, Δs^{ℓ}, defined as the link score minus the average link score of links with the same correlation value. (b) The same cluster, but with human-HMGN1 "falsely" aligned to its ortholog mouse-HMGN1, with the red links showing the poor expression overlap of this pair of genes.

4. Conclusion and outlook

In these lectures the principles of Bayesian statistics have been applied to network analysis. We have used statistical models of networks to assess quantitatively if a given part of a network deviates from a uniform global background statistics. The deviations considered here were exemplified by network clusters, network motifs, and cross-species correlations. However, these techniques have a much wider range of applicability, and we can expect further contributions of statistical mechanics at the interface between complex systems and computational biology.

Network and sequence dynamics. Bio-molecular interactions are the result of specific sequence configurations. As discussed above, regulatory interactions are mediated by binding sites for transcription factors in the regulatory region of a gene. Similarly, interactions between proteins result from matching so-called 'hotspots' on the respective protein surfaces [34,35]. In general, networks of bio-molecular interactions, however non-trivial, are a coarse-grained representation of specific information already present at the level of the sequence. An improved understanding of the dynamics of interactions at the sequence level will inform the statistical models at the heart of the scoring functions discussed here.

Interplay between regulatory interactions and expression levels. The function of regulatory interactions is to correlate the expression of genes. There is evidence for changes of the regulatory interactions leaving the expression levels conserved [36]. This leads to an ensemble of functionally equivalent regulatory networks. Including the interplay between expression level correlations and regulatory interactions into statistical models for regulatory networks will lead to new null models. As a result we may find that some of the motifs identified so far in regulatory networks are accounted for by the null-model, and conversely, new motifs may arise as statistically significant. More importantly, we may be able to address the question of which changes in regulatory networks are selectively neutral and leave expression levels unchanged, and which changes result in significantly altered expression levels and may result from an evolutionary adaptation.

These two questions are only different facets of the link between network functionality, evolutionary dynamics, and networks statistics. Treating these aspects under a unified framework remains an outstanding challenge for both theory and experiment.

Acknowledgments

This work was supported through DFG grants SFB/TR 12, SFB 680, and BE 2478/2-1 under the Emmy Noether-programme. Many thanks to Cordula Becker, Sinead Collins, and Marc Potters for discussions.

References

[1] L.D. Stein. Human genome: End of the beginning. *Nature*, 431:915–916, 2004.

[2] J.-M. Claverie. What if there are only 30,000 human genes? *Science*, 291(5507):1255–1257, 2001.

[3] D. Tautz. Evolution of transcriptional regulation. *Current Opinion in Genetics & Development*, 10:575–579, 2000.

[4] M.C. King and A.C. Wilson. Evolution at two levels in humans and chimpanzees. *Science*, 188:107–166, 1975.

[5] Sonali Mukherjee, Michael F Berger, Ghil Jona, et al. Rapid analysis of the DNA-binding specificities of transcription factors with DNA microarrays. *Nat Genet*, 36(12):1331–1339, 2004.

[6] P. Uetz, L. Giot, G. Cagney, T.A. Mansfield, and R.S. Judson et al. A comprehensive analysis of protein–protein interactions in Saccharomyces cerevisiae. *Nature*, 403:623–627, 2000.

[7] Yingming Zhao, Tom W. Muir, Stephen B.H. Kent, Ed Tischer, Jan Marian Scardina, and Brian T. Chait. Mapping protein–protein interactions by affinity-directed mass spectrometry. *PNAS*, 93(9):4020–4024, 1996.

[8] A.L. Barabási and R. Albert. Emergence of scaling in random networks. *Science*, 286(5439):509–512, 1999.

[9] A. Vazquez, A. Flammini, A. Maritan, and A. Vespignani. Modeling of protein interaction networks. *Complexus*, 1:38–44, 2003.

[10] J. Berg, M. Lässig, and A. Wagner. Structure and evolution of protein interaction networks: A statistical model for link dynamics and gene duplications. *BMC Evolutionary Biology*, 4:51, 2004.

[11] M.E.J. Newman, S.H. Strogatz, and D.J. Watts. Random graphs with arbitrary degree distributions and their applications. *Physical Review E*, 64:026118, 2001.

[12] V. Spirin and L. Mirny. Protein complexes and functional modules in molecular networks. *Proc. Natl. Acad. Sci. USA*, 100(2):12123–12128, 2003.

[13] S. Shen Orr, R. Milo, S. Mangan, and U. Alon. Network motifs in the transcriptional regulation network of Escherichia coli. *Nature Genetics*, 31:64–68, 2002.

[14] V. Mustonen and M. Lässig. Evolutionary population genetics of promoters: predicting binding sites and functional phylogenies. *Proc Natl Acad Sci U S A*, 102(44):15936–15941, Nov 2005.

[15] R. Durbin, S.R. Eddy, A. Krogh, and G. Mitchison. *Biological sequence analysis*. CUP, Cambridge, UK, 1998.

[16] T. Cover and J. Thomas. *Elements of Information Theory*. John Wiley and Sons, New York, USA, 1991.

[17] J. Bernardo and A. Smith. *Bayesian Theory*. John Wiley and Sons, New York, USA, 2001.

[18] S. Itzkovitz, R. Milo, N. Kashtan, G. Ziv, and U. Alon. Subgraphs in random networks. *Phys. Rev.*, 68:026127, 2003.

[19] M.E.J. Newman. The structure and function of complex networks. *SIAM Review*, 45:167–256, 2003.

[20] Girvan M. and M.E.J. Newman. Community structure in social and biological networks. *Proc. Natl. Acad. Sci. USA*, 99(12):7821–7826, 2002.

[21] R. Milo, S. Shen-Orr, S. Itzkovitz, N. Kashtan, D. Chklovskii, and U. Alon. Network motifs: simple building blocks of complex networks. *Science*, 298:824–827, 2002.

[22] A. Trusina, K. Sneppen, I.B. Dodd, K.E. Shearwin, and J.B. Egan. Functional alignment of regulatory networks: a study of temperate phages. *PLoS Comput. Biol.*, 7(7):e74, 2005.

[23] E.V. Koonin, A.R. Mushegian, and P. Bork. Non-orthologous gene displacement. *Trends Genet.*, 12(9):334–6, 1996.

[24] M. Stumpf, P.J. Ingram, I. Nouvel, and C. Wiuf. Statistical model selection methods applied to biological network data. *Trans. Comp. Sys. Biol. III*, 3737:65–77, 2005.

[25] C. Wiuf, M. Brameier, O. Hagberg, and M. Stumpf. A likelihood approach to analysis of network data. *Proc. Natl. Acad. Sci. USA*, 103(20):7566–7570, 2006.

[26] A.I. Su, T. Wiltshire, S. Batalov, H. Lapp, K.A. Ching, et al. A gene atlas of the mouse and human protein-encoding transcriptomes. *Proc. Natl. Acad. Sci. USA*, 101(16):6062–6067, 2004.

[27] T. Hubbard, D. Andrews, M. Caccamo, G. Cameron, Y. Chen, et al. Ensembl 2005. *Nucleic Acids Res.*, 33:D447–D453, 2005.

[28] J. Berg and M. Lässig. Local graph alignment and motif search in biological networks. *Proc. Natl. Acad. Sci. USA*, 101(41):14689–14694, 2004.

[29] J. Berg and M. Lässig. Cross-species analysis of biological networks by Bayesian alignment. *Proc. Natl. Acad. Sci. USA*, page in press, 2006.

[30] J. Berg and M. Lässig. Bayesian analysis of biological networks: clusters, motifs, cross-species correlations. In M. Stumpf and C. Wiuf, editors, *Statistical and Evolutionary Analysis of Biological Network Data*. to appear.

[31] The Gene Ontology Consortium. Gene ontology: tool for the unification of biology. *Nature Genet.*, 25:25–29, 2000.

[32] P.A. Padilla, E.K. Fuge, M.E. Crawford, A. Errett, and M. Werner-Washburne. The highly conserved, coregulated SNO and SNZ gene families in Saccharomyces cerevisiae respond to nutrient limitation. *J. Bacteriol.*, 180:5718–5726, 1998.

[33] T. Imamura, T.M. Anh, C. Thenevin, and A. Paldi. Essential role for poly (adp-ribosyl)ation in mouse preimplantation development. *BMC Molecular Biology*, 5:4, 2004.

[34] S. Jones and J.M. Thornton. Principles of protein-protein interactions. *Proc. Natl. Acad. Sci. USA*, 93:13–20, 1996.

[35] P. Aloy and R. Russell. Structural systems biology: modelling protein interactions. *Nature Reviews Molecular Cell Biology*, 7:188–197, 2006.

[36] A. Tanay, A. Regev, and R. Shamir. Conservation and evolvability in regulatory networks: The evolution of ribosomal regulation in yeast. *Proc. Natl. Acad. Sci. USA*, 2005.

Course 13

THE SLOW DYNAMICS OF GLASSY MATERIALS:
INSIGHTS FROM COMPUTER SIMULATIONS

Ludovic Berthier

Laboratoire des Colloïdes, Verres et Nanomatériaux, UMR 5587, Université Montpellier II and CNRS, 34095 Montpellier, France

J.-P. Bouchaud, M. Mézard and J. Dalibard, eds.
Les Houches, Session LXXXV, 2006
Complex Systems
© *2007 Published by Elsevier B.V.*

Contents

Une garance qui fait entendre le violoncelle
Vieira da Silva

Glassy states of matter continue to attract the interest of a large community of scientists [1–3], ranging from material physicists interested in the mechanical properties of disordered solids, to theoretical physicists who want to describe at a more fundamental level the "glass state" [4, 5]. Glassy materials can be found in a variety of materials, from soft matter (dense emulsions, concentrated colloidal suspensions, powders) to hard condensed matter (molecular liquids, polymeric glasses, disordered magnets). Several glassy phenomena unrelated to specific materials, or even outside physics, are also discussed in this book. A feature common to glassy materials is that their dynamics gets so slow in some part of their phase diagram that they appear as amorphous frozen structures on experimental timescales. The transition from a rapidly relaxing material (liquid, paramagnet...) to a frozen structure (window glass, spin glass, soft disordered solid...) is called a "glass transition". For many glassy materials, a full understanding of the microscopic processes responsible for the formation of glasses is still lacking.

In Fig. 1 we present snapshots obtained from computer simulations of three different models for materials characterized by slow dynamics. The left panel shows a binary assembly of Lennard-Jones particles with interaction parameters specifically designed to avoid crystallization, thus modelling either metallic or colloidal glasses [6]. The middle panel is taken from numerical simulations of a classical model for silica [7], the main component of most window glasses. The right panel shows the structure obtained in a soft gel [8] made of oil droplets in water connected by telechelic polymers (long hydrophilic chains ended by small hydrophobic heads). In the three cases, the dynamics of individual particles can get arrested on numerical timescales and the system essentially appears as a disordered solid—a "glass". From a statistical physics point of view, two facts are quite puzzling. First, the structural properties of liquids and glasses are essentially indistinguishable. Second, there is no clear-cut phase transition between the two, so that the standard statmech language is not obviously the most relevant one to describe the formation of these solids.

Just as in many different areas in physics, computer simulations are playing an increasing role in the field of glass formation [9, 10]. An obvious reason is

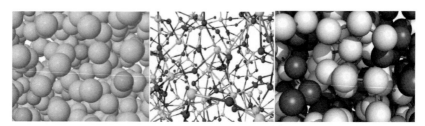

Fig. 1. Snapshots obtained from computer simulations of three different materials characterized by glassy dynamics. Left: An equilibrium configuration of a binary Lennard-Jones mixture, used as a model system for metallic or colloidal glasses [6]. Middle: Network structure of silica around 4000 K obtained from simulations of the BKS model [7]. Right: The picture shows oil droplets in white and blue, transiently connected by (red) telechelic polymers. For these parameters, a system-spanning clusters of connected particles (highlighted in white) endows the system with viscoelastic properties [8].

that, when simulating the materials shown in Fig. 1, the position of all particles is exactly known at all times—the ultimate dream for an experimentalist! Naturally, there are two immediate drawbacks. Firstly, one might wonder if it is really possible to simulate experimentally relevant materials in experimentally relevant conditions. Second question: What do we do with all this information?

The answer to the first question is positive. With present day computers, it is possible to follow for instance the dynamics of $N = 10^3$ Lennard-Jones particles shown in Fig. 1 over 9 decades of time using about 3 months of CPU time on a standard PC, thus covering a temperature window over which average relaxation timescales increase by more than 5 decades, quite a dramatic slowing down. However, at the lowest temperatures studied, relaxation is still orders of magnitude faster than in experiments performed close to the glass transition temperature. Nevertheless, it is now possible to numerically access temperatures which are low enough that many features associated to the glass transition physics can be observed: strong decoupling phenomena [11], clear deviations from fits to the mode-coupling theory [6] (which are experimentally known to hold only at high temperatures), and crossovers towards activated dynamics [12]. Of course, smaller timescales are accessed when simulating more complex systems, e.g. silica where Si and O atoms also carry charges and interact via a long-range Coulomb interaction, or more complex situations, e.g. boundary driven shear flows [13], aging phenomena [14], or gel formation [15].

The answer to the second question (what do we measure?) occupies the rest of this text. First one must make sure that the glassy dynamics one seeks to study is at least qualitatively reproduced by the chosen numerical models, which are necessarily simplified representations of the experimental complexity. One can for instance devise "theoretical models", such as the Lennard-Jones liquid shown in Fig. 1, which indeed captures the physics of glass-forming liquids [6]. One can also devise models inspired by real materials, such as the BKS model for silica and the connected micro-emulsion shown in Fig. 1. The major signatures of glassy dynamics are indeed easily reproduced in simplified models and can therefore extensively be studied in computer simulations: slow structural relaxation, sudden growth of the viscosity upon lowering the temperature, aging phenomena after a sudden quench to the glass phase, non-Debye (stretched) form of the decay of correlation functions. Kob has given an extensive account of these phenomena in the proceedings of a previous school [16].

The important topic of dynamic heterogeneity, which emerged as an important aspect of glassy materials during the 90s, is not covered in Kob's lectures, but alternative reviews exist [17]. Although different phenomena usually go under the same name, dynamic heterogeneity is generally associated to the existence, and increasing strength as dynamics gets slower, of non-trivial spatio-temporal fluctuations of the local dynamical behaviour.

Perhaps the simplest question in this context is as follows. On a given time window, t, particles in a liquid make the average displacement $\bar{d}(t)$, but the displacement of individual particles is distributed, $P(d, t)$. It is well established that $P(d, t)$ acquires non-Gaussian tails which carry more weight when dynamics is slower. This implies that relaxation in a viscous liquid must differ from that of a normal liquid where diffusion is Gaussian, and that non-trivial particle displacements exist. A long series of questions immediately follows this seemingly simple observation. Answering them has been the main occupation of many workers in this field over the last decade. What are the particles in the tails effectively doing? Why are they faster than the rest? Are they located randomly in space or do they cluster? What is the geometry, time and temperature evolution of the clusters? Are these spatial fluctuations correlated to geometric or thermodynamic properties of the liquids? Do similar correlations occur in all glassy materials? Can one predict these fluctuations theoretically? Can one understand glassy phenomenology using fluctuation-based arguments? Can these fluctuations be detected experimentally?

Although the field was initially principally driven by elegant experiments detecting indirect evidences of the existence of dynamic heterogeneity, and by a series of numerical observations in model liquids or simplified glass models, theoretical progress has been somewhat slower. It took some more time to realize that dynamic heterogeneity could be studied using a set of well-defined correlation functions that can be studied either theoretically, in computer experiments, or in real materials, thus allowing (in principle) a detailed comparison between theory and experiments [19, 20].

The main difficulty is that these correlators, unlike, say, traditional scattering functions, usually involve more than two points in space and time and represent therefore quite a challenge for computer simulations, but even more in experiments. To detect spatial correlations of the dynamics one can for instance define "four-point" spatial correlators, involving the position of two particles at two different times, a quantity which can be directly accessed in simulations. Several such measurements have been performed, and directly establish that the dynamical slowing down encountered in glassy materials is accompanied by the existence of a growing correlation lengthscales over which local dynamics is spatially correlated [19,20]. Together with theoretical developments [21–23], these results suggest that the physics of glasses is directly related to the growth of dynamic fluctuations, similar to the ones encountered in traditional phase transitions.[1]

Experimentally detecting similar multi-point quantities in, say, a molecular liquid close the glass transition would require having spatial resolution at the molecular level over timescales of the order of the second—a real challenge. Techniques have been devised to access these quantities in colloidal systems where microscopic timescales and lengthscales are more easily accessible [24]. Additionally, recent work has suggested that alternative multi-point correlation functions could be more easily studied in experiments, while containing similar physical informations [12, 25].

Despite being performed at lower temperatures and for liquids that are much more viscous than in simulations, dynamic lengthscales measured in experiments are not much larger than in simulations, Typically, one finds that relaxation is correlated over a volume containing (at most) a few hundreds of particles at low temperature. This means that even on experimental timescales, where dynamics is orders of magnitude larger than in numerical work, there is no trace of "diverging" lengthscales, as would be necessary for simple scaling theories to apply. Such modest lengthscales are, however, physically expected on general grounds: Because dynamics in glassy materials is typically thermally activated, a tiny change in activation energy (possibly related to an even smaller growth of a correlation lengthscale) translates into an enormous change in relaxation timescales [21–23, 26].

Although very few experimental results have been published, it seems that the dynamics of very many molecular liquids, and perhaps also of different types of glassy materials, could be analyzed along the lines of Ref. [25], perhaps leading to a more complete description of the time and temperature dependences of spatial correlations in a variety of materials approaching the glass transition. It remains to be seen if these correlations can successfully and consistently be ex-

---

[1] Spin glasses are one example where this behaviour is obviously realized since three-dimensional spin glasses undergo a genuine phase transition towards a spin glass phase characterized by the divergence of a correlation length measured via four-spin correlations, a static analog of the four-point dynamic functions mentioned above. During the school, however, it appeared that students did not seem to consider spin glasses as the most exciting example of "complex systems".

plained theoretically, with precise predictions that can be directly confronted to experimental results with decisive results.

I thank J.L. Barrat, G. Biroli, L. Bocquet, J.P. Bouchaud, D. Chandler, L. Cipelletti, D. El Masri, J.P. Garrahan, P. Hurtado, R. Jack, W. Kob, F. Ladieu, S. Léonard, D. L'Hôte, P. Mayer, K. Miyazaki, D. Reichman, P. Sollich, C. Toninelli, F. Varnik, S. Whitelam, M. Wyart, and P. Young for pleasant and fruitful collaborations on the topics described during these lectures. These notes were typed sitting next to a two week old little girl, whose non-glassy (but definitely complex) behaviour easily dominated over the noise of the laptop.

References

[1] *Ill-condensed matter* (Les Houches 1978), Eds.: R. Balian *et al.* (North Holland, Amsterdam, 1979).

[2] *Liquids, Freezing and the Glass Transition* (Les Houches 1989), Eds.: J.P. Hansen, D. Lévesque, J. Zinn-Justin (North Holland, Amsterdam, 1989).

[3] *Slow relaxations and nonequilibrium dynamics in condensed matter* (Les Houches 2002), Eds.: J.L. Barrat, M. eigelman, J. Kurchan, J. Dalibard (Springer, Berlin, 2003).

[4] *Spin Glasses and Random Fields*, Ed.: A.P. Young (World Scientific, Singapore, 1998).

[5] K. Binder and W. Kob, *Glassy materials and disordered solids* (World Scientific, Singapore, 2005).

[6] W. Kob and H.C. Andersen, Phys. Rev. Lett. **73**, 1376 (1994).

[7] B.W.H. van Beest, G.J. Kramer, and R.A. van Santen, Phys. Rev. Lett. **64**, 1955 (1990).

[8] P.I. Hurtado, L. Berthier, and W. Kob, *Heterogeneous diffusion in a reversible gel*, cond-mat/0612513.

[9] M. Allen and D. Tildesley, *Computer Simulation of Liquids* (Oxford University Press, Oxford, 1987).

[10] M.E.J. Newman and G.T. Barkema, *Monte Carlo methods in statistical physics* (Oxford University Press, Oxford, 1999).

[11] D.N. Perera and P. Harrowell, J. Chem. Phys. **111**, 5441 (1999); R. Yamamoto and A. Onuki, Phys. Rev. Lett. **81**, 4915 (1998); L. Berthier, Phys. Rev. E **69**, 020201 (2004).

[12] L. Berthier, G. Biroli, J.-P. Bouchaud, W. Kob, K. Miyazaki, D.R. Reichman, cond-mat/0609656; cond-mat/0609658.

[13] F. Varnik, L. Bocquet, J.-L. Barrat, and L. Berthier, Phys. Rev. Lett. **90**, 095702 (2003).

[14] W. Kob and J.-L. Barrat, Phys. Rev. Lett. **78**, 4581 (1997).

[15] P. Charbonneau and D.R. Reichman Phys. Rev. E **75**, 011507 (2007).

[16] W. Kob in Ref. [3].

[17] M.D. Ediger, Ann. Rev. Phys. Chem. **51**, 99 (2000).

[18] E. Weeks, J.C. Crocker, A.C. Levitt, A. Schofield, and D.A. Weitz, Science **287**, 627 (2000).

[19] T.R. Kirkpatrick and D. Thirumalai, Phys. Rev. A **37**, 4439 (1988); S. Franz, G. Parisi, J. Phys.: Condens. Matter **12**, 6335 (2000); S. Whitelam, L. Berthier, J.P. Garrahan, Phys. Rev. Lett. **92**, 185705 (2004); G. Biroli and J.-P. Bouchaud, Europhys. Lett. **67**, 21 (2004); C. Toninelli, M. Wyart, G. Biroli, L. Berthier, J.-P. Bouchaud, Phys. Rev. E **71**, 041505 (2005).

[20] B. Doliwa and A. Heuer, Phys. Rev. E **61**, 6898 (2000); R. Yamamoto and A. Onuki, Phys. Rev. Lett. **81**, 4915 (1998); C. Bennemann, C. Donati, J. Baschnagel, S.C. Glotzer, Nature **399**, 246 (1999); N. Lačević, F.W. Starr, T.B. Schroder, and S.C. Glotzer, J. Chem. Phys. **119**, 7372 (2003).

[21] X.Y. Xia and P.G. Wolynes, Proc. Natl. Acad. Sci. **97**, 2990 (2000).

[22] P. Viot, G. Tarjus, D. Kivelson, J. Chem. Phys. **112**, 10368 (2000).

[23] J.P. Garrahan, D. Chandler, Phys. Rev. Lett. **89**, 035704 (2002).

[24] P. Mayer, H. Bissig, L. Berthier, L. Cipelletti, J.P. Garrahan, P. Sollich, and V. Trappe, Phys. Rev. Lett. **93**, 115701 (2004); A. Duri and L. Cipelletti, Europhys. Lett. **76**, 972 (2006).

[25] L. Berthier, G. Biroli, J.-P. Bouchaud, L. Cipelletti, D. El Masri, D. L'Hôte, F. Ladieu, and M. Pierno, Science **310**, 1797 (2005).

[26] D.S. Fisher and D.A. Huse, Phys. Rev. B **38**, 373 (1988).

Course 14

EPIGENETIC LANDSCAPE AND CATASTROPHE THEORY: COMMENTARY ON A CORRESPONDENCE

Sara Franceschelli

ENS-LSH, Centre Desanti & Equipe Rehseis
frances@paris7.jussieu.fr

J.-P. Bouchaud, M. Mézard and J. Dalibard, eds.
Les Houches, Session LXXXV, 2006
Complex Systems

Contents

The paper "Une théorie dynamique de la morphogenèse" by René Thom, appeared in *Towards a theoretical Biology* I (1968), published by Conrad Hal Waddington, can be considered as the *princeps* paper of the catastrophe theory. In this paper Thom declares that embryology and, in particular, Waddington's notions of epigenetic landscape and chreod have been one of his sources in conceiving catastrophe theory. Moreover, Thom tries to show that embryology, in particular the problem of cellular differentiation as it was posed by Max Delbrück in 1949, can be a field of application of Thom's catastrophe theory, *i.e.* a mathematical theory of morphogenesis, based on the study of the property of structural stability by topology and differential analysis.

In the collection of René Thom's works *Modèles mathématiques de la Morphogenèse*, published in 1980, the French version of this paper is augmented by a correspondence between Conrad Hal Waddington and René Thom (five letters, from the 25th of January till the 23th of February 1967) about two Waddington's criticisms to the first version of Thom's paper.

The first criticism concerns the paternity of the notion of cellular differentiation. In Thom's paper Waddington argues:[1]

- "the biochemical interpretation (due to Delbrück and Szilard) of cellular differentiation".

The second one concerns the use of alternative steady states instead of time extended chreods. On this point Waddington argues:

- "every cellular specialization being – following the idea of Delbrück and Szilard – characterized by a stable regime of metabolism, *i.e.* an attractor A of the local biochemical dynamics".

And he suggests instead, as far as the first point concerns:

- "I had stated the main point as early as 1939".

And as far as the second point concerns:

- "I got it right, and spoke of alternatives between time-extended chreods (though I did not yet call them that), whereas Delbrück and Szilard had the simpler and basically inadequate idea in the context of development of alternative between steady states". Hence Thom proposes the following modifications of his text:

---

[1] All the quotations of the correspondence are from Thom (1980), 23–33.

– "The idea of interpreting cellular differentiation in terms of a "stable regime of the metabolism" *i.e.* of an attractor of the biochemical kinetics, is often attributed to Delbrück and Szilard. In fact it was stated – under its local form, which is the only correct one – in C.H. Waddington, *Introduction to Modern Genetics*, 1939."
– "every cellular specialization being – following the idea of Delbrück and Szilard – characterized by a stable regime of local metabolism".
However, concerning the second and most conceptually important point, Waddington would prefer:
• "by a stable but evolving regime of local metabolism".
Thom hence suggests:
• "an attractor of the biochemical kinetics tangent to the point under consideration".

Despite the fact that from the rest of the correspondence one can argue that even this last expression is not completely satisfactory for Waddington, it will be the one retained by Thom for the definitive version of his paper.

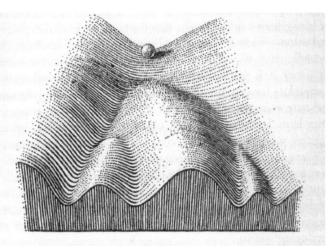

FIGURE 4
Part of an Epigenetic Landscape. The path followed by the ball, as it rolls down towards the spectator, corresponds to the developmental history of a particular part of the egg. There is first an alternative, towards the right or the left. Along the former path, a second alternative is offered; along the path to the left, the main channel continues leftwards, but there is an alternative path which, however, can only be reached over a threshold.

Waddington (1957), 29.

One can ask the question: which is the reason of the misunderstanding between the two scientists?

A first interpretation of this lack of agreement between the two scientists can be based on the taking into account of their cultural differences. To use the expression introduced by Evelyn Fox Keller in *Making Sense of Life*, they do not share the same "epistemological culture" and they do not have the same explanatory needs.

René Thom himself introduces this correspondence as an example of the difficulties in communication between a mathematician and a biologist because of the differences in their exigencies of mathematical rigour. However, following some Waddington's remarks on the peculiarity of the variable "time" in biology, I suggest another interpretation of their disagreement, based on the Waddington's unsatisfied need of representing, thanks to the metaphor of epigenetic landscape, different time scales in the process of the organism development.

The full commentary on this correspondence has been published in:

Franceschelli S. (2006). "Morphogenèse, stabilité structurelle et paysage épigénétique". In A. Lesne & P. Bourgine (eds.), *Morphogènèse. L'origine des formes.* Belin, 298–308.

Selected Bibliography

Delbrück M. (1949). *Unités biologiques douées de continuité génétique.* CNRS, Paris. In Thom (1980).

Fox-Keller E. (2002). *Making Sense of Life. Explaining Biological Development with Models, Metphors and Machines.* Harvard University Press.

Thom R. (1980). "Une théorie dynamique de la morphogenèse". In *Modèles mathématiques de la morphogenèse.* Christian Bourgeois Editeur, 1980.

Waddington C.H. (1957). *The strategy of the genes.* London, Allen and Unwin.

Acknowledgements

It is a pleasure to thank Annick Lesne for her commentaries and precious suggestions on Franceschelli (2006).

I also would thank, for interesting, discussions:

Jean Gayon, Michel Morange, Jean-Philippe Bouchaud, Silvia Demonte, Francesco D'Ovidio, Paolo Freguglia, Evelyn Fox-Keller, Charles Galperin.

Course 15

A HIKE IN THE PHASES OF THE 1-IN-3 SATISFIABILITY

Elitza Maneva[1], Talya Meltzer[2], Jack Raymond[3], Andrea Sportiello[4] and Lenka Zdeborová[5]

[1] *University of California at Berkeley, Berkeley, CA 94720, USA*
[2] *School of Computer Science and Engineering, The Hebrew University of Jerusalem, 91904 Jerusalem, Israel*
[3] *NCRG; Aston University, Aston Triangle, Birmingham, B4 7EJ*
[4] *Università degli Studi di Milano, via Celoria 16, I-20133 Milano*
[5] *CNRS; Univ. Paris-Sud, UMR 8626, Orsay CEDEX, France 91405, LPTMS*

J.-P. Bouchaud, M. Mézard and J. Dalibard, eds.
Les Houches, Session LXXXV, 2006
Complex Systems
© *2007 Published by Elsevier B.V.*

Contents

1. Introduction

At the Les Houches "Complex Systems" school we learned about spin glasses and methods by which to study their statistical properties (G. Parisi), also the applications of statistical physics to combinatorial optimization problems were introduced (R. Monasson). Following the acquired knowledges we studied the average case of 1-in-3 SAT problem. We applied rigorous methods to obtain algorithmic and probabilistic bounds on the SAT/UNSAT and Hard/Easy transitions. We also employed the cavity method to obtain a more complete picture of the problem. Our study went beyond a simple exercise, and led to several interesting results and a separate publication of three of the authors [1]. Here we summarise shortly the methods and give the main results.

1-in-3 SAT is a boolean satisfaction problem. Each formula consists of a set of variables and clauses, and each clause contains 3 literals. A clause is satisfied if exactly one of the literals is True, and the formula is *satisfiable* (SAT) if there is any assignment of variables to True or False, such that every clause is satisfied. 1-in-3 SAT has important similarities to other constraint-satisfaction and graph-theoretical problems. It is a canonical NP-complete problem [2] and thus relates to a host of practically relevant problems. In particular we consider here an ensemble of formulas parameterized by γ and ϵ, where γ is the mean connectivity of variables in the ensemble, and $\epsilon \in [0, 1/2]$ is the probability that variables appear as a negative literal in the interactions. We then study the SAT/UNSAT and Easy/Hard transition curves in this space, as the number of variables $N \to \infty$.

This parameterization is motivated by the knowledge that in many related problems there exists a sharp transition in the typical-case behaviour, from a SAT to an UNSAT regime as the parameters are varied; and a more heuristic Easy-Hard transition, where in a Easy phase many "local" algorithms work in a polynomial time, whereas in the Hard phase they need an exponential time. Previous work in (symmetric) 1-in-3 SAT ($\epsilon = \frac{1}{2}$) demonstrated that SAT/UNSAT transition is sharp at the threshold $\gamma^*_{\text{sym}} = 1$, and not accompanied by any Hard region [3]. However, for Exact Cover (i.e. positive 1-in-3-SAT, $\epsilon = 0$) this threshold is difficult to determine, with only upper [4] and lower bounds [5] to the transition being known, and the presence of a Hard region is suspected. Studying the threshold behaviour in γ, for a continuum of problems parameterized by ϵ,

495

allows us to better understand the origin of these differences and the nature of the two transitions.

2. Results

Figure 1 outlines our present knowledge on the phase diagram of the ϵ–1-in-3-SAT problem. Results are consistent with previous statements restricted to the special cases $\epsilon = 0$ [4,5], and $\epsilon = \frac{1}{2}$ [3].

A rigorous analysis of the *Unit Clause Propagation* (UC) and *Short Clause Heuristics* (SCH) algorithms [3,5] led to upper (dashed line) and lower (shifted to x-axis) bounds on the SAT/UNSAT threshold. This result identifies two regions in which the problem is known to be Easy-UNSAT and Easy-SAT, and surprisingly for $\epsilon > 0.273$ the upper and lower bounds coincide, proving the existence of a range of ϵ in which, at all values of γ, formulas are almost surely Easy. A second rigorous upper bound for the SAT/UNSAT transition is obtained by considering the First Moment Method on the 2-core [4], and a third one for the Easy-UNSAT region by an algorithmic method of embedding formulas in the less constrained 3-XOR-SAT problem [6] (out of scale in the figure). Both these bounds are an improvement w.r.t. Unit Clause for small values of ϵ, as the UC line diverges for $\epsilon \to 0$.

The other results on the phase diagram are obtained by the cavity method [7]. We worked both in the assumptions of replica symmetry (RS: existence of a single pure state) and one-step replica symmetry breaking [8] (1RSB: existence of exponentially-many pure states). Furthermore we checked the stability of the solutions thereby obtained [9].

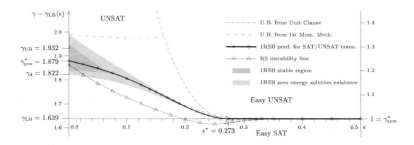

Fig. 1. Phase diagram for the ϵ–1-in-3 SAT problem. On the two axes, ϵ and γ respectively. In the plot, coordinate (ϵ, γ) is shifted down by the SCH lower bound $\gamma_{LB}(\epsilon)$, in order to make the lines appear less squeezed. As a consequence, the Easy-SAT phase w.r.t. Short Clause Heuristics is the region below the x-axis.

The RS solution is able to identify the phase transition when ϵ is large, but is proved to be unstable at smaller ϵ (solid line with triangle marks in fig. 1). The 1RSB solution we explored describes clusters which contain solutions (have zero energy) and disregards their size (entropy) [7]. A 1RSB solution of this kind exists in a region with $\epsilon \lesssim 0.21$, this gives an indication for the Easy/Hard-SAT transition. The common interpretation is that the existence of many states (clusters) is an intrinsic (i.e. algorithm-independent) reason for the average computational hardness [10, 11]. The 1RSB solution is shown to be stable in a region with $\epsilon \lesssim 0.07$, surrounding the SAT/UNSAT curve. Thus we conjecture that this portion of the curve is determined exactly. In particular, we get for the threshold of the positive 1-in-3 SAT $\gamma_{pos}^* = 1.879 \pm 0.001$, in accord with existing exhaustive search results [4, 5] but with much higher precision. There exists a region in which neither RS nor 1RSB assumptions appear to hold sway, in this region we can however guess that 1RSB, as a mean-field assumption, provides an upper bound to the SAT/UNSAT transition curve [12].

3. Concluding remarks

Several questions arise out of this study, also concerning the relation with previous works on 1-in-3 SAT, Exact Cover and K-SAT, both from Statistical-Physics and algorithmic perspectives.

The nature of the interactions, being highly constrained (only 3 out of 8 configurations satisfy a clause, and fixing 2 variables could violate a clause), leads to remarkable properties. The first of them is the success of clause-decimation methods (UC and SCH). The second one, quite unexpected, is the arising of "hard contradictions": above the light gray region in figure, the solution to 1RSB cavity equations at zero energy becomes singular.

The relative success of the SCH algorithm by comparison with the RS cavity method is also a novel result. We found a region where SCH is able to find a satisfying assignment almost surely in polynomial time, and yet from the Statistical-Physics point of view the replica symmetry is broken.

Other algorithmic issues should be investigated in the 1-in-3 SAT problem. Particularly, the performance of Belief Propagation [13] and Survey Propagation [11] (related resp. to RS and 1RSB cavity interpretation). Furthermore, it is interesting to compare our results with the behaviour of the structurally-affine $(2 + p)$-SAT problem [14].

References

[1] Raymond J, Sportiello A, Zdeborová L, *in preparation.*

[2] Cook S, Proc. third annual ACM symposium on Theory of computing, (1971), 151–158. Garey M R and Johnson D S, *Computers and Intractability: A guide to the theory of NP-Completeness,* Freeman, 1979.

[3] Achlioptas D, Chtcherba A D, Istrate G and Moore C, 2001, SODA, 721–722.

[4] Knysh S, Smelyanskiy V N and Morris E R, cond-mat/0403416.

[5] Kalapala V and Moore C, CC/0508037.

[6] Ricci-Tersenghi F, Weigt M and Zecchina R, 2001 *Phys. Rev. E* **63**, 026702. Mézard M, Ricci-Tersenghi F and Zecchina R, 2003 *J. Stat. Phys.* **111** 505.

[7] Mézard M and Parisi G, 2001 *Eur. Phys. J.* **B 20** 217. Mézard M and Parisi G, 2003 *J. Stat. Phys* **111** N. 1-2, 1–34.

[8] Parisi G, 1980 *J. Phys. A* **13** L115–L121. Mézard M, Parisi G and Virasoro M A, *Spin-glass Theory and Beyond,* World Scientific, 1987.

[9] Montanari A, Parisi G and Ricci-Tersenghi F, 2004 *J. Phys. A* **37**, 2073.

[10] Biroli G, Monasson R and Weigt M, 2000 *Eur. Phys. J.* **B 14** 551.

[11] Mézard M, Parisi G and Zecchina R, 2002 *Science* **297** 812. Mézard M and Zecchina R, 2002 *Phys. Rev. E* **66** 056126.

[12] Franz S and Leone M, 2003 *J. Stat. Phys.* **111** 535.

[13] Pearl J, *Probabilistic Reasoning in Intelligent Systems: Networks of Plausible Inference,* Morgan Kaufmann, 1988.

[14] Monasson R, Zecchina R, 1998 *J. Phys. A* **31**, 9209.